Reptile Ecology and Conservation

Techniques in Ecology and Conservation Series

Series Editor: William J. Sutherland

Bird Ecology and Conservation: A Handbook of Techniques
William J. Sutherland, Ian Newton, and Rhys E. Green

Conservation Education and Outreach Techniques
Susan K. Jacobson, Mallory D. McDuff, and Martha C. Monroe

Forest Ecology and Conservation: A Handbook of Techniques
Adrian C. Newton

Habitat Management for Conservation: A Handbook of Techniques
Malcolm Ausden

Conservation and Sustainable Use: A Handbook of Techniques
E.J. Milner-Gulland and J. Marcus Rowcliffe

Invasive Species Management: A Handbook of Principles and Techniques
Mick N. Clout and Peter A. Williams

Amphibian Ecology and Conservation: A Handbook of Techniques
C. Kenneth Dodd, Jr.

Insect Conservation: A Handbook of Approaches and Methods
Michael J. Samways, Melodie A. McGeoch, and Tim R. New

Remote Sensing for Ecology and Conservation: A Handbook of Techniques
Ned Horning, Julie A. Robinson, Eleanor J. Sterling, Woody Turner, and Sacha Spector

Marine Mammal Ecology and Conservation: A Handbook of Techniques
Ian L. Boyd, W. Don Bowen, and Sara J. Iverson

Carnivore Ecology and Conservation: A Handbook of Techniques
Luigi Boitani and Roger A. Powell

Primate Ecology and Conservation: A Handbook of Techniques
Eleanor J. Sterling, Nora Bynum, and Mary E. Blair

Conservation Education and Outreach Techniques Second Edition
Susan K. Jacobson, Mallory D. McDuff, and Martha C. Monroe

Reptile Ecology and Conservation: A Handbook of Techniques
C. Kenneth Dodd, Jr.

Reptile Ecology and Conservation

A Handbook of Techniques

Edited by
C. Kenneth Dodd, Jr.

OXFORD
UNIVERSITY PRESS

OXFORD
UNIVERSITY PRESS

Great Clarendon Street, Oxford, OX2 6DP,
United Kingdom

Oxford University Press is a department of the University of Oxford. It furthers the University's objective of excellence in research, scholarship, and education by publishing worldwide. Oxford is a registered trade mark of Oxford University Press in the UK and in certain other countries

First Edition published in 2016

Impression: 1

Published in the United States of America by Oxford University Press
198 Madison Avenue, New York, NY 10016, United States of America

British Library Cataloguing in Publication Data

Data available

Library of Congress Control Number: 2015956170

ISBN 978–0–19–872613–5 (hbk.)
ISBN 978–0–19–872614–2 (pbk.)

Printed and bound by
CPI Group (UK) Ltd, Croydon, CR0 4YY

In memory of Jonathan Scott Loftis

Preface

As this volume is completed, more than 10,200 non-avian reptile species have been recognized (6175 lizards and amphisbaenians, 3496 snakes, 341 turtles, 25 crocodilians, and 1 Tuatara), with new taxa being described nearly every day (Uetz, P. and Hošek, J. (eds), *The Reptile Database*, http://www.reptile-database.org, accessed 17 August 2015). The life histories and conservation status of most of these species are imperfectly understood or completely unknown except for a few of the more charismatic or popular larger species. Because of a host of threats such as habitat loss and fragmentation, trade, toxic and endocrine disrupting chemicals, emerging infectious diseases, and global climate change, it seems likely that reptiles may be declining at rates approaching those of amphibians (Gibbons et al., 2000; Böhm et al., 2013) and that many species will disappear by the end of the twenty-first century (Alroy, 2015; Ceballos et al., 2015; McCallum, 2015). At the same time, powerful economic interests have united to adversely influence decisions affecting the environment, even rejecting well-established scientific research such as the origin and extent of climate change. There is thus an urgent need for field research on reptile species and their community interactions.

There are a great many techniques available for ecological and conservation-based research on reptiles. Journals such as *Herpetological Review, Herpetological Conservation and Biology*, and *Chelonian Conservation and Biology* frequently contain techniques papers. Specialized books and papers, such Henle and Veith (1997), Gent and Gibson (1998), Hachtel et al. (2009), Eekhout (2010), McDiarmid et al. (2012), Cacciali (2013), and Graeter et al. (2013), offer additional summaries that are as applicable today as when they were published. Although not strictly a techniques book, many of the chapters in Lutterschmidt (2013) offer excellent guides as to the dynamic state of research in reptile biology currently underway. The current volume is meant not to supplant earlier works, but to supplement them and add new areas not previously summarized, such as statistical modelling, landscape ecology, genetics, disease-biosecurity, and human dimensions. Our objectives have been to delineate important new developments, to give an idea as to what the techniques tell or do not tell a researcher, to focus attention on biases and data inference, and to get readers to appreciate sampling as an integral part of their science, rather than just a means of capturing animals. The techniques used will set the boundaries within which results can or should be interpreted.

No one volume can include all techniques, nor can the techniques included be discussed in more than passing detail. Because of space limitations, chapters on growth (Andrews, 1982), behaviour, advanced spatial modelling, social science, relocation/reintroduction (see *Animal Conservation*, Volume 17, Supplement 1, 2015), and the use of stable isotopes (Fry, 2006) could not be included. Authors of individual chapters were advised to incorporate some of these topics as best they could into existing chapters, and to point readers to important references where more detailed information may be obtained. In addition, many of the topics covered in the amphibian volume of this series also are pertinent to reptiles (Dodd, 2010). Lack of space also precludes the addition of a comprehensive glossary. Readers should consult Lillywhite (2008) or online sources when unfamiliar terms are encountered.

Publishing results is an integral component of research undertaken on reptile ecology and conservation. It is simply not good enough to bury research findings in unpublished reports or university theses. Although not all investigations result in 'high impact' publications, there are myriad outlets available for disseminating research results. At the same time, researchers should avoid so-called 'predatory' or 'pay to publish' journals, as publication in journals of dubious scholarly reputation will lead to a questioning of research reliability and interpretation. A list of questionable publishers can be found at: http://scholarlyoa.com/publishers/.

The editor thanks the following for taking their valuable time assisting with various aspects of this project: Matthew Allender, George Balazs, Jaime Bertoluci, Bayard Brattstrom, Steven J.B. Carter, Michael Cherkiss, William Cooper, Ben Croak, Wei-Guo Du, Ruth Elsey, Kevin Enge, Neil Ford, Mercedes Foster, Frank Glaw, Gurutzeta Guillera-Arroita, April Goodman Hall, John Iverson, Kate Jackson, Ambika Kamath, Joshua Kapfer, Amy Lathrop, Harvey Lillywhite, Peter Lindeman, Scott Loarie, Victor Loehr, Erin Marnocha, Jonathan Mawdsley, Shai Meiri, Damian Michael, Donald B. Miles, Debra Miller, Joe Mitchell, Akira Mori, Paul Ouboter, Ted Papenfuss, Michael Plummer, Thomas Rainwater, John H. Roe, Jose Rosado, Christopher Rowe, Andrew Royle, Raul F.D. Sales, Benedikt Schmidt, Coleman Sheehy, Rick Shine, Cameron Siler, Javier Manjarrez Silva, Lora Smith, Phillip Q. Spinks, James Spotila, Laurie Vitt, Jayme Waldron, Dan Warner, James Watling, Kimberley M. Watson, Grahame Webb, Scott Weir, Allan Woodward, and Amy Yackel. I greatly appreciate the support from Ian Sherman and Lucy Nash at Oxford University Press, and thank series editor, Bill Sutherland, for inviting me to edit the reptile volume. Alan Skull and Indumadhi Srinivasan deserve special thanks for their efforts to improve the text and illustrations. A special thanks to Marian Griffey, Morganna Fairchild (feline, not the actress), Allen K. Fearless, and the rest of the cat pride. This volume is dedicated to all the biologists who take up the challenge of reptile ecology and conservation.

C. Kenneth Dodd, Jr.

References

Alroy, J. (2015). Current extinction rates of reptiles and amphibians. *Proceedings of the National Academy of Science, USA*, **DOI: 10.1073/pnas.1508681112**.

Andrews, R.M. (1982). Patterns of growth in reptiles. In C. Gans (ed) *Biology of the Reptilia, Volume 13. Physiology D: Physiological Ecology*. London: Academic Press, pp. 273–320.

Böhm, M., Collen, B., Baillie, J.E.M., et al. (2013). The conservation status of the world's reptiles. *Biological Conservation*, **157**, 372–85.

Cacciali, P. (2013). *Colecta y Preparación de Anfibios y Reptiles*. Saarbrücken, Germany: Editorial Académica Española, AV Akademikerverlag GmbH & Co.

Ceballos, G., Ehrlich, P.R., Barnosky, A.D., et al. (2015). Accelerated modern human-induced species losses: entering the sixth mass extinction. *Science Advances*, **1**, e1400253.

Dodd, C.K., Jr. (ed). (2010). *Amphibian Ecology and Conservation: A Handbook of Techniques*. Oxford: Oxford University Press.

Eekhout, X. (2010). Sampling amphibians and reptiles. *ABC Taxa*, **8**, 530–57.
Fry, B. (2006). *Stable Isotope Ecology*. New York: Springer Science.
Gent, T., and Gibson, S. (eds). (1998). *Herpetofauna Worker's Manual*. Peterborough, UK: Joint Nature Conservation Committee.
Gibbons, J.W., Scott, D.E., Ryan, T.J., et al. (2000). The global decline of reptiles, déjà vu amphibians. *Bioscience*, **50**, 653–66.
Graeter, G.K., Buhlmann, K.A., Wilkinson, L.R., et al. (eds). (2013). *Inventory and Monitoring: Recommended Techniques for Amphibians and Reptiles*. Technical Publication IM-1. Birmingham, AL: Partners in Amphibian and Reptile Conservation.
Hachtel, M., Schlüpmann, M, Thiesmeier, B., et al. (eds). (2009). *Methoden der Feldherpetologie*. Bielefeld, Germany: Laurenti Verlag.
Henle, K., and Veith, M. (eds). (1997). *Naturschutzrelevante Methoden der Feldherpetologie*. Deutsche Gesellschaft für Herpetologie und Terrarienkunde.
Lillywhite, H.B. (2008). *Dictionary of Herpetology*. Malabar, FL: Krieger Publishing.
Lutterschmidt, W.I. (ed). (2013). *Reptiles in Research. Investigations of Ecology, Physiology, and Behavior from Desert to Sea*. New York: Nova Biomedical.
McCallum, M.L. (2015). Vertebrate biodiversity losses point to a sixth mass extinction. *Biodiversity and Conservation*, DOI: 10.1007/s10531-0940-6.
McDiarmid, R.W., Foster, M.S., Guyer, C., et al. (eds). (2012). *Reptile Biodiversity. Standard Methods for Inventory and Monitoring*. Berkeley, CA: University of California Press.

Contents

Part 3. Sampling Reptiles

List of Contributors

Giovanni Amori CNR-Institute of Ecosystem Studies, viale dell'Università 32, 00185 Rome, Italy. E-mail: giovanni.amori@uniroma1.it

Christine Bishop Environment Canada, Science and Technology Branch, Wildlife Research Division, 5421 Robertson Road, Delta, BC V4K 3N2, Canada. E-mail: cab.bishop@canada.ca

Monika Böhm Institute of Zoology, Zoological Society of London, Regent's Park, London NW1 4RY, UK. E-mail: monika.bohm@ioz.ac.uk

Xavier Bonnet CEBC-CNRS, 79360 Villiers en Bois, France. E-mail: bonnet@cebc.cnrs.fr

Matthew Brien IUCN-SSC Crocodile Specialist Group, 6 Fitzmaurice Drive, Bentley Park, QLD 4869, Australia. E-mail: matthew_brien@hotmail.com

François Brischoux CEBC-CNRS, 79360 Villiers en Bois, France. E-mail: francois.brischoux@gmail.com

Keith A. Christian Research Institute for the Environment and Livelihoods, Charles Darwin University, Darwin, NT 0909, Australia. E-mail: keith.christian@cdu.edu.au

Indraneil Das Institute of Biodiversity and Environmental Conservation, Universiti Malaysia Sarawak, 94300 Kota Samarahan, Sarawak, Malaysia. E-mail: idas@ibec.unimas.my

Tiffany M. Doan Department of Biology, University of Central Florida, 4000 Central Florida Boulevard, Orlando, FL 32816, USA. E-mail: tiffperu@yahoo.com

C. Kenneth Dodd, Jr. Department of Wildlife Ecology and Conservation, University of Florida, Gainesville, FL 32611, USA. E-mail: Terrapene600@gmail.com

Karen L. Eckert Wider Caribbean Sea Turtle Conservation Network, 1348 Rusticview Drive, Ballwin, MO 63011, USA. E-mail: keckert@widecast.org

John W. Ferner Department of Biology, Thomas More College, Crestview Hills, KY 41017, USA. [Mailing address: Casa de los Lagartos, 2966 E. Placita Santa Lucia, Tucson, AZ 85716, USA.] E-mail: fernerjw@gmail.com

Robert N. Fisher U.S. Geological Survey, Western Ecological Research Center, San Diego Field Station, 4165 Spruance Road, Suite 200, San Diego, CA 92101, USA. E-mail: rfisher@usgs.gov

Nancy N. FitzSimmons Australian Rivers Institute, Griffith School of Environment, Griffith University, Nathan, QLD 4111, Australia. E-mail: n.fitzsimmons@griffith.edu.au

Steve W. Gotte U.S. Geological Survey, Patuxent Wildlife Research Center, National Museum of Natural History, Museum Support Center, 4210 Silver Hill Road, Suitland, MD 27046, USA. E-mail: sgotte@usgs.gov

Brian Gratwicke Smithsonian's National Zoological Park, Smithsonian Conservation Biology Institute, 3001 Connecticut Avenue NW, Washington, DC 20009, USA. E-mail: gratwickeb@si.edu

Robert W. Henderson Milwaukee Public Museum, 800 W. Wells St., Milwaukee, WI 53233, USA. E-mail: rh@mpm.edu

Brian T. Henen Post Office Box 1676, Twentynine Palms, CA 92277, USA. E-mail: bthenen@yahoo.com

Margaretha D. Hofmeyr Biodiversity and Conservation Biology Department, University of the Western Cape, Private Bag X17, Bellville 7535, South Africa. E-mail: mdhofmeyr@uwc.ac.za

Jeremy F. Jacobs Department of Vertebrate Zoology, National Museum of Natural History, Museum Support Center, 4210 Silver Hill Road, Suitland, MD 27046, USA. E-mail: jacobsj@si.edu

Elliott R. Jacobson College of Veterinary Medicine, University of Florida, Gainesville, FL 32610, USA. E-mail: jacobsone@ufl.edu

Bruce A. Kingsbury Department of Biology, Indiana University-Purdue University Fort Wayne, Fort Wayne, IN 46805, USA; E-mail: Bruce.Kingsbury@ipfw.edu

Gunther Köhler Senckenberg Forschungsinstitut und Naturmuseum, Senckenberganlage 25, 60325 Frankfurt a.M., Germany. E-mail: gkoehler@senckenberg.de

Pilar Lopez Departamento de Ecologia Evolutiva, Museo Nacional de Ciencias Naturales, CSIC, Jose Gutierrez Abascal 2, 28006 Madrid, Spain. E-mail: pilar.lopez@mncn.csic.es

Luca Luiselli Environmental Studies Centre Demetra, via Olona 7, I-00198 Rome, Italy. E-mail: lucamlu@tin.it

Darryl I. MacKenzie Proteus Wildlife Research Consultants, P.O. Box 7, Outram 9062, New Zealand. E-mail: darryl@proteus.co.nz

Marco Mangiacotti Museo Civico di Storia Naturale di Milano, Corso Venezia, 55, 20121 Milan, Italy. E-mail: marco.mangiacotti@gmail.com

Charlie Manolis Wildlife Management International, P.O. Box 530, Karama, NT 0813, Australia. E-mail: cmanolis@wmi.com.au

Jose Martín Departamento de Ecologia Evolutiva, Museo Nacional de Ciencias Naturales, CSIC, Jose Gutierrez Abascal 2, 28006 Madrid, Spain. E-mail: Jose.Martin@mncn.csic.es

Lindsay Renick Mayer Global Wildlife Conservation, P.O. Box 129, Austin, TX 78767, USA. E-mail: renickmayer@gmail.com

Earl D. McCoy Department of Integrative Biology, University of South Florida, Tampa, FL 33620, USA. E-mail: edm@mail.usf.edu

Stephen J. Mullin Department of Biology, Stephen F. Austin State University, Nacogdoches, TX 75962, USA. E-mail: sjmullin@sfasu.edu

Henry R. Mushinsky Department of Integrative Biology, University of South Florida, Tampa, FL 33620, USA. E-mail: mushinsk@usf.edu

Matthew Neff Smithsonian's National Zoological Park, Reptile Discovery Center, 3001 Connecticut Avenue NW, Washington, DC 20009, USA. E-mail: NeffM@si.edu

David A. Pike College of Marine and Environmental Sciences, James Cook University, Townsville, QLD 4811, Australia. [Mailing address: 5108 Holland Avenue, Tampa, FL 33617, USA.] E-mail: david.pike22@gmail.com

Michael V. Plummer Department of Biology, Box 12251, Harding University, Searcy, AR 72149, USA. E-mail: plummer@harding.edu

Viorel D. Popescu Ohio University, Biological Sciences, 107 Irvine Hall, Athens, OH 45701, USA. E-mail: popescu@ohio.edu

Robert Powell Department of Biology, Avila University, 11901 Wornall Road, Kansas City, MO 64145, USA. E-mail: Robert.Powell@avila.edu

Arne R. Rasmussen School of Conservation Esplanaden 34, 1263 København K, Denmark. E-mail: arr@kadk.dk

Nathan J. Robinson The Leatherback Trust, Golding-Gund Marine Biology Station, Playa Grande, Guanacaste, Costa Rica. E-mail: nathanjackrobinson@gmail.com

J. Andrew Royle USGS Patuxent Wildlife Research Center, 12100 Beech Forest Road, Laurel, MD 20708, USA. E-mail: andy_royle@usgs.gov

Sharon Ryan Smithsonian Tropical Research Institute, Apartado Postal 0843–03092, Panamá, República de Panamá. E-mail: RyanSM@si.edu

Roberto Sacchi Dipartimento di Scienze della Terra e dell'Ambiente, Università di Pavia, 27100 Pavia, Italy. E-mail: roberto.sacchi@unipv.it

Marco Sannolo CIBIO, Research Centre in Biodiversity and Genetic Resources, InBIO, Universidade do Porto, Campus Agrário de Vairão, Rua Padre Armando Quintas N.7, 4485-661 Vairão, Vila do Conde, Portugal. E-mail: marco.sannolo@gmail.com

Stefano Scali Museo Civico di Storia Naturale di Milano, Corso Venezia, 55, 20121 Milan, Italy. E-mail: stefano.scali@comune.milano.it

Richard A. Seigel Department of Biological Sciences, Towson University, 8000 York Road, Towson, MD 21252, USA. E-mail: rseigel@towson.edu

Jennifer Sevin Center for Leadership in Global Sustainability, 900 North Glebe Road, Arlington, VA 22203, USA. E-mail: jennifer.sevin@gmail.com

Seth Stapleton Jumby Bay Hawksbill Project, Jumby Bay, P.O. Box 243, St. John's, Antigua, West Indies. [Mailing address: 1317 Juno Ave., St. Paul, MN 55116, USA.] E-mail: seth@jbhp.org

Joanna Sumner Museum Victoria, GPO Box 666, Melbourne, VIC 3001, Australia. E-mail: jsumner@museum.vic.gov.au

Chris Sutherland Department of Environmental Conservation, University of Massachusetts Amherst, Amherst, MA 01379, USA. E-mail: csutherland@umass.edu

Christopher R. Tracy Department of Biological Sciences, California State University Fullerton, 800 N. State College Blvd., Fullerton, CA 92831, USA. E-mail: ctracy@fullerton.edu

C. Richard Tracy Department of Biology, MS-315, University of Nevada, Reno, Reno, NV 89557, USA. E-mail: dtracy@unr.edu

Laurie J. Vitt Sam Noble Museum, University of Oklahoma, 2401 Chautauqua Avenue, Norman, OK 73072, USA. E-mail: vitt@ou.edu

Richard C. Vogt INPA/CBIO, Av. André Araújo, nº 2.936, Petrópolis, CEP 69.067–375, Manaus, Amazonas, Brazil. E-mail: vogt@inpa.gov.br

John D. Willson Department of Biological Sciences, University of Arkansas, Fayetteville, AR 72701, USA. E-mail: jdwillson@uark.edu

Marco A.L. Zuffi Museo Storia Naturale di Pisa, Università di Pisa, Via Roma 79, 56011 Calci (Pisa), Italy. E-mail: marco.zuffi@unipi.it

George R. Zug Department of Vertebrate Zoology-MRC162, National Museum of Natural History, P.O. Box 37012, Washington, DC 20013, USA. E-mail: zugg@si.edu

Part 1
Introduction

1

Reptile diversity and life history

Laurie J. Vitt

1.1 Introduction

The two most fundamental pieces of information necessary to begin developing ecological studies and conservation strategies for reptiles (or any organisms) are identifying the species and knowing what a species does in its natural habitat. Identifying species requires knowledge of other species or populations, particularly those that are closely related and may appear similar to the species or population of concern. The advent of Linnaean taxonomy provided a useful means of cataloguing species such that estimates of the number of species in various 'taxonomic' groups could be determined. When *Systema Naturae* was published in 1735 (Linnaeus, 1735) establishing a taxonomic hierarchy of names (classes, orders, genera, and species), the notion that evolution by natural selection occurred was 124 years in the future (Darwin, 1859), and the mechanism underlying evolution by natural selection would not appear until 131 years later (Mendel, 1866). We now recognize that Linnaean taxonomy in its basic form does not accurately represent evolution and it has been replaced with evolutionary taxonomies.

Learning what a species or population does in its natural habitat can be a challenge, particularly for species that are rare. I briefly review both reptile 'diversity' and life histories as a baseline for the following chapters that provide specific information on how to approach studies of these fascinating animals.

1.2 Reptile 'diversity'

1.2.1 'Diversity'

The term 'diversity' has had several meanings in evolutionary biology and ecology and is often used loosely by professional biologists, land managers, and the general public. Technically speaking, when we are talking about the number of species in a given region, on a global level, or within a particular clade (evolutionary group with a common ancestor), we are really talking about species richness (see Chapter 21). Species diversity, as used in ecology, incorporates relative abundance of each species.

1.2.2 Evolutionary history and numbers of reptile species

Extant reptiles are amniotes (i.e. they have an amniotic membrane in their embryo). Mammals are also amniotes and birds are nested within reptiles (Figure 1.1). Turtles

Reptile Ecology and Conservation. Edited by C. Kenneth Dodd, Jr. © Oxford University Press 2016.
Published 2016 by Oxford University Press.

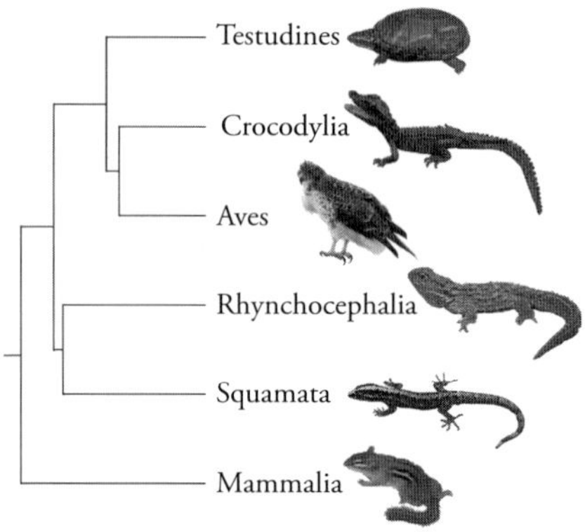

Figure 1.1 Phylogeny of extant amniote tetrapod vertebrates.

and tortoises are now considered as sister to the Crocodylia–Aves (bird) clade, and the positions of some turtle families have changed (Figure 1.2). Based on a large number of molecular studies, it has become clear that both snakes and amphisbaenians are monophyletic clades nested within 'lizards' and thus the classic division of squamates into three suborders (Amphisbaenia, Serpentes, and Lacertilia) does not reflect the evolutionary history of these animals (Figure 1.3).

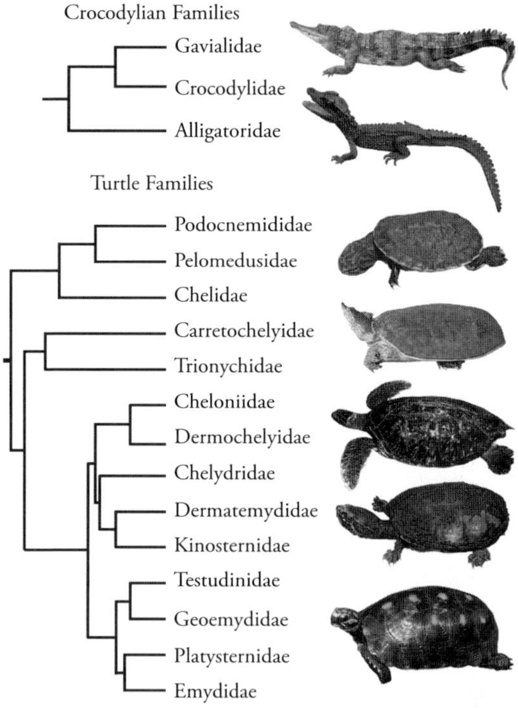

Figure 1.2 Phylogeny of extant crocodilian and turtle families based on molecular studies (see Vitt and Caldwell, 2014, for original references).

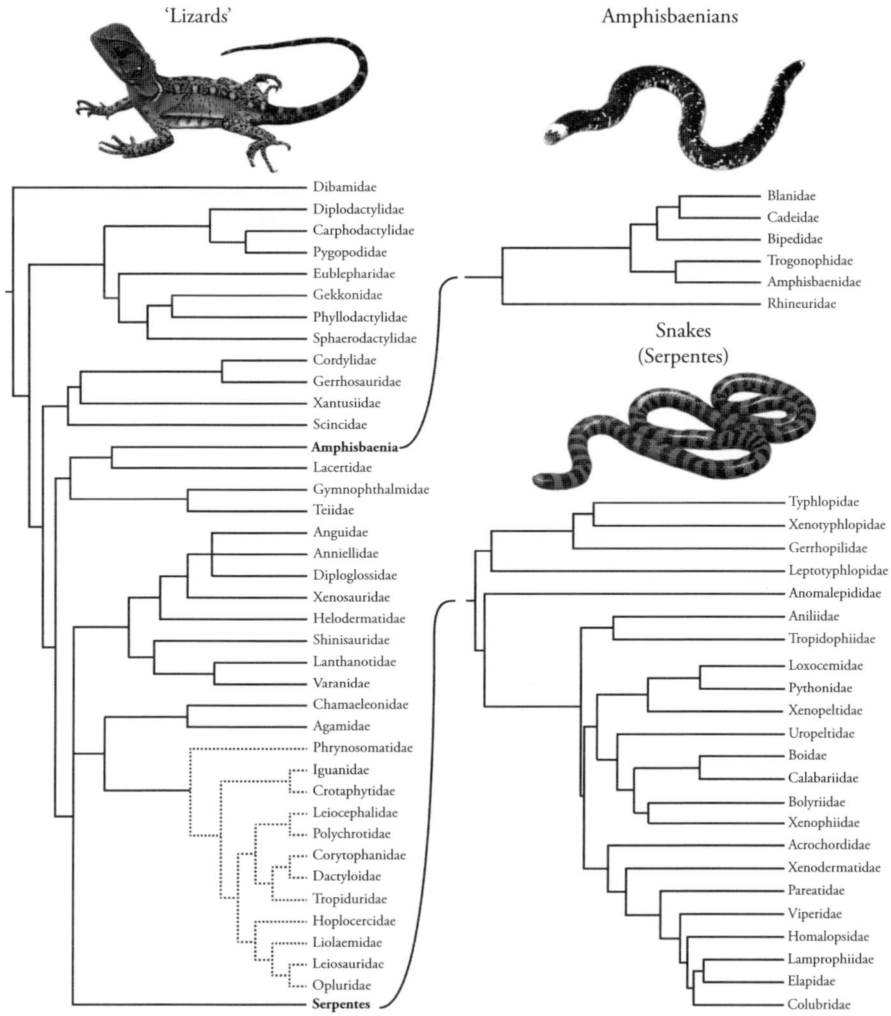

Figure 1.3 Phylogeny of extant lizard, amphisbaenian, and snake families based on molecular studies (see Vitt and Caldwell, 2014, for original references).

As of March 2015, 10,178 species of living non-avian reptiles were known, and new species are being described regularly (Uetz and Hošek, 2014). Of these, 341 are turtles or tortoises, 25 are crocodilians, 1 is a rhynchocephalian (some consider *Sphenodon* to consist of two species), and the remainder are squamates (Table 1.1). Among squamates, 6040 are what we typically refer to as lizards, 3522 are snakes, and 190 are amphisbaenians.

1.2.3 Morphological and ecological diversity

So, what are these animals that we typically call reptiles, how do the major groups differ, and how variable morphologically, behaviourally, and ecologically are species within

Table 1.1 *Taxonomic diversity of reptiles*

Major clade	Families	Genera	Species
Turtles	14	95	341
Crocodilians	3 (4)	9	25
Squamate 'lizards'	36	533	6040
Squamate amphisbaenians	6	20	190
Squamate 'snakes'	27	522	3522
Rhynchocephalians	1	1	1 (2)

Notes: Numbers of families, genera, and species are based on The Reptile Database *(http://www.reptile-database.org/). The potentially additional crocodilian family (Caimanidae) is based on Willis (2009) and some consider* Sphenodon *(Rhynchocephalia) to consist of two species.*

each major group? Most people would easily distinguish a turtle from a crocodilian, and both of these from other extant reptiles.

All turtles have shells consisting of a dorsal carapace and a ventral plastron. Relative size of the carapace and plastron varies greatly, and shell shape varies from very flat to high domes. Most shells are hard with distinct keratinized scales (called scutes), but some, like softshell turtles and the marine Leatherback Turtles, have softer shells lacking scutes. Turtles lack teeth, but have sharp keratinized ridges resting on their jawbones. Crocodilians are lizard-like in overall body form, but can easily be distinguished from all lizards by the absence of lips, an elongate snout, and exposed teeth that interlock. They also have a cloacal slit that runs parallel to the body, whereas lizards have a cloacal slit that runs perpendicular to the body. To most people, snakes would be easily distinguishable from lizards because they have no limbs. However, loss of limbs has occurred independently a number of times within squamates, and only one of those times was in the ancestor to snakes (Wiens et al., 2006). For example, five of the six families of amphisbaenians lack limbs and the other family is represented by only three species, each of which has only small, mole-like front limbs. Consequently, absence of limbs is not the best trait for distinguishing a snake from a lizard. All snakes have spectacles over the eyes and no eyelids, but some lizards (e.g. certain gecko families, xantusiids, and others) also lack eyelids and have spectacles. Snakes have long forked tongues used to detect chemicals in the environment, but many lizards (e.g. varanids and helodermatids) also have long, forked tongues. Snakes lack external ear openings as do a few lizards (e.g. *Cophosaurus* and *Holbrookia*). Characters that distinguish lizards from snakes are for the most part skeletal, and thus not obvious. Nevertheless, the *combination* of long-forked tongue, spectacle over the eyes, lack of an external ear opening, and lack of limbs distinguishes all snakes from nearly all lizards and amphisbaenians. Amphisbaenians are easily distinguished from other squamates by the ring-like pattern of scales around their bodies. The Tuatara is distinguished from all other extant reptiles by the combination of lack of a male copulatory organ, a cloacal slit that runs parallel to the body, and a beak-like structure in the front of the jaws.

As a group, turtles have the widest global distribution among reptiles, largely because all species in two turtle families live in oceans. Freshwater and terrestrial turtles occur on all continents except Antarctica, and a number of species occur at high latitudes (Vitt and Caldwell, 2014). Crocodilians for the most part are restricted to freshwater and estuarine habitats associated with tropical and subtropical landmasses, but the Estuarine Crocodile (*Crocodylus porosus*) can venture long distances into the oceans off Southeast Asia, the Philippines, and Papua New Guinea. In terms of climatic zones, squamates have the widest distribution occurring on all major continents except Antarctica, as well as many oceanic islands. A few species, such as the Viviparous Lizard (*Zootoca vivipara*), enter the Arctic Circle. Sea snakes (subfamily Hydrophinae) occupy continental shelf regions of the Pacific and Indian Oceans with one species reaching tropical areas along the western coast of Central and South America. Tuataras are restricted entirely to about 30 islands off the northeast coast of North Island and western Cook Strait of New Zealand.

Turtles are diverse ecologically. Many species live in freshwater and feed primarily on invertebrates, although some emydid, carettochelyid, and chelid turtles are herbivorous. Map turtles (*Graptemys*) specialize on molluscs that they capture on river bottoms. The large terrestrial tortoises (Testudinidae) are herbivorous. Sea turtles tend to be dietary specialists as adults, and different species specialize on different organisms. For example, *Eretmochelys imbricata* feeds on sponges and soft corals, *Dermochelys coriacea* feeds largely on gelatinous invertebrates such as jellyfish and salps, whereas *Chelonia mydas* feeds on marine grasses or algae.

Crocodilians live primarily in water, all are relatively large bodied, and all feed on animals. The type and size of animals eaten depend a lot on body size—juvenile crocodilians often feed on invertebrates as well as small vertebrates, whereas adults feed on vertebrates, some as large as they are, which are dismembered and torn apart. Crocodilians with long narrow snouts, such as the gavial, feed largely on fish.

Squamates are so diverse ecologically that any generalizations are filled with exceptions (Greene, 1997; Pianka and Vitt, 2003; Lillywhite, 2014). Snakes and lizards can be found in nearly every imaginable microhabitat. Many are arboreal, others terrestrial, others subterranean, some aquatic, and yet others are marine. Many are diurnal, whereas many others are nocturnal. Some are highly specific in their dietary choices, whereas others will eat nearly anything that moves. Much of the variation in ecological traits is clade or species specific (Cadle and Greene, 1993; Vitt et al., 2003; Vitt and Pianka, 2005). For example, snakes in the genus *Dipsas* are thin-bodied, typically arboreal, nocturnal, and specialize on snails and slugs, whereas snakes in the family Leptotyphlopidae live in ant and termite nests feeding on larvae. Lizards in the genera *Phrynosoma, Plica*, and *Uracentron* feed almost entirely on ants. *Phrynosoma* are strictly terrestrial whereas *Plica* and *Uracentron* are arboreal. With the exception of herbivorous lizards, nearly all lizards and all snakes swallow their prey whole. Predaceous lizards usually feed on arthropods or other invertebrates, although some of the large species, particularly in the families Helodermatidae, Varanidae, and Teiidae, frequently feed on vertebrates (Pianka and King, 2004; Beck, 2005). The Komodo Dragon (*Varanus komodoensis*) can mortally wound large prey including water buffalo, wait for them to die and begin rotting, and dismember them, swallowing large portions of the dead prey (Auffenberg, 1981). All snakes swallow their prey whole, but mechanisms for capturing and killing prey vary from strike + constriction, simply biting and swallowing alive, to strike + envenomation.

1.3 Diversity of life histories

1.3.1 Definitions

The term 'life history' generally refers to a combination of the following traits: number of offspring, frequency of reproduction, egg or offspring size, female size or age at first reproduction, reproductive life span, longevity, and age-specific survivorship. Additional variables, including nesting ecology, juvenile development, senescence, and parental investment, are often included (Williams, 1966; Stearns, 1992). Data on age-specific survivorship, reproductive life span, and senescence are not available for the vast majority of reptile species. Without this critical information, population reproductive rates (referred to as *r*, the intrinsic rate of population growth), potential replacement rates (*R*, in which a value of 1 indicates a stable population), and reproductive value (RV = current reproduction + residual reproductive value) cannot be calculated. These are among the most critical variables necessary to develop realistic long-term management programmes for species. Nevertheless, insight into the many different life histories exhibited by reptiles can serve as useful starting points.

1.3.2 General observations

All turtles, crocodilians, and the Tuatara deposit eggs (none are live-bearing) and are relatively long-lived and late maturing when compared with most squamates. All reproduce sexually, and most have environmental sex determination (ESD) often referred to as temperature-dependent sex determination (TSD) (Valenzuela and Lance, 2004). Parental care exists in most if not all crocodilians and has been reported in several turtle species. Among squamates, a remarkable diversity of life histories exist, including both sexual and asexual reproduction, ESD in some, multiple independent origins of live-bearing, extended parental care in some, and a near-continuum from short-lived, early maturing species to long-lived and late maturing species.

1.3.3 Turtle life histories

All turtles reproduce sexually, do not guard or attend nests, most deposit eggs in nests dug into the ground, and species may or may not have sex chromosomes (Table 1.2). Heterogamety (presence of sex-determining chromosomes) occurs in some species (e.g. Trionychidae, *Siebenrockiella crassicollis*), but most lack sex chromosomes and exhibit TSD. TSD is believed to be ancestral in turtles, with gonadal sex determination (GSD) evolving at least six times independently (Warner and Janzen, 2010; Warner, 2011). Parental care of offspring is rare in turtles, reported in only three of the known 341 species.

1.3.4 Crocodilian life histories

All crocodilians reproduce sexually and females deposit eggs in nests constructed of vegetative debris (Graham and Beard, 1973; Thorbjarnarson, 1996). Parental care in

Table 1.2 *Mechanisms of sex determination in reptiles*

	Genetic sex determination			Temperature-dependent sex determination
	Heterogamety in males (XY/XX)	Heterogamety in females (ZZ/ZW)	Homogamety	
Turtles	Chelidae, Geomydidae, Kinosternidae	Geomydidae, Trionychidae	Chelidae	Pelomedusidae, Podocnemididae, Geomydidae, Carettochelyidae, Cheloniidae, Chelydridae, Dermatemydidae, Dermochelyidae, Emydidae, Kinosternidae, Testudinidae, Trionychidae
Crocodilians	None	None	None	Alligatoridae, Crocodylidae, Gavialidae
Tuataras	None	None	None	Sphenodontidae
Squamates	Iguania, Gekkonoidea, Teiidae, Scincidae	Gekkonoidea, Lacertidae, Amphisbaenia, Varanidae, Boidae, Colubridae, Elapidae, Viperidae	Iguania, Gekkonoidea, Lacertidae, Teiidae, Scincidae, Colubridae, Elapidae	Agamidae, Diplodactylidae, Eublepharidae, Gekkonidae, Scincidae

Notes: Original sources can be found in Vitt and Caldwell (2014).

the form of nest attendance, freeing hatchlings from eggs, and carrying (in the mouth) hatchlings to water occurs in most species studied. Smaller species and individuals produce smaller clutches of eggs than larger species or individuals. For example, the relatively small (1.3 m adult snout–vent length (SVL)) *Paleosuchus trigonotus* typically produces about 15 eggs, whereas the large (2.7 m SVL) *Crocodylus porosus* can produce, on average, about 50 eggs. All studied crocodilians have ESD (Valenzuela and Lance, 2004). Nest temperatures near 31°C produce equal numbers of males and females. Temperatures of 30°C or lower produce females and nest temperatures of 32–33°C produce males. Age at sexual maturity and longevity vary among species, but relative to most squamates, all can be considered late maturing. For example, Gavials (*Gavialis gangeticus*) reach maturity in 15–18 years and American Alligators (*Alligator mississippiensis*) reach maturity in 7–10 years.

1.3.5 Squamate life histories

Squamates (lizards and snakes) exhibit remarkable life history variation, and much of the overall variation is historical (e.g. correlated with phylogeny) (Dunham et al., 1988; Mesquita and Colli, 2010). Variation in some traits, such as egg or offspring size and number, is often associated with species body size, and egg or offspring number is usually associated with female size within species. Nevertheless, in some clades, such as Dactyloidae, Phyllodactylidae, and Sphaerodactylidae, clutch size is one egg and eggs are produced in rapid succession. In some other clades, such as Anniellidae, Gekkonidae, Diplodactylidae, Carphodactylidae, and Eublepharidae, females produce clutches or litters of only two. Larger varanids (e.g. *Varanus bengalensis*) can produce clutches of 40 or more eggs. The largest species of pythons and boas can produce upwards of 100 eggs or 60 offspring, respectively. Most squamates produce eggs (oviparity), but 20% of squamate species are live-bearing (viviparous).

Frequency of reproduction varies greatly across clades, but appears to be constrained by temperature, at least at higher elevations and latitudes. For example, some tropical *Anolis* can produce an egg every seven days whereas most Temperate-zone squamates produce one or two clutches per year. Most viviparous squamates produce a single litter per year, and some tropical egg-laying squamates (e.g. *Polychrus acuirostris*) also produce only a single clutch each year.

Because many of the more interesting aspects of squamate life histories cut across taxa, I briefly summarize these rather than providing details on a taxon-by-taxon basis (but see Fitch, 1970, and Vitt and Caldwell, 2014, for additional summaries).

- Parthenogenesis is the production of offspring by females without involvement of males and is cloning given that offspring produced are genetically identical to their mothers. Parthenogenesis was first discovered in the lizard *Darevskia saxicola* by Illya Darevsky (1958). Parthenogenesis has since been confirmed by (1) lack of males in populations, (2) rearing females from birth until they produced offspring, (3) immunocompatibility (skin grafting) experiments, (4) chromosome studies, and (5) genetic studies. More than 50 squamate species are parthenogenetic, and in nearly all cases, parthenogenetic species arose via hybridization between two sexual species or back-crossing between a parthenogenetic species and a sexual one (Reeder et al., 2002). Diploid and triploid parthenogens are most common, but one (*Aspidoscelis neavesi*) is tetraploid (Cole et al., 2014). The xantusiid lizard *Lepidophyma flavimaculatum* appears to be an exception in that it is not a hybrid species (Sinclair et al., 2010). Parthenogenesis occurs in eight lizard families (Agamidae, Chamaeleonidae, Gekkonidae, Gymnophthalmidae, Teiidae, Lacertidae, Xantusiidae, and Scincidae) and one snake family (Typhlopidae). In addition, females of a small number of normally sexually reproducing species (e.g. Nile Monitors, Komodo Dragons, Indian Pythons, and Sierra Gartersnakes) have produced offspring in captivity in the absence of males. The ecological significance of parthenogenesis in reptiles remains a mystery (Kearney et al., 2009), but the short-term advantage is that it increases reproductive rate because every individual produced is a reproducing female.

- Viviparity refers to live birth, and among extant reptiles, occurs only in squamates. Although relatively uncommon even in squamates, viviparity has arisen at least 114 times, which is by far the greatest number of independent origins known for tetrapod vertebrates (e.g. Blackburn, 1982, 1985, 2000). The primary advantage of viviparity is that it allows the parental female to exert some control over incubation temperatures via behavioural thermoregulation. Because most viviparous species live in areas with cool climates (high elevations or latitudes), adaptation to cold climates appears to best explain most origins of viviparity (Shine and Berry, 1978; Shine, 1985, 2002). The primary disadvantage is that because females must carry their offspring during the entire incubation period, and because the additional weight of the embryos impacts locomotion, females can incur increased risk during pregnancy. Because of the many origins of viviparity, it does not occur in the same way in all viviparous species. Many species simply hold fully yolked eggs in the oviduct until development is complete (e.g. the snake *Opheodrys vernalis*). Others provide various degrees of embryonic nourishment to developing offspring. Depending upon species, squamate placentae vary both morphologically and in the degree to which they supply nutrients to developing embryos (Blackburn et al., 1984; Qualls et al., 1997).
- Parental care is rare, or at least poorly known in squamates. Forty-one species of lizards and 47 species of snakes exhibit some form of parental care. Parental care in squamates varies from aiding neonates out of placental membranes at birth, creating nests and attending or brooding eggs until they hatch, to defending nests or allowing juveniles to remain in adult territories. A remarkable example of brooding occurs in some pythons, which not only brood their eggs until they hatch, but elevate temperatures of the nest and eggs by shivering thermogenesis (Somma, 2003).
- Among squamates, at least four mechanisms of sex determination are recognized (Table 1.2; Warner and Janzen, 2010; Warner, 2011). Three of these are genetic (i.e. GSD): heterogamety in males (XY/XX), heterogamety in females (ZZ/ZW), and homogamety (sex chromosomes undifferentiated). The fourth is TSD, in which narrow ranges of temperature variation can shift the hatchling sex ratio. Although TSD is most common in species lacking differentiated sex chromosomes, it occurs in some species with heteromorphic sex chromosomes. Moreover, elements of GSD and TSD are known to occur within species and populations, and sex reversal has even been shown to trigger the rapid transition from GSD to TSD (Holleley et al., 2015). Teasing out the adaptive significance of TSD (or ESD) has been elusive, but experiments with the agamid lizard *Amphibolurus muricatus* revealed that fitness of each sex was maximized by the incubation temperature that produced each sex, just as theory had predicted (Warner and Shine, 2008).

1.3.6 Tuatara life history

Tuataras reproduce sexually, females deposit 5–15 eggs in shallow nest cavities that they dig, and hatching occurs 11–16 months later with development arrested during winter. Females nest once every four years and eggs are carried in the oviducts for 6–8 months prior to deposition (Cree et al., 1992; Cree, 2002, 2014). This unique reproductive pattern is thought to be an adaptation to the cool environment on islands of New Zealand where the Tuatara occurs. Unlike other extant reptiles, male tuataras do not have a copulatory organ. Sperm transfer is accomplished by cloacal apposition, similar to birds. TSD occurs, with developmental temperatures of 21°C producing equal numbers of males and females, 22°C producing mostly males, and 20°C producing mostly females. Nevertheless, both genetic and environmental factors can be involved in tuatara sex determination (Cree et al., 1995). Tuataras also grow more slowly than other reptiles, continuing to grow in size during the first 35 years of their lives. Tuataras are extremely long-lived, with individuals known to reach more than 100 years in age (Anonymous, 2009). Tuataras reach sexual maturity at an age of 10–20 years.

1.4 Summary

Extant reptiles constitute a taxonomically, morphologically, and ecologically diverse group of tetrapods. They occur on all continents except Antarctica and can be found in nearly every possible habitat, including some that might be considered extreme for animals that depend on behavioural adjustments to gain and lose heat. A few occur within the Arctic Circle, an entire clade of snakes (sea snakes) occurs in continental shelf regions of the oceans, and many other species live nearly exclusively underground. The diversity of life histories among reptiles is also impressive, although most of that diversity occurs only in squamates.

From an ecological and conservation perspective, it seems clear that no single model can be applied to reptiles. Rather, detailed natural history data are necessary to determine factors that might lead to persistence or extirpation of local populations. Ideally, age-specific survivorship schedules determined over long time periods (at least one cohort's lifespan) should provide data necessary to determine whether populations are stable, increasing, or decreasing. Unfortunately, these data exist for but a handful of species. Often observations on a single life-history trait can influence development of conservation strategies. For example, if freshwater turtles deposit their eggs 200 m from the ponds they live in, then those nesting areas must be included in the turtle's management programme. Likewise, head-start programmes for sea turtles should incorporate incubation temperatures that maximize fitness for each sex. The reactionary approach to conservation that is often tied to funding (if it isn't broken, don't fix it, but when it breaks, throw a large amount of money at it) is unrealistic when it comes to reptile conservation. In most cases we don't have pre-disturbance data, rendering it nearly impossible to (a) determine whether a population is actually declining and (b) what to do about it. Because we lack long-term monitoring data for a vast majority of reptile species, we often cannot be certain whether it is normal for populations to vary dramatically through time. We already know this is true for North American amphibians, but comparable data do not exist for most reptile species. Moreover, conservation funding is often biased towards large, highly visible species such as sea turtles, crocodilians, and the tuatara, even though large numbers of small-bodied squamates are on the decline (e.g. Sinervo et al., 2010).

References

Anonymous. (2009). Tuatara becomes a father for the first time, aged 111. *New Zealand Herald* Available at: http://www.nzherald.co.nz/nz/news/article.cfm?c_id=1&objectid=10553616.

Auffenberg, W. (1981). *The Behavioral Ecology of the Komodo Dragon*. Gainesville, FL: University Presses of Florida.

Beck, D.D. (2005). *Biology of Gila Monsters and Beaded Lizards*. Berkeley, CA: University of California Press.

Blackburn, D. (1982). Evolutionary origins of viviparity in the Reptilia. I. Sauria. *Amphibia-Reptilia*, **3**, 185–205.

Blackburn, D. (1985). Evolutionary origins of viviparity in the Reptilia. II. Serpentes, amphisbaenia, and ichthyosauria. *Amphibia-Reptilia*, **6**, 259–91.

Blackburn, D.G. (2000). Reptilian viviparity: past research, future directions, and appropriate models. *Comparative Biochemistry and Physiology Part A*, **127**, 391–409.

Blackburn, D.G., Vitt, L.J., and Beuchat, C.A. (1984). Eutherian-like reproductive specializations in a viviparous reptile. *Proceedings of the National Academy of Sciences USA*, **81**, 4860–3.

Cadle, J.E., and Greene, H.W. (1993). Phylogenetic patterns, biogeography, and the ecological structure of Neotropical snake assemblages. In R.E. Ricklefs and D. Schluter (eds) *Species Diversity in Ecological Communities: Historical and Geographical Perspectives*. Chicago, IL: University of Chicago Press, pp. 281–93.

Cole, C.J., Taylor, H.L., Baumann, D.P., et al. (2014). Neaves' Whiptail Lizard: the first known tetraploid parthenogenetic tetrapod (Reptilia: Squamata: Teiidae). *Breviora*, **539**, 1–20.

Cree, A. (2002). Tuatara. In T. Halliday and K. Adler (eds) *The New Encyclopedia of Reptiles and Amphibians*. Oxford: Oxford University Press, pp. 210–11.

Cree, A. (2014). *Tuatara. Biology and Conservation of a Venerable Survivor*. Christchurch, NZ: Canterbury University Press.

Cree, A., Cockrem, J.F., and Guillette, L.J., Jr. (1992). Reproductive cycles of male and female tuatara (*Sphenodon punctatus*) on Stephens Island, New Zealand. *Journal of Zoology*, **226**, 199–217.

Cree, A., Thompson, M.B., and Daugherty, C.H. (1995). Tuatara sex determination. *Nature*, **375**, 543.

Darevsky, I.S. (1958). Natural parthenogenesis in certain subspecies of rock lizard, *Lacerta saxicola* Eversmann. *Doklady Akademii Nauk SSSR Biological Science*, **122**, 877–9.

Darwin, C. (1859). *On the Origin of Species by Means of Natural Selection*. London: John Murray.

Dunham, A.E., Miles, D.B., and Reznick, D.N. (1988). Life history patterns in squamate reptiles. In C. Gans and R.B. Huey (eds) *Biology of the Reptilia, Volume 16. Ecology B. Defense and Life History*. New York: A. R. Liss, pp. 441–552.

Fitch, H.S. (1970). Reproductive cycles in lizards and snakes. *Miscellaneous Publications of the Museum of Natural History of the University of Kansas*, **52**, 1–247.

Graham, A., and Beard, P. (1973). *Eyelids of Morning. The Mingled Destinies of Crocodiles and Men*. Greenwich, CT: New York Graphic Society, Ltd.

Greene, H.W. (1997). *Snakes. The Evolution of Mystery in Nature*. Berkeley, CA: University of California Press.

Holleley, C.E., O'Meally, D.O., Sarre, S.D., et al. (2015). Sex reversal triggers the rapid transition from genetic to temperature-dependent sex. *Nature*, **523**, 79–82.

Kearney, M., Fujita, M.K., and Ridenour, J. (2009). Lost sex in the reptiles: constraints and correlations. In I. Schön, K. Martens, and P. van Dijk (eds) *Lost Sex, The Evolutionary Biology of Parthenogenesis*. Dordrecht: Springer, pp. 447–74.

Lillywhite, H.B. (2014). *How Snakes Work. Structure, Function, and Behavior of the World's Snakes*. Oxford: Oxford University Press.

Linnaeus, C. (1735). *Systema Naturae, sive regna tria naturae systematice proposita per classes, ordines, genera, & species*. Leiden: Haak, pp. 1–12.

Mendel, G. (1866). Versuche über Plflanzen-hybriden.Verhandlungen des naturforschenden Vereins. *Brünn*, **IV**, 3–47.

Mesquita, D.O., and Colli, G.R. (2010). Life history patterns in tropical South American lizards. In O.H. Gallegos, F.R.M. Cruz, and J.F.M. Sánchez (eds) *Reproducción en Reptiles: Morphologia, Ecología, y Evolución*. Mexico: Universidad Autónoma del Estado de México, pp. 45–71.

Pianka, E.R., and King, D.R. (eds) (2004). *Varanoid Lizards of the World*. Bloomington, IN: Indiana University Press.

Pianka, E.R., and Vitt, L.J. (2003). *Lizards. Windows to the Evolution of Diversity*. Berkeley, CA: University of California Press.

Qualls, C.P., Andrews, R.M., and Mathies, T. (1997). The evolution of viviparity and placentation revisited. *Journal of Theoretical Biology*, **185**, 129–35.

Reeder, T.W., Cole, C.J., and Dessauer, H.C. (2002). Phylogenetic relationships of whiptail lizards of the genus *Cnemidophorus* (Squamata: Teiidae): a test of monophyly, reevaluation of karyotypic evolution, and review of hybrid origins. *American Museum Novitates*, **3365**, 1–61.

Shine, R. (1985). The evolution of viviparity in reptiles: an ecological analysis. In C. Gans and F. Billett (eds) *Biology of the Reptilia*. New York: John Wiley and Sons, pp. 677–80.

Shine, R. (2002). An empirical test of the 'predictability' hypothesis for the evolution of viviparity in reptiles. *Journal of Evolutionary Biology*, **15**, 553–60.

Shine, R., and Berry, J.F. (1978). Climatic correlates of live-bearing in squamate reptiles. *Oecologia*, **33**, 261–8.

Sinclair, E.A., Pramuk, J.B., Bezy, R.L., et al. (2010). DNA evidence for nonhybrid origins of parthenogenesis in natural populations of vertebrates. *Evolution*, **64**, 1346–57.

Sinervo, B., Méndez-de-la-Cruz, F., Miles, D.B., et al. (2010). Rapid erosion of lizard diversity at global scales: altered thermal niches due to climate change. *Science*, **328**, 894–9.

Somma, L.A. (2003). *Parental Behavior in Lepidosaurian and Testudinian Reptiles*. Malabar, FL: Krieger Publishing Company.

Stearns, S.C. (1992). *The Evolution of Life Histories*. Oxford: Oxford University Press.

Thorbjarnarson, J. (1996). Reproductive characteristics of the order Crocodylia. *Herpetologica*, **52**, 8–24.

Uetz, P., and Hošek, J. (eds) (2014). *The Reptile Database*. Available at: http://www.reptile-database.org/.

Valenzuela, N., and Lance, V. (eds) (2004). *Temperature-Dependent Sex Determination in Vertebrates*. Washington, DC: Smithsonian Books.

Vitt, L.J., and Caldwell, J.P. (2014). *Herpetology: An Introductory Biology of Amphibians and Reptiles*. 4th ed. San Diego, CA: Academic Press.

Vitt, L.J., and Pianka, E.R. (2005). Deep history impacts present-day ecology and biodiversity. *Proceedings of the National Academy of Sciences USA*, **102**, 7877–81.

Vitt, L.J., Pianka, E.R., Cooper, W.E., Jr., et al. (2003). History and the global ecology of squamate reptiles. *American Naturalist*, **162**, 44–60.

Warner, D.A. (2011). Sex determination in reptiles. In D.O. Norris and K. Lopez (eds) *Hormones and Reproduction of Vertebrates, Volume 3. Reptiles*. San Diego, CA: Academic Press, pp. 1–38.

Warner, D.A., and Janzen, F.J. (2010). Diversity of sex-determining mechanisms. In D. Westneat and C.W. Fox (eds) *Evolutionary Behavioral Ecology*. Oxford: Oxford University Press, pp. 81–3.

Warner, D.A., and Shine, R. (2008). The adaptive significance of temperature-dependent sex determination in a reptile. *Nature*, **451**, 566–8.

Wiens, J.J., Brandley, M.C., and Reeder, T.W. (2006). Why does a trait evolve multiple times within a clade? Repeated evolution of snakelike body form in squamate reptiles. *Evolution*, **60**, 123–41.

Williams, G.C. (1966). *Adaptation and Natural Selection: A Critique of Some Current Evolutionary Thought*. Princeton, NJ: Princeton University Press.

Willis, R.E. (2009). Transthyretin gene (TTR) intron 1 elucidates crocodylian phylogenetic relationships. *Molecular Phylogenetics and Evolution*, **53**, 1049–54.

2

Planning and setting objectives in field studies

Robert N. Fisher

2.1 Introduction

The objective of this chapter is to engage the researcher in the logic steps required to design and plan field studies on the ecology and conservation of reptiles. I previously described the approaches needed to design inventory and monitoring programmes for reptiles (Fisher and Mitrovich, 2012; see also Cogălniceanu and Miaud, 2010); most of the information provided in these publications is applicable here. Here, however, I expand the focus from ecology to conservation techniques, many of which become management actions designed to recover or maintain at-risk species or populations. Often these techniques involve a higher level of uncertainty and risk than inventory and monitoring techniques, which vary from observational to highly quantitative, but involve the tenet of 'do no harm'. Setting goals for management may be straightforward, but developing objectives can be less straightforward as it often involves decisions that are uncomfortable to make and difficult to implement as a result of this additional risk. It is often much easier to make no decision than to make one that might result in harm to an at-risk species. Unfortunately, even 'no decision' is a decision when it comes to conservation (Meek et al., 2015). First, I'll review the difference between goals and objectives and then discuss how we implement field studies and/or management actions.

2.2 Planning: goals versus objectives

Goals are descriptions of what one intends to accomplish. These are typically not directly actionable, but serve as the larger vision for the study or set of actions that will follow. For instance, goals might be to:

- Determine the diversity of reptiles of some geographically defined area.
- Conserve *x, y*, or *z* species and their habitats.
- Identify the covariates driving the distribution of certain species.
- Increase the abundance of a certain species.

These goals seem relatively simple and straightforward, but they give no guidance on the steps to be taken or how to implement them, track progress, or follow timelines. This is the function of objectives. Objectives are the measurable steps required to achieve the established goals. They are the strategic activities developed during the planning

Table 2.1 *Examples and definitions of goals and objectives*

	Goal	Objective
Meaning	General purpose of the endeavour	Something that can be attained or accomplished; typically a target
Example	We want to maintain biodiversity in the Canary Islands	We need to remove all invasive California Kingsnakes from the Canary Islands
Action	Outcome for which we strive to attain	Specific action that supports goal accomplishment
Measure	Goals might not be measurable	Must be tangible
Time frame	Longer term	Short term

of a project and are tied to available resources and a specific time frame. Objectives should be SMART—specific, measurable, attainable, realistic/relevant, and timed/time bound (Schroeder, 2009). Many online tools are available to help one define objectives and evaluate whether they are SMART or not (example URLs are listed at end of this chapter). Meeting a specific goal may require many objectives. It may not be possible to define some of them until certain experiments have been conducted; often evaluations of sampling protocols are needed to increase certainty in the biological results. Examples of objectives might be:

- Test traps to verify that they effectively sample all life classes of a certain species during the first year.
- Review and, if necessary, develop marking techniques for uniquely identifying a certain species during the first month.
- Measure covariates of trapping stations using quantitative sampling during the first spring.

Additional definitions and examples are in Table 2.1.

2.3 Design

Time and/or money are never sufficient to conduct a complete, ideal project. Projects almost always are more complex than originally realized. Biological assumptions are typically flawed and need to be reassessed throughout a project. The best way to deal with uncertainties is to maximize the scope and intent when formulating a project and setting goals so that a certain level of failure can be absorbed. This allows one to push beyond safe boundaries (a conservative design) that might not ultimately lead to project success. A key design element is identifying the minimum accomplishment required to achieve success. Phases that equate to milestones that meet some success criteria should be clearly built into the objectives. Having more objectives than possible to accomplish with the available time and resources will create complexity, challenging the researcher to evaluate and prioritize actions continually, but it will also enhance the significance of a project in the end. Conceptual models used to visualize this complexity and identify critical uncertainties are discussed in Section 2.4.

An additional key element of a successful project is balancing elegant design and functional design. Involving statisticians in the project design is critical to success, but can also lead to a study that is biologically irrelevant or impossible to carry out. Because so many uncertainties (i.e. knowing the number of replicates, sample sizes, or repeat visits required, or logistical difficulties) exist at the onset of a project, designing it may seem difficult (or even impossible) and thus lead to design paralysis. It is possible to model some parameters initially and run power analyses to estimate what some sample sizes might need to be, and how many replicates are required, as a first way to deal with uncertainty. At this point it is important to consider functional design and adaptive design. The concept of functional design is that as the data are collected in a way that there are multiple independent elements, such that once a design is implemented if it needs to be changed or modified, or some sampling is unfeasible, not all parts of the design are impacted. Adaptive design is where changes in sampling are informed by the data being collected. This approach typically serves to make studies more efficient (such as fewer days, shorter transects) while meeting or exceeding study objectives. Knowing the type of study design at the beginning will allow a researcher to better understand what parts of the design are flexible, and what parts are fixed, such that if they change, the study will not be successful. Asking the initial question '*What is really important in this study?*' followed by '*Is it a statistical trend? Is it successful reproduction? Is it demographic?*' will better define the relevant design approach. A study that determines that there are other ways to measure and implement sampling that meets the objectives and still has a reasonable chance of success would be adaptive, but not necessarily elegant. Sometimes simpler sampling procedures with more replication will result in a more straightforward dataset to interpret and analyse.

One critical way to think about design is in the dichotomy between verifying trends that are obvious versus forcing statistics to extract trends from patterns that are unclear. Utilizing robust sampling frames with statistical inference is not the same as utilizing complex measurements on the ground. An example of this contrast might be having many plots stratified across habitat types that are coarsely defined utilizing existing GIS layers. Let's say the scale of the vegetation layer is 1 ha, which means that there has been a category of vegetation assigned to each 1 ha pixel of the GIS layer. If the sampling plots are 50 m wide and the region being sampled is 500 km^2, then you can have a very robust sampling frame with replication across vegetation categories and random plot placement. Statistically robust trends might be determined from this design regarding vegetation and species of interest, and this might well answer one set of questions. If the species or community is responding to soil texture (rock size) and there is a general relationship between vegetation and soil texture, however, then this relationship might be missed. If complex measurements are taken of vegetation and soil at the plot scale (50 m) to describe the local heterogeneity within and between plots, then a different set of relationships might emerge, answering a different set of questions. Thinking through which of these questions is more important to managers or management will inform which approach to utilize when conducting the study, even though both might be statistically robust. That said, researchers want to eliminate observer bias as much as possible when sampling; incorporating procedures that minimize ambiguity in species identification or measurement (during trapping or other sampling, as vouchers) is thus an important consideration.

2.4 Conceptual models

One way to address uncertainty when dealing with complex issues is to use conceptual models to direct goals, objectives, and, ultimately, study design (Atkinson et al., 2004). Conceptual models are based on literature reviews and serve as visual depictions of the relationships between variables that describe a system of interest. Example systems could be a vernal pool ecosystem or the life history of a given species of reptile. For a vernal pool, the model might include all of the species using the pool for recruitment, the predator–prey relations between them, the timing of water availability, and the upland context of the pool system in driving diversity or species density. For a single species, a model might focus on critical reproductive traits, various mortality factors at different life stages, and individual movement requirements for territoriality or mate selection.

Conceptual models vary in complexity and rigour depending on the questions being asked and the amount of pre-existing knowledge of the system. Substantial literature on development and use of conceptual models, which 'simultaneously need to embrace and deconstruct contextual complexity', is available (Margoluis et al., 2009). When a system is minimally known or primarily understood through expert opinion, the models can initially be quite simplistic and diagrammatic in nature. As more parameters are resolved and relationships between variables quantified, the researcher revises models to make them more robust. A good example of this process is the progression of the parameterization of the models and the adaptive revisions made during a study of the occupancy of threatened lizards, and the reptile communities in general, in xeric sand dune habitats in southern California in order to understand habitats and their physical response to anthropogenic land use changes (Barrows, 1996; Griffiths et al., 2002; Barrows and Allen, 2010). The result was a more robust study design and better management options.

Studies often are designed to test some future action, such as the changes in a riverine system after a dam is constructed upstream or predictions of responses from climate change. Conceptual models can be developed for future scenarios based on the literature and knowledge from comparative systems, and the uncertainties in adapting these to the current situation can be described. When sampling is punctuated in time, especially involving different observers over time, appropriate methods and observer bias must be clearly understood. Sampling variability can be part of the conceptual model, as any sources of data collection bias should inform the uncertainty of the results of the study. There are good examples of robust studies in which sampling intervals are long and others in which they are short. Sampling methods and placement of sampling sites must be well documented to limit ambiguity when sampling repeatedly over time (see Chapter 3).

2.5 Sampling considerations

Tools that can be used to enhance sampling include digital databases of vouchered specimens and geospatial covariate climate data. An example from the Global Biodiversity Information Facility (GBIF; gbif.org) illustrates the contrast in density of specimen records (all taxa, not just reptiles) or observations from western Europe versus northern Africa (Figure 2.1). Such data can be used either to locate global or local gaps in biodiversity sampling or to provide a historic framework for a study. An illustration of this is the change in composition of observations or records for reptiles and amphibians from southern California over the last 20 years. A graph of specimens archived in museums per year over the last 170 years shows some peaks where large numbers were accumulated one year or several years in sequence, and large gaps where few or no specimens were registered (Figure 2.2). In the last 20 years, almost all recent records used for modelling and identifying trends resulted from a large pit-fall trap study; very few specimens were archived in museum collections because the majority of animals were released alive after capture (Franklin et al., 2009; Figure 2.2). This means that future researchers will have to accept the accuracy of field pit-fall trap identifications when comparing datasets, whereas they can verify the identification of archived specimens. Future datasets of species distributions (presence only, typically) will increasingly be composed of records from citizen scientists and in-field voucher photos, assuming accurate identification. An example of such a robust online database is the Herpetological Education & Research Project (http://www.naherp.com/), which currently has 2840 contributors and more than 220,000 records of reptiles and amphibians (August 2015).

Another useful resource is *The Reptile Database* (http://www.reptile-database.org/; Uetz, 2014), which provides an up-to-date list of all currently recognized reptile species. This is a great resource of information and it illustrates well how fast our knowledge of the diversity of reptiles is increasing. For instance, when I finished my postdoctoral

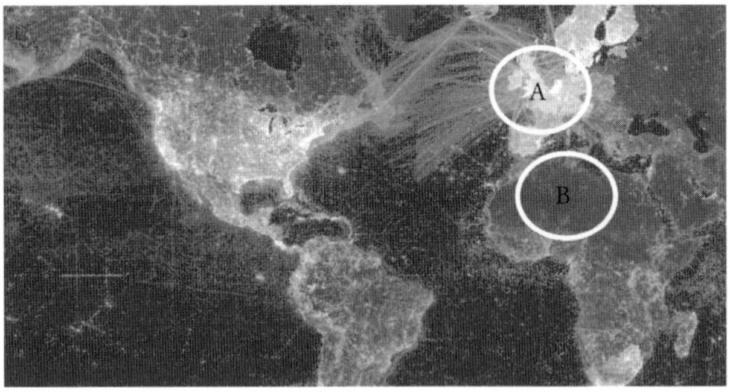

Figure 2.1 Distribution of 466,839,088 georeferenced global occurrence records from GBIF (gbif.org) covering all taxa. White and grey shading/stippling represent individual record locations. Dark areas lack any records for any species in the GBIF database. The contrast in records from western Europe (A) to northern Africa (B) is very striking and illustrates the biodiversity knowledge inequality between these regions. Someone asking a research question in western Europe would have much baseline data available to think through in study design, whereas someone in northern Africa might not.

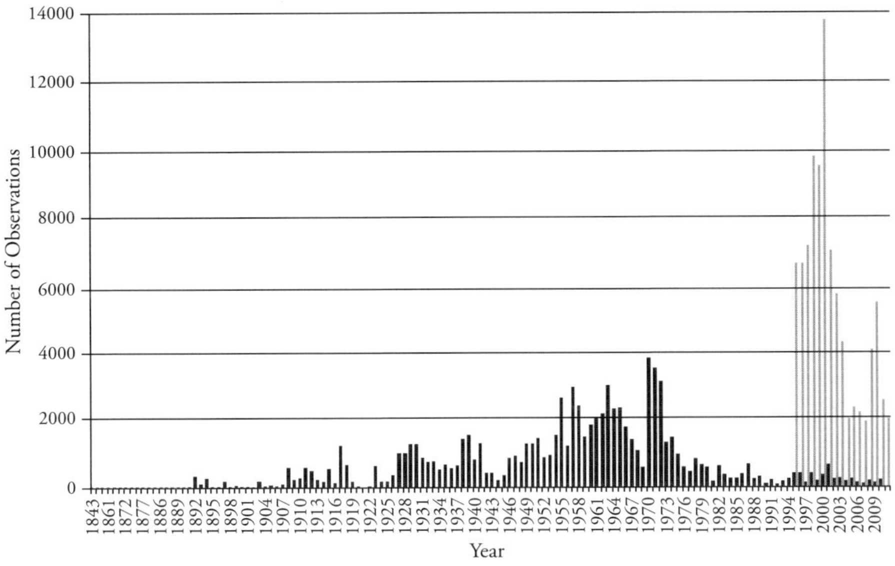

Figure 2.2 Museum records (black bars) for five southern California counties for the last 170 years (*N* = 90,918) showing clear peaks in vouchers collected. On the right side of the graph are records from the U.S. Geological Survey pitfall trap sites (grey bars) from these same five counties from 1995 to present (*N* = 91,541). The majority of the later samples were field collected and released, illustrating the shifting types of data available for discerning future trends and comparing distributions.

research in 1998, 8000 species of reptiles were known; currently more than 10,000 species are known, a 20% increase in less than 20 years. Using up-to-date taxonomy is critical to prevent future confusion in the literature regarding which species were studied. Moreover, many researchers fail to publish geospatial coordinates for sampling locations or report them at a coarse resolution, making it difficult to determine which species was collected, especially when species complexes are sampled and vouchers are not saved. This is extremely important as over time, many taxonomic changes will be made. Changes in species lists for a particular location over time may be more indicative of name changes than of changes in occurrence.

Species names can also be tracked through the Integrated Taxonomic Information System (ITIS; www.itis.gov; Hardisty et al., 2013). This system assigns a taxonomic serial number (TSN) to each named entity such that it has a unique identifier that allows changes in taxonomy to be tracked; the TSN can be stored as a value in the database with the species names. There are currently (30 April 2015) 684,369 scientific names (not just species, but across all taxonomic ranks) with TSNs.

The value of this system can be illustrated by a large standardized inventory/monitoring programme carried out across coastal and desert transitional southern California habitats. When I started this programme in 1995, I developed a list of 49 reptile species that I might expect to detect within this geographic area (Table 2.2).

Table 2.2 *The focal 49 species of reptiles of within five counties of southern California. Column 1 represents their scientific name in common use in the literature in 1995. Column 4 indicates their current name in common use in the literature in 2015. Bold names in Column 4 represent nomenclature changes over these 20 years. Column 2 represents the names still valid in 2015 as in common use in 1995, but indicates that these taxa that as currently defined no longer occur in the study area. Column 3 represents the names still valid in 2015 as in common use in 1995, but indicates that these taxa as currently defined no longer occur in the USA and are now restricted to Mexico.*

	Original name	Original name	
1995 Taxonomy	Still valid only extra-limital	No longer occurs in USA	2015 Taxonomy
Turtle			
Clemmys marmorata	Y		***Actinemys pallida***
Lizards			
Anniella pulchra	Y		*Anniella* ***stebbinsi***
Callisaurus draconoides			*Callisaurus draconoides*
Cnemidophorus hyperythrus			***Aspidoscelis hyperthra***
Cnemidophorus tigris			***Aspidoscelis*** *tigris*
Coleonyx variegatus abbotti			*Coleonyx variegatus abbotti*
Crotaphytus insularis bicinctores	Y (*insularis*)	Y (*insularis*)	*Crotaphytus* ***bicinctores***
Crotaphytus insularis vestigium	Y (*insularis*)	Y (*insularis*)	*Crotaphytus* ***vestigium***
Eumeces gilberti			***Plestiodon*** *gilberti*
Eumeces skiltonianus			***Plestiodon*** *skiltonianus*
Gambelia wislizenii			*Gambelia wislizenii*
Gambelia wislizenii			*Gambelia* ***copeii***
Gerrhonotus multicarinatus			***Elgaria*** *multicarinatus*
Petrosaurus mearnsi			*Petrosaurus mearnsi*
Phrynosoma coronatum	Y	Y	*Phrynosoma* ***blainvillii***
Phrynosoma platyrhinos			*Phrynosoma platyrhinos*
Phyllodactylus xanti	Y	Y	*Phyllodactylus* ***nocticolus***
Sceloporus graciosus	Y		*Sceloporus* ***vandenburgianus***
Sceloporus magister			*Sceloporus magister*
Sceloporus occidentalis			*Sceloporus occidentalis*

Sceloporus orcutti			*Sceloporus orcutti*
Urosaurus microscutatus			*Urosaurus microscutatus*
Uta stansburiana			*Uta stansburiana*
Xantusia henshawi			*Xantusia henshawi*
Xantusia vigilis			*Xantusia vigilis*
Xantusia vigilis			*Xantusia* ***wigginsi***
Snakes			
Arizona elegans			*Arizona elegans*
Charina bottae	Y		*Charina* ***umbratica***
Coluber constrictor			*Coluber constrictor*
Crotalus mitchellii	Y	Y	*Crotalus* ***pyrrhus***
Crotalus ruber			*Crotalus ruber*
Crotalus viridis	Y		*Crotalus* ***oreganus***
Diadophis punctatus			*Diadophis punctatus*
Hypsiglena torquata	Y	Y	*Hypsiglena* ***ochrorhyncha***
Lampropeltis getula	Y		*Lampropeltis* ***californiae***
Lampropeltis zonata			*Lampropeltis zonata*
Leptotyphlops humilis			***Rena*** *humilis*
Lichanura trivirgata	Y		*Lichanura* ***orcutti***
Masticophis flagellum fuliginosus			***Coluber fuliginosus***
Masticophis flagellum piceus			***Coluber*** *flagellum piceus*
Masticophis lateralis			***Coluber*** *lateralis*
Pituophis melanoleucas	Y		*Pituophis* ***catenifer***
Rhinocheilus lecontei			*Rhinocheilus lecontei*
Salvadora hexalepis			*Salvadora hexalepis*
Tantilla planiceps			*Tantilla planiceps*
Thamnophis elegans			*Thamnophis elegans*
Thamnophis hammondii			*Thamnophis hammondii*
Thamnophis sirtalis			*Thamnophis sirtalis*
Trimorphodon biscutatus	Y	Y	*Trimorphodon* ***lyrophanes***

I created a four-digit code for each one (the first two letters of the genus followed by the second two letters of the species) and entered it into the database with its ITIS TSN. This ensured that even if a name changed over time, I would maintain the same species code over time and the TSN would serve in most cases as a crosslink back to the current taxonomy. Since 1995, the species, genus, or both names of 25 species (53%) in this fauna have changed. For 30.6% or our species, the species epithet used in 1995 is now restricted to an entity not found in our study area. Thus, even though I and colleagues published papers with one name, someone reading an old paper may assume that the names we used are still current, and not realize that the species now has a different species epithet. Even more concerning is that 14.3% of the species on our digital list no longer occur in the United States but now are restricted to Mexico. The individuals that we initially studied did not migrate, nor did their descendants, but someone looking at these data now or in the future might interpret the name changes as range changes or local extirpations. With clear documentation of the geographic scope and locations of sampling areas, a researcher should be able to determine the current name for any species.

2.6 Covariates

One important part of study design that is often under considered is the measurement of covariates, of which there may be two types, depending on the type of study. If sampling locations are fixed and sampling events are repeated over time, then both study-specific covariates and sampling-specific covariates should exist. Study-specific variables are often defined once for each study site and do not change between sampling bouts. These may include vegetation type, canopy cover, geographic coordinates, slope, distance to roads, and other factors that are relatively stable at the study site over time. Although they may seem straightforward, researchers must clearly document how they were measured and the units of measurement. For vegetation, there might be several different ways of measurement (Chapter 19), or standardized GIS layers might already be available (Chapter 22). Whatever covariates are used, they should be well documented and thought through as to relevance, cost, logistics, and necessity. Sampling-specific variables are those that may change between visits, for example, date, time, weather, wind, moon phase, water depth, or presence of invasive species. Each variable and its character states should have standardized definitions and measurements. Covariates should be carefully considered and tied to trap or capture success, detectability, or life history parameters of interest. Perhaps the most aggravating issue for a researcher when analysing data is missing data or realizing that additional data should have been collected (see Chapter 3).

2.7 Timescales

Timescales are another critically important study design consideration, because life spans of reptiles vary so greatly. A decade-long study of small geckos would likely encompass several generations encompassing variation in recruitment and survivorship among generations, whereas an equally long study of Tuataras (*Sphenodon*) might involve a single cohort of individuals in the absence of recruitment or mortality. An investigator must decide on an interval of repeated sampling (e.g. annually, monthly) according to the population under study. In some cases, significant population demographic events can be obtained through genetic tools and analysis. For instance, if one finds low genetic diversity within a large population, it may signal of an historic population bottleneck. Analysis of the genetic data may allow a researcher to establish, with confidence intervals, how many generations ago this occurred and in turn tie the change to some environmental driver, such as a flood or fire. On the Galapagos island of Pinzon, no hatchling tortoises (*Chelonoidis ephippium*) were recorded for 150 years, even though the island has been a park since 1959 (Aguilera et al., 2015). Following the removal of Black Rats (*Rattus rattus*) in 2012, hatchling tortoises were recorded in 2014 (Aguilera et al., 2015). Clearly, a generational study of tortoises will take centuries; during one scientific career since the park was established, subsets of the same individual tortoises would have been counted every year.

2.8 Permits

Permitting can be one of the most time intensive parts of planning a study. Although the need for permits is generally acknowledged, their importance to a study tends to be under appreciated, especially when an endangered species is involved. Often three types of permits are required: (1) one issued by the investigator's institution allowing him or her to conduct the research; (2) those from outside institutions allowing the investigator to conduct work on their properties; and (3) those issued by regulatory agencies allowing the investigator to conduct work on the species of interest. If the collection and export of specimens are involved, the number of permits required increases significantly. Although permits are critical to a study, the permitting process receives little to no attention in the literature on study planning. Local (city, county), national (federal, state, province), special (parks, wildlife refuges and reserves, protected forests, traditional native lands), and international (e.g. European Union, CITES) permits may be required. To illustrate the complexity of the permitting process, I will step through an actual example of what it takes to conduct critical research on one endangered species. The time required for obtaining permits and the order in which they must be obtained makes some types of research or conservation activities temporally beyond the scope of short-term projects, such as graduate theses.

To start a project, an academic institution or resource agency often requires a study or research plan that outlines the background, questions, hypotheses, and methods of the study. If a grant proposal for funding the project exists, it can often be used. In addition, methods for research involving vertebrates need to be reviewed by an Institutional Animal Care and Use Committee (IACUC) in the United States and some other countries to ensure compliance with all federal regulations and verify that the researcher is trained to carry them out. IACUC approval is required before research can be initiated or funding released. Many IACUCs have specific submission deadlines and review meeting schedules (e.g. monthly or quarterly), so researchers must plan accordingly.

Researchers must determine whether similar protocol reviews are required by other national or international (e.g. European Union) entities.

Permits may be required for conducting research in a particular country. Going over the steps I needed to take to legally conduct research on iguanas (*Brachylophus* spp.) in Fiji will illustrate some of the complexities. Fiji requires a research permit to conduct research. An application, including a research proposal, letter of support, résumé, and copies of published research papers, is submitted to the Ministry of Education, Heritage and Arts for review and ministry approval. My first step was to obtain a letter from a governmental or other local institution (e.g. university) to support the research; I worked with the National Trust of Fiji. The letter might be obtained quickly provided that the researcher has already taken the time to develop a relationship with the sponsoring institution. Review and approval of the application could take 3–6 weeks, depending on the committee's schedule. Reports are due intermittently through the life of the permit to the Ministry.

Once the research permit was obtained, I had to apply to the Ministry of Immigration for a research visa. The application had to include payment of substantial fees, the research permit, a police report, medical assessment, résumé, and research proposal. This process could take another one to two months. Once the visa was obtained, research could be conducted legally in Fiji. Research visas are valid for 18 months, but must be renewed every 6 months; the application to the Ministry for renewal requires submission of a progress report covering the previous six months. Another permit, from the Biosecurity Authority of Fiji (BAF), is required to collect and transport research samples. They restrict, with major fines for non-compliance, holding species non-native to Fiji (e.g. the invasive Green Iguana, *Iguana iguana*). They also regulate all movement of live native species between islands and must be notified of such potential events. Their goal is to reduce the spread of invasive species or native species outside their native ranges in Fiji.

Once these permits are all in order, the researcher must seek permission to access lands where the research will be conducted. Approximately 87% of the land in Fiji is Native Land that belongs to a village group or 'land-owning unit' under the iTaukei Land Trust Act. The Provincial Council, which represents all of the chiefs from the province, must approve access to Native Lands. The sponsoring institution in Fiji sends a letter to the Provincial Council at least a month before access is required identifying the time frame and specific places where research will take place. The Provincial Council informs the traditional land owners (Mataqali) and seeks their approval. When the actual field visit takes place, the researcher must go first to the Provincial office to meet with the *Roko Tui*, who is the appointed head of the Provincial Council. They then perform a 'Qaloqalovi' ceremony, which is the traditional welcome ceremony that consists of a 'sevusevu' during which one asks permission to enter the province and do the research. The 'sevusevu' consists of presenting a gift of 'yaqona' to the *Roko Tui*. Yaqona is the dried root of the kava plant (*Piper methysticum*), and it is ground into a powder and mixed with water for drinking. You present it as a root, then it is ground and you drink it as part of the ceremony with the Provincial officers. Any previous reports or papers are presented during this ceremony, as well as educational and outreach materials. When this traditional ceremony is completed the *Roko Tui* invites you to enter the province to do your work, and proceed on to villages.

Once at a village, one needs to find the Village Headman or Turaga *ni Koro* (who works for the Provincial Administration) to explain what is needed from the village. They will then find the village Chief (Turaga) and you will go with them to present the 'sevusevu'. The Headman will present you to the Chief, explain what you are going to do, and ask for their permission to enter village lands to conduct work. Sometimes the Turaga *ni Koro* is not in the village. You have to wait for him to return, which could take hours. You are not allowed to leave where you were taken, typically the Turaga *ni Koro's* house, or move around the village. I have been placed in bures (traditional Fijian houses) previously for hours, even after it became dark and with no lights, waiting for the Turaga *ni Koro* to return. Then, the Headman will proceed to present you to the Chief to conduct the 'sevusevu', as was completed in the Provincial office.

After work is completed in a village, one then needs to present the 'i tatau' prior to leaving the village. This is the farewell ceremony when someone leaves the community. Again the Turaga *ni Koro* will present you to the Chief and you will give a report of the work completed while in the village, and present yaqona. After work is completed in all of the villages you visit in that province, you go to the Provincial office to present the 'i tatau', which represents that you will now be leaving the Province, and give them a report of what work was accomplished, including another presentation of yaqona to the *Roko Tui*.

Once work is completed in the country, permits need to be obtained from the Department of Environment (CITES Management Authority; the signatory authority is listed here: http://www.cites.org/) if exportation of samples will take place. There are typically three types of export permits. Export permits for non-CITES listed species is one type. Some of the species covered under this permit may be listed as endangered by a country, yet not be CITES listed. For these a CITES permit is not required, but other national permits might be required initially to study these species, and different regulations might exist for their export. Typically for export with a non-CITES permit, import into another country is less complex. For import into the United States, for example, one needs to complete eDecs with the U.S. Fish and Wildlife Service (https://edecs.fws.gov/edecshome.cfm) prior to import, if possible; otherwise, the import documentation (3–177 forms) can be filled out manually or digitally and delivered at the port of entry. For other countries, various national or local biosecurity restrictions may apply. The United States also has a list of injurious wildlife that cannot be imported without special permits (http://www.fws.gov/injuriouswildlife/). Separate from the federal import laws, each U.S. state may have its own importation rules and lists of species that are considered restricted and not importable.

The other two types of permit relate to CITES listed species: the CITES Appendix II and the CITES Appendix I export permits. The CITES Appendix II permit covers all species listed in Appendix II of the Convention (www.cites.org) and typically does not require a CITES import permit from the country receiving the samples or specimens, although the CITES permit must be submitted to the representatives of the CITES management authority when arriving at a port of entry. There are specific formats for this permit, and if any elements of the permit are not correct, it will be invalid. Appendix II covers several entire groups of reptiles, e.g. boas, iguanas, and tortoises.

The process for obtaining the CITES Appendix I export permit is the most rigorous and it is only valid for six months. These permits cover species listed in Appendix I of CITES, including some of the rarest species, as well as some not so rare that are regulated to control trade. There is a requirement to obtain a CITES import permit from the country that will be receiving the specimens, prior to export. Sometimes, the CITES import permit will take 90 days to receive, so the overlap in the two permits is relatively short and any export-import must take place within this window. For all permits in Fiji, there is an additional requirement for BAF to check samples for count and species accuracy; they sign off on that as you exit the country. Under CITES and other federal laws, some biological materials do not require permits. Faecal materials are one example; if the country of origin does not regulate these, then even CITES Appendix I samples can be exported or imported with just a simple letter or statement from the country of origin that this material is not restricted.

Overall, it could easily take 6–9 months for permitting certain research projects. Some permits require documentation of the grant of funding and a bond. Some research grants require documentation of all necessary permits prior to funding. If the use of radio telemetry is required, then additional permits might be needed for certain radio frequencies. If chemicals are needed to process samples, there might be another set of permits required. As the Fiji example illustrates, permit requirements may be complex and must be integrated into the overall timing and costs for any research project.

2.9 Ethical considerations

In addition to the IACUC assessments discussed in the previous section, there needs to be an assessment of the risk of the study to the system itself and to other organisms, or more generally, an assessment of the environmental impact of the study. This impact assessment could be as simple as evaluating what impact seining for turtles has on benthic communities of invertebrates in small ponds, or what effect repeatedly walking to field traps has on vegetation or soil compaction. For example, there is often trail development, reduction in vegetation along the trail, and access to the study site by non-researchers when repeated walking takes place for months or years. These impacts might change the responses of the target species to the habitat or sampling plots through a feedback loop. Objective consideration of such impacts on study design is relevant and should be considered. For example, when conducting 1 ha total removal plots, the effect could be important; if this is the only way to conduct the work at determining absolute species density, this effect can be described but not be mitigated (Rodda et al., 2001). Study designs can be thoroughly examined and discussed as to whether the value of the study overrides the non-target impacts to other species or the environment.

In some cases, the sampling method might be illegal or heavily regulated, such as the use of mouse sticky traps or glue boards in the United Kingdom, Australia, and

New Zealand. They are restricted in these countries when applied for home use because of animal cruelty concerns towards the targeted mammalian species. When used properly, these traps are an excellent method for standardizing sampling effort in tropical woodland habitats for lizards, in particular for skinks and geckos (see Chapter 11); however, if used irresponsibly, they can result in high levels of mortality.

2.10 Biosecurity

Biosecurity is often the last thing remembered when designing field studies, although critically important (see Chapter 28). Key issues with biosecurity involve aquatic invertebrates, plants, and pathogens such as amphibian chytrid fungus. The simple act of moving sampling equipment or traps between sites may bring risk to habitats or other species. Traps must be cleaned and seeds or parasites removed between uses. When sampling amphibians, we treat equipment and clothing between sites, and the same care should be taken when someone moves turtle traps or as researchers move between sites when capturing water snakes. Some aquatic invasive invertebrates become ecosystem engineers, such as New Zealand Mudsnails (*Potamopyrgus antipodarum*). Mudsnails average 5 mm and can reach densities of 500,000 per m^2 (http://nas.er.usgs.gov/queries/factsheet. aspx?SpeciesID=1008). They can hide anywhere—in seams of waders, in webbing on traps, in handles of nets—and can survive out of water for extended periods of time. Having site-specific equipment and clothing increases study costs significantly, so potential environmental benefits and risks need to be carefully considered beforehand.

For terrestrial biosecurity, shoes, clothes, traps, and collecting bags or containers are often identified as the mechanism for between-site movements of invasive seeds and the spread of disease. For example, cloth collecting bags could be significant risk factors since bringing snake bags from labs where live animals are kept could expose field captures to mites or diseases that might put wild snakes at risk. To prevent these risks, cloth bags should be washed in hot water and sterilized between uses. The recent outbreaks of fungal skin disease in reptiles is a cause for serious concern, as some of the strains identified are the same in captive and wild snakes; infections are typically contagious and fatal (Sigler et al., 2013). The same concern might be relevant to tubes used for venomous snakes, and these should be sterilized between uses.

Biosecurity also concerns the inadvertent spread of invasive reptiles and their parasites during conservation programmes. Currently the use of rat poisons, as mentioned for eradicating rats on the Galapagos (Section 2.7), is part of a management strategy for restoring the tortoises and other reptiles at risk of extinction. These same rat poisons, when sitting in plastic containers or on pallets in ports in the tropics, can become vectors for the movement of invasive species such as the House Gecko (*Hemidactylus frenatus*) that place their eggs in small crevices such as under bucket lids (Fisher, 2011). This species continues to disperse a constant 'rainfall' of propagules (eggs) around the tropical Pacific, as identified by genetic markers (Tonione et al., 2011). This last example illustrates some of the complexity involved with biosecurity planning and implementing strategies to 'do no harm'.

2.11 Conclusion

In planning and designing field studies, there are many questions that need to be considered. Critical to this is having an understanding what success might look like for the research. For instance, will some management action be informed by the results of this research? Some parts of field studies typically are underappreciated (i.e. permits) in both their complexity and the amount of time they take. Thinking logically about these considerations before heading directly into research design should streamline the planning process and increase the likelihood of a successful research project.

2.12 Example URLs for SMART objectives

http://www.cdc.gov/dhdsp/programs/spha/evaluation_guides/docs/smart_objectives.pdf
http://iom.nationalacademies.org/About-IOM/Making-a-Difference/Community-Outreach/Smart-Bites-Toolkit/~/media/17F1CD0E451449538025EBFE5B144 1D3.pdf
https://www.projectsmart.co.uk/smart-goals.php

References

Aguilera, W.T., Malaga, J., and Gibbs, J.P. (2015). Giant tortoises hatch on Galapagos island. *Nature*, **517**, 271.

Atkinson, A.J, Trenham, P.C., Fisher, R.N., et al. (2004). Designing monitoring programs in an adaptive management context for regional multiple species conservation plans. U.S. Geological Survey Technical Report, USGS Western Ecological Research Center, Sacramento, CA. Available at: http://www.scwa2.com/home/showdocument?id=1010.

Barrows, C.W. (1996). An ecological model for the protection of a dune ecosystem. *Conservation Biology*, **10**, 888–91.

Barrows, C.W., and Allen, M.F. (2010). Patterns of occurrence of reptiles across a sand dune landscape. *Journal of Arid Environments*, **74**, 186–92.

Cogălniceanu, D., and Miaud, C. (2010). Setting objectives in field studies. In C.K. Dodd, Jr. (ed) *Amphibian Ecology and Conservation. A Handbook of Techniques*. Oxford: Oxford University Press, pp. 21–35.

Fisher, R.N. (2011). Considering native and exotic terrestrial reptiles in island invasive species eradication programmes in the tropical Pacific. In C.R. Veitch, M.N. Clout, and D. Towns (eds) *Island Invasives: Eradication and Management*. Gland, Switzerland: IUCN, pp. 51–5.

Fisher, R.N., and Mitrovich, M. (2012). Study design and sampling: overview. In R.W. McDairmid, M.S. Foster, C. Guyer, et al. (eds) *Reptile Biodiversity: Standard Methods of Inventory and Monitoring*. Berkeley, CA: University of California Press, pp. 27–9.

Franklin, J., Wejnert, K.E., Hathaway, S.A., et al. (2009). Effect of species rarity on the accuracy of species distribution models for reptiles and amphibians in southern California. *Diversity and Distributions*, **15**, 167–77.

Griffiths, P.G., Webb, R.H., Lancaster, N., et al. (2002). Long-term sand supply to Coachella Valley fringe-toed lizard habitat in the northern Coachella Valley, California. U.S. Geological Survey Water-Resources Investigations Report 02–4013. Available at: http://pubs.usgs.gov/wri/wri024013/pdf/WRIR4013.pdf.

Hardisty, A., Roberts, D., and The Biodiversity Informatics Community (2013). A decadal view of biodiversity informatics: challenges and priorities. *BMC Ecology*, **13**, 16.

Margoluis, R., Stem, C., Salafsky, N., et al. (2009). Using conceptual models as a planning and evaluation tool in conservation. *Evaluation and Program Planning*, **32**, 138–47.

Meek, M.H., Wells, C., Tomalty, K.M., et al. (2015). Fear of failure in conservation: the problem and potential solutions to aid conservation of extremely small populations. *Biological Conservation*, 184, 209–17.

Rodda, G.H., Campbell, E.W.I., and Fritts, T.H. (2001). A high validity census technique for herpetological assemblages. *Herpetological Review*, **32**, 24–30.

Schroeder, R. (2009). Evaluating the quality of biological objectives for conservation planning in the National Wildlife Refuge System. *The George Wright Forum*, **26**, 22–30.

Sigler, L., Hambleton, S., and Pare, J.A. (2013). Molecular characterization of reptile pathogens currently known as members of the *Chrysosporium* Anamorph of *Nannizziopsis vriesii* Complex and relationship with some human-associated isolates. *Journal of Clinical Microbiology*, **51**, 3338–57.

Tonione, M.A., Reeder, N., and Moritz, C.C. (2011). High genetic diversity despite the potential for stepping-stone colonizations in an invasive species of gecko on Moorea, French Polynesia. *PLoS One*, **6**, e26874.

Uetz, P. (2010). The original descriptions of reptiles. *Zootaxa*, **2334**, 59–68.

3

Data collection and storage

Richard A. Seigel

No battle plan survives contact with the enemy.

—*Quote attributed to German military strategist Helmuth von Moltke, ca. 1861*

3.1 Introduction

I want to start this chapter by asking you to do what Einstein and others have called a 'thought experiment' and what sports psychologists call 'visualization'. Pretend that you are an undergraduate or graduate student (the prime targets for this text) who has successfully submitted your research proposal (see Chapter 2) and you have just arrived at your study site to begin your thesis work on some aspect of reptile conservation biology. You pull your vehicle off the road at the edge of your study site, shut off the engine, get out of the car, and . . . now what? Your research mentor has approved your sampling plan, but are you ready to record data for your study? If so, how are you going to record those data? Do you have a field notebook? If so, what kind? Do you have data sheets ready to record data on? Have these been field-tested under adverse weather conditions? Do you have a clipboard for storing these sheets? Do you have an appropriate writing instrument? Spares? Camera? Extra batteries and flash cards? Once you have recorded data on your data sheets, how and when will these paper sheets be transcribed into an electronic database? When will this occur? How will these data be backed up?

For many field herpetologists, getting the right answers to these questions was not something that happened the first time out in the field. In many cases (certainly in mine) there were last-minute trips to local stores to get the missing items or frantic calls to fellow grad students or major professors to get items shipped overnight to the study site. In some cases, these oversights resulted in no more than some embarrassment or a short-term cost in data. However, I have known researchers who have lost significant amounts of data due to poor planning or who had to spend many hours trying to recover information that was lost due to not performing this kind of thought experiment.

The primary goal of this chapter is to provide a beginning field herpetologist a basic overview of the *logistics* involved in conducting a field study on reptile conservation. This chapter will not cover *how* to collect these data; such information can be found elsewhere

Reptile Ecology and Conservation. Edited by C. Kenneth Dodd, Jr. © Oxford University Press 2016.
Published 2016 by Oxford University Press.

in this text (see Chapters 2, 4, and 6). Instead, this chapter focuses on the logistical preparation and planning needed to successfully conduct a field study on reptiles, with emphasis on the need for flexibility when working with such wonderful (but often uncooperative) animals under conditions that are always unpredictable and frequently frustrating.

3.2 Flexibility: the research proposal versus the real world

Without trying to duplicate materials elsewhere in this text (Chapter 2), it is important to emphasize just how critical a well-thought-out research proposal is when doing field work in conservation (see Congdon and Dunham, 1999, for a good overview). While many of the comments you will get from your research professor or supervisor for such a document will, of necessity, focus mainly on your overall hypotheses and statistical design, a critical part of your proposal can and should provide a detailed overview of your sampling procedures. I cannot count the number of times I have read statements in student research proposals that say something such as 'snakes will be weighed, measured, marked, and released at the site of capture' without noting *how* they will be weighed, using *what* instruments, and what the *accuracy* will be when doing so.

Ideally, anyone reading your proposal should be able to use the text to repeat the study, i.e. correctly identify the equipment needed to do your field work as well as what information will be recorded on your data sheets. In the real world, however, the opposite is often true, and the data recorded in the field are often either less than what the proposal called for or you wind up recording data that you never planned on collecting. These outcomes can stem from three causes. First, you may have simply not given your planned research enough thought and your mentor or thesis committee may have not provided you with a sufficiently critical review. It happens. Second (and more commonly), you find that data that made perfect sense to collect when you or your professor were planning your research in your nice comfortable office do not make a lot less sense when you are actually in the field and are constrained by time, resources, and safety. When I was starting research on Massasauga Rattlesnakes in Missouri for my Ph.D., my major professor (Henry Fitch) strongly encouraged me to track fang replacement rates by opening the mouth of each Massasauga and counting the number of fangs. That made great sense when he proposed it to me and a lot less sense when working by myself in the field, trying to keep a struggling venomous snake from hurting me or itself. I quickly learned that the investment of time, loss of safety, and increased stress to the snake were not worth the data being collected, so this was dropped from my research plan (to Henry's dismay, I should add).

The third cause for the deviation between your proposal and your actual data collection occurs when you realize that you have opportunities to collect additional or different data than you had planned when writing your proposal. Here is where the concept of *flexibility* (as espoused by General von Moltke) becomes paramount. Without losing sight of your primary goals, having the flexibility to realize that data you never planned on collecting could be as or more important than what you and your committee originally envisioned could be the difference between an ordinary study and one that makes a truly original contribution to the field.

Voices of Experience I What data to collect

A chapter such as this can easily become something of an editorial essay, rather than a summary of divergent views. To counter this, I reached out to my colleagues in the form of an online poll and asked them a series of 14 questions regarding recording field data for conservation biology. A total of 40 eminent figures in herpetology (including all of the authors of this text) were asked to participate and 28 of these kindly lent their ideas and comments to this poll. A summary of these anonymous responses will be presented throughout this chapter under the heading 'Voices of Experience'.

One of the questions posed to these experts focused on the concept of how much data to collect, as noted previously. Fifty-six per cent of those polled reported having collected either too much or too little data and having to adjust their plans while in the field. Interestingly, they were almost equally split between feelings that they had initially planned on recording too much extraneous information (30%) versus too little (26%). Some example comments include the following:

- Once started, data collection can be streamlined/shortened after a preliminary analysis that reveals some of your data items are overkill and likely preventing you from concentrating on your field experiments.
- Decide what data you are collecting and why before beginning your field work.
- Don't collect data simply because others have done so. Many turtle biologists record multiple shell measurements (carapace length, depth, width) without realizing that all these measures are so strongly interrelated that they are not independent variables, thus wasting precious field time.
- My advice to students is to be consistent and thorough in recording data and incidental observations. Think hard before you start a project about the questions/hypotheses and the way you hope to analyse your data.
- Some faculty members in my department said that students should collect only the minimum amount of data necessary to get a significant result, which I strongly disagree with and is why I have been able to publish as many papers as I have.

3.3 Field notes

Field notes have been recorded by biologists for centuries, with the detailed notes of John James Audubon, Meriwether Lewis, and Charles Darwin being perhaps the most famous examples (e.g. Lewis and Clark, 2002). However, the need for anyone working in the field to keep a record of their daily work activities is something to get into the habit of doing early in your career. Of the 28 biologists polled for 'Voices of Experience', 89% said that keeping accurate field notes was critically important and 11% said it was moderately important, so it is very likely that your research mentor will be asking you to record notes during your field work.

3.3.1 Mechanics of field notes

Although the need for recording accurate field notes was considered a priority by all the biologists polled, *how* you record these notes produced somewhat more variable responses. Three key decisions dictate how you will record these notes: (1) the kind of notebook you use, (2) the way (format) you record those notes, and (3) what kinds of information you record. Each of these is reviewed briefly in the following subsections.

3.3.2 The field notebook

By far the most popular field notebook in my 'Voices of Experience' poll was the Rite in the Rain series produced by J.L. Darling LLC (http://www.RiteintheRain.com). Available in a wide variety of sizes and formats, some variant of these notebooks were used by 46% of the biologists polled and 81% of those who mandated a specific kind of notebook. These notebooks have the advantages of being relatively inexpensive (generally less than US$8 from online suppliers), not damaged by exposure to water or rain, and easy to transport. The bright yellow covers that these notebooks usually employ also have advantages, as they can often be recovered when lost in the field (see the following discussion). Three-ring binders are also commonly used as field notebooks, as are a variety of other small notebooks from commercial suppliers such as Office Depot or Staples in the United States.

The advent of small, portable tablet computers has introduced another option to the choice of a field notebook. As I noted in a recent paper (Seigel, 2013), these devices have at least two significant advantages to field biologists when compared with standard notebooks. First, since notes are recorded in a standard format (Microsoft Word or compatible), notes can be stored in the cloud and then easily searched for specific entries or events even when the notebooks are not physically present. This may not seem like much of an advantage when you are first starting out, but after a number of years in the field, attempting to find information recorded in a notebook a decade or more ago is much easier when the notes are available in a searchable format. This also makes notes immediately available to groups of workers collaborating on a project. Second, since the information can be backed up both in the cloud and on portable media, physical loss of a notebook and all the information it contains is much less likely. Again, this is not a hypothetical possibility. One of my former students had his apartment robbed recently and thieves took all of his field gear, including all of his field notebooks. I have also been in an academic building when it was struck by lightning and caught fire and my first thought was that I was going to lose 10 years of data if the fire reached my office. Certainly, making xerographic copies of field notebooks is possible, but I doubt many biologists are doing this on a routine basis. Finally, new commercial programs (e.g. http://www.neukadye.com/mobile-applications/field-journal/) allow automatic geolocation and time-stamping of field notes as well as automatic cloud storage.

Despite these advantages, electronic notebooks are not yet widely accepted by those in the 'Voices of Experience' poll (see Voices of Experience II). The main concerns centre on having tablets malfunction under field conditions or losing them in the field. The first issue can be a cause for concern when working in wet or dusty conditions, although there are modified (albeit expensive) tablets that are designed to be used in demanding environments (e.g. Samsung Galaxy Tab Active). The second caution involves loss in the field, but this can happen to any notebook, electronic or otherwise, and I would suggest that if you back up your tablet daily, the loss of data will only be for that day, not for the entire project as would be true when losing a standard field notebook (assuming you are not photocopying it daily).

The biggest drawback with electronic tablets as a replacement for field notebooks has to do with costs, which are obviously much greater for tablets. However, tablet prices as of 2015 are down to less than US$100–150 for 7" field-portable devices, and these could theoretically be used for many years, unlike standard notebooks, which fill fairly rapidly and need constant replacement.

A useful supplement to both hard-copy field notebooks and tablet computers are digital voice recorders, either as stand-alone units or by using an app on smartphones equipped with a microphone for such purposes. Audio recordings can be especially useful when the investigator does not have a hand free for writing or typing, such as while walking a line transect or while watching behaviours of fast-moving lizards. Stand-alone units are easily field-transportable (masses between 70 and 180 g), relatively inexpensive (US$50 in the United States, £25–50 in the United Kingdom), and some offer the option of direct downloads onto a hard drive for convenient storage as well as automatic 'time stamps' for recording date and time of the recording. Obviously, information from such audio files must eventually be transferred to hard-copy or electronic data sheets for analysis and should be archived for later retrieval (Section 3.6.1).

3.3.3 What to record and how to record it

The information you should record in your field notes is, of course, a reflection of the goals and aims of your project, so coming up with a single, standardized list of information to be recorded would not be practical or in keeping with the flexibility concept underlying this chapter (see also Greaves, 2012). However, there are certain key pieces of information that most field biologists would likely want to make a priority for recording. A basic, but non-exhaustive list, would include: date (always written as 9 March 2015 and NOT as '3–9–15' which can be read differently in different cultures), specific locations visited (include state or province, county, and GPS coordinates), time of starting and stopping specific field activities (using military time [00:00–24:00] only), and basic weather conditions including shaded air temperature (Celsius), wind speed and direction, and approximate cloud cover (sunny, partly sunny, cloudy). There are a number of relatively inexpensive and field-portable devices that record basic weather conditions quickly, with the Kestrel Pocket Weather Meter being especially well known (http://kestrelmeters.com/). Other information that can be recorded includes who is with you in the field (which can be helpful in calculating sampling effort) and any unusual occurrences or observations that are not going to be recorded in your data sheet. Stebbins (2003) has additional ideas for what information to record.

How one goes about recording field data in a notebook is highly personalized and there is no 'right' or 'wrong' way to do this. That being said, there are two basic styles, each with advantages and disadvantages. Perhaps best known is the 'diary style' of recording field notes, long associated with Joseph Grinnell (Grinnell, 1912). An example of this style of recording field notes is reproduced in Stebbins (2003: 21). This style takes longer to record, but is certainly more interesting to read. The second basic style can be referred to as the 'bullet style', which eschews the details of the diary style for a terser format that focuses more narrowly on the basic information noted at the start of this subsection.

For any hard-copy field notes you record, do not forget that all notes must be written using a writing instrument that will not run or smear if the notebook gets wet. I have seen students try and use ballpoint pens whose ink ran all over the page the first time it rained in the field. See comments under Voices of Experience II for more on this. Finally, about half of the colleagues polled indicated that students must leave their field notebooks behind when they leave the lab, so give some careful thought about any extraneous (non-scientific) information you might record.

Voices of Experience II Field notebooks

Kinds of notebooks

- In general, I still have students use paper, lined, bound notebooks with hardbacks. I am open to students presenting alternatives that we can discuss and decide on for specific projects.
- Field notes must be written in indelible ink.
- We do work in the water so it is best to have a waterproof book.
- Rite in the Rain is the standard I have always followed, but of course this is as much habit as anything else.
- Rite in the Rain is the standard. You know that it will hold up to any conditions. Loose-leaf pages get lost.
- Hard bound Rite in the Rain field books NO exceptions, must also write with permanent black resistant ink. No ballpoint pens, etc.
- Don't forget to put your name and contact information on whatever device you use for field notes, in case they become lost.

Electronic versus hard copy

- Obviously hard copy. I am opposed to electronic, which is acceptable as a supplement as long as it does not deprive the hard copy from essential items.
- For some studies, electronic data collection might be useful, but I've never done it.
- I don't trust electronic storage devices, plus they can get destroyed in field conditions.
- I have also used/had students use an iPad in the field. I believe this cuts down on errors in data entry and transcribing.

Format and style

- Student choice, but 'bullet' format is faster.
- For general field notes I require that the 'bullet' be the time of day that the observation or record was observed.
- Not a stickler on actual format but desire a consistent format that contains all the essential items.
- I am old school and most field notes are in diary format. The key thing is to record observations we may or may not enter these into a project database.
- I think the best way to record field notes is using a checklist, because you can obtain data on species absence/non-detection that way. Effort must be recorded as well.
- I like seeing some text largely, so that there will be context for the specific information in field notes.
- We use both a diary format to document the day's activities and bullet format for the data collected; you need both.

3.4 Data sheets

3.4.1 General considerations

Although your field notes are highly important, the data that will be most critical to your thesis or publication are likely going to be that information recorded on your data sheets. When done properly and with some careful thought, your data sheets should accomplish three things: (a) acts as a mechanism of 'quality control' to ensure that you have recorded everything you and your research mentor agreed needed to be recorded; (b) allows recording of your field data in a quick and efficient way; and (c) facilitates the transcription of your data from the data sheets to the computer program used for later data analysis.

Creating data sheets that fulfil all three of these criteria takes time, careful consideration, and appropriate testing in the field. How often this happens is unknown, but too often I see students struggling in the field with data sheets that simply are not suited to the study at hand. Indeed, I would argue that you should have your data sheets reviewed by at least two colleagues who are familiar with your study and then taken to a field site for a trial run before being deployed for actual data collection. One of the most common things my students learn from such trial runs is whether the spaces available on the data sheets are large enough for the information they need to record. I have seen data sheets where the space provided for recording UTM information from a GPS unit is so small that even someone with excellent handwriting would be hard-pressed to find enough room to fit the necessary data. In addition, you should give careful attention to the order in which your data sheet calls for information to be entered. This depends on the nature of the reptiles you are processing. For example, when processing venomous snakes using a hotbox or tube, the last data point to record might well be body mass, since it is easier to go from the tube into a bag than the other way around.

3.4.2 Mechanics of data sheets

As is true of field notes, data sheets can be prepared via either traditional hard copies or electronically using a laptop, tablet computer, or even a smartphone. Although most of my colleagues and students still use hard copies for data sheets, I have argued (Seigel, 2013) that a key advantage of recording data directly into a spreadsheet or database is the reduction of error rates incurred when transcribing information from a hard copy to the computer, particularly if constraints are placed on the data that can be entered, perhaps through pull-down menus with pre-selected choices. In addition, although loss of a tablet in the field is an important consideration, the same can happen with hard copies. I had one graduate student who kept multiple pages of data sheets in an aluminium cover (see next paragraph), but, despite my repeated requests, never photocopied these sheets and kept at least the copies at the lab. Unfortunately, he lent his binder to another student who promptly lost it in a local river, taking with it at least a month of hard-won field data. See the section on 'backing up your data' for more on this theme.

Likely the most widely used materials for data sheets in the field are hard-copy data sheets using 'Rite in the Rain' paper and a closable aluminium clipboard for both writing and storing and protecting data sheets. These products are available in the United

States from Forestry Suppliers, which has an international sales department (http://www.forestry-suppliers.com), and other online retailers, and Rite in the Rain products are available in other countries as well (from Amazon in the United Kingdom, Damen/Papier Royaal in the Netherlands, and the Mercado Livre chain in Brazil, just as a few examples). I recommend tying a length of fluorescent flagging to clipboards to facilitate finding them when (not if) you leave yours in the field when moving between sites.

Voices of Experience III Data sheets

- Must write with permanent black resistant ink.
- Establish a protocol at the beginning of each study and then design data sheets (or spreadsheets) that organize the data to be gathered in an efficient manner. Leave some room or time in each sample for additional comments or records that may not have been anticipated in your planning. Always do a quick preliminary sample to test the effectiveness of your record taking and adjust before starting your project.
- Where possible, data sheets should incorporate units of measure and lists of categorical data. Having a prepared list of choices so that one can be circled will minimize variation in how data are collected.
- Be sure to test your data sheets under simulated field conditions BEFORE starting 'real' data collection. This often reveals issues with the structure of the data sheets or items you forgot to include.
- One of the most important is to transfer the data to Excel as quickly as possible so they can fix things before they forget them.

3.5 Documenting the field site: photographs, GIS, and environmental data

In addition to data sheets and field notebooks, recording additional information on your field site can be an essential compliment to your data set. In the 'Voices of Experience' poll, 79% of those responding routinely documented their study sites using digital photographs at intervals ranging from monthly (4% of those taking photographs) to simply 'whenever possible' (79%). Keep in mind that if you are conducting a long-term study, it is highly possible that you will eventually represent the longest 'institutional memory' of that area and photographs that show changes in the site over time may prove invaluable. Files containing such photographs need to be labelled clearly and georeferenced where possible. Backing up these files is also critical (see Section 3.6).

I did my first undergraduate research at the Great Swamp National Wildlife Refuge in New Jersey (USA) in the mid-1970s. In the early 1990s, I corresponded with the refuge biology staff about re-starting my research, aiming to look at long-term changes in turtle populations. When I went for a field tour with the staff biologist, I was especially eager to see a red maple swamp that contained the largest population of Spotted Turtles

(*Clemmys guttata*) on the site. When I told the biologist this, he said that no such area existed. Going to the site, I found it had been converted to farmland, with all the turtles clearly now gone. None of the current staff even knew that such a site had existed 20 years before, so were unaware that their largest population of Spotted Turtles had been extirpated. They did (eventually) believe me about this, but photographs of the site from the 1970s would have made a big difference.

In addition to photographs, recording landscape (GIS/GPS) information and environmental data could prove highly important. In the 'Voices of Experience' poll, 32% of those responding regretted not having recorded additional data on environmental variables such as constant recording of air and water temperatures, 31% would have recorded more data on habitat/landscape variables, and 18% would have recorded more detailed locality data using GPS units. Given the wide availability of both GPS units (both standalone and as free apps on many smartphones) and data loggers such as iButtons (Maxim Integrated, San Jose, California) or HOBO TidBits (Onset Computer Corporation, Bourne, Massachusetts), recording such information today in a more detailed and rigorous manner seems fairly apparent (see Chapter 24 as well).

Voices of Experience IV Additional field data

- These data may come in handy beyond the original study goals. For example, I did not anticipate rapid climate change in early field studies and now going back to these sites wish I had better environmental and vegetation data.
- Hindsight is always 20:20. Early in my career I did not use GPS, now it is habitual. At times I have been inconsistent in collecting weather/temperature data. Other species present and relative abundance are often missing from my earlier work, but there is a trade-off because recording this type of extraneous data does take time.
- Survey temperature and canopy cover are measures I wish I had routinely taken. I would still only record more labour-intensive measures, such as specific vegetation communities, if they are directly relevant to a motivating research question. Having a thorough 'Notes' field in databases has proved useful on occasion during data proofing but have never been used formally. Being more systematic about what kinds of observations are taken may require less time and thought.
- It would be nice to have more detailed environmental data for use in climate change analyses.

3.6 Data: backing up and archiving

Once you have collected your field data, careful thought needs to go into how you are going to back up and archive these data. First, we must realize that backing up data and archiving them are not the same thing. As multiple online services will tell you, *backups* are relatively short-term mechanisms for keeping data sets that are in active use from being lost, whereas *archives* are for long-term storage of data sets that are no longer being added to or used. Historically, biologists have paid more attention to backups than to archiving, but new technologies are changing that situation.

3.6.1 Data backups

Assuming that most field biologists have not yet adopted tablet computers for field recording of data, the most basic way of backing up data is by photocopying or scanning data sheets and leaving these copies in a safe location. This is especially important if there is a significant delay in transcribing hard-copy data sheets into an electronic file (see comments about loss of data sheets in Section 3.4.2). In the 'Voices of Experience' poll, only 4% of respondents reported converting data to an electronic format on the same day as being generated and only 29% did so within a week. The rest transcribed the data either within a month (43%) or at intervals longer than 30 days (25%).

Once data are in an electronic database, these files need to be backed up to prevent loss from electronic corruption, theft, or other disasters. Current technology offers a wide variety of backup mechanisms, including 'thumb drives' (flash drives or Secure Digital cards), external hard disks, and cloud storage. Costs for these range widely, from free for many cloud services such as DropBox (http://www.dropbox.com) or Google Drive (http://www.google.com/drive), to less than US$15 for a small (8 GB) flash drive to less than US$75 for a medium-sized external hard drive (750 GB). Given the low costs for these technologies versus the catastrophic consequences of losing data, I insist that my students make routine backups on a portable hard disk in our lab in addition to copies on flash drives or the cloud, and on their main computer hard disks (three copies). I find that cloud backups are perhaps the easiest to use, as it is easy to 'synch' copies between a hard disk and the cloud, making it less likely to overwrite a newer file with an older one.

In the 'Voices of Experience' poll, the most common mechanism used for backing up data was an external hard drive (41%), followed by cloud storage (19%) and a flash drive (17%). Surprisingly, 24% of those surveyed reported they made no backups of their data, with all information stored on their main hard drive.

3.6.2 Data archiving and metadata

A fairly recent development is the advent of online data archiving services that allow authors of published papers to place the data from a published study on a cloud-based server where the information contained is available permanently to the author or other researchers. Dryad (http://datadryad.org/) is a well-known repository for such data, and it counts among its members the Ecological Society of America, The British Ecological Society, the American Society of Naturalists, and the U.S. Fish & Wildlife Service. Depending on whether the journal you are publishing in is a member, costs for the service can be free or up to US$90 (as of March 2015). Given that many studies of conservation biology of reptiles may have long-term importance, this seems like a reasonable price to pay for having a safe and secure place for data storage and retrieval.

An important consideration when archiving data sets concerns *metadata* (literally, 'data about data'). Metadata are essential in guiding a user through the structure and definitions used in a data set (Michener et al., 1997). As Michener et al. (1997) put it:

'metadata comprise all information that is necessary . . . to enable long-term secondary use (reuse) of the data set . . .' For example, it is common in data sheets and spreadsheets to use acronyms and abbreviations such as CL or BM to stand for variables such as 'carapace length' or 'body mass'. However, what seems obvious to you at the time you are creating these data files may not be quite so obvious to you when you are 10 or 20 years older or when someone else is trying to interpret your data. Space constraints prevent a more detailed examination of metadata and data management in general, but useful websites include DataONE (https://www.dataone.org/data-management-planning) and a U.S. Fish & Wildlife Service site devoted to biological metadata (http://www.fws.gov/stand/standards/pr_biomet_WWW.html).

3.7 Conclusions

Recording data in field notebooks and on data sheets are some of the most fundamental activities in herpetological ecology and conservation. Done well, these activities make publishing well-received papers much easier and can reduce staffing time and costs. Done poorly, critical data can be missed, research opportunities can be lost, and time spent in the field or office significantly increased. Modern technologies are providing more diverse methods of recording and preserving such hard-won field data, but adoption has been slow and more basic technologies remain a fixture in our field.

Acknowledgements

I thank James D. Anderson for first teaching me the value of field notes. Thanks also go to my colleagues who participated in the 'Voices of Experience' poll, and to Nathan Byers, Scott Martin, and two anonymous reviewers for constructive comments on the text.

References

Congdon, J.D., and Dunham, A.E. (1999). Defining the beginning: the importance of research design. In K.L. Eckert, K.A. Bjorndal, F.A. Abreu-Grobois, et al. (eds) *Research and Management Techniques for the Conservation of Sea Turtles*. IUCN/SSC Marine Turtle Specialist Group Publication No 4. IUCN/SSC, pp. 83–7.

Greaves, S. (2012). Making, maintaining, and using serious field notes. Citizens Science League. Available at: http://citizenscientistsleague.com/2012/02/09/making-maintaining-and-using-serious-field-notes/.

Grinnell, J. (1912). An afternoon's field notes. *The Condor*, **14**, 104–7.

Lewis, M., and Clark, W. (2002). *The Journals of Lewis and Clark*. F. Bergon (ed). New York: New American Library.

Michener, W.K., Brunt, J.W., Helly, J.J., et al. (1997). Nongeospatial metadata for the ecological sciences. *Ecological Applications*, 7, 330–42.

Seigel, R.A. (2013). Applicability of 'tablet' computers for use by field biologists. *Herpetological Review*, 44, 82–5.

Stebbins, R.C. (2003). *A Field Guide to Western Reptiles and Amphibians*. 3rd ed. New York: Houghton Mifflin.

Part 2
The Individual

4
Marking and measuring reptiles

John W. Ferner and Michael V. Plummer

4.1 Introduction

This chapter reviews marking and measuring techniques for most reptiles (sea turtle tagging is covered in Chapter 15 and crocodilian tagging in Chapter 16). Field and behavioural studies often require marking individuals for studies of growth and age, survivorship, movements, and other phenomena that require repeated identification of individual animals. Ethical issues relative to certain techniques may be of concern and thus, some older techniques such as radioactive tags and mutilation procedures should be discouraged. On the other hand, major advancements in digital identification (Chapter 5), telemetry (Chapter 9), and passive integrated transponder (PIT) tags have added flexibility in the choice of marking methods.

Recent reviews of marking and identification techniques for reptiles include those by Henle et al. (1997), Baker and Gent (1998), Ferner (2007), and Plummer and Ferner (2012). Some criteria to consider for ideal marks or tags are as follows (after Ferner, 2007, and Plummer and Ferner, 2012):

1. They should not affect the survivorship or behaviour of the organism.
2. They should allow the animal to be as free from stress and pain as possible.
3. They should identify the animal as a particular individual or member of a cohort, if desirable.
4. They should last indefinitely or at least through the duration of the study.
5. They should be easily read and/or observable by all informed individuals.
6. They should be adaptable to organisms of different sizes.
7. They should be easy to use in both laboratory and field, and use easily obtained material at minimal cost.
8. They should be tested to meet these listed criteria before being put into widespread use.
9. They should prevent marking application tools from being reused without first being thoroughly disinfected and cleaned.

No techniques satisfy all of these criteria in all circumstances. Criteria 1, 5, and 8 can be especially challenging to satisfy. Selecting a technique, then, requires deciding which criteria are most important for any particular study. Often, two or more marks can be used in order to meet more of the criteria (e.g. toe-clipping and photographing dorsal pattern), and techniques described for use with one group of reptiles may be adapted for a different group. Therefore, investigators might benefit by considering all available techniques. Standardization of marking techniques can be advantageous in comparative studies or studies that might be continued by future researchers.

In this chapter, we emphasize the use of marking techniques for adult reptiles that have proved most practical in major studies. We include information on the advantages and disadvantages of each technique when available. As far as possible, technical information is provided for investigators to consider without consulting original publications. Once a technique has been selected, however, we advise investigators to consult the primary literature if time and facilities permit for important details and information on special problems. The use of more complex techniques, such as PIT tags, will require a review of original references. Sources for marking materials needed for many of the techniques discussed here are listed in Table 8.1 of Ferner (2010), and all are still valid as of this printing; availability and contact information for those and others included in the following may change over time.

4.2 Toe-clipping

Toe-clipping with small sharp scissors or fingernail clippers is the most common technique used for marking lizards (Figure 4.1). Tinkle (1967) suggested clipping up to four toes per individual, but never more than two per foot and no adjacent toes. Numbering systems or codes for the toes clipped have been described by Tinkle (1967), where each toe is given a number from 1 to 20, by Medica et al. (1971), where numbers assigned are added together (Figure 4.1), and by Waichman (1992), where each limb has a letter (A through D) and each digit on the limb a number (1 to 5). Lizard A2D3 thus would have the second digit on the left forelimb and the third digit on the right hindlimb excised. Because natural toe loss can reach as great as 20% (Hudson, 1996), it is important to carefully inspect each lizard before clipping and clip only one toe per limb when possible to eliminate the problem of misidentification from toe loss after marking.

In general, toe-clipping does not appear to have a detrimental effect on lizards (see review by Ferner, 2007). Dodd (1993) found that toe-clipping in *Aspidoscelis sexlineata* had no immediate or permanent impact on sprint performance when only two toes were removed per individual. Australian skinks were found to lose toes naturally at a high rate with no major impact on survivorship (Hudson, 1996), indicating that toe-clipping may be an acceptable marking technique. Huey et al. (1990) reported that the sprint speed of individual *Sceloporus merriami* was not compromised when up to four toes were removed. Similarly, Borges-Landaez and Shine (2003) found no effect of toe-clipping on the average or maximum speeds of *Eulamprus quoyii* in Australia. On the other hand, Bloch and Irschick (2005) reported that toe-clipping on the arboreal *Anolis carolinensis* resulted in a 40% decrease in clinging ability with two toes excised, and 60% with four toes clipped. Paulissen and Meyer (2000) recommended no more than one toe per foot be removed in arboreal or wall-climbing species. By measuring stress via monitoring plasma corticosterone levels in *Eulamprus heatwolei*, Langkilde and Shine

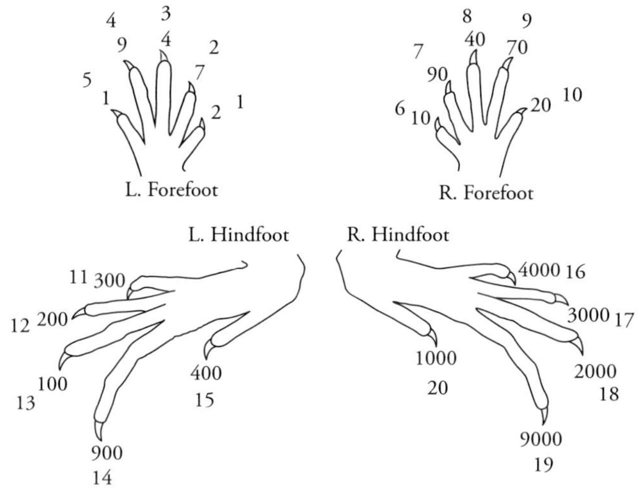

Figure 4.1 Dorsal views of lizard feet with examples of numbered digits. The outer series of numbers follow Tinkle's (1967) system and the inner series represent Medica et al.'s (1971) system. A unique individual number can be obtained by recording or adding the numbers of the clipped toes. For example, clipping the toes 3, 7, and 19 would yield the code 3–7–19 in Tinkle's system and the code number 9094 in Medica et al.'s system.

(2006) found toe-clipping had less impact on corticosterone levels than implanted PIT tags (when blood levels remained elevated for two weeks).

4.3 Scale/scute-clipping

4.3.1 Snakes

Most researchers mark individual snakes by clipping either the subcaudal or ventral scales (Plummer and Ferner, 2012). A ventral scale-clipping system (Figure 4.2) developed by Brown and Parker (1976) with a serial enumeration designation is now most commonly used and eliminates the more problematic use of subcaudal scales. Clipping the ventrals may traumatize the snake, so caution should be used (Fitch, 1987). To clip ventrals, Brown and Parker (1976) used the tips of small, sharp-pointed surgical scissors, describing the procedure as follows:

1. Insert a tip of the scissors under the caudal edge of the scute using sterile technique, push it forward under the entire scute and cut. Make two such incisions, one on each side of the block of tissue to be removed.
2. Insert the scissor tip under and across the top (cranial edge) of the scute and make another cut transversely to remove the entire portion (recommended to be about half of the ventral). The excision should be through the entire skin to expose the ventral musculature. Numbers in the code using adjacent scutes should be omitted from the series as adjacent scales tend to invade excised areas.

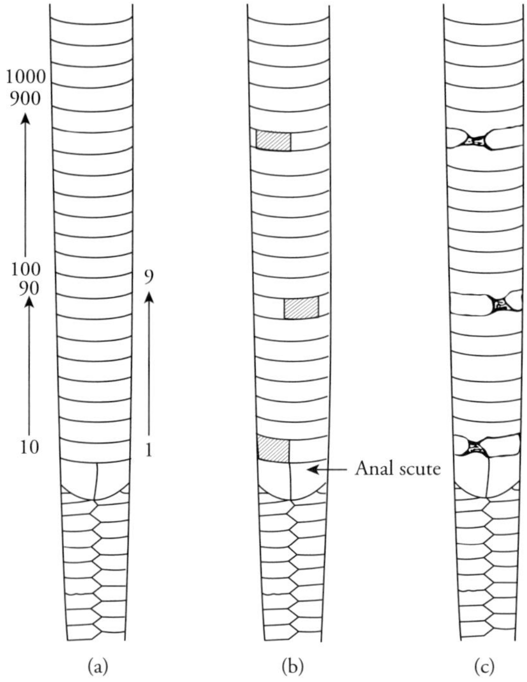

Figure 4.2 Ventral scute-clipping system for identifying snakes. Views of caudal portion of the body of a North American Racer (*Coluber constrictor*). (a) Enumeration of ventrals proceeding cranially from the anal scute: series of scales designated 10s, 100s, and 1000s on the left; scales designating units one through nine on the left. (b) Snake freshly marked as described in text with the identifying number 718. (c) Snake 718 showing resulting scars three years after marking. (From Brown and Parker, 1976; © *Journal of Herpetology*, redrawn with permission, as seen in Plummer and Ferner, 2012).

Brown and Parker (1976) found these marks to last at least four years. Moreover, shed skins from marked *Coluber* could be identified more than 90% of the time (Brown and Parker, 1976). Clipped snake scales may need to be reclipped periodically as regeneration can obscure the scars after several years and make identification difficult (Fitch, 1987). Plummer (1980) found that scute patterns may sometimes be anomalous (e.g. having a ventral duplicated on one side) or result from injuries, and suggested these conditions should be incorporated into any scute-marking system.

4.3.2 Lizards

Rodda et al. (1988) clipped and removed three dorsal crest scales with scissors on *Iguana iguana* to delineate three sections on the crest, spaced 11 scales apart. The 10 scales within each of these sections were each designated numbers 0 through 9, and one scale was removed from each of the sections. Therefore, up to 999 individuals could be identified, adjusting to accommodate any naturally missing scales. The authors reported that subsequent scale loss occurred, but that confusion was minimized by having additional notes on tail regeneration and other characteristics. This technique was not reliable with hatchlings because scales quickly regenerated.

4.4 Branding and painting

4.4.1 Turtles

A white-hot wire was used to brand a line on the carapace of *Gopherus agassizii* where scutes were given numbers or letters to yield individual recognition codes by Woodbury and Hardy (1948). Burning scutes too deeply resulted in complete regeneration and too lightly allowed the scar to wear off in a few years. Woodbury and Hardy (1948) also marked tortoises with a variety of colours of paint, but these marks are less permanent than branding and other morphological modifications as would be expected. In general, painting codes have been used by researchers for short-term visual identification of individuals along with a more permanent mark such as notches on the scutes.

Figure 4.3 (a) *Cyclura cychlura cychlura* on South Andros Island, Bahamas, marked with bright yellow bead (circle) and paint mark. (b) A turtle being notched on its marginal scutes. Photo by Kent Bekker. (c) *Leiocephalus carinatus* marked with bright red spots (arrow) for individual recognition. (d) Another *L. carinatus* coated with fluorescent green powder (light dorsum in photograph; arrow) allowing daytime observation; at night, the lizard's pathways can be determined using a blacklight to follow the lizard's daytime trail. (e) A PIT being inserted under the skin of a *C. c. cychlura*. Lizard photos by C.K. Dodd, Jr.

Paint markings for free-ranging aquatic cooters (*Pseudemys concinna*) were employed in a short-term study by Kornilev et al. (2012). White (contrasting with the dark shell), oil-based, non-toxic paint markers (563 Speedry, Diagraph, Marion, Illinois, USA; cost ca. US$3/marker) were used to make numbers with lines up to 1–2 cm wide. Researchers removed algae and loose scutes from the carapace, thoroughly dried the shell, applied wide number markings at the most appropriate location on the carapace, and released the turtles at point of capture after a paint drying time of ca. 10 min. Within 60 days of resampling 82% of the 41 marked individuals were identified at least once.

4.4.2 Lizards

Heated wire branding was used successfully on *Anolis carolinensis* and *Phrynosoma cornutum* by Clark (1971) who preferred it to toe-clipping because locomotion appeared to be less impacted and brands were easier to read. Freeze branding has also been found to be successful with iguanas according to R.K. Farrell (personal communication). Any brand on a reptile may need to be reapplied after several skin sheddings.

Adult *Uta stansburiana* were painted with coloured insignias (dots, longitudinal stripes, transverse bands, plus signs, and arrows) to allow individual recognition without recapture (Tinkle, 1967). Periodic recapture and marking were needed after ecdysis to avoid confusion; young individuals did not have enough surface area to allow adequate marking for individual recognition. Enamel model paint in a variety of colours was used by Medica et al. (1971) with thinned out nail polish brushes to increase precision. Similarly, bright colours of fingernail polish worked successfully to mark *Leiocephalus carinatus* in a short-term study in the Bahamas (Figure 4.3(c)).

Using a quick drying paint, Jenssen (1970) placed coloured numbers on the dorsum of *Anolis carolinensis* with females marked as yellow and males orange. Jenssen used combinations of four dorsal numbers (1, 2, 4, and 7) to obtain all numbers between 1 and 9; these four numbers were located on the neck, mid-body, pelvic region, and base of the tail. He then designated the tens column for each identification number by painting the tips of the tails different colours (e.g. white for 10s and green for 20s). As with all painting systems on reptiles, skin shedding and, in this case, tail loss need to be monitored and paint reapplied.

Other studies using paint marking include use of reflective paint spots on *Iguana iguana* (Rodda et al., 1988); minimally paint marking only on the tails of *Sceloporus undulatus* to limit possible impact on survivorship (Vinegar, 1975); use of purple indelible pencil on *S. occidentalis* by Stebbins and Cohen (1973); and use of a black felt-tip pen on *I. iguana* (Henderson, 1974). Boone and Larue (1999) recommended not using xylene-based paint to mark animals, as they found significant mortality when marking *Uta stansburiana*.

4.4.3 Snakes

Plummer and Ferner (2012) noted that tattooing and branding have not been regularly used for snakes by most researchers because scale clipping has proven more successful. A historical summary of branding techniques for snakes can be found in Ferner (2007). More recently, battery-powered medical cautery units were used to brand over 200 snakes from 15 species by Winne et al. (2006). Each snake brand included the anterior portion of the ventral scale and extended dorsally onto adjoining lateral scales and lasted at least two years. The units ('Aaron Medical Change-A-Tip cautery units') are available from Aaron Medical (St. Petersburg, FL 33710, USA; http://www.aaronmed.com).

Paint marking has found little use in snakes with the exception of rattlesnakes. For example, Brown et al. (1984) used an acrylic paint to put an identifying number on the basal rattle of *Crotalus horridus* that lasted for up to four years. Paint marking of crotalids should be augmented with a permanent ventral scale clip.

4.5 Shell notching

The carapace of terrestrial and aquatic turtles has been notched by cutting, sawing, grinding, filing, and drilling on selected scutes to obtain various codes for individual recognition as seen in Figure 4.3(b) (Plummer and Ferner, 2012). Since natural scarring also can occur, it is important to incorporate such injuries into any code and be alert for any new marks that may occur between captures.

Cagle (1939) proposed a scute non-sequential numbering system that used a combination of notching the carapace and plastron and clipping toes. More recently, scute sequential numbering systems have been commonly used. For example, Ernst et al. (1974) provided a sequential system for hard-shelled Pond Sliders (*Trachemys scripta*) (Figure 4.4). Moving from the cranial end, the right marginal scutes were designated 2,

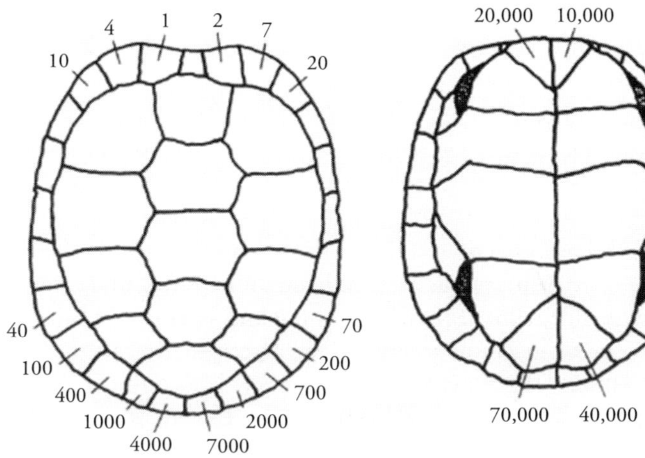

Figure 4.4 Numerical coding system for notching hard-shelled turtles as shown for Pond Sliders (*Trachemys scripta*). Left view: carapace with numerical code for each marginal scute. Right view: plastron with code for each gular and anal scute. Unique identifying numbers are created by adding the numerical values given to each notched scute. For example, specimen 4721 would have notches in marginals numbered 4000, 700, 20, and 1. For turtles with only 11 marginals (e.g. kinosternids), this numbering system may be modified. (after Ernst et al., 1974).

7, 20 (then skipping the 4th–7th where the carapace to plastron bridge occurs), 70, 200, 700, 2000, and 7000. The left marginals were coded as 1, 4, 10 (then skipping again the 4th–7th), 40, 100, 400, 1000, and 4000. The plastron gular and anal scutes were also coded: right gular (20,000), right anal (70,000), left gular (10,000), left anal (40,000). By notching a combination of these scutes and adding the numbers, thousands of individuals can be marked.

A simple sequential system using letters rather than numbers to mark *Kinosternon sonoriense* was used by J. Congdon (personal communication). The marginals were lettered A to M (which includes those species with up to 13 scutes) cranial to caudal on the right side of the turtle and N to Z on the left. As with the Ernst et al. (1974) method, the marginals along the juncture of the carapace and plastron were not notched but were included in the lettering sequence, that is, marginals D to F on the right and R to T on the left were not used. Congdon notched two to four scutes and these were then read and recorded sequentially (e.g. AB, CH, BHI, JKWY). This lettering scheme can be applied to all turtle species regardless of the number of marginals and, according to Congdon, can reduce misread identifications compared to numbering systems.

Softshell turtles, which lack epidermal scales and dermal bone at the edge of the carapace, can be notched by removing a small V-shaped piece of the carapace edge with a scalpel or punching a small hole with a paper punch (Doody and Tamplin, 1992). These marks may heal quickly, but leave distinctive white scars which are best seen from a ventral perspective. With no scutes to designate numbers, coding the notches can be difficult. Plummer (2008) created numbered positions around the carapace 1 through 12 as if on a clock face to allow combinations for individual identification codes.

4.6 Tagging and banding

4.6.1 Lizards

Lizards have been tagged with beads (Figure 4.3(a)), tape, foil, aluminium rings, glowing tubes (for nocturnal study), plastic bird bands, elastomers, and more. Items have been attached with glue, wire, thread, and fishing line, as reviewed by Ferner (2007). Some examples of more successful techniques are described here.

Zwickel and Allison (1983) marked *Emoia physicae* with pressure sensitive rip-stop nylon tape (Coghlan's Ltd., 121 Irene St., Winnipeg, Canada R3T 4C7; http://www.coghlans.com) in addition to permanent toe clips because the tape was lost with each skin shed. They first wiped the dorsum with 95% alcohol before attaching a $5 \times 10\ mm^2$ piece of tape. The tape was then colour coded with acrylic paint and allowed to dry before each lizard was released.

Rodda et al. (1988) tagged individuals of *Iguana iguana* on their mid-dorsal flap of integument with beads approximately 2 mm in diameter strung on nylon monofilament line. The dorsal crest was pierced using a hypodermic needle so the line could be threaded through it and then secured by melting the ends of the nylon after the beads were inserted. The tag was tested for strength by pulling on the beads to be sure they would not slip off and that enough slack was left in the line to allow for growth. Other than some very small hatchlings losing these tags, they proved to be very durable.

Johnson (2005) marked four species of lizards (*Anolis carolinensis, A. gundlachi, A. krugi, Sceloporus undulatus*) for short-term studies with bee marking kits (The Bee Works of Orillia, Ontario, Canada L3V 6M2; http://www.beeworks.com). The relatively inexpensive kits include numbered cardboard dots in five colours, glue, and applicator. Johnson found the highly visible dots to be useful in about 85% of the lizards in the study over a three-week period. The dots had to be replaced after each ecdysis.

Daniel et al. (2006) injected visible implant elastomer (VIE) under the skin around the leg joints of *Hemidactylus turcicus*. They found high retention of VIE during shedding and concluded that even though VIE requires more cost and effort than surface marks, the disadvantages were out-weighed by the long retention time.

4.6.2 Freshwater and terrestrial turtles

Various types of plastic and metal tags have been affixed to hard-shelled turtles with the preferred site on the caudal carapace (Plummer and Ferner, 2012). Coloured or numbered tags have been attached with glue, screws, or wires through holes drilled in the shell.

Pough (1970) suggested using a Buttoneer tool (Dennison Manufacturing Co., Framingham, MA 01701, USA; http://www.averydennison.com) to fasten plastic plugs in a 3/32-inch (2.5 mm) hole drilled through a marginal plate. The plugs (used to fasten buttons on clothing) come in various sizes and colours and have a stem with a bulb at one end and a crossbar at the opposite end. Pough found this technique good for use with juvenile turtles where regeneration can be a problem when marking with notches.

Apalone spinifera was marked by Dreslik (1997) using low cost 'spaghetti tags' made from a 3 m length of Romex cable wire containing eight different coloured wires within its sheath. After cutting the eight wires into 10 cm pieces, the coloured plastic sheath around each was removed from the copper core providing 240 tags. Dreslik used one to four per turtle to create a total of 4680 unique colour combinations. A small hole was made at the edge of the soft carapace and the spaghetti tag was slipped through and knotted at both ends, leaving 1.5 cm between the knot ends and the turtle shell for potential growth. Excess plastic from the ends of the tag was then removed. Most tags were retained for three years, some much longer (M. Dreslik, personal communication). Plummer and Ferner (2012) stressed that tags on softshell turtles should be loosely attached due to the tendency of the fleshy shell to become necrotic at the point of any firm attachment of foreign material.

Tortoises have been marked with numbered titanium disks fastened with a metal–resin adhesive into depressions drilled into the keratin scutes on the carapace (Gaymer, 1973). Gaymer reported a very high retention rate with this technique. The carapace of *Clemmys guttata* was marked by Ward et al. (1976) using an adhesive tag which bore an identification number.

4.6.3 Snakes

External tags have been rarely used on snakes (Ferner, 2007). Plummer and Ferner (2012) recommended against their use due to the potential problems of entanglement in habitat vegetation and reduced access to narrow burrows.

4.7 Trailing devices

A variety of trailing devices have been used with turtles. These are typically fastened to the caudal portion of the carapace and used for short-term monitoring of movements and behaviour. Thread trailing is the most common technique used. A continuous thread is pulled from a spool fastened to the carapace or a device attached and pulled behind the turtle.

Wilson (1994) used encapsulated thread bobbins (Culver Textiles, P.O. Box 360, West New York, NJ 07093, USA; 800.526.7188), rather than a traditional spool, to mark and trail *Kinosternon baurii.* After pulling about one metre of thread from the bobbin (small at 1.8 g, large at 4.5 g), Wilson covered it in clear plastic wrap which was twisted at the bottom and taped to the sides. This package was then dipped in Plastic Dip® (used for dipping tool handles) and dried on waxed paper. After drying, the edge on the paper was flat enough to allow easy attachment to the carapace (Wilson, 1994). This technique was also adapted by Jennings (2007) to follow juvenile Box Turtles (*Terrapene carolina*).

Blankenship et al. (1990) used fluorescent powder (JS-DH3020, type 300 from Radiant Color, 2800 Radiant Avenue, Richmond, CA 94804, USA, 415.233.9119) to track adult *Gopherus polyphemus*. Five cc of the powder was placed in a fine mesh nylon pouch and attached to the posterior marginal scutes with cotton twine so the pouch would drag along the ground behind the tortoise. Daily movements were tracked at night using a portable UV light ('Woods Light', from Henry Schein Inc., 5 Harbor Park Drive, Port Washington, NY 11050, USA, 800.872.4346). The trails generated were marked with flagging for mapping over the period of their six-day study (Blankenship et al., 1990). This technique was also used in tracking the Six-lined Racerunner, *Aspidoscelis sexlineata* (Dodd, 1992), and *Leiocephalus carinatus* in the Bahamas (Figure 4.3(d)).

4.8 Passive integrated transponder (PIT) tags

The PIT tag is a radiofrequency glass-encased device that is injected into the animal with a hypodermic syringe. It transmits a specific alphanumeric code when it is inductively powered by a portable reader with a hand-held wand within a few cm of the tag. This technology allows identification of each animal without requiring recapture or handling. The tag is biologically inert, requires no battery, and may function for as long as 75 years. Other advantages of PIT tagging include avoidance of morphological modifications which may impact the clarity of code reading. Most PIT tags used for reptiles transmit at 125 kHz and cost approximately US$5 each with a portable reader priced about US$300. In addition to cost, other disadvantages of these tags can be movement of tags, potential high rate of tag loss, and finding appropriate injection sites (Ferner, 2007; Plummer and Ferner, 2012).

4.8.1 Turtles

PIT tags are recommended for use in both ecological studies and when tracking turtles that may become trafficked in illegal commercial trade. Tag retention appears to be high in all species, although difficulty in detecting tags may be a problem in large individuals. While placement of the tags in turtles is not standardized, the body cavity in the cranial inguinal regions parallel to the shell bridge works well for freshwater and terrestrial species, and intramuscular or subcutaneous locations in the shoulder work well for sea turtles. We do not recommend the older technique of placing PIT tags in predrilled holes in the shell because of high loss and breakage (Ferner, 2007; Plummer and Ferner, 2012).

4.8.2 Lizards and snakes

Dimensions of 125 kHz PIT tags (8.4–10 mm × 1.4–2.1 mm) may preclude their use with smaller lizards. Intra-abdominal injection has been reported most often in the literature and retention of the tags has been high. PIT tags can be an ideal marking technique for many snake species (Camper and Dixon, 1988). Greatest retention reported by investigators is for intraperitoneal placement at mid body, as shown in Figure 4.3(e) (Plummer and Ferner, 2012).

4.9 Taking measurements

In order to place individuals in an age or size class (i.e. hatchling, juvenile, or adult) and to document growth rates with subsequent recaptures, it is important that snout–vent length (SVL) and tail length (TL) be recorded for snakes and lizards, and carapace length (CL) and plastron length (PL) for turtles. Snakes and lizards may be placed on flat transparent surfaces or placed inside transparent Plexiglas® tubes with an attached metric scale for measuring these lengths. Callipers may be used on small specimens while being hand held. SVL is taken from the tip of the snout to the caudal end of the cloacal aperture, and TL from the caudal end of the aperture to the tip of the tail. In addition, regenerated portions of the tail should be measured and recorded. The CL of a turtle can be measured straight-line with callipers from the mid-cranial edge to the mid-caudal edge, as CL in turtles is difficult to determine due to the body form and neck flexibility. Straight-line PL length can typically be measured without resorting to callipers.

For some species additional measurements or counts may be helpful with age estimation. For example, the number of individual units of the caudal rattle in *Crotalus* and *Sistrurus* may be somewhat correlated with age even though terminal segments may break off easily. Similarly, the growth rings on the keratin scutes of turtles are sometimes helpful with aging specimens (Reed and Tucker, 2012).

After patting specimens dry and cleaning off any substrate particles, their body mass should be measured in grams. A specimen bag or appropriate customized container can be used to contain the animal that then can be placed on or suspended from a scale (such as Pesola®). It is recommended that the body mass be obtained to the nearest 0.10 g except for the largest specimens. If a species, such as tortoises, is prone to evacuate the cloaca or bladder upon capture, researchers should allow this to happen before weighing.

Measurement error is of concern in any study. It may be more so in some cases, such as taxonomic descriptions of new species, than in others, as in determining age classes in population studies. We recommend using the most accurate devices in taking measurements and practising your technique with an experienced mentor before initiating a study. Can all members of the research team replicate measurements within a tolerable margin of error? What is the tolerable margin of error in your data? When possible, we have found it beneficial to have one member of a team take all the measurements in order to have the highest consistency. If the researcher taking a measurement is reciting this to another person recording the data, it is essential that the scribe repeat the information orally and have the data confirmed by the person who took the measurement.

4.10 Recommendations

In keeping with our criteria to select the best marking technique for a particular situation (see Section 4.1), we make the following recommendations (adapted from Ferner, 2010, and Plummer and Ferner, 2012):

- Shell notching, toe-clipping, and scale-clipping are the most commonly used, reliable, and cost-effective methods of marking reptiles.
- PIT tags are effective and increasingly popular for providing positive individual recognition in reptiles used in long-term studies.
- All external tags should be tested for possible interference with behaviour.
- Colour marking used in conjunction with permanent marks can reduce the need for frequent recaptures by facilitating field identification.
- Measurements of all captured reptiles should be made and recorded, even if those data are not essential to the immediate research goal.
- Measurement error is to be minimized by using the best equipment and techniques.

In conclusion, researchers should pay particular attention to the criteria given in Section 4.1 and the recommendations listed here when deciding on a particular marking technique. With hundreds of references in the literature describing marking techniques for reptiles, we have selected those most commonly used and practical to include in this review. Careful choice of a technique is essential to the success of research, so we strongly recommend careful deliberation in the choice of a marking technique prior to initiating field research.

References

Baker, J., and Gent, T. (1998). Marking and recognition of animals. In T. Gent and S. Gibson (eds) *Herpetofauna Worker's Manual*. Peterborough, UK: Joint Nature Conservation Committee, pp. 45–54.

Blankenship, E.L., Bryan, T.W., and Jacobson, S.P. (1990). A method for tracking tortoises using fluorescent powder. *Herpetological Review*, **21**, 88–9.

Bloch, N., and Irschick, D.J. (2005). Toe-clipping reduces clinging performance in a pad-bearing lizard (*Anolis carolinensis*). *Journal of Herpetology*, **39**, 293–8.

Boone, J., and Larue, E. (1999). Effects of marking *Uta stansburiana* (Sauria: Phyrnosomatidae) with xylene-based paint. *Herpetological Review*, **30**, 33–4.

Borges-Landaez, P.A., and Shine, R. (2003). Influence of toe-clipping on running speed in *Eulamprus quoyii*, an Australian scincid lizard. *Journal of Herpetology*, **37**, 592–5.

Brown, W.S., and Parker, W.S. (1976). A ventral scale clipping system for permanently marking snakes (Reptilia, Serpentes). *Journal of Herpetology*, **10**, 247–9.

Brown, W.S., Ganon, V.P.J., and Secoy, D.M. (1984). Paint marking the rattle of rattlesnakes. *Herpetological Review*, **15**, 75–6.

Cagle, F.R. (1939). A system of marking turtles for future identification. *Copeia*, **1939**, 170–3.

Camper, J.D., and Dixon, J.R. (1988). Evaluation of a microchip marking system for amphibians and reptiles. *Texas Parks and Wildlife Department, Research Publication*, **7100–159**, 1–22.

Clark, D.R. (1971). Branding as a marking technique for amphibians and reptiles. *Copeia*, **1971**, 148–51.

Daniel, J.A., Baker, K.A., and Bonine, K.E. (2006). Retention rates of surface and implantable marking methods in the Mediterranean house gecko (*Hemidactlyus turcicus*), with notes on capture methods and rats of skin shedding. *Herpetological Review*, **37**, 319–21.

Dodd, C.K., Jr. (1992). Fluorescent powder is only partially successful in tracking movements of the six-lined racerunner (*Cnemidophorus sexlineatus*). *Florida Field Naturalist*, **20**, 8–14.

Dodd, C.K., Jr. (1993). The effects of toe-clipping on sprint performance of the lizard *Cnemidophorus sexlineatus*. *Journal of Herpetology*, **27**, 209–13.

Doody, J.S., and Tamplin, J.W. (1992). An efficient marking technique for soft-shelled turtles. *Herpetological Review*, **23**, 54–6.

Dreslik, M.J. (1997). An inexpensive method for creating spaghetti tags for marking trionychid turtles. *Herpetological Review*, **28**, 33.

Ernst, C.H., Hershey, M.F., and Barbour, R.W. (1974). A new coding system for hard-shelled turtles. *Transactions of the Kentucky Academy of Science*, **35**, 27–8.

Ferner, J.W. (2007). A review of marking and individual recognition techniques for amphibians and reptiles. *Society for the Study of Amphibians and Reptiles Herpetological Circular*, No. **35**.

Ferner, J.W. (2010). Measuring and marking post-metamorphic amphibians. In C.K. Dodd Jr. (ed) *Amphibian Ecology and Conservation. A Handbook of Techniques*. Oxford: Oxford University Press, pp. 123–41.

Fitch, H.S. (1987). Collecting and life-history techniques. In R.A. Seigel, J.T. Collins, and S.S. Novak (eds) *Snakes: Ecology and Evolutionary Biology*. New York: Macmillan Publishing Company, pp. 143–64.

Gaymer, R. (1973). A marking technique for giant tortoises and field trials in Aldabra. *Journal of Zoology*, **169**, 393–401.

Henderson, R.W. (1974). Aspects of the ecology of the juvenile common iguana (*Iguana iguana*). *Herpetologica*, **32**, 304–10.

Henle, K, Kuhn, J., Podloucky, R., et al. (1997). Individualerkennung und Markierung mitteleuropaischer Amphibien and Reptilien: Ubersicht und Bewertung der Methoden. *Empfehlungen aus Natur- und Tierschutzsicht Mertensiella*, 7, 133–84.

Hudson, S. (1996). Natural toe loss in southeastern Australian skinks: implications for marking lizards by toe-clipping. *Journal of Herpetology*, **30**, 106–10.

Huey, R.B., Dunham, A.E., Overall, K.L., et al. (1990). Variation in locomotor performance in demographically known populations of the lizard *Sceloporus merriami*. *Physiological Zoology*, **63**, 845–72.

Jennings, A.H. (2007). Use of habitats and microenvironments by juvenile Florida box turtles, *Terrepene carolina bauri*, on Egmont Key. *Herpetologica*, **63**, 1–10.

Jenssen, T.A. (1970). The ethoecology of *Anolis carolinensis* (Sauria, Iguanidae). *Journal of Herpetology*, **4**, 1–38.

Johnson, M.A. (2005). A new method of temporarily marking lizards. *Herpetological Review*, **36**, 277–9.

Kornilev, Y.V., Dodd, C.K., Jr., and Johnston, G.R. (2012). Retention of paint markings for individual identification of free-ranging basking aquatic turtles (Suwannee Cooters, *Pseudemys concinna suwanniensis*). *Herpetological Review*, **43**, 61–4.

Langkilde, T., and Shine, R. (2006). How much stress do researchers inflict on their study animals? A case study using a scincid lizard, *Eulamprus heatwolei*. *Journal of Experimental Biology*, **209**, 1035–43.

Medica, P.A., Hoddenbach, G.A., and Lannom, J.R. (1971). Lizard sampling techniques. *Rock Valley Miscellaneous Publication*, **1**, 1–55.

Paulissen, M.A., and Meyer, H.A. (2000). The effects of toe-clipping on the gecko *Hemidactylus turcicus*. *Journal of Herpetology*, **34**, 282–5.

Plummer, M.V. (1980). Ventral scute anomalies in a population of *Opheodrys aestivus*. *Journal of Herpetology*, **14**, 199.

Plummer, M.V. (2008). A notching system for marking softshell turtles. *Herpetological Review*, **39**, 64–5.

Plummer, M.V., and Ferner, J.W. (2012). Marking reptiles. In R.W. McDiarmid, M.S. Foster, C. Guyer, et al. (eds) *Reptile Biodiversity: Standard Methods for Inventory and Monitoring*. Berkeley, CA: University of California Press, pp. 143–50.

Pough, F.H. (1970). A quick method for permanently marking snakes and turtles. *Herpetologica*, **26**, 428–30.

Reed, R.N., and Tucker, A.D. (2012). Determining age, sex and reproductive condition. In R.W. McDiarmid, M.S. Foster, C. Guyer, et al. (eds) *Reptile Biodiversity: Standard Methods for Inventory and Monitoring*. Berkeley, CA: University of California Press, pp. 151–63.

Rodda, G.H., Bock, B.S., Burghardt, G.M., et al. (1988). Techniques for identifying individual lizards at a distance reveal influences of handling. *Copeia*, **1988**, 905–13.

Stebbins, R.C., and Cohen, N.W. (1973). The effect of parietalectomy on the thyroid and gonads in free-living western fence lizards, *Sceloporus occidentalis*. *Copeia*, **1973**, 662–8.

Tinkle, D.W. (1967). The life and demography of the side-blotched lizard, *Uta stansburiana*. *Miscellaneous Publications, Museum of Zoology, University of Michigan*, **132**, 1–182.

Vinegar, M.B. (1975). Life history phenomena in two populations of the lizard *Sceloporus undulatus* in southwestern New Mexico. *American Midland Naturalist*, **93**, 388–402.

Waichman, A.V. (1992). An alphanumeric code for toe clipping amphibians and reptiles. *Herpetological Review*, **23**, 19–21.

Ward, F.P., Hohmann, C.J., Ulrich, J.F., et al. (1976). Seasonal microhabitat selections of spotted turtles (*Clemmys guttata*) in Maryland elucidated by radioisotope tracking. *Herpetologica*, **32**, 60–4.

Wilson, D.S. (1994). Tracking small animals with thread bobbins. *Herpetological Review*, **25**, 13–14.

Winne, C.T., Willson J.D., Andrews, K.M., et al. (2006). Efficacy of marking snakes with disposable medical cautery units. *Herpetological Review*, **37**, 52–4.

Woodbury, A.M., and Hardy, R. (1948). Studies of the desert tortoise, *Gopherus agassizii*. *Ecological Monographs*, **18**, 145–200.

Zwickel, F.C., and Allison, A. (1983). A back marker for individual identification of small lizards. *Herpetological Review*, **14**, 82.

5

Digital identification and analysis

Roberto Sacchi, Stefano Scali, Marco Mangiacotti, Marco Sannolo, and Marco A.L. Zuffi

5.1 Introduction

The identification of individual animals in a population (usually achieved through marking) aids in acquiring accurate estimates of population size in ecological studies. In reptiles, marking usually involves scale clipping, branding, tattooing, subcutaneous elastomer injections, PIT tags, painting, and toe-clipping (see Chapter 4). However, marking techniques involve capturing and handling animals, which can alter survival probability or other life-history traits leading to serious violations of the assumptions of demographic models (Bloch and Irschick, 2004).

Digital identification (i.e. photographic identification) of individual animals is a non-intrusive way to obtain information on life-history traits that is based on distinctive individual morphological characteristics (such as patterns of ornamentation or morphology; Figure 5.1) that remain unchanged over time despite skin moulting. With respect to other techniques, digital identification has the advantage of being relatively inexpensive (entailing only a digital camera and computer), requiring only basic expertise to manage image analysis tools, and allowing large numbers of individuals to be re-identified. Typically, it employs catalogues of images that are matched against new images in order to assess if they are re-sightings of previously photographed individuals (i.e. 'recaptures') or still unknown individuals (i.e. 'captures'), which would then be included in the catalogue.

A major constraint involved in this process is the large amount of time required for image acquisition and comparison, which is likely to increase exponentially with increasing sample size. Indeed, for small catalogues (i.e. <50 individuals), image matching is still feasible by eye (Stenhouse, 1985). When catalogue items exceed 100 or so, however, the identification process becomes hugely time-consuming and misclassification errors occur frequently (Gamble et al., 2008; Sacchi et al., 2010; Bolger et al., 2012). The increasing use of photo-identification techniques coincided with advances in image analysis tools and pattern recognition algorithms, which have accelerated the image matching process. For this reason, photographic identification has become a popular technique in reptile capture–recapture studies

Reptile Ecology and Conservation. Edited by C. Kenneth Dodd, Jr. © Oxford University Press 2016.
Published 2016 by Oxford University Press.

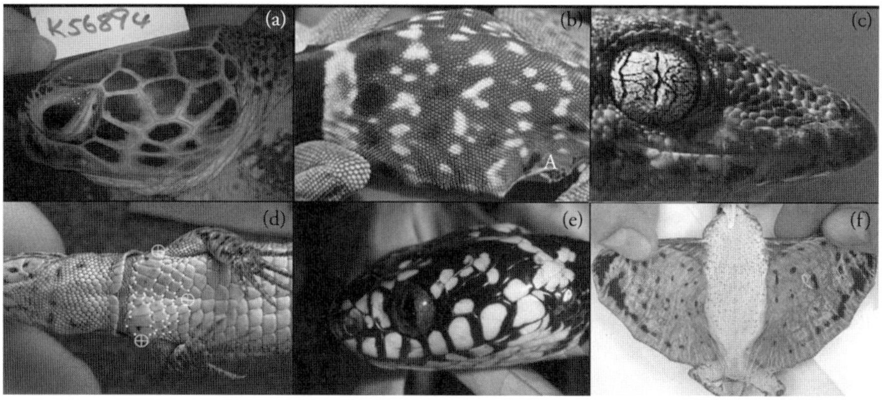

Figure 5.1 Examples of species with ornamentation or lepidosis patterns that are useful for photographic identification. (a) Facial scutes in Green Turtle (*Chelonia mydas*, from Carter et al., 2014). (b) Lateral dots in Dickerson's Collared Lizard (*Crotaphytus dickersonae*, from Faber et al., 2013). (c) Iris pattern in Boettger's Wall Gecko (*Tarentola boettgeri*, from Rocha et al., 2013). (d) Scale morphology in Common Wall Lizard (*Podarcis muralis*, from Sacchi et al., 2010). (e) Head ornamentation in Mangrove Snake (*Boiga dendrophila*, from Speed et al., 2007). (f) Patagium ornamentation of Indian Gliding Lizard (*Draco dussumieri*, from Sreekar et al., 2013).

only in the last decade despite the fact that the ability to identify animals through photographing distinctive traits was well known in the 1970s (e.g. in cetaceans, Hammond et al., 1990).

Computer algorithms decrease the time required for comparing photos, as they rapidly select from the catalogue the subset of images that most closely match the one to be identified, thus drastically reducing the number of comparisons made by the operator. The algorithms do this by computing a standardized similarity score for all pairwise combinations of the new image with those in the catalogue, and ranking the catalogued images from the closest to the most distant with respect to the new image. 'Match' or 'no-match' decisions are made by the user who must visually inspect the highest scored images defined by an a priori selected threshold. In this sense, digital identification should be referred to as a computer-assisted system for digital identification, as the algorithms actually do not make any decision in establishing the 'truth' between potential matches in place of the user.

Computer-assisted identification systems rarely check for efficiency or comparisons with other marking techniques. In cases when this has occurred, the results demonstrated highly accurate performance. For example, computer matching has been estimated to lead to a 38-fold time savings relative to a completely manual matching process in avian studies (Morrison et al., 2011), although no estimates are available for reptiles. Moreover, costs related with digital identification are sensibly lower compared with conventional methods allowing for larger sample sizes (Bolger et al., 2012).

Consequently, digital identification may outperform other marking techniques such as visible implant elastomers (Bendik et al., 2013).

During the last decade, a range of computer-assisted systems for digital identification have been developed for both general and species-specific applications. Irrespective of their own algorithm, photo-identification software can be classified in two main groups according to the way the individually distinctive information is extracted from the digital images. We will refer to the first group as *partially automated* digital identification (PADI) systems, which include all the programs of the family of the Interactive Individual Identification System (I^3S, Van Tienhoven et al., 2007) software, and to the second as *fully automated* digital identification (FADI) systems, including Wild-ID (Bolger et al., 2012), MYDAS (Carter et al., 2014), and the Automatic Photo Identification Suite (APHIS, Moya et al., 2015). In PADI systems, each image is pre-processed by the user, who manually finds and extracts all the distinctive image features that are used by the algorithm to compute the distance between image pairs and the relative ranking with respect to the target image. By contrast, the algorithms in FADI systems automatically perform all steps, from the extraction of the distinctive features of images to the computation of pair distances and ordination of the catalogued images with respect to the target image. The FADI systems have the remarkable advantage of reducing the user's work, but do not allow for a complete control of the information involved in identification. They also are more sensitive than the PADI systems to the quality of images and to the non-informative noise in the image background.

A typical photographic recognition system has three main stages of operation: the enrolment stage, the identification stage, and the validation stage. In the enrolment stage, the user selects the best suitable trait for individual recognition and captures a digital image of it. This stage is totally under the user's control, and it completely depends on the operator's skill. During the identification stage, the salient features of the images are extracted, stored, and compared with those of the templates stored in the reference database. This second stage is computer assisted, whereas the user's intervention is needed for image pre-processing (only in the PADI system) and image matching. During the validation stage, both performance and reliability of the classification obtained through algorithms are assessed. Unfortunately, this last stage is often omitted or underestimated, primarily when relying with species or traits that have been successfully used in identification in previous studies.

Here, we review these three main stages, mainly focusing on the practical implications when applying the technique in field work. We then present a short review of the main studies in which a photo-identification has been applied successfully to reptiles in order to supply case studies useful for future research planning. Finally, we highlight future prospects for photographic identification in reptiles.

5.2 Collecting images

The first step for photographic identification is, of course, the photo acquisition. This phase is particularly tricky because all the identification processes depend on a picture's quality, so careful attention should be made to photo planning. In general, we can set five points for image acquisition:

- Identification of distinctive features
- Set-up of a photographic shoot
- Photo shooting
- Photo coding
- Photo elaboration

5.2.1 Identification of distinctive features

Many anatomical features have been used for the photo-identification of reptiles, but their choice is subject to certain constraints: they must always occur in each individual; they must be stable for the entire lifespan or, at least, over the duration of the study period; and they must be unambiguously photographed under differing field conditions (Bolger et al., 2012), especially when capturing target animals is not always feasible (e.g. sea turtles, Buonantony, 2008). Ornamentation patterns have been used to identify individual sea turtles (Buonantony, 2008), lizards (Faber et al., 2013), and snakes (Vaughan, 1999), but these features are applicable only to those species that do not have plain or uniform colouration. Cases of changes in ornamentation patterns have been documented for chelonians (Tichý and Kintrová, 2010) and lizards (Henle et al., 1997; Sacchi et al., 2006), so the stability of spots and patches should be verified before using these characteristics for identification. Furthermore, ontogenetic pattern changes have been documented, so caution should be provided in long-term studies. Lepidosis of reptiles greatly aids researchers, because scale pattern is usually constant within a species, but scale shapes and combinations are different among individuals, so the vertexes and the edges can be used profitably in photo-identification. Many studies have been conducted using these features on sea turtles (Carter et al., 2014; Dunbar et al., 2014), tortoises (Tichý and Kintrová, 2010), and lizards (Perera and Perez-Mellado, 2004; Li et al., 2009; Sacchi et al., 2010; Sreekar et al., 2013). Individual identification could be particularly difficult for some taxa, such as geckos because of their uniform morphology, although some peculiar traits can be used for recognition, e.g. the iris pattern in *Tarentola* sp. (Rocha et al., 2013).

5.2.2 Set-up of a photographic shoot

Photographs should be taken under standardized conditions to facilitate matching. This can be a problem during field sessions when light conditions may vary strongly from shot to shot. For small- and medium-sized species, the easiest way to standardize images is to capture and photograph individuals indoors at fixed distances using artificial lights. However, photos also can be taken in the field and corrections applied to the images afterwards. If possible, cameras should be mounted on a tripod to avoid involuntary movements and to standardize the distance from the subject. Handheld pictures are discouraged and should be taken when capture and immobilization is hard to achieve, as in the case of crocodilians or large monitors. Backgrounds should be uniform and contrasting in order to easily locate important features. A ruler or a millimetre-lined paper sheet should be included next to the animal to set the image scale, together with a label indicating date, identification (ID) number, and sex. A certified colour chart would be desirable for colour calibration in successive analyses.

5.2.3 Photo shooting

The choice of camera and lens depends on the subject's size, species, and fieldwork conditions. However, the greatest possible image resolution is always preferable. Since camera technology is evolving rapidly, suggestions on camera selection is beyond the scope of this book. When possible, the photo should be restricted only to the areas that will be used for the identification, and light reflection should be avoided. The photo image should be taken perpendicular to the subject to avoid parallax deformation that could be problematic when the shooting angle is too great. For example, images skewed by more than 30° cause severe problems with individual identification using I^3S software (Speed et al., 2007). When feasible, users should mount the camera to a copy stand and keep it orthogonal to the subject.

5.2.4 Photo coding

Images should be stored in a database using a predefined name coding to facilitate search and information recovery. Many ways of coding can be used for this purpose, but we suggest always including the individual's locality in the file name, the identification number, and the sex.

5.2.5 Photo enhancement

Picture quality can be enhanced before analysis for photo-identification. In some cases, the images should be cropped in order to isolate the area of interest (Bendik et al., 2013). In other cases, brightness and contrast may be enhanced to highlight spots, patches, and edges that were used as identification features (Arzoumanian et al., 2005; Li et al., 2009). In case of reptiles with very flexible bodies, forcing them to maintain a straight pose can be difficult, so a rectification can be applied to the images by the operator (Gamble et al., 2008). Furthermore, corrections for scaling, rotation, translation, and correction of perspective can be applied to the pictures (e.g. Van Tienhoven et al., 2007).

5.3 Software and algorithms

Thanks to the development of computers and digital cameras, a large number of software programs dedicated to photo-identification have been developed over the years. In this section, we briefly present some of the most recent software programs, which are distinguished by their availability, ease of use, and effectiveness. We also give a short description of the procedures and algorithms that allow them to discriminate among different individuals.

5.3.1 I^3S, Interactive Individual Identification System

I^3S is a family of software originally designed for the recognition of whale sharks and sand tiger sharks (Van Tienhoven et al., 2007). In all versions, the first operation to be performed for each photo is the identification of three anatomical homologous points, and the preparatory step allows researchers to superimpose photos, even if taken at different angles (Speed et al., 2007). It does this through a mixture of scaling, shifting, and stretching (using the algorithm of Gauss–Jordan; Press et al., 1992; Van Tienhoven et al., 2007), and obtains an aligned set of images suitable for the computation of the dissimilarity indexes. This last metric depends on the software version: I^3S Classic, the earliest version, is based on a point-to-point comparison, where the identification is obtained by computing the level of matching of a variable number of points rather than the three homologous ones. These 'points of interest' (or key-points) must be digitized previously on the photos by the operator and can be different both in order and in number; two points are considered as a correct match when the first alternative point is at least double the distance calculated for the first two points. At the end of the process, the software returns a ranking of distance between the target photo and each of the images in the database; the smaller this distance, the greater the probability of a correct match. I^3S Manta, now improved in I^3S Spot, adds the ability to draw ellipses around a point of interest. The program is particularly appropriate for species with distinct and well-shaped spots (Sreekar et al., 2013). I^3S Contour allows researchers to compare profiles, and was compiled for the comparison of whale fin photos (Caci et al., 2013). I^3S Pattern is the latest release and detaches completely from the previous programs in terms of the recognition algorithm. It employs an improved version, SURF (speeded-up robust features) with the algorithm SIFT (scale invariant feature transform), originated in the field of computer vision where it is widely used for comparing similar scenes (Lowe, 2004; Bay et al., 2007). Compared to previous software, the extraction of the points of interest is automatic and not carried out by the operator, who is instead free to set several advanced options. This software was originally developed for analysing head scales in sea turtles, although it performs better with species with an undefined ornamentation pattern. Detailed information on the various versions of I^3S software can be found at http://www.reijns.com/i3s/index.html.

5.3.2 Wild-ID

Wild-ID has been widely used for individual recognition of giraffes (Bolger et al., 2012). As with I^3S Pattern, it hosts the SIFT algorithm, although the photo must be properly orientated and cropping the area of interest must be accomplished by the user. Indeed, contrary to I^3S Pattern, Wild-ID does not supply a selection tool to identify the anatomical region used for the image superimposition. This software has been applied successfully to salamander larvae (Bendik et al., 2013), terrestrial box turtles (Cross et al., 2014), and has application to lizards. It can be downloaded from http://dartmouth.edu/faculty-directory/douglas-thomas-bolger.

5.3.3 MYDAS

Recently, Carter et al. (2014) developed a novel software, MYDAS, which is a PADI system based on a neural network, suitable for Green Turtles (*Chelonia mydas*). This software uses a neural network toolbox already available in MatLab, and individual animals are identified by their distinctive post-ocular scute patterns. The success rate in correctly determining whether a new photo matches a photo in the database is better than 95%. Thus, the methodology used by MYDAS is promising and has the potential to be applied to other species of marine turtles as well as other animals. At the present, the program can be obtained only by contacting the authors, but a worldwide distribution is planned for 2016.

5.3.4 APHIS

This software was developed at the Population Ecology Group of the Institut Mediterrani d'Estudis Avançats (IMEDEA) with particular reference to individual lizard recognition (Moya et al., 2015). It consists of two different routines. The first is a comparison of user-defined spot pattern (spots pattern matching, SPM) that uses the same algorithm of I^3S Classic and processes the images using the same procedure previously described. The second is a pixel-based comparison of a given body region (image template matching, ITM) and is based on algorithms developed in open computer vision (OpenCV). APHIS can be downloaded at: http://www.imedea.uib-csic.es/bc/ecopob/.

5.4 How they work

Compared to the algorithm used in I^3S Classic and to the others developed in the first years of the digital-identification era, those designed recently, especially those borrowed from computer vision, are much more complex (Fukunaga, 1990; Howse, 2013). All are based on the use of interest point detectors for finding the same features in a given image under different viewing conditions. Several scale-invariant interest point detectors have been proposed and tested that reliably recognize distinctive locations in the image, such as corners, blobs, and T-junctions. However, the performances sensibly differ among algorithms, depending on the descriptor used in image recognition and the advances in algorithm improvement that attempt to reconcile speed computation with accuracy.

Apart from the mathematical considerations, which fall outside the scope of this book, SIFT and SURF algorithms share the same method of image processing that has proved to be best performing in terms of stability and repeatability; they considerably differ, however, in the approximations they use to improve computation speed. SURF uses weaker approximations than SIFT, thus obtaining a faster speed at a low cost in terms of lost accuracy. Both algorithms detect key-points within images that can be repeatedly assigned under differing views of the same image, and both have the properties to be invariant to scale change and in plane rotation of the image itself. In other words, they identify within each image a list of descriptors that are individually distinctive while at the same time being robust to noise, detection displacement, and geometric and photometric deformation (Bay et al., 2007). The comparison between two images then is achieved by arranging descriptors in vectors that are matched by means of distance metrics (e.g. Mahalanobis or Euclidean distance; see Chapter 21). SIFT and SURF carry out key-point detection by searching for stable features across all possible scales using a Gaussian function (with different approximations). They progress through four main steps for comparing two images after their superimposition using reference points: (1) a scan over all scales and image locations to identify all the candidate interest points (such as corners, blobs, and T-junctions) that are invariant to scale and orientation; (2) each candidate location is evaluated for stability by fitting an appropriate model, and key-points are selected based on measures of their stability; (3) one or more orientations are assigned to each key-point location based on local image gradient directions; all further operations are performed on image data that have been transformed relative to the assigned orientation, scale, and location for each feature, thereby providing invariance to these transformations; (4) the local image gradients are measured at the selected scale in the region around each key-point. These are transformed into a representation that allows for significant levels of local shape distortion and change in illumination. For image matching and recognition, features are first extracted from a set of reference images and stored in a database. A new image is matched by individually comparing each feature from the new image to the database previously created and finding candidate matching features based on Euclidean distance of their feature vectors (Lowe, 2004; Bay et al., 2007).

The approach of the ITM algorithm is quite different, since it needs two reference points to align the images, applying afterward a fully automated recognition procedure basing on six non-overlapping contiguous templates extracted automatically from the lower half of the pattern. The six resulting templates extracted from a single sample are individually compared with the candidate pattern and the scores resulting from the six comparisons are added up to produce the final sample-candidate matching score (Moya et al., 2015).

5.5 Validation

Since the objective of a marking technique is to allow individual identification, a serious error that should be minimized is the misidentification of individuals. The incidence of such errors has been emphasized as one of the possible limitations to photo-identification (Bolger et al., 2012), at least when computer-assisted procedures are used (Caorsi et al., 2012), and may lead to serious errors (e.g. biased demographic parameters estimation; Morrison et al., 2011). Given the importance of this issue, a non-trivial step in setting up a photo-identification protocol is to supply a validation procedure to estimate the magnitude of the misidentification rate and decide if the method is adequate to study objectives (Bendik et al., 2013). This is a fundamental and necessary step when an already existing protocol is applied for the first time to a particular structure or pattern (e.g. Caci et al., 2013), to a new taxon (e.g. Sacchi et al., 2010), or when a novel protocol is proposed (e.g. Bolger et al., 2012). On the other hand, the validation procedure may appear unnecessary when a well-established technique is used with a reliable previously derived error rate estimate. Researchers should consider this assumption cautiously since the results of a validation test may depend on a combination of factors that are unlikely to be replicable from one study to another.

In order to illustrate the problem, it may be useful to distinguish between two groups of factors that could potentially affect the misidentification rate. The first group relates to all those features that contribute to the final error by defining the quality of the information on which the photo-identification process acts. For example, we may expect an increasing misidentification rate when the marking technique changes over time, when low between-individual variation is apparent, when low quality photos with low resolution are used, and when photos are taken in non-standardized ways (i.e. with varying angle, distance, exposure; Bendik et al., 2013). The second group deals with the photo-identification process itself, including both user and algorithm effects. Even the same algorithm applied to the same structure of the same species may produce different misidentification rates depending on the remnant factors (not always easy to check and control *a priori*). This suggests that validation tests always should be performed (Morrison et al., 2011, Bendik et al., 2013).

The validation procedure could be summarized in three steps:

- assembling a test database of images of individuals unambiguously identified by an independent procedure (i.e. the individual identification has been obtained in a different way from the photo-identification procedure that we want to test);
- choosing a threshold of the similarity score that defines the maximum number of candidate images proposed to the user for the comparison;
- computing the misidentification (error) metrics associated to the chosen threshold.

The first step allows researchers to set up a true reference for validation. The database should simulate the actual data (e.g. number of images, photo quality) and does not contain an unbalanced number of replicates for each individual (i.e. few individuals with many replicates and many individuals with few replicates) in order to avoid a biased estimation of the error metrics (Bendik et al., 2013). The second step is a matter of optimization; choosing a threshold that leaves many candidates increases the user's processing time, but may decrease misidentification likelihood. On the other hand, a restrictive threshold reduces the user's processing time, but may increase the error rate. A good compromise would be testing different thresholds and comparing the associated costs (error rate) and benefits (time saved). In this way, researchers will be aware of the error rate associated to a particular effort.

The third step quantifies the identification error rate; two non-overlapping error measures are usually estimated in biometric recognition (Jain, 2007). The false rejection rate (FRR, or false negative rate) measures the probability that a recapture event is wrongly classified as a new capture. In photo-identification terms, a false rejection occurs when we fail to match two photos of the same individual. The false acceptance rate (FAR, or false positive rate) measures the probability that a new capture is wrongly classified as a recapture; a false acceptance occurs when two photos of different individuals are incorrectly matched. Computationally, FRR is defined as the ratio between the number of false rejections and the total number of identification attempts. Similarly, FAR is defined as the ratio between the number of false acceptances and the total number of identification attempts. Be sure to note that 'number of attempts' is the number of classified images—e.g. capture/recapture—and not the number of pairs of photos that have been compared in order to obtain the classification. FAR and FRR values range from 0 (good) to 1 (bad).

Considering that computer-assisted photo-identification (Figure 5.2) gives the user the final decision (i.e. the user must choose the matching photograph, if any, among a preselected list of candidates), the processes that produce false acceptance (FA) and false rejection (FR) are not balanced. Indeed, FR may be due to algorithm faults if the algorithm fails in scoring the true matching image among the top ones, preventing the user from inspecting it and recognizing the correspondence. In any case, FR also could be due to the user's fault if a true matching image is not recognized as such, even if it is listed by the algorithm among the favourites. On the other hand, FA may be due only to the user's error in matching two non-corresponding images (independently by the low or high ranking in the algorithm list). If it is assumed that human error is negligible with respect to the algorithm, FA is expected to be less frequent than FR and, consequently, of less concern. This simplification of the problem is supported by studies that have explicitly estimated both FAR and FRR. FAR values are 80-fold lower than FRR (range 0.007–0.25) and practically nil ($p < 0.001$) (see Table 1 in Bolger et al., 2012).

Figure 5.2 A typical computer-assisted photo-identification process illustrated with the ventral scales pattern in the Common Wall Lizard (from Sacchi et al., 2010): (a) set-up of a photographic shoot using a tripod to avoid involuntary movements and to standardize the distance from the subject; (b) perpendicular shot of the salient features (i.e. the ventral scales) used for digital identification. The paper sheet next to the lizard shows the date and individual code; (c) image processing using I^{3}S Classic (the three reference points encompass the ventral scales used for photo-identification); (d) a match with an image stored within the database: the user has to make the decision in establishing if this potential match is 'true'.

5.6 Photo-identification in reptiles: present and future

5.6.1 The state of the art of photo-identification in reptiles

Here, we present a short review of the most relevant applications of computer-assisted systems for the digital identification of reptiles over the last decade (Table 5.1).

Hatchlings and juveniles of the Spur-thigh Tortoise (*Testudo graeca ibera*) were recognized with a plastron photo-identification system (time-series) mediated by a dissimilarity–similarity index (Tichý and Kintrová, 2010). The combined dissimilarity measure was calculated from the relative differences in seam distances between two plastron photographs using multivariate analyses.

Baker et al. (2012) demonstrated that near-infrared (NIR) imaging with an alternate light source (ALS) and digital photography were useful tools for revealing and documenting original dorsal skin patterns found on dyed snake leather products in the wildlife trade. Documented species included pythons, anacondas, and water snakes (e.g. genera *Homalopsis* and *Enhydris*). Although this application of photo-identification fell outside of its use in field surveys and experiments, we found this approach particularly intriguing and stimulating. In the same way, photographing individuals also can be used to track protected species (Bender, 2001).

For marine turtles, photo-identification is based on specific physical characteristics (position and structure) of the scutes from both facial profiles located posterior to the eye towards the neck and from the line of the upper jaw to the top of the turtle's head on either side of the head (Jean et al., 2010). This approach enabled researchers to separate individuals having very similar patterns and helped in understanding the behavioural patterns of foraging Hawksbill and Green Turtles (Chassagneux et al., 2013). Similarly, Goodman Hall and Braun McNeill (2013) applied a photographic and visual recognition technique to sea turtles, where the most reliable information was derived from scale shape and distribution on the dorsum of the head, the easiest part to recognize when turtles were encountered underwater or captured in fishing nets. Photo-identification has been used to avoid the repeated satellite tagging of individual sea turtles, to revise remigration estimates, to identify the number of individuals potentially injured by boat strikes, to examine the regression of fibropapillomas, to assess behavioural interactions of conspecifics, to estimate tag loss, and to determine operational population sex ratios and breeding periodicity (reviewed in Goodman Hall and Braun McNeill, 2013).

An application of photographic identification that used the individual colouration pattern of the dorsal skin, invariable between years, in collared lizards was described recently by Faber et al. (2013). The procedure is applicable to small samples without automated integration, similar to I^3S software, and provided significant results in individual recognition among years. Sreekar et al. (2013) pointed out the importance of photographic capture–recapture technique in the Indian Gliding Lizard, focusing the target spots of the ventral region of the patagium. The authors used I^3S Manta to assess individual recognition, furthermore finding a significant sexual dimorphism in the pattern and size of markings.

Although originally developed for amphibians, Wild-ID has been used to identify other organisms, such as Leatherback Turtles, by analysing the different coloured area of the pink pineal spot on the head (Buonantony, 2008; Bendik et al., 2013).

Table 5.1 *Representative examples of studies in which photographic identification techniques have been used to study reptiles over the last 15 years*

Family	Species	Year	Methods–software	Features	Validation	Reference
Cheloniidae	*Chelonia mydas* *Eretmochelys imbricata*	2013	Visual identification	Facial scutes	Yes	Chassagneux et al. (2013)
Cheloniidae	*Caretta caretta* *Chelonia mydas* *Lepidochelys kempii*	2013	Visual identification	Head and facial scutes	No	Goodman Hall and Braun McNeill (2013)
Cheloniidae	*Chelonia mydas*	2014	MYDAS	Post-ocular scutes	Yes	Carter et al. (2014)
Cheloniidae	*Eretmochelys imbricata*	2014	I^3S Classic	Scale morphology	Yes	Dunbar et al. (2014)
Dermochelyidae	*Dermochelys coriacea*	2008	Wild-ID	Head pineal spots	Yes	Buonantony (2008)
Testudinidae	*Testudo graeca*	2010	Similarity–dissimilarity algorithms	Plastron seams	Yes	Tichý and Kintrová (2010)
Colubridae	*Natrix natrix*	1999	Visual identification	Facial spots	No	Vaughan (1999)
Crotaphytidae	*Crotaphytus dickersonae*	2013	Visual identification	Spots and markings, lateral body side	Yes	Faber et al. (2013)
Agamidae	*Draco dussumieri*	2013	I^3S Manta	Spots of ventral region of patagium	Yes	Sreekar et al. (2013)
Lacertidae	*Podarcis muralis*	2010	I^3S Classic	Scale morphology	Yes	Sacchi et al. (2010)
Lacertidae	*Podarcis muralis*	2006	Visual identification	Black spots on ventral scales	Yes	Sacchi et al. (2006)
Lacertidae	*Lacerta perspicillata*	2004	Visual identification	Ventral scale morphology	Yes	Perera and Perez-Mellado (2004)
Scincidae	*Tiliqua adelaidensis*	2009	Edge detection algorithm	Scale pattern	No	Li et al. (2009)
Sphaerodactylidae	*Tarentola boettgeri bischoffi*	2013	I^3S Classic	Iris pattern	Yes	Rocha et al. (2013)

5.6.2 Where should we go from here?

The future of any software dedicated to photo-identification, especially in capture–mark–recapture studies, resides in the development of tools based on algorithms already developed in the context of computer vision in order to compare digital scenes. These algorithms often are freely available and have proved to be powerful, fast, and reliable. More effort should be therefore made to bridge the gap that currently exists between computer scientists who develop these tools and field biologists that intend to use them in order to make computer programs more able to consistently pick up the different characteristics that are specific to a species and that allow for the accurate identification of individuals.

References

Arzoumanian, Z., Holmberg, J., and Norman, B. (2005). An astronomical pattern-matching algorithm for computer-aided identification of whale sharks *Rhincodon typus*. *Journal of Applied Ecology*, **42**, 999–1011.

Baker, W.B., Reinholz, A.D., and Espinoza, E.O. (2012). Digital near-infrared photography as a tool in forensic snake skin identification. *Herpetological Journal*, **22**, 79–82.

Bay, H., Ess, A., Tuytelaars, T., et al. (2007). Speeded-up robust features (SURF). *Computer Vision and Image Understanding*, **110**, 346–59.

Bender, C. (2001). *Photodocumentation of Protected Reptiles*. Rheinbach, Germany:German Society for Herpetology and Terrarium Keeping.

Bendik, N.F., Morrison, T.A., Gluesenkamp, A.G., et al. (2013). Computer-assisted photo identification outperforms visible implant elastomers in an endangered salamander, *Eurycea tonkawae*. *PLoS One*, **8**, e59424.

Bloch, N., and Irschick, D.J. (2004). Toe-clipping dramatically reduces clinging performance in a pad-bearing lizard (*Anolis carolinensis*). *Journal of Herpetology*, **37**, 293–8.

Bolger, D.T., Morrison, T.A, Vance, B., et al. (2012). A computer-assisted system for photographic mark–recapture analysis. *Methods in Ecology and Evolution*, **3**, 813–22.

Buonantony, D. (2008). *An analysis of utilizing the leatherback's pineal spot for photo-identification*. Master of Environmental Management Thesis, Nicholas School of the Environment and Earth Sciences, Duke University, Durham, NC, USA.

Caci, G., Biscaccianti, A.B., Cistrone, L., et al. (2013). Spotting the right spot: computer-aided individual identification of the threatened cerambycid beetle *Rosalia alpina*. *Journal of Insect Conservation*, **17**, 787–95.

Caorsi, V.Z., Santos, R.R., and Grant, T. (2012). Clip or snap? An evaluation of toe-clipping and photo-identification methods for identifying individual southern red-bellied toads, *Melanophryniscus cambaraensis*. *South American Journal of Herpetology*, **7**, 79–84.

Carter, S.J.B., Bell, I.P., Miller, J.J., et al. (2014). Automated marine turtle photograph identification using artificial neural networks, with application to green turtles. *Journal of Experimental Marine Biology and Ecology*, **452**, 105–10.

Chassagneux A., Jean C., Bourjea, J., et al. (2013). Unraveling behavioral pattern of foraging hawksbill and green turtles using photo-identification. *Marine Turtle Newsletter*, **137**, 1–4.

Cross, M.D., Lipps, G.J., Jr., Sapak, J.M., et al. (2014). Pattern-recognition software as a supplemental method of identifying individual eastern box turtles (*Terrapene c. carolina*). *Herpetological Review*, **45**, 584–6.

Dunbar, S.G., Ito, H.E., Bahjiri, K., et al. (2014). Recognition of juvenile hawksbills *Eretmochelys imbricata* through face scale digitization and automated searching. *Endangered Species Research*, **26**, 137–46.

Faber, L., Plasman, M., Duchateau, M.J. (2013). The use of photo identification on Dickerson's collared lizard (*Crotaphytus dickersonae*). *Herpetological Review*, **44**, 600–3.

Fukunaga, K. (1990). *Introduction to Statistical Pattern Recognition*. 2nd ed. San Diego, CA: Academic Press.

Gamble, L., Ravela, S., and McGarigal, K. (2008). Multi-scale features for identifying individuals in large biological databases: an application of pattern recognition technology to the marbled salamander *Ambystoma opacum*. *Journal of Applied Ecology*, **45**, 170–80.

Goodman Hall, A., and Braun McNeill, J. (2013). Inferring sea turtle recapture rates using photographic identification. *Herpetological Review*, **44**, 561–9.

Hammond, P.S., Mizroch, S.A., and Donovan, G.P. (1990). *Individual Recognition of Cetaceans: Use of Photo-Identification and Other Techniques to Estimate Population Parameters*. Special Issue 12. Cambridge, UK: International Whaling Commission.

Henle, K., Kuhn, J., Podloucky, R., et al. (1997). Individualerkennung und Markierung mitteleuropäischer Amphibien und Reptilien: Ubersicht und Bewertung der Methoden; Empfehlungen aus Natur- und Tierschutzsicht. *Mertensiella*, 7, 133–84.

Howse, J. (2013). *OpenCV Computer Vision with Python*. Birmingham, UK:Packt Publishing Ltd.

Jain, A.K. (2007). Biometric recognition. *Nature*, **449**, 38–40.

Jean, C., Ciccione, S., Talma, E., et al. (2010). Photo-identification method for green and hawksbill turtles and first results from Reunion. *Indian Ocean Turtle Newsletter*, **11**, 8–13.

Li, J.S.J., Tohl, D., Randhawa, S., et al. (2009). Non-invasive lizard identification using signature curves. In *TENCON 2009, IEEE Region 10 Conference*, pp. 1–5.

Lowe, D.G. (2004). Distinctive image features from scale-invariant key-points. *International Journal of Computer Vision*, **60**, 91–110.

Morrison, T.A., Yoshizaki, J., Nichols, J.D., et al. (2011). Estimating survival in photographic capture–recapture studies: overcoming misidentification error. *Methods in Ecology and Evolution*, **2**, 454–63.

Moya, Ó., Mansilla, P.L., Madrazo, S., et al. (2015). APHIS: a new software for photo-matching in ecological studies. *Ecological Informatics*, **27**, 64–70.

Perera, A., and Perez-Mellado, V. (2004). Photographic identification as a noninvasive marking technique for lacertid lizards. *Herpetological Review*, **35**, 349–50.

Press, W.H., Teukolsky, S.A., Vetterling, W.T., et al. (1992). *Numerical Recipes in C, the Art of Scientific Computing*. Cambridge: Cambridge University Press.

Rocha, R., Carrilho, T., and Rebelo, R. (2013). Iris photo-identification: a new methodology for the individual recognition of *Tarentola* geckos. *Amphibia-Reptilia*, **34**, 590–6.

Sacchi, R., Scali, S., Fasola, M., et al. (2006). The numerical encoding of scale morphology highly improves photographic identification in lizards. *Acta Herpetologica*, **2**, 27–35.

Sacchi, R., Scali, S., Pellitteri-Rosa, D., et al. (2010). Photographic identification in reptiles: a matter of scales. *Amphibia-Reptilia*, **31**, 489–502.

Speed, C.W., Meekan, M.G., and Bradshaw, C.J.A. (2007). Spot the match: wildlife photo-identification using information theory. *Frontiers in Zoology*, **4**, 1–11.

Sreekar, R., Purushotham, C.B., Saini, K., et al. (2013). Photographic capture–recapture sampling for assessing populations of the Indian gliding lizard *Draco dussumieri*. *PLoS One*, **8**, e55935.

Stenhouse, S.L. (1985). Migratory orientation and homing in *Ambystoma maculatum* and *Ambystoma opacum*. *Copeia*, **1985**, 631–7.

Tichý, L., and Kintrová, K. (2010). Specimen identification from time-series photographs using plastron morphometry in *Testudo graeca ibera*. *Journal of Zoology*, **281**, 210–17.

Van Tienhoven, A.M., Den Hartog, J.E., Reijns, R.A., et al. (2007). A computer-aided program for pattern-matching of natural marks on the spotted ragged tooth shark *Carcharias taurus*. *Journal of Applied Ecology*, **44**, 273–80.

Vaughan, R. (1999). Provisional results from study of facial features as a means of individual identification in *Natrix natrix*. *British Herpetological Society Bulletin*, **68**, 39–46.

6
Preserving reptiles for research

Steve W. Gotte, Jeremy F. Jacobs, and George R. Zug

6.1 Introduction

What are voucher specimens and why do we collect them? Voucher specimens are animals and/or their parts that are deposited in a research museum to document the occurrence of a taxon at a specific location in space and time (Pleijel et al., 2008; Reynolds and McDiarmid, 2012). For field biologists, vouchers are the repeatable element of a field study as they allow other biologists, now and in the future, to confirm the identity of species that were studied. The scientific importance of a voucher specimen or series of specimens is that other people are afforded the opportunity to examine the entire animal and confirm or correct identifications. A photographic record is somewhat useful for recording the occurrence of a species, but such records can be insufficient for reliable confirmation of specific identity. Even if a photo shows diagnostic characters of currently recognized taxa, it may not show characters that separate taxa that may be described in the future. Substantial cryptic biodiversity is being found in even relatively well-known herpetofaunas (Crawford et al., 2010), and specimens allow researchers to retroactively evaluate the true diversity in a study as understanding of taxonomy evolves. They enable biologists to study the systematic relationships of populations by quantifying variation in different traits. Specimens are also a source of biological data such as behaviour, ecology, epidemiology, and reproduction through examination of their anatomy, reproductive and digestive tracts, and parasites (Suarez and Tsutsui, 2004).

Preserving reptiles as vouchers is not difficult, although doing it properly requires care, effort, and time. Poorly preserved vouchers can invalidate the results and conclusions of your study because of the inability to confirm the identity of your study animals. Good science requires repeatability of observations, and the *absence* of vouchers or poorly preserved ones prevents such confirmation.

Due to space restrictions, we are unable to go into as much detail as we would like in this chapter. A number of publications give more details on some topics discussed in this chapter, such as Pisani (1973), Pisani and Villa (1974), Etheridge (1996), Karns (1986), McDiarmid (1994), Cortez et al. (2006), Foster (2012) (and subchapters therein), Reynolds and McDiarmid (2012), and Simmons (2015). Although some of these works focus on amphibians, they also apply to reptiles in many aspects.

Reptile Ecology and Conservation. Edited by C. Kenneth Dodd, Jr.
Published 2016 by Oxford University Press.

6.2 Planning and permits

Collecting vouchers for your research project must be in accordance with your institution's animal care regulations and the laws and regulations of the various levels of government where your research site is located (see Chapter 2). Scientists are not above the law and have an ethical obligation to perform their research within the boundaries of the legal code of their country of research. Vouchering specimens usually requires collecting and/or research permits from local authorities (including state or provincial governments) in addition to national permits. Some countries also require permission from indigenous peoples. Transporting voucher specimens across internal political boundaries may require permits and transport across international borders definitely requires government permission (export, import, and potentially CITES permits). Failure to obtain legal permission potentially subjects you to fines and/or jailing, will not be accepted by responsible depositories, and will prevent you from publishing your research in primary scientific journals.

Because you are euthanizing a living organism and removing an individual from the population's gene pool, it is important to carefully evaluate the number of individuals preserved and *equally important* to preserve sufficient specimens to meet your research goals. You will usually have to provide the maximum number of specimens per taxa to be collected to obtain permits, which will be determined by the parameters of your research. If your goal is simply to document the species studied in your research, then two individuals, an adult female and an adult male, may be adequate. However, if morphological examination is part of the research project, a sample that yields statistically valid results will be necessary. In most instances, this can be achieved with a sample of about 20 individuals per species per sex from each locality that is surveyed. For a survey of a large site with multiple habitat types, additional samples may be required due to the increased chance that cryptic taxa may be encountered that are indistinguishable in the field. Studies of food habits, reproductive biology, and parasite loads, for example, may require larger samples over defined time increments. See Reynolds and McDiarmid (2012) for an additional discussion of sample size. It is usually desirable or required by permits that some specimens be deposited in the country of origin to support local research and training, so adjust your sample size accordingly.

Following collection, live individuals should be temporarily and humanely stored in a preparation area that is quiet, well ventilated, out of direct sunlight, and away from human interference so as to help reduce any additional trauma or injury to the individuals. Specimen preparation should begin as soon after capture as practical; rapid preparation is essential from a humane perspective as well as to help ensure high quality specimen preparation. The preservation station must have stable work surfaces and should be comfortable for working for extended periods of time. Further, it should be kept clean to reduce potential accidents with chemicals or sharp instruments and to reduce hiding places if an individual escapes during handling. Table 6.1 provides a list of supplies and equipment needed for preservation.

Table 6.1 *Basic supplies and equipment for proper preservation of voucher specimens*

Chemicals	Formalin
	Paraformaldehyde
	Buffers
	Ethanol
	Euthanizing agents (sodium pentobarbital, MS222, chlorobutanol)
Documentation	Permits, IACUC, safety data sheets (SDSs) for the chemicals used
	Field catalogue
	Tags with string attached (sequentially numbered; blanks)
	Pens with indelible ink
	Pencils (medium lead)
Prep kit	Syringes, 1 ml, 5 ml, 10 ml, 30 ml, 60 ml
	Syringe needles, #16, #18, #22, #27 (ca. 30 mm long)
	Thread (cotton or linen, #16 three ply)
	Tissue tubes and etcher
	Small pointed forceps
	Bottle forceps
	Scalpel handle and blades
	Small and large scissors
	Metric tape measure
	Preserving trays and non-embossed paper towels
Storing, packing, and shipping	Plastic jugs (2 l, 4 l, for mixing formalin and for storing/floating specimens)
	Small plastic jars or vials (for particularly delicate specimens, removed organs, parasites)
	Tight sealing plastic trays
	Cheese cloth
	Large plastic bags (such as garbage bags, for secondary containment)
	Plastic bags (multiple sizes for secure specimen packing)
	Zip-ties, twist ties, or other secure bag sealers
Personal protective gear (PPE)	Nitrile gloves
	Eye protection
	Apron, dust mask, or respirators are optional

6.3 Euthanasia

Euthanasia is the act of killing an animal with the least amount of trauma and as painlessly as possible, while at the same time not damaging the scientific value of the voucher. For reptiles, this generally requires an intracardiac injection or an intracoelomic injection of one of several chemical agents. Whatever the chemical agent is and its method of use, it should always conform to Institutional Animal Care and Use Committee (IACUC) protocols or similar requirements imposed by national or international agencies, such as those of the European Union. Many of the chemicals mentioned in this chapter are toxic or suspected carcinogens. It is important that safety data sheets be consulted for safe handling and disposal procedures of each of the chemicals and should be carried with you in the field and during transport.

Historically, Nembutal (aqueous sodium pentobarbital) has been used as a very effective euthanizing agent. It quickly stops respiration and cardiovascular activity rendering the individual a quick death with minimal trauma and relaxed musculature. Relaxed musculature is essential when it comes time to fix and position your specimen. The problem with using Nembutal is that it is highly regulated in many countries. For instance, it is a Class II drug in the United States and Federal Drug Enforcement Agency and state permits are required to possess it. Additionally, detailed procedures must be followed: strict usage logs must be kept and the drug must be kept under lock and key. In some countries the possession of Nembutal is strictly illegal. Basic dosage is 100 mg/kg body weight. Some researchers dilute 1:5 to 1:10 for very small individuals, but this can cause bloating particularly when the preservative is injected.

Several other euthanizing agents have been successfully used that are less regulated. The chemical MS222 (tricaine methanesulphonate, also called TMS, tricaine, tricaine mesylate, and Finquel) has been used on reptiles for over 50 years (Karlstrom and Cook, 1955; Conroy et al., 2009). It comes in a crystalline form that dissolves readily in water and will dispatch a reptile with little to no discernible distress. Injection is either cardiac or coelomic. Use about half a teaspoon (ca. 0.3 g) in 20 ml water and make fresh solutions regularly as it loses its effectiveness after several days (Simmons, 2015). See Figure 1 of Conroy et al. (2009) for detailed instructions for preparing 1% buffered solution and dosing.

Chlorotone (hydrous chlorobutanol) has been shown to be effective. Berry (2012) suggests 0.5–1.0 ml of a saturated solution for individuals less than 50 g, with more for larger individuals. The anaesthetic effect is quick, but death may require up to 30 minutes.

Other possibilities include lidocaine, procaine (prescription may be needed, but controls are not as restrictive as Nembutal), and benzocaine; procaine is used at 0.05 cc of a 10% solution per gram body weight (Livezey, 1958; Simmons, 2015), whereas benzocaine (available as a 20% gel or liquid at many drugstores for tooth ache pain under names such as Orajel) has been used for small reptiles, either orally or topically (Simmons, 2015).

Inhalants such as acetone, ether, and chloroform as well as CO_2 have also been used with some success, but these techniques are often difficult to use or transport safely under field conditions. Further, they yield unsatisfactory muscle relaxation. Freezing has been used to kill reptiles, but sometimes is considered inhumane and may be unacceptable as part of an Animal Care and Use (IACUC) protocol (but see Shine et al., 2015). Also, freezing ruptures cells thereby producing significantly lower quality specimens (Scott and Aquino-Shuster, 1989).

Death of an individual is recognized by a total relaxed and limp state. No muscular reaction should be observed if one pinches or pricks a limb. If you see a muscular, reaction further exposure to the euthanizing agent may be necessary. Preservation should not begin until one is certain that the animal is dead. Please note that with venomous snakes, one must be especially alert during the euthanizing process because the reflex action of a nearly dead snake can produce a serious or fatal bite. Always immobilize the snake's head when testing its state of relaxation and during the initial preservation stages. Also be careful not to impale oneself on a fang of a venomous snake before and during preservation even after it is clearly dead, as the venom is still 'hot' until it has been preserved. Turtles are difficult to kill and even when they appear dead, they may only be deeply anaesthetized. A dead turtle will be totally relaxed! Pick up the turtle by the plastron, and holding horizontal, gently shake to and fro. If dead, the head, neck, and legs will all fall outward and swing limply.

6.4 Specimen preservation and data collection

6.4.1 Record keeping

The minimum data that must be recorded and maintained in association with each voucher specimen include: the full name of the collector(s); the unique field number of the specimen; the date the specimen was captured (dd/Mmm/yyyy, spell out the month); and the locality where it was found (name of location or distance and compass direction to nearest town) and geo-coordinates. A tentative identification, even listing a specimen as a colubrid snake or an agamid lizard should be made. Record whether the specimen was photographed or if parasites or tissues were taken. For a more comprehensive approach, the following data should also be recorded: elevation, time of day, weather, site description, microhabitat, behaviour, colour patterns, and measurements. These additional data facilitate understanding the biology of the species. Record date/time of preservation if more than a few hours from that of capture; this provides context for food habits and reproductive condition that may be determined by future examination of internal organs. There are various ways to maintain these data. A hard copy (Figure 6.1) is preferred, as it is less susceptible to data loss or corruption. If data are taken electronically and the final museum depository for your specimens is known, coordinate the data structure with the depository to reduce post-collection processing time and transcription errors. There are various types of field books, that is, loose leaf in a ringed binder, bound ledgers, and preprinted field datasheets (see Chapter 3). Regardless of type, archival quality paper, legibility, and permanence of the written entry are paramount. Use only indelible ink or number 2 lead pencil; ball-point pen ink is unsuitable for permanent field records because it is water soluble and easily obliterated. Field notes become part of the voucher collection and will accompany specimens into a museum for permanent conservation and access by other researchers. Retention of specimens and their records in a researcher's lab or personal museum is acceptable while the research is active, but permanence of vouchers and their supporting data is possible only in a research museum with a historical record of commitment of perpetuity of voucher specimen maintenance and making the specimens available to the scientific community.

Best record keeping practice requires that each specimen bears a uniquely numbered field tag (Figure 6.1(a)). Tags need to be water- and alcohol-proof heavy stock rag-content paper. Metal tags corrode over time and become unreadable and disintegrate, and most plastic tags become soft and/or melt over time as well as leaching plasticizers into the preservation fluid. Plastic embossing tape such as Dymo tape and similar products in particular should be avoided. Not only does the tape leach colour into the preservation fluid and specimens, but the glue reacts adversely with preservatives potentially

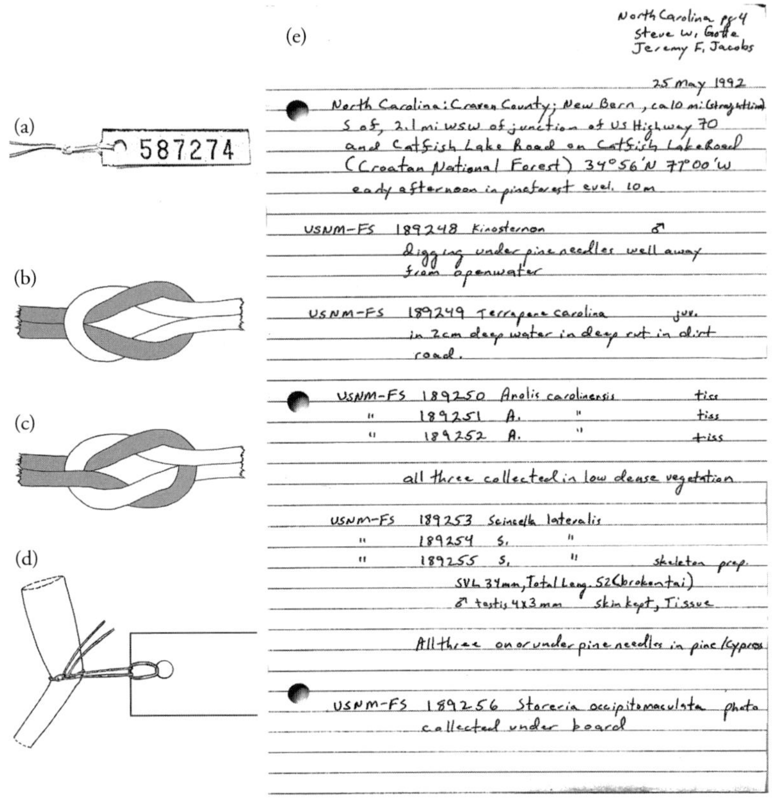

Figure 6.1 Essential items for maintenance of voucher data. Field tag with string (a); note the manner of attachment of the string to tag (ca. 1 cm from tag to knot), providing flexibility in the movement of the tag but limited to avoid entanglement with other tags and specimens. Use uncoloured linen or cotton string (not a synthetic); tie with a square knot (b) not a 'granny' (c), as they become untied over time with handling. Cut the loose ends of the string at least 1 cm but no more than 3 cm from the second knot to maintain the integrity of the knot while reducing tangling (d). Sample page from a loose-leaf field book (e). Note that locality information is presented from broadest to narrowest divisions, starting with country.

creating a sticky slime on specimens. To prevent the incorrect association of collecting data with a specimen, attach the tag to the specimen as it is preserved. Tie the tag to the right hindlimb just below the knee or around the waist just above the hind limbs of very small specimens. For snakes and legless lizards, tie the tag about one fifth the body length behind the head (in the vicinity of the heart is usually a good place; do not go much posterior of the heart so that the tag does not impede examination of internal organs). Alternatively, the tag can be tied one or two head lengths behind the head for snakes with distinct necks. Some collectors use suture needles to thread field tag strings through the body of the snake and tie the string using the vertebral column as an anchor. Using a square knot (Figure 6.1(b)) (granny knots (Figure 6.1(c)) loosen over time and are not acceptable), tie the tag tight enough that it will not slide off with handling. Cut the trailing ends of the tag string ca. 1 to 2 cm beyond the square knot. This field number must be recorded in your field book in direct association with the capture date and locality data. Data recording is performed simultaneously with preservation.

6.4.2 Preservation and positioning

Collection of tissue for subsequent molecular studies has become a common feature of voucher preparation (Schulte, 2012; Gamble, 2014). Harvest tissue immediately after euthanasia but prior to preservation, and process them in an area removed from where preservation is occurring to prevent accidentally compromising tissues with specimen preservative. Liver and muscle tissue are the standard for subsequent molecular analysis. Both can be removed with minimum damage to specimens. Any organs that come out of a body incision during the tissue removal process should be pushed back inside the specimen prior to preservation. Clean blood off the surface of the specimen prior to preservation as it can become permanently affixed to the skin thus obscuring or falsely indicating a pattern.

There are two primary 'preservatives' used for reptile preservation, 10% formalin (3.7% formaldehyde) and 70% ethanol. There are important differences between the two. Formalin is a fixative that crosslinks proteins and is mostly irreversible, whereas ethanol is a preservative (by dehydration) that is reversible; therefore, specimens preserved in ethanol can rot if the ethanol concentration gets too low at some point in the future (Simmons, 2014). Specimens fixed in formalin are much more suitable for later histological examination. One argument for preservation in ethanol is that it preserves DNA; however, the entire specimen does not need to be a source of DNA, and the specimen should be maintained at a different concentration (70%) than the tissue sample (≥95%).

Formalin is a toxic chemical that must be used carefully. Use only in well-ventilated areas and avoid breathing the fumes as much as possible when working with formalin. Use neoprene, nitrile, or other gloves rated for formalin when preserving; latex gloves are easily penetrated by formaldehyde (Raloff, 1985). Formalin can cause skin damage, and prolonged exposure can cause hypersensitivity and result in contact dermatitis. Mix a 10% formalin solution from full strength formalin stock (37% formaldehyde), buffer with sodium phosphate monobasic (4 g) and sodium phosphate dibasic anhydrous (6.5 g) per litre (other buffers that have been suggested have various disadvantages, including clearing (becoming translucent) of specimens; Simmons, 2015). Buffering is important because if the pH is too high, specimens will clear, and if too low, decalcification can occur. You can premix 1 l aliquots of the buffers in plastic bags to be added to the 10% formalin in the field.

Paraformaldehyde when properly prepared works as well as pre-mixed formalin and can be easier to transport. Paraformaldehyde is an irritant if it gets on your skin. It is sometimes a very light powder and easily blown about; it will cause considerable damage if inhaled or blown into your eyes. Before beginning fieldwork, make up double-sealed bags of paraformaldehyde to make 1 l (36 g) or 4 l (143 g) of 10% formalin to reduce handling in the field. Paraformaldehyde dissolves slowly in 'room-temperature' water so it should be mixed with water heated to 50–60°C or prepared two or three days before use and agitated regularly. Ehmann (1989) provides a detailed but easy procedure for preparing and mixing paraformaldehyde into a neutral 10% formalin solution. Water should be as clean and free of contaminates as possible.

Although we strongly encourage fixation with formalin, there are times when it is unavailable. Ethanol preservation is the preferred alternative. The ethanol should not be 'denatured', that is, chemically contaminated to prevent human consumption. Denaturing agents often pose health risks, degrade DNA, and likely adversely affect long-term viability of the vouchers. Specimens should be preserved in 70% (140 proof) ethanol. High-proof unflavoured vodka, clear rum or gin, or even higher proof spirits such as grain alcohol can be used (diluted) as a source of ethanol if necessary. Isopropanol and methanol have been mentioned as preservatives, but their use is strongly discouraged (Simmons, 2014).

Preservation or fixation (henceforth preservation) is the critical step in producing voucher specimens that are useful for identification and for long-term use and retention in a museum collection. Preservation should begin as soon as the animal has died. The following instructions are specifically for formalin-fixed specimens, but also apply to ethanol preserved individuals as well. First affix a tag to the specimen. Because of their relatively impermeable skin (particularly hydrophobic in some groups of small/tiny lizards), all reptile specimens, no matter how small, must have the formalin injected into the body cavity and muscles. Be careful not to inject air into specimens from syringes that are not completely full of preservative. Have multiple size hypodermic needles and syringes to deal with various size specimens.

Lizards less than 50 mm snout–vent length (SVL) are injected by inserting the hypodermic needle into and through the cloacal wall into the body cavity; inject sufficient formalin to return the animal to the volume of a well-fed specimen, not until it looks like an over-inflated balloon. For lizards over 50 mm SVL, also inject at one or more points mid-body to make sure the formalin is uniformly distributed in the body cavity. You should inject the feet and legs, starting distally with the sole of the foot (which often will help spread the digits) and then larger muscle masses of the legs (Figure 6.2). Starting distally, inject formalin into the tail as soon as it is thick enough to receive the needle, then move forward in increments to make sure that there is a good distribution of formalin throughout the tail. Where the tail is too narrow for injection, use a needle or scalpel tip to perforate the ventral surface or use very small sharp scissors to cut along the venter of the tail to allow absorption of formalin. Be very careful, as many lizards have fragile and/or autotomizing tails. With practice, gentle pressure with your thumb followed by careful injection near the base of the tail can evert the hemipenes of males (see Raxworthy, 2012). Crocodilians are injected like lizards, though perfusion is preferable for very large individuals. Perfusion is basically using the circulatory system to disperse the formalin by replacing the blood (see Simmons, 2015, and references therein for methods of perfusion). Because of the elongate bodies of snakes (and some lizards), you will need to move the needle and inject sequentially along the body to ensure that all parts of the body cavity and organs receive an adequate volume of formalin. As you inject snakes, you will see the body swell as the fluid enters, and generally you will need

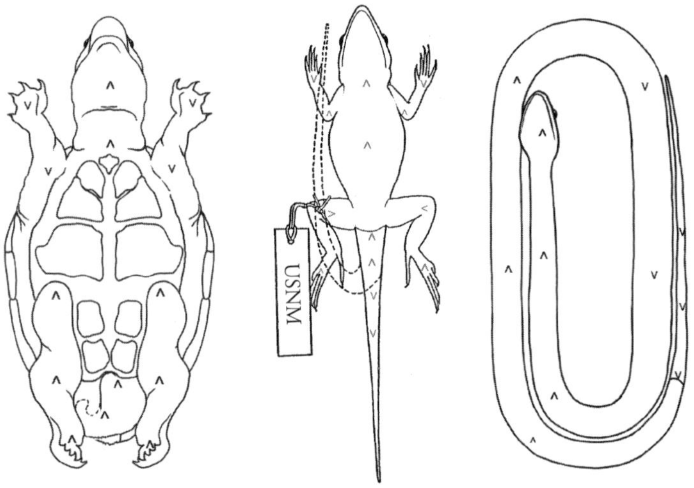

Figure 6.2 Preferred positioning of specimens of reptiles to permit ease and accuracy of subsequent data-gathering and long-term storage. Lizard tails are sometimes too thick or brittle to wrap forward, in such cases, the tail should be preserved straight. Turtles should have head and limbs outstretched to allow examination of colour patterns and soft part morphology. Snakes are best positioned in a single plane (not coiled spring-like in a jar), either circularly or looped (the authors prefer the latter owing to accuracy of subsequent measurements and for storage in standard museum jars). 'V' (showing insertion direction) indicates injection sites.

to move the needle at least every 10–20 cm (the goal is uniform distribution of the formalin, so more smaller injections are often preferable to a few large injections, particularly in individuals with large food items or that are gravid); inject from the cloaca towards the head, and inject and perforate the tail as in lizards. For particularly large snakes and lizards, you also need to inject the jaw muscles and dorsally into the muscles along the spine and tail. Do this with multiple small injections so as to avoid pockets of formalin. It is sometimes necessary to manually massage the injection sites to help distribute the formalin.

Turtles, because of their shell, offer a special challenge in preservation and require a large volume of formalin to preserve them properly. Estimate the volume of the shell and insert at least half that amount of formalin into the body cavity. For all turtles, inject formalin by inserting the needle into the body cavity at the insertion of all four limbs to distribute the formalin uniformly (Figure 6.2). The legs, feet, and neck all require separate injections. Insert the needle in the sole of the foot (carefully because skin is thick and the needle may bend or slip potentially causing you injury); then inject the remainder of the limb (this may require multiple sites of injection in larger turtles). The head-neck requires multiple small injections, first at base of neck at the juncture of the carapace then moving forward with a final injection at the base of the skull, massaging any pockets of formalin to maintain uniform distribution. For the tail, only an injection at its base is typically required for females; in males (which have larger tails), additional injections in the middle of the tail and towards the tip may be required to ensure formalin penetration for satisfactory preservation. Turtles should be preserved with their mouths held completely open by inserting an appropriate size cork or other object into the buccal cavity.

Injection is not the final step in preservation. Positioning is also required, usually in sealable flat-bottomed plastic trays lined with non-textured undyed paper towels saturated with formalin. Do not use metal trays if possible, as they can cause chemical reactions that have long-term consequences to specimen conservation (Simmons, 2015). Tupperware™ and Rubbermaid™ rectangular food storage containers about 25 cm × 40 cm × 5–10 cm tall work well for most specimens and are inexpensive and easily available (and can be used for packing and transport). After the specimens have been positioned, cover them with formalin-saturated paper towels or cheese cloth. The illustrations (Figures 6.2 and 6.3) show the various layout postures for reptiles. The posture becomes permanent during preservation. By carefully adjusting the body and limbs immediately after injection, the voucher will have a posture that permits easy use for identification and study. It is important to take the time to carefully lay out specimens; the additional minutes spent at this stage greatly enhance the usefulness of the specimens. Straightening and spreading toes allows easy determination of toe length proportions, lamellae counts, and morphology; propping mouths open provides access to tooth counts and distributions, nares shape and position, tongue shape and ornamentation, rhamphotheca structure, and mouth lining colour; everting hemipenes facilitates access to a number of taxonomically important characters.

Oversized individuals can pose problems, particularly when resources are limited. Specimens to be preserved whole can be positioned on a board, gently wrapped in saturated cheesecloth, and sealed in a plastic garbage bag to harden. Very large specimens also can be skinned, leaving the head, limbs, and tail attached to the skin, and preserved by injection and submersion in formalin. The resulting specimen will take up much less space while still maintaining many important characters. Take notes on reproductive condition, gut contents, and parasites if they are not preserved. Sometimes it is desirable to preserve vouchers as skeletons. In this case, record measurements (SVL, total length, and weight) and sex because they cannot be recovered after preparation. Remove skin, viscera, and larger muscle masses. When possible, preserve the skin and viscera. Dehydrate the carcass either in the sun or preferably in 50–70% ethanol so that it can be cleaned by dermestid beetle larvae in the laboratory. Dermestids (Dermestidae) can be purchased commercially from a variety of internet sources.

Removal of visceral organs, including gonads and gut contents, of standard wet specimens in the field is not recommended, although on rare occasions it may be required by research design. It is important that whenever a specimen has associated material, all parts are labelled with the same field number so that they can be easily and unambiguously associated with the voucher. We discourage the practice of assigning different numbers to individual parts, including tissues.

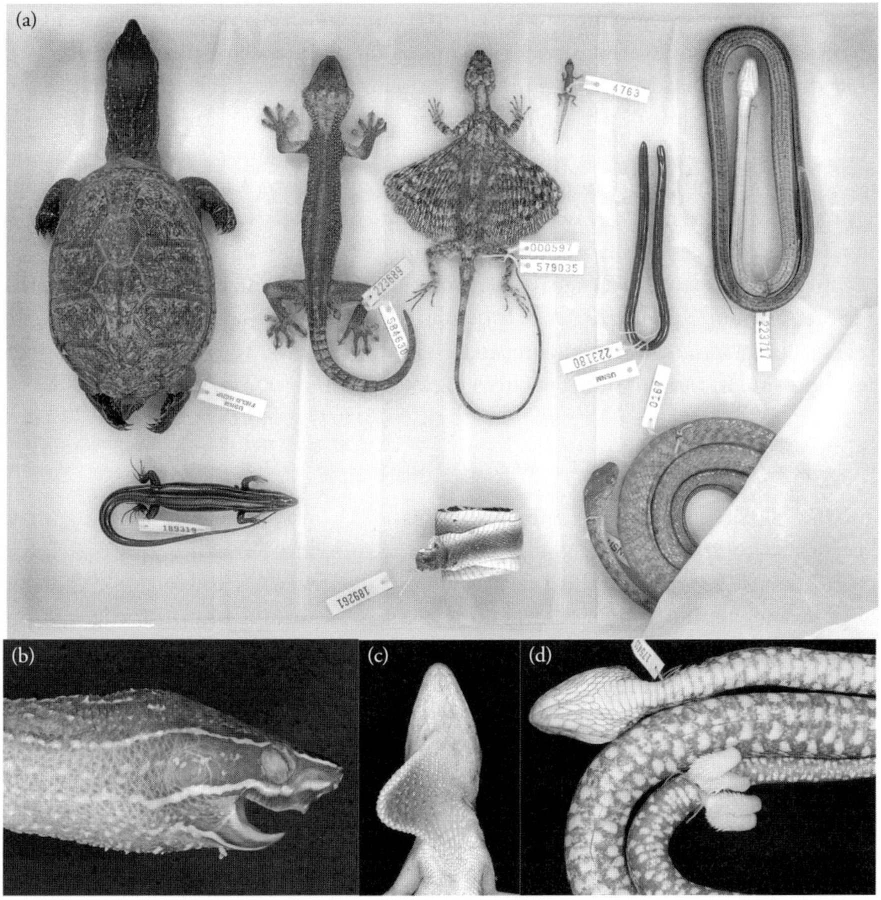

Figure 6.3 Examples of reptile voucher specimens. (a) Specimens laid out in a preservation tray, clockwise from upper left *Kinosternon subrubrum* (USNM-FS 176073), *Gekko* sp. (USNM 584638), *Draco timorensis* (USNM 579035), *Tropiocolotes somalicus* (BSFS 4763), *Ramphotyphlops braminus* (USNM 571866), *Dendrelaphis* sp. (USNM 577693), *Boiga irregularis* (USNM 580052), rolled skin of specimen prepared as a skeletal prep *Opheodrys aestivus* (USNM 332751), *Plestiodon fasciatus* (USNM 332704, note tick on body behind left forelimb). (b) *Kinosternon subrubrum* (USNM-FS 176073) showing open mouth, and pattern and dermal ornamentation on the neck that would be obscured had the specimen been preserved with its neck retracted into the shell. (c) *Anolis sagrei* (USNM 578281) showing spread dewlap. (d) *Bothrops atrox* (USNM 286957) showing everted hemipenes. If you choose to tie-off an everted hemipenis, do it as close to the body as possible to avoid obscuring ornamentation in the knot, and use only a single overhand knot so that it can be easily undone if additional manipulation of the hemipenis is needed later. The ventral incisions visible on the specimen were made so gut contents could be examined after the specimen was deposited in the museum.

Properly preserved specimens will harden (maintain their posture) fairly quickly and can be removed from the tray after 12–24 hours. Once the specimens are well hardened, we recommend submerging them in 10% formalin for the rest of the trip (or up to two weeks in the laboratory). This maximizes the contact with the formalin and frees-up often limited tray space. When possible, have several containers so that different sizes and types of specimens can be segregated (containers such as 4 l wide mouth Nalgene™ jugs or 9.5 l Rubbermaid 'Easy Find Lids'™ food containers work well for many specimens). Do not force specimens into a submerging container; you do not want to distort specimens, but instead float them loosely so as not to press against the container or other specimens. Submerging is particularly helpful for specimens that have been cut open for organ or tissue removal since the injected formalin can leak. Keep the preserved specimens cool and in the dark until ready to pack for shipping.

6.5 Specimen transport and shipping

Both formaldehyde and ethanol are regulated as hazardous materials for shipping; always check local laws and shipping regulations prior to shipping. Some shipping policies and regulations can be found on the International Air Transport Association (IATA) (http://www.iata.org/) and Natural Sciences Collection Association (http://www.natsca.org/) web pages as well as from specific shippers. Formalin preserved specimens should be soaked in water for 24–48 hours prior to packing. Ethanol preserved specimens, however, should *never* be put into water. Prior to transport, specimens should be removed from their field storage containers and wrapped in preservative moistened cheesecloth. Smaller specimens may be wrapped in a series but separated from adjoining specimens by a layer of preservative moistened cloth to help prevent abrasion and desiccation. Wrapped specimens are placed inside a plastic bag, and the bag is sealed to prevent leakage. It may be necessary to wrap feet of large lizards and turtles to prevent claws from perforating bags. The bagged specimens are then placed inside a second plastic bag to further help prevent leakage. Note that no free liquid should be visible within the first bag. For airline travel or for commercial shipping, the double plastic bag should be placed in a third bag containing an absorbent material such as vermiculite, paper towels, or even disposable diapers.

6.6 Useful resources

- American Society of Ichthyologists and Herpetologists (http://www.asih.org) provides sources for supplies and *Guidelines for Use of Live Amphibians and Reptiles in Field and Laboratory Research,* Second Edition, Revised by the Herpetological Animal Care and Use Committee (HACC) of the American Society of Ichthyologists and Herpetologists, 2004. (Committee Chair: Steven J. Beaupre, Members: Elliott R. Jacobson, Harvey B. Lillywhite, and Kelly Zamudio).
- CITES (the Convention on International Trade in Endangered Species of Wild Fauna and Flora; http://www.cites.org/) provides appendices of listed populations and CITES Authorities.

- Natural Sciences Collection Association (http://www.natsca.org/) and The Society for the Preservation of Natural History Collections (http://www.spnhc.org/) provide information on natural history collections including supply sources and legal issues.
- International Air Transport Association (IATA; http://www.iata.org/) provides up to date information on shipping regulations.

Acknowledgements and notice

We thank Robert Reynolds, James Poindexter, Jens Vindum, Amy Lathrop, Ted Kahn, and an anonymous reviewer for their help in improving earlier drafts of this chapter. Any use of trade, product, or firm names is for descriptive purposes only and does not imply endorsement by the United States government.

References

Berry, S.J. (2012). Preparing scientific specimens. In R.W. McDiarmid, M.S. Foster, C. Guyer, et al. (eds) *Reptile Biodiversity: Standard Methods for Inventory and Monitoring*. Berkeley, CA: University of California Press, pp. 96–106.

Conroy, C.J., Papenfuss, T., Parker, J., et al. (2009). Use of tricaine methanesulfonate (MS222) for euthanasia of reptiles. *Journal of American Association for Laboratory Animal Science*, **48**, 28–32.

Cortez, C., Suárez-Mayorga, A.M., and López-López, F.J. (2006). Preparación y preservación de material científico. In A. Angulo, J.V. Rueda-Almonacid, J.V. Rodríguez-Mahecha, et al. (eds) *Técnicas de Inventario y Monitireo para los Anfibios de la Región Tropical Andina*, Serie Manuales de Campo No. 2. Conservación International, pp. 173–220.

Crawford, A.J., Lips, K.R., and Bermingham, E. (2010). Epidemic disease decimates amphibian abundance, species diversity, and evolutionary history in the highlands of central Panama. *Proceedings of the National Academy of Sciences USA*, **107**, 13777–82.

Ehmann, H. (1989). Neutral paraformaldehyde solution for field use. *Herpetological Review*, **20**, 44–5.

Etheridge, R.E. (1996). Collections: preservation techniques (adapted for the WWW and updated by M. O'Brien and G. Schneider, May 1996). Available at: http://www.lsa.umich.edu/ummz/herps/collections/herp-prep.asp

Foster, M.S. (2012). Preparing reptiles as voucher specimens. In R.W. McDiarmid, M.S. Foster, C. Guyer, et al. (eds) *Reptile Biodiversity: Standard Methods for Inventory and Monitoring*. Berkeley, CA:University of California Press, pp. 95–125.

Gamble, T. (2014). Collecting and preserving genetic material for herpetological research. *Society for the Study of Amphibians and Reptiles Herpetological Circular*, No. 41.

Karlstrom, E.L., and Cook, S.F., Jr. (1955). Notes on snake anesthesia. *Copeia*, **1955**, 57–8.

Karns, D.R. (1986). *Field Herpetology: Methods for the Study of Amphibians and Reptiles in Minnesota*. Occasional Paper 18. Minneapolis, MN: James Ford Bell Museum of Natural History.

Livezey, R.L. (1958). Procaine hydrochloride as a killing agent for reptiles and amphibians. *Herpetologica*, **13**, 280.

McDiarmid, R.W. (1994). Preparing amphibians as voucher specimens. In R.W. Heyer, M.A. Donnelly, R.W. McDiarmid, et al. (eds) *Measuring and Monitoring Biological Diversity: Standard Methods for Amphibians*. Washington, DC:Smithsonian Institution Press, pp. 289–97.

Pisani, G.R. (1973). *A Guide to Preservation Techniques for Reptiles and Amphibians*. Herpetological Circular No. 1. Society for the Study of Amphibians and Reptiles.

Pisani, G.R., and Villa, J. (1974). *Guia de Tecnicas de Preservacion de Anfibios y Reptiles*. Herpetological Circular No. 2. Society for the Study of Amphibians and Reptiles.

Pleijel, F., Jondelius, U., Norlinder, E., et al. (2008). Phylogenies without roots? A plea for the use of vouchers in molecular phylogenetic studies. *Molecular Phylogenetics and Evolution*, **48**, 369–71.

Raloff, J. (1985). Unwanted formaldehyde 'breakthrough'. *Science News*, **127**(**19**), 297.

Raxworthy, C.J. (2012). Hemipenis preparation. In R.W. McDiarmid, M.S. Foster, C. Guyer, et al. (eds) *Reptile Biodiversity: Standard Methods for Inventory and Monitoring*. Berkeley, CA:University of California Press, p. 102.

Reynolds, R.P., and McDiarmid, R.W. (2012). Voucher specimens. In R.W. McDiarmid, M.S. Foster, C. Guyer, et al. (eds) *Reptile Biodiversity: Standard Methods for Inventory and Monitoring*. Berkeley, CA:University of California Press, pp. 89–94.

Scott, N.J., and Aquino-Shuster, A.L. (1989). The effects of freezing on formalin preservation of specimens of frogs and snakes. *Collections Forum*, **5**, 41–6.

Shine, R., Amiel, J., Munn, A.J., et al. (2015). Is 'cooling then freezing' a humane way to kill amphibians and reptiles? *Biology Open*, DOI: 10.1242/bio.012179.

Schulte, J.A. (2012). Collecting and preserving tissues for biochemical analyses. In R.W. McDiarmid, M.S. Foster, C. Guyer, et al. (eds) *Reptile Biodiversity: Standard Methods for Inventory and Monitoring*. Berkeley, CA:University of California Press, pp. 121–5.

Simmons, J.E. (2014). *Fluid Preservation: A Comprehensive Reference*. Lanham, MD:Rowman & Littlefield.

Simmons, J.E. (2015). Herpetological collecting and collection management, third edition. *Society for the Study of Amphibians and Reptiles Herpetological Circular*, No. 42.

Suarez, A.V., and Tsutsui, N.D. (2004). The value of museum collections for research and society. *BioScience*, **54**, 66–74.

7
Reproduction

Gunther Köhler

7.1 Introduction

Reproduction is a principal component of the life history of an organism, and knowledge of reproductive biology is crucial to understand the ecology and natural history of a species. Nevertheless, such information is minimal to non-existent in the published literature for a majority of reptile species. Important traits in the life history of reptiles include age and size at sexual maturity, size and frequency of clutches or litters, egg or offspring size, and reproductive phenology (Aldridge and Brown, 1995; Kuchling, 1999; Balestrin and Di-Bernardo, 2005). Complex trade-offs between fecundity and energy costs become evident when studying parameters such as age and size at maturity, clutch (or litter) size, and number of offspring per season in a given species (Shine, 1980; Qualls and Shine, 1998; Kunz and Orrell, 2004). Information on the reproductive biology of reptiles can be generated using a wide array of methods applied either to live individuals or preserved specimens. These data can include the number of offspring (counting follicles, eggs, or foetuses), minimum and average size of adults, temporal variation of blood chemistry parameters and hormones, and the temporal patterns of gonadal data, all of which shed light on the reproductive ecology of the species studied. Research that includes both sexes can uncover reproductive phenomena such as sexual size dimorphism, sperm storage, or asynchronous reproduction (e.g. Seigel and Ford, 1987; Almeida-Santos et al., 2004).

7.2 A brief description of the genital tract in reptiles

The male reproductive system in reptiles consists of the testes, the testicular ducts, the sexual segment of the kidney, and the cloaca. Copulatory organs (e.g. hemipenes in squamate reptiles; a penis in turtles and crocodilians) and secondary sexual characteristics (e.g. enlarged femoral pores, body appendages) are not treated in this contribution. As in all amniotes, the testis in reptiles is enclosed in a connective tissue capsule. The testes are intra-abdominal and positioned anterior to the kidneys and attached to the dorsal body wall. Except for the more elongate shape of the testes in snakes and some limbless, elongate lizards, most reptile testes are round to slightly oval in shape (Gribbins and Rheubert, 2015), with the right gonad anterior to the left (Figure 7.1). They are mostly dirty white to cream in external appearance, although in some taxa, such as the

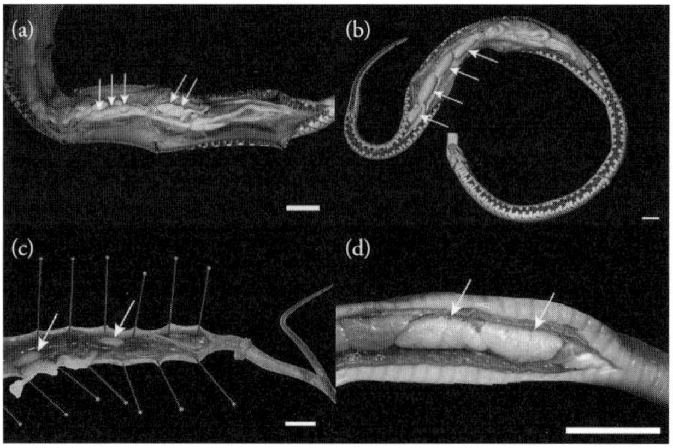

Figure 7.1 Dissections of (a) *Macroprotodon cucullatus* SMF 99019, female with vitellogenic follicles (arrows); (b) *Macroprotodon cucullatus* SMF 99025, female with oviductal eggs (arrows); (c) *Sibon sanniolus* GK-5161, male (arrows point to testes); (d) *Ninia sebae* GK-5141, male (arrows point to testes). Scale bars equal 10.0 mm.

Chamaeleontidae, they can be heavily pigmented. Seasonal variation in size has been reported with testes reaching double their inactive size during peaks of reproductive activity, producing high intratesticular pressure and thus giving the testes added firmness (Gribbins and Rheubert, 2015). The sperm develop in the germinal epithelium of the seminiferous tubules in the testis. The testicular ducts drain testicular products, the spermatozoa and seminal fluids, to the exterior. They comprise the rete testis, ductuli efferentes, epididymis, and ductus deferens (Sever, 2010; Rheubert et al., 2015). *In situ*, the ductus deferens is convoluted, and in this form occupies approximately one-third of the body length in snakes; for example, when freed from the peritoneal membrane and manually distended, its length is about six times the snout–vent length (SVL) in *Crotalus durissus terrificus* (Almeida-Santos et al., 2004).

The female reproductive system in reptiles consists of the ovary, the oviduct, and the cloaca. Reptiles have paired ovaries and oviducts. The ovaries are attached to the dorsal body wall via a thin mesovarium. The oocytes are produced within germinal beds, small regions located on the dorsal side of the ovary. Only a single germinal bed per ovary is present in species that produce a single egg from each ovary (monoallochronic ovulation) alternately as in anoles (family Dactyloidae). In highly fecund species with multiple eggs per clutch (polyauthochronic ovulation), usually several germinal beds are observed per ovary (for exceptions to this trend, see Jones et al., 1982; Radder and Shine, 2007), suggesting a correlation with clutch size and breeding frequency (Radder et al., 2008). On the surface of an active ovary, follicles of different sizes (= stages of development) as well as corpora lutea and atretic follicles can be observed (Figure 7.2). The oviducts enter the urodaeum of the cloaca either through a single, medial urogenital papilla or through pores of bilateral urogenital papillae (although the absence of oviducal papillae was reported in some lizards and snakes; see Siegel et al., 2011, 2015; Trauth and Sever, 2011). The lumina of the two oviducts remain separate before

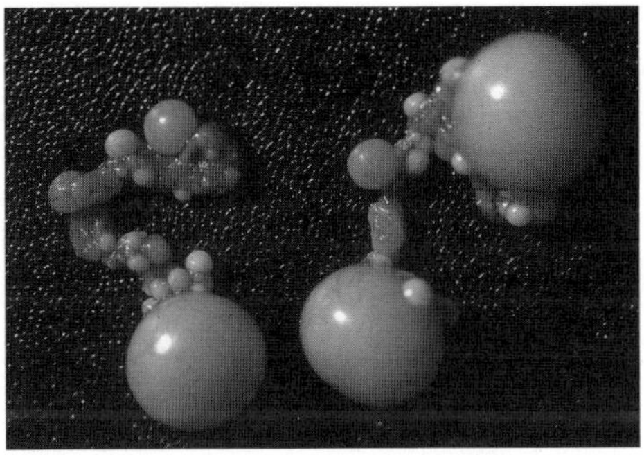

Figure 7.2 Corpora luteum and follicles of *Terrapene ornata*. The different sizes of the follicles indicate that multiple clutches are developing. Photo courtesy of John Iverson.

emptying into the urodaeum in the majority of squamate reptiles (Siegel et al., 2015). The oviduct can be divided into three portions: the infundibulum; the magnus (which secretes albumin); the glandular uterus (often termed uterus); and the non-glandular uterus (often termed vagina). The infundibulum is the most cranial portion of the oviduct and can be divided into two regions: the caudal portion with distinct infundibular glands where sperm storage occurs in some taxa and the more cranial portion that is characterized by the absence of such glands (Sever and Hamlett, 2002). Uterine glands secrete proteins that are involved in shell membrane formation (Blackburn, 1998) and considerable variation in uterine gland morphology has been documented among different squamate reptile taxa, often correlated with different parity modes (Siegel et al., 2011, 2015; Rheubert et al., 2015). In numerous lizard taxa, the non-glandular uterus contains sperm storage tubules (Sever and Hamlett, 2002). In practice, the three portions of the oviduct are often difficult to differentiate. In gravid females these portions are easily observed as 'functional regions': the portion carrying the eggs is the glandular uterus; the portion cranial to the eggs is the infundibulum; and the portion caudal to the eggs is the non-glandular uterus (Blackburn, 1998).

7.3 Dissections

Direct inspection of gonads, fat bodies, and gonadal products is an effective means of producing data on the reproductive ecology of species (Balestrin and Di-Bernardo, 2005; Figure 7.1). This can be done either in live individuals through laparotomy or endoscopy, or in freshly dead or preserved specimens (e.g. material from museum collections). Road-killed specimens can be a valuable source of study material, yielding important biological data on reproduction. In the course of a four-year study on the impact of traffic on the local snake populations along a 40 km road transect, about 450 individual snakes of 31 species were collected (at survey intervals of 14 days; Köhler, unpublished data).

Of course, all relevant collecting permits have to be acquired and any restrictions must be followed complying with local, regional, and national laws and policies. A road-killed reptile is collected and placed in a plastic bag with a paper label that contains the collecting information (field number, species name, sex, time, name of place or distance to a reference point, GPS coordinates, elevation above sea level, weather and moon conditions, condition of specimen, and SVL or carapace length when fresh) and then placed in a container with ice to slow down the autolysis process. The specimen should be dissected for internal inspection as soon as possible after collecting. If this is not possible within a few hours after collecting, then preservation of the reptile should be considered. However, the internal organs are much easier to differentiate and evaluate in a non-preserved reptile compared with a preserved specimen. Nevertheless, an autolytic specimen is completely useless! For preservation, I recommend the injection of a fixation solution (i.e. 5–10 ml absolute [i.e. 36%] formalin in 1 l of 96% ethanol) into the body cavity and thighs; the injected reptile can then be placed directly in 70% ethanol for permanent maintenance (see Chapter 6). Although this preservation protocol is not widely used by herpetologists, at the Senckenberg Research Institute this technique has been used for about 100 years now—yielding preserved specimens of excellent quality!

Relying on chance mortalities (road-killed specimens) presents two possible drawbacks: the time span necessary for full seasonal representation usually requires a long time and individuals are often found after they have been dead for hours before processing or preservation, making histological analysis of some tissues difficult.

To access the organs of the body cavity, a longitudinal cut is made along the ventral surface of the body of the reptile. After cutting through the ventral scales, the body cavity is opened. In turtles, the use of a drilling saw with circular saw discs is recommended to open the shell along the bridge on both sides. Care must be exercised to avoid damaging internal organs. In snakes, the right gonad is positioned anteriorly of the left one, as is true for all paired organs in snakes (Figure 7.1). Often, the intestinal tract and the liver need to be removed to access the gonads, which are situated dorsally near the vertebral column and anterior to the kidneys.

In males, the longitudinal and transverse diameters of each testis should be measured, preferably using callipers. The testes should be inspected for convoluted tubules in the epididymis region. Each testis should then be removed from the body cavity and placed into 10% buffered formalin, prepared using standard paraffin infiltration techniques, sectioned at 7–8 μm, and stained with haematoxylin (some additionally use Biebrich scarlet, orange G, and fast green; see Aldridge and Brown, 1995) for histological examination, especially to evaluate the ducts for the presence of spermatozoa. Histological preparation measurements may include: seminiferous tubule diameter, stage of spermatogenesis, diameter of the sexual segment of the kidney, and presence of sperm in the ductus deferens.

In females, the longitudinal and transverse diameters of each ovary are measured with callipers. The ovaries can then be examined for the presence of previtellogenic and vitellogenic follicles (Figure 7.1(a)), and/or corpora lutea. The size of all vitellogenic and of the largest 3–5 previtellogenic follicles is measured with callipers. If oviducal eggs are present (Figure 7.1(b)), the length, width, and weight of each egg is determined and the number of eggs is recorded for each oviduct separately. The ovaries are then removed from the body cavity and fixed with 10% buffered formalin for histological examination. Histologically, the type of follicle, previtellogenic or vitellogenic, and stage of atresia (resorption of the follicle—early, middle, and late) can be determined based upon the degree of connective tissue invasion of the follicle (Aldridge and Semlitsch, 1992). Counts and size determinations of corpora lutea should be made, recognizing that females of species that produce multiple annual clutches/litters may have multiple sets of corpora lutea.

Generally, males can be considered as mature when they have enlarged, turgid testes or opaque and convoluted deferent ducts—indicating the presence of sperm—whereas females can be regarded as mature when bearing oviducal eggs, embryos, or ovarian vitellogenic follicles (Shine, 1977a, 1977b; Bizerra et al., 2005; Marques et al., 2009). For all specimens used, a minimum of morphometric data should be obtained, including SVL and tail length (see Chapter 4).

Since the coelomic adipose tissue mass varies with the stage of the reproductive cycle, this parameter can yield additional useful data about the reproductive cycle of a given species. The relationship between date of collection and size-standardized abdominal fat body mass can then be compared in reproductive (individuals with vitellogenic follicles or oviducal eggs) and non-reproductive reptiles by months or season.

The usefulness of cloacal smears for the detection of sperm has been demonstrated in various studies (e.g. Aldridge and Brown, 1995). For a cloacal smear, a small amount of fluid is forced from the cloaca onto a microscope slide and examined under high power (400×) for the presence of sperm. Another source of semen is from the contents of the vas deferens in freshly dead specimens. The vasa deferentia are removed from the body cavity and stretched out on a glass slide for semen extraction. For this purpose, a compression over the whole extension of the two vasa deferentia is applied. The semen obtained can then be used for spermatozoa concentration counts and for analysis of motility and survival time (Almeida-Santos et al., 2004).

In case of a fresh road kill that contains mature eggs in its oviducts, an attempt can be made to incubate the eggs. The oviducts are opened with a longitudinal incision and the eggs are removed from the oviducts. Any adhesive blood or tissue should be removed with a moist paper towel. The eggs can then be placed in an incubation box filled with slightly moist vermiculite or perlite and maintained in an incubator for artificial incubation. The temperature and moisture conditions appropriate for the incubation of reptile eggs varies considerably species specifically. Köhler (2005) provided an overview of incubation techniques and a summary of incubation data for numerous reptile species. Reptile eggs should not be turned during incubation; the embryo should always be on the top of the egg. To determine the position of the embryo, the egg can be candled with a light. To candle eggs, a bright light that produces only minimal heat should be used, so that the embryo is not damaged by the heat.

7.4 Endoscopy

Endoscopy is a minimally invasive surgery method that allows a visual examination of internal organs in the body cavity of live reptiles (Schildger et al., 1999; Kuchling, 2009). For endoscopy, the reptile must be anaesthetized, which is usually done by inhalation anaesthesia (Köhler, 2006; Chapter 28). The choice of lens is a compromise between the desire for good visibility (large diameter endoscope) and a small incision (small-diameter endoscope). For reptiles under 1000 g mass, small-diameter (2.7–3.0 mm), rigid endoscopic equipment has been used successfully for direct visual inspection of the gonads, the gonadal ducts, and abdominal fat bodies (Divers, 2010). Reptiles should not be fed for a few days prior to endoscopic examination in order to avoid access problems caused by a full digestive tract. Reduced content in the digestive tract also reduces the risk of intestinal perforation and ensures that there is sufficient space between the organs and the lens (Schildger et al., 1999).

7.5 External examination and palpation

In lizards and snakes, the presence of oviducal eggs can be determined in gravid females by gentle palpation. In species with thin, almost transparent abdominal skin, such as in many geckoes, the eggs can be seen through the skin by external inspection of the posterior region of the venter (Figure 7.3(a)), with or without the aid of a small lamp.

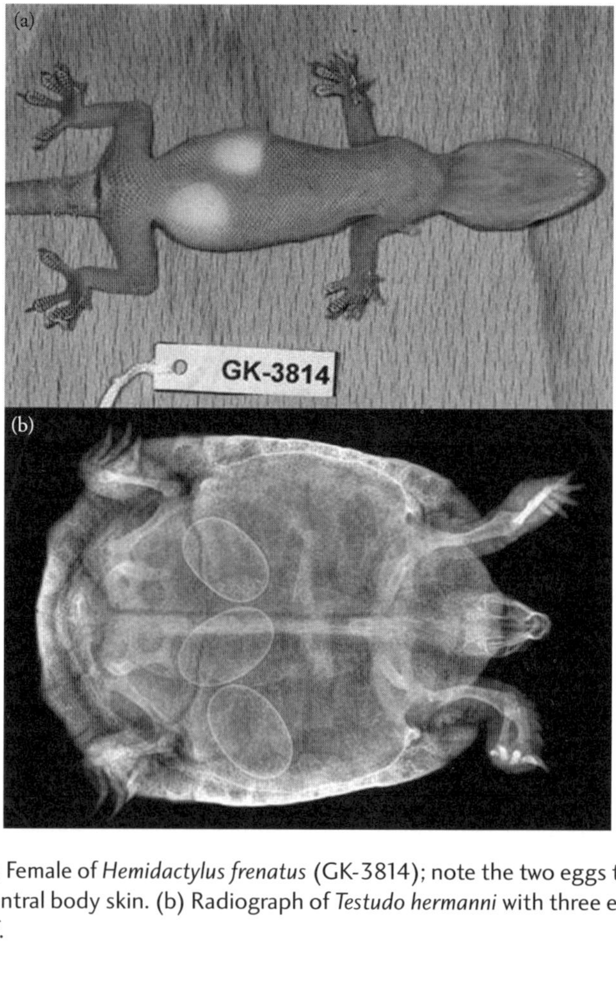

Figure 7.3 (a) Female of *Hemidactylus frenatus* (GK-3814); note the two eggs that are visible through the ventral body skin. (b) Radiograph of *Testudo hermanni* with three eggs. Photo by Birgit Rüschoff.

In gravid snakes, the number of eggs can often be counted by palpation. Data derived from palpation can yield information on clutch size, size at maturity, and reproductive season (Pizzatto and Marques, 2002). In turtles, calcified oviducal eggs can be detected by inguinal palpation (Tinkle et al., 1981; Bertolero and Marín, 2005; Zuffi et al., 2005). However, this can be difficult in species such as mud turtles (genus *Kinosternon*) that have moveable hinges that allow almost complete closure of the shell.

7.6 Imaging methods

Radiology, ultrasound, and computerized tomography are non-invasive methods that allow the observation of internal structures, including soft tissue organs and gonadal products such as ovarian follicles and oviducal eggs (Silverman and Janssen, 1996; Zuffi et al., 2005). The diameters of testes, follicles, and eggs can then be measured using the calliper measurement function for distances on the computer screen or with the aid of callipers on a printout of the image.

Most reptiles can be placed directly on the film box to obtain an image. Venomous reptiles should be left in plastic containers while the X-rays are taken. Shelled eggs are seen as rounded, soft tissue opaque structures in the caudal abdomen in radiographs, often arranged in a linear pattern resembling an overlapping string of pearls. In chelonians, the egg shells may be mineralized so that they appear as radiopaque structures (Figure 7.3(b)); repeated radiographs during long-term studies of turtles have not demonstrated harm to females or their developing embryos (Hinton et al., 1997).

Ultrasound is particularly useful for examining soft tissues (Köhler, 2006; Kuchling, 2015). For ultrasonography of most small to medium-sized reptiles, high-resolution equipment with 3.5 MHz for small and 5.0–7.5 MHz transducers for large reptiles is often used, although small footprints are recommended (Schildger et al., 1994; Silverman and Janssen, 1996). Linear scanners work well for lizards and snakes, whereas sector scanners have proven useful for chelonians (Kramer and Gerwing, 1991; Sainsbury and Gili, 1991). An acoustic coupling gel is applied to the skin 10–15 minutes prior to ultrasonograph examination. With ultrasound, ovaries in early follicular growth are identified as round hypoechogenic masses (small diameter, non-vitellogenic follicles), whereas vitellogenesis is determined by the presence of hyperechogenic follicles (large diameter, vitellogenic follicles; Bertona and Chiaraviglio, 2003). Clutch or litter size can be determined based on the number of vitellogenic follicles or oviducal eggs.

7.7 Blood chemistry

The determination of hormones in the blood plasma can yield valuable information on reproductive cycles in reptiles (Weil and Aldridge, 1981; Wibbels et al., 1992). In reptiles, blood samples have often been obtained by cardiac puncture (Aldridge et al., 1990). However, I recommend collecting blood samples from the ventral caudal vein in lizards and snakes, from the occipital blood sinus in crocodilians, and from the dorsal caudal vein or the right jugular vein in turtles (Jenkins, 1996; Köhler, 2006; also see Chapter 28). However, whereas using the ventral caudal vein works great in large iguanians and teiids, it has the potential risk of tail autotomy in some species lizards such as lacertids and geckoes. In these more delicate species I recommend using small capillary tubes to collect blood from occipital sinus which is easy and effective in lizards, and rarely causes permanent damage. Fresh blood samples are then centrifuged, and stored at −20°C. Plasma can be assayed for total androgen by radioimmunoassay (Aldridge et al., 1990; Kuchling, 2011).

7.8 Hormonal induction of egg laying

Egg laying can be induced in gravid reptiles with mature eggs by the intramuscular injection of the hormone oxytocin (Cree et al., 1991; Köhler, 2006). However, the status of pregnancy of the female should be checked with ultrasound scanning prior to oxytocin injection. In lizards and snakes, a dosage of 1–5 units oxytocin per kg usually works well whereas 2–20 units oxytocin per kg have been used in chelonians (Köhler, 2006; Tucker et al., 2007).

7.9 Conclusions

Elucidating reproductive characters is essential for understanding a reptile's life cycle. However, the reproductive biology of many reptile species is poorly known, particularly for tropical and subtropical species. The large collections of reptiles in museums worldwide are a potential source of valuable information on the reproductive biology of reptiles. One drawback of a mass-dissection approach is the destructive nature of this method that will not meet with approval by many collection curators. This is understandable given that collection specimens are irreplaceable because of accompanying collecting data for each specimen that often gives the particular specimen a historical value aside from its value for reproductive or taxonomic study. Nevertheless, these collections have a great potential for the generation of natural history data that until now has been minimally exploited (but see Aldridge and Brown, 1995). Road-killed specimens are an excellent source of material for such studies, especially since the individuals were already dead when collected. With a little effort, long-term studies can be designed that may result in a meaningful series of individuals for inclusion in projects on natural history, internal anatomy, external morphology, and on the evolutionary biology of reproduction.

References

Aldridge, R.D., and Brown, W.S. (1995). Male reproductive cycle, age at maturity, and cost of reproduction in the timber rattlesnake (*Crotalus horridus*). *Journal of Herpetology*, **29**, 399–407.

Aldridge, R.D., and Semlitsch, R.D. (1992). Female reproductive biology of the southeastern crowned snake *Tantilla coronata*. *Amphibia-Reptilia*, **13**, 209–18.

Aldridge, R.D., Greenhaw, J.J., and Plummer, M.W. (1990). The male reproductive cycle of the rough green snake (*Opheodrys aestivus*). *Amphibia-Reptilia*, **11**, 165–72.

Almeida-Santos, S.M., Laporta-Ferreiraa, I.L., Antoniazzib, M.M., et al. (2004). Sperm storage in males of the snake *Crotalus durissus terrificus* (Crotalinae: Viperidae) in southeastern Brazil. *Comparative Biochemistry and Physiology, Part A*, **139**, 169–74.

Balestrin, R.L., and Di-Bernardo, M. (2005). Reproductive biology of *Atractus reticulatus* (Boulenger, 1885) (Serpentes, Colubridae) in southern Brazil. *Journal of Herpetology*, **15**, 195–9.

Bertolero, A., and Marín, A. (2005). Efficacy of inguinal palpation for detecting oviductal eggs in Hermann's tortoise, *Testudo hermanni*. *Amphibia-Reptilia*, **26**, 523–6.

Bertona, M., and Chiaraviglio, M. (2003). Reproductive biology, mating aggregations, and sexual dimorphism of the Argentine boa constrictor (*Boa constrictor occidentalis*). *Journal of Herpetology*, **37**, 510–16.

Bizerra, A., Marques, O.A.V., and Sazima, I. (2005). Reproduction and feeding of the colubrid snake *Tomodon dorsatus* from south-eastern Brazil. *Amphibia-Reptilia*, **26**, 33–8.

Blackburn, D.G. (1998). Structure, function, and evolution of the oviducts of squamate reptiles, with special reference to viviparity and placentation. *Journal of Experimental Zoology*, **282**, 560–617.

Cree, A., Daugherty, C.H., Schafer, S.F., et al. (1991). Nesting and clutch size of tuatara (*Sphenodon Guntheri*) on North Brother Island, Cook Strait. *Tuatara*, **31**, 9–16.

Divers, S.J. (2010). Reptile diagnostic endoscopy and endosurgery. *Veterinary Clinics of North America: Exotic Animal Practice*, **13**, 217–42.

Gribbins, K.M., and Rheubert, J.L. (2015). The architecture of the testis, spermatogenesis, and mature spermatozoa. In J.L. Rheubert, D.S. Siegel, and S.E. Trauth (eds) *Reproductive Biology and Phylogeny of Lizards and Tuatara*. Boca Raton, FL:CRC Press, pp. 340–424.

Hinton, T.G., Fledderman, P.D., Lovich, J.E., et al. (1997). Radiographic determination of fecundity: is the technique safe for developing turtle embryos? *Chelonian Conservation and Biology*, **2**, 409–14.

Jenkins, J.R. (1996). Diagnostic and clinical techniques. In D.R. Mader (ed) *Reptile Medicine and Surgery*. Philadelphia, PA:Saunders, pp. 264–76.

Köhler, G. (2005). *Incubation of Reptile Eggs*. Malabar, FL:Krieger.

Köhler, G. (2006). *Diseases of Amphibians and Reptiles*. Malabar, FL:Krieger.

Kramer, M., and Gerwing, M. (1991). The basics of sonography in reptiles. In *Proceedings des 4. Internationalen Symposiums der Pathologie und Therapie der Reptilien und Amphibien*, Bad Nauheim, pp. 135–38.

Kuchling, G. (1999). *The Reproductive Biology of the Chelonia*. Berlin:Springer.

Kuchling, G. (2009). Letter to the editor: differences among studies on sex identification of hatchling turtles. *Journal of the American Veterinary Medical Association*, **235**, 932.

Kuchling, G. (2011). Hormones and reproductive cycles in turtles. In D. Norris and K. Lopez (eds) *Hormones and Reproduction of Vertebrates. Volume 3. Reptiles*. Oxford: Academic Press, pp. 277–303.

Kuchling, G. (2015). Ultrasound scanning as an effective tool in the conservation of chelonians. *International Zoo Yearbook*, **49**, DOI: 10.1111/izy.12081.

Kunz, T.H., and Orrell, K.S. (2004). Energy costs of reproduction. In C. Cleveland (ed) *Encyclopedia of Energy*, Vol. 5. Oxford: Elsevier, pp. 423–42.

Jones, R.E., Guillette, L.J., Norman, M.F., et al. (1982). Corpus luteum–uterine relationships in the control of uterine contraction in the lizard *Anolis carolinensis*. *General and Comparative Endocrinology*, **48**, 104–12.

Marques, O.A.V., Almeida-Santos, S.M., Rodrigues, M.G., et al. (2009). Mating and reproductive cycle in the Neotropical colubrid snake *Chironius bicarinatus*. *South American Journal of Herpetology*, **4**, 76–80.

Pizzatto, L., and Marques, O.A.V. (2002). Reproductive biology of the false coral snake *Oxyrhopus guibei* (Colubridae) from southeastern Brazil. *Amphibia-Reptilia*, **23**, 495–504.

Qualls, C.P., and Shine, R. (1998). Costs of reproduction in conspecific oviparous and viviparous lizards, *Lerista bougainvillii*. *Oikos*, **82**, 539–51.

Radder, R.S., and Shine, R. (2007). Germinal bed condition in a polyautochronic single-clutched lizard, *Bassiana duperreyi* (Scincidae). *Amphibia-Reptilia*, **28**, 159–62.

Radder, R.S, Pizzatto, L., and Shine, R. (2008). Morphological correlates of life-history variation: is lizard clutch size related to the number of germinal beds in the ovary? *Biological Journal of the Linnean Society*, **94**, 81–8.

Rheubert, J.L., Sever, D.M., Siegel, D.S., et al. (2015). Male reproductive anatomy: the gonadoducts, sexual segment of the kidney, and cloaca. In J.L. Rheubert, D.S. Siegel, and S.E. Trauth (eds) *Reproductive Biology and Phylogeny of Lizards and Tuatara*. Boca Raton, FL:CRC Press, pp. 253–339.

Sainsbury, A.W., and Gili, C. (1991). Ultrasonographic anatomy and scanning technique of the coelomic organs of the Bosc monitor (*Varanus exanthematicus*). *Journal of Zoo and Wildlife Medicine*, **22**, 421–33.

Schildger, B.J., Casares, M., and Kramer, M., et al. (1994). Technique of ultrasonography in lizards, snakes and chelonians. *Seminars in Avian and Exotic Pet Medicine*, **3**, 147–55.

Schildger, B., Haefeli, W., Kuchling, G., et al. (1999). Endoscopic examination of the pleuroperitoneal cavity in reptiles. *Seminars in Avian and Exotic Pet Medicine*, **8**, 130–8.

Seigel, R.A., and Ford, N.B. (1987). Reproductive ecology. In R.A. Seigel, J.T. Collins, and S.S. Novak (eds) *Snakes: Ecology and Evolutionary Biology*. New York:McMillan, pp. 210–52.

Sever, D.M. (2010). Ultrastructure of the reproductive system of the black swamp snake (*Seminatrix pygaea*). VI. Anterior testicular ducts and their nomenclature. *Journal of Morphology*, **271**, 104–15.

Sever, D.M., and Hamlett, W.C. (2002). Female sperm storage in reptiles. *Journal of Experimental Zoology*, **292**, 187–99.

Shine, R. (1977a). Reproduction in Australian elapid snakes. I. Testicular cycles and mating seasons. *Australian Journal of Zoology*, **25**, 647–53.

Shine, R. (1977b). Reproduction in Australian elapid snakes. II. Female reproductive cycles. *Australian Journal of Zoology*, **25**, 655–66.

Shine, R. (1980). 'Costs' of reproduction in reptiles. *Oecologia*, **46**, 92–100.

Siegel, D.S., Miralles, A., Chabarria, R.E., et al. (2011). Female reproductive anatomy: cloaca, oviducts, and sperm storage. In R.D. Aldridge and D.M. Sever (eds) *Reproductive Biology and Phylogeny of Snakes*. Boca Raton, FL:CRC Press, pp. 347–409.

Siegel, D.S., Miralles, A., Rheubert, J.L., et al. (2015). Female reproductive anatomy: cloaca, oviduct, and sperm storage. In J.L. Rheubert, D.S. Siegel, and S.E. Trauth (eds) *Reproductive Biology and Phylogeny of Lizards and Tuatara*. Boca Raton, FL:CRC Press, pp. 144–95.

Silverman, S., and Janssen, D.L. (1996). Diagnostic imaging. In D.R. Mader (ed) *Reptile Medicine and Surgery*. Philadelphia, PA:Saunders, pp. 258–64.

Tinkle, D.W., Congdon, J.D., and Rosen, P.C. (1981). Nesting frequency and success: implications for the demography of painted turtles. *Ecology*, **62**, 1426–32.

Trauth, S.E., and Sever, D.M. (2011). Male urogenital ducts and cloacal anatomy. In R.D. Aldridge and D.M. Sever (eds) *Reproductive Biology and Phylogeny of Snakes*. Boca Raton, FL:CRC Press, pp. 411–75.

Tucker, J.K., Thomas, D.L., and Rose, J. (2007). Oxytocin dosage in turtles. *Chelonian Conservation and Biology*, **6**, 321–4.

Weil, M.R., and Aldridge, R.D. (1981). Seasonal androgenesis in the male water snake, *Nerodia sipedon*. *General and Comparative Endocrinology*, **44**, 44–53.

Wibbels, T., Owens, D.W., Light, P., et al. (1992). Serum gonadotropins and gonadal steroids associated with ovulation and egg production in sea turtles. *General and Comparative Endocrinology*, **87**, 71–8.

Zuffi, M.A.L., Citi, S., Giusti, M., et al. (2005). Assessment of reproductive frequency in the European pond turtle (*Emys orbicularis*) using manual palpation, ultrasonography, and radiography. In N. Ananjeva and O. Tsinenko (eds) *Herpetologia Petropolitana*, pp. 295–6. [SEH Meeting 2003].

8
Diet

Luca Luiselli and Giovanni Amori

8.1 Introduction

Diet studies are crucial for understanding disparate aspects of the evolutionary biology, ecology, and life-history traits of reptiles (e.g. Pianka, 1986), as well as prey–predator interactions and even energy fluxes within and across ecosystems (e.g. Brown and Gillooly, 2003). For example, topics such as niche relationships, competitive processes, and predation can be analysed by investigating the dietary traits of a given species. It is crucial, therefore, for investigators to develop efficient and practical methodologies to analyse the dietary habits of their target animals. Considering that many reptile species are in decline (Reading et al., 2010; Böhm et al., 2013), and taking ethical issues into consideration, herpetologists must adopt experimental procedures that minimize disturbances and survival risks of target organisms. For brevity, coverage in this chapter is limited to discussions of the source of samples, to elucidating analytical techniques through macroscopic and microscopic means and briefly, to addressing certain analytical issues so that the role of diet composition, selectivity, and preference can be studied (Pianka, 1986; Mushinsky et al., 2003).

8.2 Sources of material

Three main sources of individuals are used for analysing the dietary ecology of reptiles: dead specimens that are available (i) in museum collections (e.g. Shine et al., 1996, 2006) or (ii) as roadkills (e.g. Daltry et al., 1998), and (iii) live animals captured during field surveys (e.g. Rugiero et al., 2012).

From a practical point of view, the availability of specimens in museum collections may considerably help reduce the costs of a research project. Indeed, it is frequently much less expensive for a scientist (in terms of both time and funds) to examine collections of preserved specimens than to carry out field research. An exception to this rule might be when local scientists work in regions with reduced operating costs. For instance, it is generally much less expensive for local scientists in tropical Africa to collect reptile specimens in the field than to organize and plan visits to museum collections, since most museums with large numbers of specimens are located in Europe, North America, or Australia. Museum collections may house specimens collected over a wide range of time (over several decades) and throughout a large portion of a species'

Reptile Ecology and Conservation. Edited by C. Kenneth Dodd, Jr. © Oxford University Press 2016.
Published 2016 by Oxford University Press.

geographical range, thus allowing researchers to explore temporal, geographic, and habitat dietary shifts.

Nevertheless, there are some serious disadvantages that may compromise aspects of dietary research using museum specimens: (i) uncertainty in the identification of species; (ii) uncertainty in the capture location (as labelling is often not very precise, especially when specimens were collected several decades in the past); (iii) absence of information about how animals were killed, and how much time has passed before they were killed after capture. If individuals remain in captivity before death, gut contents may have digested, thus overestimating the proportion of empty stomachs in the sample or underestimating easily digestible (soft-bodied) prey; (iv) lack of information about the capture method (active search, pitfall trap, glue trap), which may have a confounding effect in the analyses when animals collected by different methods are included in a single sample, thus biasing conclusions; (v) sampling biases. These latter may include biases in populations and localities, over- or under-representation of a gender or a size-class in a given collection, and over- or under-representation of prey remains in the stomach and gut contents of dissected specimens. For example, there are some species in which the males are more active (hence more likely to be captured and preserved) than the females during spring, when they do not feed (for instance, *Vipera aspis*; see Luiselli and Agrimi, 1991). Thus, it is highly likely that the museum specimens of this species will be male biased and that most of the males will not have any prey in the gastrointestinal tract.

Another noteworthy shortcoming of using museum specimens is that it is taxon biased. Indeed, although preserved lizards and snakes are often available for dietary studies (dissection is easy; many entire specimens might be available), the same is not true for chelonians and crocodilians, which typically are stored in museum and public collections in relatively lower numbers and without their guts available for dissection (many crocodilians are stuffed, whereas many chelonians are available only as shells or skeletons).

Reptiles often cross roads (Rahman et al., 2013), and may become victims of car traffic. Roadkills can be a source of information, particularly in tropical areas where it is very easy to collect a large number of specimens killed on roads; this is especially true where snakes are concerned and during certain phases of a species' annual activity cycle (e.g. Rahman et al., 2013). The collection of roadkills should be planned using standardized methodology, but it is a relatively inexpensive technique and easy to accomplish, especially by inexperienced researchers. Thus, roadkills are an efficient source of data in many instances. As already pointed out for the use of museum specimens, however, this source of material is severely taxon biased. Indeed, whereas snakes and lizards are frequent victims of vehicle traffic and can hence be used for dietary studies, crocodilians and aquatic chelonians are generally rare or unavailable.

In contrast, field studies of live individuals/populations can be planned in order to minimize the biases that were pointed out in the case of museum specimens and in order to maximize data collection (e.g. by adding data on microhabitat at capture site and fresh body mass). Field studies are based on more or less standardized surveys throughout the natural habitat of a target species, and on the capture, handling, and processing of the individuals encountered (e.g. Luiselli and Agrimi, 1991; Weaver, 2010; Rugiero et al., 2012). A particular strength of the method is that it is not taxon biased. The dietary habits of all reptile taxa can be investigated using appropriately designed field surveys and with the application of convenient methodologies to obtain the gut contents (see Section 8.3.2). The logistical difficulties are essentially financial (many field surveys are relatively expensive) and/or temporal (obtaining adequate data may require years of field investigation; e.g. Rugiero et al., 2012).

All the abovementioned sources of material can be used conveniently and effectively to obtain reliable and rigorous datasets on the diet of all extant reptile groups. When planning a research study, however, it is always necessary to budget the costs of applying a given method.

8.3 Methods for examining diet and trophic interactions

Once it has been decided which source(s) of material should be used to gather data, a scientist has several alternative methodologies for examining the diet of a target species. In addition, trophic interactions can also be complemented with field studies that include a list of potential species available locally.

8.3.1 Direct observation

Direct observation of feeding events, especially if with minimal or no human intervention, can contribute to our understanding of reptile trophic ecology (Sáez and Traveset, 1995), particularly in poorly known species (Figure 8.1(b)). Direct observations are generally opportunistic and rarely involve an adequate replication of samples (e.g. Taylor and Gardner, 2014). Stochastic observations are published in peer-reviewed journals specializing in natural history notes, thus illustrating that direct observation of reptile prey/predation is still important in herpetological studies. The majority of quantitative studies on reptile diets, however, do not come from direct natural history observations of single individuals, but from a prescribed determination of food eaten in stomach contents (by dissection or flushing), faecal analysis, or even by using advanced techniques such as doubly labelled water.

8.3.2 Dissection of stomachs

Dissection of stomachs or of entire guts can be carried out on museum specimens (e.g. Shine et al., 1996, 2006) or on road-killed specimens (e.g. Daltry et al., 1998); killing live animals should be discouraged for both ethical and conservation reasons (Böhm et al., 2013). Valuable dietary studies have been conducted by ecologists dissecting hundreds of individuals of a given species that were originally sacrificed for genetic, systematic, or distribution purposes (i.e. as vouchers) and later examined for ecological studies (e.g. Pérez-Mellado and Corti, 1993). Dissection of stomachs and guts of preserved specimens or roadkills is easy, inexpensive, and convenient, as it allows the experimenter to obtain whole food items contained in an individual's gut (Figure 8.1(a)). Especially when examining lizards and snakes, the identification of ingested prey is usually easy using appropriate identification keys (when necessary), particularly when the bolus is

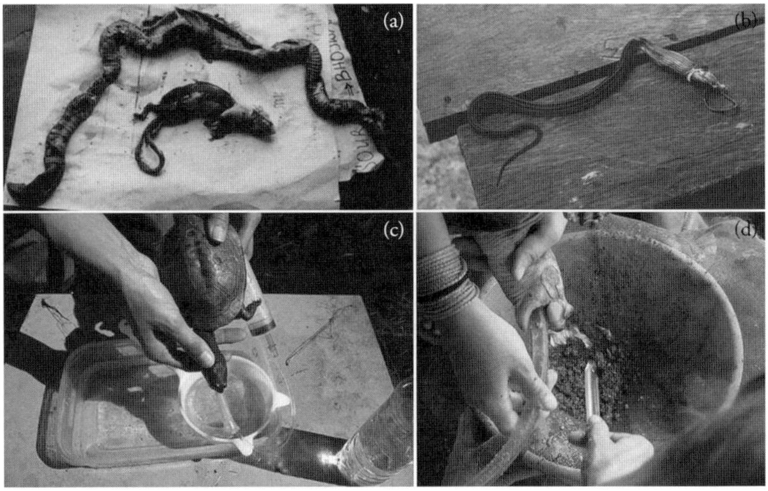

Figure 8.1 (a) A freshly killed, dissected specimen of a Spitting Cobra (*Naja nigricollis*) from the environs of Abraka (Niger Delta, Nigeria) with its prey, a large rodent. Photo by Luca Luiselli. (b) A poorly known colubrid (*Bothrophthalmus lineatus*) regurgitating a rodent in the Upper Orashi Forest Reserve (Niger Delta, southern Nigeria). Recently ingested prey can be gently palpated anteriorly where the snake will regurgitate it. Photo by Luca Luiselli. (c) Stomach flushing *Acanthochelys spixii* in southern Brazil. The tube is inserted down the throat into the stomach, and water is gently flushed in using a squeeze bottle. Photo by Gabriel de Freitas Horta. (d) Stomach contents of *Podocnemis unifilis* are washed into a screen for sorting in the Amazon Basin. Photo by Richard Vogt.

contained within the stomach of the animal. Soft tissues of the body of the ingested prey also occasionally may be conserved and easily identified. Apart from potential biases surrounding the collection of museum specimens (see Section 8.2), the main shortcoming of dissecting museum specimens is that the voucher can be considerably damaged by exposing the gastrointestinal tract. Thus, researchers should adopt approaches to minimize damage to preserved specimens.

8.3.3 Stomach flushing

Stomach flushing is one of the most popular techniques to obtain ingested food from lizards and chelonians (Table 8.1; see also Legler and Sullivan, 1979; Figure 8.1(c, d)), and it has even been used with crocodilians (Fitzgerald, 1989). Basically, the method consists of gently inserting a Teflon™ probing tube of adequate diameter, depending on the animal's size, into the mouth and down inside the gastrointestinal tract of the restrained, immobilized animal. Water is gently forcedly introduced into the stomach, then exercising some pressure to open the pyloric sphincter especially in adult individuals. This method is easily applied to most small- to medium-sized species, but it may become difficult with large and powerful species, including large turtles, monitor

Table 8.1 *Brief synopsis of the literature available on the diet of reptiles, showing the methodology used to collect the data.*

Species	Order	Methods	References
Crocodylus moreletii	Crocodylia	stomach flushing	Platt et al. (2006)
Crocodylus niloticus	Crocodylia	stomach flushing	Wallace and Leslie (2008)
Sphenodon punctatus	Sphenodontia	stomach flushing	Ussher (1999)
Sphenodon punctatus	Sphenodontia	faecal analysis	Walls (1981)
Trogonophis wiegmanni	Amphisbaenia	faecal analysis	Martin et al. (2013)
Lacerta bilineata	Sauria	faecal analysis	Angelici et al. (1997)
Niveoscincus palfreymani	Sauria	stomach flushing	Brothers et al. (2003)
Podarcis hispanica	Sauria	dissection of specimens	Pérez Mellado and Corti (1993)
Podarcis tiliguerta	Sauria	dissection of specimens	Pérez Mellado and Corti (1993)
Podarcis lilfordi	Sauria	dissection of specimens	Pérez Mellado and Corti (1993)
Calloselasma rhodostoma	Serpentes	dissection of specimens, faeces, squeezing	Daltry et al. (1998)
Hierophis viridiflavus	Serpentes	squeezing	Lelièvre et al. (2012)
Zamenis longissimus	Serpentes	squeezing	Lelièvre et al. (2012)
Vipera aspis	Serpentes	squeezing and faeces	Rugiero et al. (2012)
Atractaspis spp.	Serpentes	dissection of specimens	Shine et al. (2006)
Python reticulatus	Serpentes	dissection of specimens	Shine et al. (1996)
Hypsiglena chlorophaea	Serpentes	squeezing	Weaver (2010)
Chelonia mydas	Testudines	stomach flushing	Amorocho and Reina (2007)
Chelonia mydas	Testudines	stable isotopes	Arthur et al. (2008)
Phrynops rufipes	Testudines	stomach flushing and faecal analysis	Caputo and Vogt (2008)
Chelonia mydas	Testudines	dissection of specimens	López-Mendilaharsu et al. (2005)
Dermochelys coriacea	Testudines	stable isotopes	Seminoff et al. (2009)
Trachemys scripta	Testudines	stable isotopes	Seminoff et al. (2007)

lizards, and large crocodilians. The use of ketamine may reduce stress and increase the effectiveness of the technique in large reptiles (Fields et al., 2000). This procedure has the advantage of making soft parts of the food available for identification, and this may be important when the species being examined is a predator of soft-bodied organisms (Pincheira-Donoso, 2008). Thus, Pincheira-Donoso (2008) suggested that stomach flushing should be the preferred method for analysing reptile diets.

Ethical and conservation concerns have been raised because stomach flushing may not be entirely safe to the animals processed. For instance, De Lima et al. (1997), working on the turtle *Phrynops rufipes*, wrote: 'Initially, we tried to pump the stomach contents from individuals. However, we were unable to dislodge seeds that could be felt through the body wall, and one animal died as a result of puncturing the stomach wall'.

The same problems also may occur when flushing lizard stomachs. For example, flushed individuals of *Podarcis carbonelli* and *P. hispanica* were clearly injured by the procedure, even if the Teflon™ tube was gently introduced in the gut (Pérez-Mellado et al., 2011). Injuries included ruptures of the gut wall, severe impairment of subsequent locomotion making the lizard unable to escape after release (100% of captured lizards), and even the rapid death of some individuals. In addition, post mortem inspection of lizard stomachs and intestines revealed that in at least 30% of the samples from *Podarcis muralis, Psammodromus algirus*, and *Psammodromus hispanicus*, soft and hard-bodied prey were retained in the stomach after flushing (Pérez-Mellado et al., 2011). Obviously, this would be an additional limitation to the reliability of this method. Mortality of stomach-flushed lizards also was greater for smaller lizards, with as many as 8.7% of the lizards killed by flushing procedures in some cases.

Stomach flushing also may have negative impacts on subsequent survival. For example, Luiselli et al. (2011) demonstrated that a flushed sample of lizards (*Agama agama*) exhibited lower survivorship than non-flushed lizards of all sex and age classes, with the negative effects of stomach flushing being stronger in adult lizards than in subadults. This result appears counterintuitive, since the authors predicted that smaller and more delicate individuals would have suffered greater survival risks. Luiselli et al. (2011) suggested that several reasons may explain this result. Large lizards might be more stressed than smaller individuals because of their body vigour, forcing the experimenters to be less delicate when handling them for flushing their stomachs. It is also possible that the stronger pressure exerted when using the probing tube to open the pyloric sphincter in adult individuals could increase the probability of rupturing gut walls (Luiselli et al., 2011).

Gastric suction is considered an alternative technique to obtain diet contents in small lizards (Barreto-Lima, 2009). In this method, a syringe dampened with saline solution is carefully inserted through the mouth into the stomach pylorus or stomach, and the stomach material is sucked out by softly retracting the syringe pump as the syringe is slowly withdrawn from the animal. There may be some mortality associated with this method, as 10.5% of the processed specimens died during the procedure (Barreto-Lima, 2009).

8.3.4 Faecal pellets

Collection of faecal pellets is an extremely easy and safe technique for exploring the dietary habits of reptiles. The method can be applied with all groups of reptiles. The simplest way to obtain a faecal capsule from a reptile is to keep the animal in a cage until it defecates. This method is safe and works very well, but may be logistically problematic when many individuals need to be examined within a short period of time. Logistically, it is necessary to keep each individual separate from the others in order to avoid confusion of the faeces' identity and to minimize inter-individual aggression and stress. In snakes, faeces can also be easily obtained by gentle massaging the side of the body towards the posterior (e.g. the tail) (Rugiero et al., 2012). In addition, many species of lizards and snakes, as well as several turtles, immediately defecate as a response to being captured (for instance, lizards of the genus *Podarcis* and snakes of the genus *Natrix* and *Nerodia*, just to cite a few). Lizard scats can also be collected from walls and other perches (Luiselli et al., 2011), but of course this method does not allow for the identification of the individual who has deposited the single scat.

Food contents from both stomach flushing and faecal pellets can be identified using similar procedures. Dietary contents should be spread on Petri dishes with water or a disinfectant agent (5% solution of Lysol), and the fragments (chitin of insects, hairs, scales) should be separated and identified to the lowest possible taxonomic level, counted, and measured for width and length (for determination of volumetric prey composition) using digital callipers (to the nearest 0.01 mm in some cases; see Pincheira-Donoso, 2008). A reference collection of local animals that are potential prey can be assembled, as it can be used for comparisons of residues of food items and hence enhance the probability of correctly identifying the remains. Identification is usually easier with remains derived from flushing than from faecal pellet analysis. Having a species list also allows for a comparison of the potential prey with prey ingested.

Although some authors have pointed out that faeces do not provide a complete assessment of the dietary spectrum of a given species (because the soft body parts of the prey tend to be less well preserved in remains; see Pincheira-Donoso, 2008), other authors have disagreed. For instance, Luiselli et al. (2011) demonstrated that the diet composition of rainbow lizards (*Agama agama*) was very similar (>98% identity) whether by employing stomach flushing or simply by collecting faeces from the soil; similar conclusions were reached by Angelici et al. (1997) and Pérez-Mellado et al. (2011). For this reason, Luiselli et al. (2011) suggested that stomach flushing should be avoided when studying threatened species or populations. In the case of threatened species, we suggest researchers employ use faecal analysis, especially with insectivorous lizards.

8.3.5 Forced regurgitation

A frequently used field technique to analyse the dietary habits of snakes is the forced regurgitation of the ingested bolus (e.g. Luiselli and Agrimi, 1991; Lelièvre et al., 2012). This technique is facilitated by the fact that snakes ingest the prey whole, and that digesting snakes are often slow and torpid, thus enabling capture and handling by the researchers. Most snakes clearly exhibit an enlarged body after eating, thus allowing researchers to determine if the captured snakes have fed recently. Abdominal palpation also may reveal the presence of food items that are not immediately seen. Once food is identified to occur in the gut of a snake, it may be squeezed up to the mouth and identified. This procedure is easy and safe for most snakes, and the ingested bolus even can be reinserted into the snake's gut, after prey identification, with the help of tools such as forceps. Some snake species are reluctant to re-ingest the disgorged bolus (e.g. the viperid snakes *Vipera aspis, V. berus, Causus maculatus*), but other species tend to naturally re-ingest the prey when handled in this way (e.g. *Hierophis viridiflavus* and *Psammophis phillipsii*; unpublished observations). The disgorged bolus can be easily weighed when recently ingested, and this type of measurement can provide very valuable information.

Large and vigorous species (e.g. large pythons) can be processed only with difficulty, but with experience it is still possible to work with them. Researchers should be very careful with large-sized venomous snakes (e.g. cobras, mambas, large viperids) because they can easily bite during processing. With these species, we suggest squeezing the prey item up to the neck while still holding the snake's neck with the fingers, and then releasing hold of the neck just when the food comes into the mouth. In this way, the snake is usually impeded by the food if attempting to bite. Nevertheless, it should be remembered that this procedure is not 100% safe, and a dangerous bite can always be inflicted by a snake when the food arrives in the mouth. This technique should be used with care, given that strong pressure on the belly may risk damage to some internal organs such as the heart. In the course of our research, however, we have handled thousands of snakes of very different sizes (from small 20 cm vipers to large pythons of more than 3 m in length), without a single mortality.

An alternative to forced regurgitation was described by Kjaergaard (1981), and is particularly valuable with venomous snakes. This method consists of placing specimens under study in cages at temperatures <8°C, thus causing a cessation of digestion with consequent regurgitation of prey (Kjaergaard 1981). The main shortcoming of this method is that it is logistically difficult, as refrigerated cages are energy and space-consuming, especially if large snake species are studied.

8.3.6 Stable isotopes

Stable isotopes have been used in marine and freshwater turtle and in lizard dietary studies (e.g. Seminoff et al., 2007; Arthur et al., 2008; Table 8.1), although mammals and birds have been the traditional subjects of these analyses (Kelly, 2000). Stable isotope analysis is particularly useful when an organism's diet is difficult to establish with conventional techniques (Seminoff et al., 2006). Indeed, traditional ways of analysing diets may fail to adequately resolve temporal patterns in diet use, since these techniques reflect small and non-random samples where pseudo-replication may be a problem (Darimont and Reimchen, 2002). Measuring stable isotopes in animal tissues can provide important data since isotopes reflect average dietary records and thus eliminate some of the shortcomings of more traditional diet analyses (e.g. Dalerum and Angerbjörn, 2005). In general, carbon and nitrogen isotopes are the main elements used in dietary analyses (reviewed by Kelly, 2000). Their utility derives from two properties, that is, some sources of dietary carbon or nitrogen have distinct isotope signatures and the isotope signature of a food item is incorporated into the consumer's tissues (Kelly, 2000). Hence, the carbon- or nitrogen-isotope composition of a consumer should be a direct reflection of its diet (DeNiro and Epstein, 1978).

The main shortcomings of this technique are that (i) determination of prey types is too coarse (i.e. imprecise) at the species level; (ii) the temporal patterns of dietary habits cannot be revealed (seasonal patterns may remain hidden, for example); and (iii) it often requires a series of complex and labour-intensive procedures compared to the abovementioned field techniques (Kelly, 2000).

8.3.7 Doubly labelled water

Field energy budgets (including metabolic rates, feeding rates, and growth rates) can be studied using the doubly labelled water technique, which has been used intensively with squamate reptiles in the laboratory (e.g. Andrews and Pough, 1985) and the field

(Peterson et al., 1998). However, this method does not give an indication of the diet of an organism, and hence is not mentioned further in this chapter.

8.4 Diet by volume or mass vs. diet by prey number

The dietary composition of a species can be interpreted differently depending upon whether the volume or the number of ingested prey is considered. For example, Rugiero and Luiselli (1991) found that water snakes may feed on much larger numbers of tadpoles than adult frogs, although adult frogs accounted for a much greater proportion of the biomass ingested. Hence, the techniques discussed in Section 8.3 are not equally useful when volumetric calculations are needed. For volumetric calculations, it is necessary to have well-preserved food items, and whole items are not common in the digestive tracts of most reptiles, particularly small lizards (Pérez-Mellado and Corti, 1993).

Similar problems are encountered for volumetric calculations using prey remains from faecal and gut samples (Luiselli et al., 2011). Such problems are unlikely to occur when working with prey regurgitated by snakes. An alternative method, at least for some prey, is the calculation of total prey size based on the measurement of particular anatomical pieces, and then using regression equations to estimate prey body sizes or biomass (Hódar, 1996). An alternative approach is to assign an 'importance value' to two or three food variables (mass of the item + volume item + item number).

Obviously, the precision of volumetric measurements is a direct function of the completeness of the remains examined, which in turn depends on gut clearance times (i.e. on the physiological performance of digestion). Gut clearance times in reptiles depend on external conditions (i.e. temperature, rainfall) other than on species-specific physiological performance, individual-specific physiological performance, and the prey/predator ratio (e.g. Naulleau, 1983). Thus, variation in gut clearance times may represent a complication for understanding of the precise prey composition of given species or population of individuals. This is true especially when populations in different climatic conditions are compared or when certain species feed on a wide variety of organisms that require different digestion rates. In such cases, stable isotope analysis may represent a valuable technique for minimizing potential problems.

8.5 Gut clearance times

Gut clearance time is an important variable in reptile dietary studies, given that the percentage of 'fed' individuals in a given sample is directly influenced by the rate of gut clearance. Gut clearance time is dependent on species (some species are more efficient in digestion than others; McKinon and Alexander, 1999), external conditions (ambient temperatures) where the animals live, and diet quality (Alexander et al., 2012). For example, the gut passage time was 25% faster at 27°C than at 20°C in an African elapid snake, but it was not affected by food type or snake body mass; larger meals took longer to digest than smaller meals (Alexander et al., 2012).

In general, the relevance of gut clearance time for understanding the dietary patterns of reptiles is more important in those groups with occasional feeding events (snakes) than in those groups with nearly daily ingestion of food (lizards, chelonians); not surprisingly, most studies have focused on lizards (e.g. McKinon and Alexander, 1999).

Digestive efficiency (= digestion coefficient, DC) is generally measured as the number of calories removed from the food relative to the number of calories consumed (Zug, 2008):

$$\mathrm{DC} = (\text{consumed} - \text{defecated})/(\text{consumed}).$$

The energy content of both meals and faeces are measured using bomb calorimetry (e.g. Davenport et al., 1989). Digestive efficiencies are generally much greater in carnivorous lizards than they are in herbivorous lizards (DC ranging from 30 to 80%, McKinon and Alexander, 1999). Experimentation on gut clearance times in reptiles requires laboratory treatments, where diets of different quality are provided to the same individuals under different values of controlled temperatures (e.g. see Alexander et al., 2012). The gut clearance time is then evaluated as the time passing between the feeding event and the first defecation event.

8.6 Quantitative analyses of diet

Typically, dietary data can be analysed with appropriate statistical tools that are valid for other types of data. For instance, frequencies of utilization of distinct prey categories can be analysed using a χ^2 test and, where appropriate, different kinds of factor analyses, dendrograms, or multivariate sets of analyses. In addition, researchers have developed a suite of indices to summarize the richness of prey species and their relative abundances into a single value condensing both types of information. These indices are defined as 'diversity indices' (see Chapter 21). In herpetology, the most used formula describing the *food niche breadth* of a given species has been Simpson's (1949) diversity index:

$$\text{Diet breadth}\,(B) = 1 \Big/ \sum_{i=1}^{n} p_i^2$$

where p_i is the frequency of utilization of each food category in the dietary spectrum. For calculating the similarity in food types used by two species, the symmetric niche overlap index of Pianka (1986) can be used:

$$O_{jk} = \frac{\sum_{i=1}^{n} p_{ij} \times p_{ik}}{\sqrt{\sum_{i=1}^{n} p_{ij}^2 \sum_{i=1}^{n} p_{ik}^2}}$$

where p_{ij} and p_{jk} are the frequency of utilization of each food category in the dietary spectrum of the species j and k. In this formula, the values yield from 0 (no overlap) to 1 (total overlap).

Although Pianka's overlap formula was originally designed to assess niche overlap between two potential competitors, nevertheless it is merely a similarity index, and hence can be readily used as a univariate measure of similarity in the dietary spectrum between two different populations of animals (Pianka, 1986). Pianka's index is similar to MacArthur and Levins' (1967) index, with a denominator that has been normalized to make it symmetric but with the stability properties that were unchanged.

An important tool for improving comparisons of diversity indices for determining assembly rules between potential competitors has been the use of null models with Monte Carlo simulations (Gotelli and Graves, 1996). In this case, when the diet of two species is compared, the original species utilization matrices of food types from which Pianka's overlap is calculated are randomized by shuffling the original values among the resource states, with two randomization algorithms (defined as RA2 and RA3) that are preferably used, as they are particularly robust for niche overlap studies (Gotelli and Graves, 1996). The software EcoSim Professional (Acquired Intelligence Corp., Kesey-Bear; http://garyentsminger.com/ ecosim/index.htm) is particularly useful and widely used to calculate overlap indices and to generate Monte Carlo simulations. There also are some indices that combine in a single value the numeric and volumetric contributions of a prey category. For instance, the importance index (*I*) is an arithmetic average of the numeric proportion, volumetric proportion, and frequency of occurrence of each prey category on the population's diet (Mesquita and Colli, 2003).

References

Alexander, G.J., Hanrahan, S.A., and Mitchell, D. (2012). Assimilation efficiency and gut passage time in an African elapid snake, *Hemachatus haemachatus*. *African Journal of Herpetology*, **61**, 3–13.

Amorocho, D.F., and Reina, R.D. (2007). Feeding ecology of the East Pacific green sea turtle *Chelonia mydas agassizii* at Gorgona National Park, Colombia. *Endangered Species Research*, **3**, 43–51.

Andrews, R.M., and Pough, F.H. (1985). Metabolism of squamate reptiles: allometric and ecological relationships. *Physiological Zoology*, **58**, 214–31.

Angelici, F.M., Luiselli, L., and Rugiero, L. (1997). Food habits of the green lizard, *Lacerta bilineata*, in central Italy and a reliability test of faecal pellet analysis. *Italian Journal of Zoology*, **64**, 267–72.

Arthur, K.E., Boyle, M.C., and Limpus, C.J. (2008). Ontogenetic changes in diet and habitat use in green sea turtle (*Chelonia mydas*) life history. *Marine Ecology Progress Series*, **362**, 303–11.

Barreto-Lima, A.F. (2009). Gastric suction as an alternative method in studies of lizard diets: tests in two species of *Enyalius* (Squamata). *Studies on Neotropical Fauna and Environment*, **44**, 23–9.

Böhm, M., Collen, B., Baillie, J.E.M., et al. (2013). The conservation status of the world's reptiles. *Biological Conservation*, **157**, 372–85.

Brothers, N., Wiltshire, A., Pemberton, D., et al. (2003). The feeding ecology and field energetics of the Pedra Branca skink (*Niveoscincus palfreymani*). *Wildlife Research*, **30**, 81–7.

Brown, J.H., and Gillooly, J.F. (2003). Ecological food webs: high-quality data facilitate theoretical unification. *Proceedings of the National Academy of Sciences of the USA*, **100**, 1467–8.

Caputo, F.P., and Vogt, R.C. (2008). Stomach flushing vs. fecal analysis: the example of *Phrynops rufipes* (Testudines: Chelidae). *Copeia*, **2008**, 301–5.

Dalerum, F., and Angerbjörn, A. (2005). Resolving temporal variation in vertebrate diets using naturally occurring stable isotopes. *Oecologia*, **144**, 647–58.

Daltry J.C., Wüster W., and Thorpe, R.S. (1998). Intraspecific variation in the feeding ecology of the crotaline snake *Calloselasma rhodostoma* in Southeast Asia. *Journal of Herpetology*, **32**, 198–205.

Darimont, C.T., and Reimchen, T.E. (2002). Intra hair stable isotope analysis implies seasonal shift to salmon in grey wolf diet. *Canadian Journal of Zoology*, **80**, 1638–42.

Davenport, J., Antipas, S., and Blake, E. (1989). Observations of gut function in young green turtles *Chelonia mydas* L. *Herpetological Journal*, **1**, 336–42.

De Lima, A.C., Magnusson, W.E., and Da Costa, V.L. (1997). Diet of the turtle *Phrynops rufipes* in central Amazonia. *Copeia*, **1997**, 216–19.

DeNiro, M.J., and Epstein, S. (1978). Influence of diet on the distribution of carbon isotopes in animals. *Geochimica et Cosmochimica Acta*, **42**, 495–506.

Fields, J.R., Simpson, T.R., Manning, R.W., et al. (2000). Modifications to the stomach flushing technique for turtles. *Herpetological Review*, **31**, 32–3.

Fitzgerald, L.A. (1989). An evaluation of stomach flushing techniques for crocodilians. *Journal of Herpetology*, **23**, 170–2.

Gotelli, N.J., and Graves, G.R. (1996). *Null Models in Ecology*. Washington, DC:Smithsonian Institution Press.

Hódar, J.A. (1996). The use of regression equations for estimation of arthropod biomass. *Acta Oecologica*, **17**, 421–33.

Kelly, J.F. (2000). Stable isotopes of carbon and nitrogen in the study of avian and mammalian trophic ecology. *Canadian Journal of Zoology*, **78**, 1–27.

Kjaergaard, J. (1981). A method for examination of stomach content in live snakes and some information on feeding habits in common viper (*Vipera berus*) in Denmark. *Natura Jutlandica*, **19**, 45–8.

Legler, J.M., and Sullivan, L.J. (1979). The application of stomach-flushing to lizards and anurans. *Herpetologica*, **35**, 107–10.

Lelièvre, H., Legagneux, P., Blouin-Demers, G., et al. (2012). Trophic niche overlap in two syntopic colubrid snakes (*Hierophis viridiflavus* and *Zamenis longissimus*) with contrasted lifestyles. *Amphibia-Reptilia*, **33**, 37–44.

López-Mendilaharsu, M., Gardner, S.C., Seminoff, J.A., et al. (2005). Identifying critical foraging habitats of the green turtle (*Chelonia mydas*) along the Pacific coast of the Baja California peninsula, Mexico. *Aquatic Conservation: Marine and Freshwater Ecosystems*, **15**, 259–69.

Luiselli, L., and Agrimi, U. (1991). Composition and variation of the diet of *Vipera aspis francisciredi* in relation to age and reproductive stage. *Amphibia-Reptilia*, **12**, 137–44.

Luiselli, L., Akani, G.C., Ebere, N., et al. (2011). Stomach flushing affects survival/emigration in wild lizards: a study case with rainbow lizards (*Agama agama*) in Nigeria. *Amphibia-Reptilia*, **32**, 253–60.

Macarthur, R., and Levins, R. (1967). The limiting similarity, convergence, and divergence of coexisting species. *American Naturalist*, **101**, 377–85.

Martin J., Ortega J., Lopez P., et al. (2013). Fossorial life does not constrain diet selection in the amphisbaenian *Trogonophis wiegmanni*. *Journal of Zoology*, **291**, 226–33.

McKinon, W., and Alexander, G.J. (1999). Independence of digestive efficiency an experimental artifact in lizards? A test using the common flat lizard (*Platysaurus intermedius*). *Copeia*, **1999**, 299–303.

Mesquita, D.O., and Colli, G.R. (2003). The ecology of *Cnemidophorus ocellifer* (Squamata, Teiidae) in a neotropical savanna. *Journal of Herpetology*, **37**, 498–509.

Mushinsky, H.R., Stilson, T.A., and McCoy, E.D. (2003). Diet and dietary preference of the juvenile gopher tortoise (*Gopherus polyphemus*). *Herpetologica*, **59**, 475–83.

Naulleau, G. (1983). The effects of temperature on digestion in *Vipera aspis*. *Journal of Herpetology*, **17**, 166–70.

Pérez-Mellado, V., and Corti, C. (1993). Dietary adaptations and herbivory in lacertid lizards of the genus *Podarcis* from western Mediterranean islands (Reptilia: Sauria). *Bonner Zoologische Beitraege*, **44**, 193–220.

Pérez-Mellado, V., Pérez-Cembranos, A., Garrido, M., et al. (2011). Using faecal samples in lizard dietary studies. *Amphibia-Reptilia*, **32**, 1–7.

Peterson, C.C., Walton, B.M., and Bennett, A.F. (1998). Intrapopulation variation in ecological energetics of the garter snake *Thamnophis sirtalis*, with analysis of the precision of doubly labeled water measurements. *Physiological Zoology*, **71**, 333–49.

Pianka, E.R. (1986). *Ecology and Natural History of Desert Lizards*. Princeton, NJ:Princeton University Press.

Pincheira-Donoso, D. (2008). Testing the accuracy of fecal based analyses in studies of trophic ecology of lizards. *Copeia*, **2008**, 322–5.

Platt, S.G., Rainwater, T.R., Finger, G.A., et al. (2006). Food habits, ontogenetic dietary partitioning and observations of foraging behaviour of Morelet's crocodile (*Crocodylus moreletii*) in northern Belize. *Herpetological Journal*, **16**, 281–90.

Rahman, S.C., Rashid, S.M.A., Das, K., et al. (2013). Monsoon does matter: annual activity patterns in a snake assemblage from Bangladesh. *Herpetological Journal*, **23**, 203–8.

Reading, C.J., Luiselli, L., Akani, G.C., et al. (2010). Are snake populations in widespread decline? *Biology Letters*, **6**, 777–80.

Rugiero, L., and Luiselli, L. (1991). Food niche partitioning by water snakes (Genus *Natrix*) at a freshwater environment in central Italy. *Journal of Freshwater Ecology*, **6**, 443–7.

Rugiero, L., Milana G., Capula, M., et al. (2012). Long-term variation in small mammal composition of a snake diet do not mirror climate change trends. *Acta Oecologica*, **43**, 158–64.

Sáez, E., and Traveset, A. (1995). Fruit and nectar feeding by *Podarcis lilfordi* (Lacertidae) on Cabrera Archipelago (Balearic Islands). *Herpetological Review*, **26**, 121–3.

Seminoff, J.A., Bjorndal, K.A., and Bolten, A.B. (2007). Stable carbon and nitrogen isotope discrimination and turnover in pond sliders *Trachemys scripta*: insights for trophic study of freshwater turtles. *Copeia*, **2007**, 534–42.

Seminoff, J.A., Jones, T.T., Eguchi, T., et al. (2006). Stable isotope discrimination ($\delta^{13}C$ and $\delta^{15}N$) between soft tissues of the green sea turtle *Chelonia mydas* and its diet. *Marine Ecology Progress Series*, **308**, 271–8.

Seminoff, J.A., Jones, T.T., Eguchi, T., et al. (2009). Stable carbon and nitrogen isotope discrimination in soft tissues of the leatherback turtle (*Dermochelys coriacea*): insights for trophic studies of marine turtles. *Journal of Experimental Marine Biology and Ecology*, **381**, 33–41.

Shine, R., Branch, W.R., Harlow, P.S., et al. (1996). Sexual dimorphism, reproductive biology, and food habits of two species of African filesnakes (*Mehelya*, Colubridae). *Journal of Zoology*, **240**, 327–40.

Shine, R., Branch, W.R., Harlow, P.S., et al. (2006). Biology of burrowing asps (Atractaspididae) from southern Africa. *Copeia*, **2006**, 103–15.

Simpson, E.H. (1949). Measurement of diversity. *Nature*, **163**, 649.

Taylor, B., and Gardner, C. (2014). Nectar feeding by the day gecko *Phelsuma mutabilis* (Squamata: Gekkonidae) on the mangrove tree *Sonneratia alba* (Lythraceae) in southwest Madagascar. *Herpetology Notes*, 7, 85–7.

Ussher, G.T. (1999). Tuatara (*Sphenodon punctatus*) feeding ecology in the presence of kiore (*Rattus exulans*). *New Zealand Journal of Zoology*, **26**, 117–25.

Wallace, K.M., and Leslie, A.J. (2008). Diet of the Nile crocodile (*Crocodylus niloticus*) in the Okavango Delta, Botswana. *Journal of Herpetology*, **42**, 361–8.

Walls, G.Y. (1981). Feeding ecology of the tuatara (*Sphenodon punctatus*) on Stephens Island, Cook Strait. *New Zealand Journal of Ecology*, **4**, 89–97.

Weaver, R.E. (2010). Diet and reproductive ecology of the desert nightsnake (*Hypsiglena chlorophaea*) in central Washington State. *Journal of Herpetology*, **44**, 148–52.

Zug, G.R. (2008). *Herpetology: An Introductory Biology of Amphibians and Reptiles*. New York: Academic Press.

9

Movement patterns and telemetry

Bruce A. Kingsbury and Nathan J. Robinson

9.1 Introduction

Movement is among the most basic and conspicuous characteristics of life and is central to the ecology of all vertebrates. By tracking the movements of animals in nature, we gain valuable insights into their habitat requirements, foraging strategies, social and reproductive interactions, and physiological tolerances (e.g. Doody et al., 2002; James et al., 2005; Kay, 2005; Knapp and Owens, 2005; Waldron et al., 2006; Ferrara et al., 2013). Furthermore, information on movement patterns and habitat associations can help identify critical habitat needs and thus guide management efforts.

Researchers have tracked reptile movements for over a century. In recent decades, much of this has been achieved through telemetry—the wireless transmission of data between transmitters on an animal and a receiver in hand or attached to automated systems. As telemetry technologies improve, an ever-expanding array of movement data are becoming available. The primary benefit to telemetry over visual tracking and mark–recapture techniques is the reliability with which animals can be relocated coupled with remote tracking. Telemetry devices allow relocation of even highly cryptic individuals, enable tracking of free-ranging animals as they move undisturbed through their environment, and allow animals to be tracked as they move over long distances or terrain that is largely inhospitable to humans (e.g. marine environments).

In this chapter, we offer a methodological 'road-map' to assist researchers contemplating telemetry projects to help determine which approach is best for them and to maximize the benefits of such endeavours. We discuss the advantages and limitations of different telemetry devices (also see Guilhon et al., 2011); outline the most common-place statistical approaches for analysing telemetry data; discuss taxa-specific considerations that should be kept in mind; and look towards future opportunities for tracking reptiles. Telemetry is not the only way to keep track of individuals and associate location with an individual. Marking and trapping techniques are covered elsewhere in this book, but users of these techniques may find useful information in this chapter when interpreting movement patterns.

9.2 Common considerations for telemetry studies

There are many issues that must be considered before beginning a telemetry study. The most important is that any animal-borne telemetry device has the potential to affect the behaviour and movement patterns of the animal (McMahon et al., 2011), thus compromising the data collected and even endangering the animal itself. The transmitter should be both compact and lightweight enough not to incur any negative effects. A commonly used yet relatively arbitrary weight limit suggested for terrestrial reptiles is 5%. The source of this threshold is not clear. Nevertheless, carrying additional mass will at some point affect behaviour and, in lieu of alternatives, the 5% maximum provides a starting point for deciding how to proceed. The form of the transmitter may also have a greater impact than its weight. Every effort should be made to ensure that the transmitter does not restrict or interfere with the animal's movements, such as becoming snagged on objects in the environment.

Telemetry devices can cost anywhere between US$150 and US$5000. Consequently, sample sizes are often dictated by funding availability rather than the statistical requirements of the study. Since the individual rather than the location is the sample unit, any division of individuals into groups (e.g. sex, size) quickly introduces a statistical challenge for projects with modest budgets. Another consideration is labour. Unless the device is a satellite transmitter, affiliated with a data logger, or is automatically monitored with an antenna array, researchers must triangulate to the animal every time a location is desired. Cost–benefit analysis of different approaches may help assess trade-offs, such as spending more on initial expenses of remote monitoring equipment versus the labour costs for personnel walking or boating when looking for telemetered animals.

9.3 Telemetry devices

9.3.1 VHF transmitters

Most transmitters used in reptile research are VHF (very high frequency) devices, with operating frequencies commonly ca. 150–170 KHz. Such units have low power consumption and do not require large batteries, and thus are suitable for smaller animals. Ultra high frequency (UHF) devices might have appeal in that the higher the frequency used, the smaller the antennas involved, although transmission range suffers. Transmitters may be very small (less than a gram), but that comes with a sacrifice of longevity. Battery mass correlates with amp/hrs that the battery can deliver; the larger the battery, the longer the transmitter will provide a signal. Transmitters that weigh less than a gram may only last for weeks, whereas larger ones may last for several years.

Transmitters have unique frequencies to allow differentiation among animals close together, and may have temperature-sensitive components such that the pulse rate changes as transmitter temperature changes. This attribute can allow body or environmental temperature monitoring, depending upon placement. Pulse interval is then matched to temperature using a calibration curve. Temperature sensitivity may also allow determination if the animal is inactive underground: a slow pulse rate is caused by cooler soil temperatures, whereas a high rate occurs above ground where ambient temperatures are much greater. Transmitter battery life dramatically increases during overwintering, since battery life is dependent on how many pulses are produced by the unit.

The signal from a radio transmitter is acquired using a radio receiver and antenna tuned to the same frequency as the transmitter. The signal emitted from the transmitter is omnidirectional; thus, all that can be detected is the strength of the signal. Directionality is achieved by adding wires or tubes (elements) positioned perpendicular to a central supporting strut to the receiving antenna. It is possible to have an effective antenna with only two elements, but reception and directional performance improve with additional elements. Trade-offs involve weight and size (e.g. navigating through thick brush with an antenna of even three elements can be frustrating). Antennas may be hand-held with data recorded manually, or they may be built into stationary platforms that store data for later analysis. Omnidirectional receiving antennae are also available and are useful for mounting on a vehicle when searching large areas.

Many varieties of receivers are available. Portability is critical when tracking on foot, and the ability to control signal gain and volume is valuable. Gain is the amplification of the signal relative to background noise. Researchers evaluate the input by optimizing the output from the receiver by manipulating volume and gain. Gain is best maximized for distant animals, then attenuated/reduced when close for greater ability to discern signal strength and directionality. Some trackers will find that simply listening to the speaker output is best for them; others will opt to wear headphones (which have the benefit of blocking sound not coming from the receiver); while others will examine a visual output available with some receivers, since the receiver is better than we are at audibly distinguishing signal strength.

Tracking reptiles in the field is an art. When tracking animals, it is important to develop a triangulation strategy that does not involve a direct approach that might result in clumsily stepping upon, flushing, or otherwise disturbing the subject. Approaching the presumed location 'imprecisely' rather than directly and checking periodically to obtain approximations perpendicular to one's trajectory, then adjusting direction of travel, allows approach to an animal's vicinity with less chance of harming or disturbing it. The result is a 'spiralling' movement to the location rather than a direct approach. Aiming the antenna at the ground in front of you periodically also alerts you to an imminent encounter. Researchers must balance the value of finding a precise location with the consequences of (repeatedly) disturbing the animal. Many spatial ecology questions do not require sub-metre accuracy; if being unsettled causes unusual movements, subsequent data points may not accurately reflect normal behaviour.

An alternative to manual tracking is the use of antenna arrays, receivers, and databases to store locality information. While considerably more expensive than manual tracking, arrays can be established to collect data automatically as frequently as desired as long as the subject is in range. An extensive dataset can be collected with a reduced labour cost. Kays et al. (2011) provided a good explanation of how automated radio-telemetry system (ARTS) can be used for tracking animals; examples for reptiles include those reported by Ward et al. (2013) and Tucker et al. (2015).

Researchers should be cognizant of the accuracy limitations of the more affordable hand-held Global Positioning System (GPS) units. Without post-differential correction, locations reported by the GPS 'wander' over time, depending upon weather and satellite availability. Consequently, trackers should use a compass and measuring tape for measuring short moves rather than the GPS to document shifts relative to an established point, such as an initial GPS point measurement; an averaged position relocated over time might also be considered unless using a GPS unit that already provides submetre accuracy. This messiness in location positioning leads to inaccuracies in measuring distances moved. Further, it argues for habitat assignment in the field as opposed to later assignment based on maps. The combination of the inherent inaccuracy in the map and in the GPS location leads to incorrect habitat assignments, particularly in fine-grained habitats or when animals are using ecotones.

9.3.2 Acoustic telemetry

Acoustic telemetry is mechanistically very similar to VHF/UHF telemetry, but the signal only travels well through water. Acoustic telemetry is thus only suitable for tracking aquatic and marine species such as crocodiles or turtles. Acoustic transmitters can relay a signal up to 1 km. It is possible to use acoustic transmitters to track the movements of animals using a triangulation method by suspending a receiver over the side of a boat, but this is relatively challenging in many situations. It is more common to deploy arrays of acoustic receivers to function as 'listening posts' within areas of interest. These listening posts are generally omnidirectional to maximize the chance of picking up a signal from nearby transmitters, yet may allow tracking an animal's directional movements based on the arrangement of the array relative to the movement of the animal. The use of acoustic telemetry remains uncommon, but Thums et al. (2013) and Campbell et al. (2010) provide examples.

9.3.3 Satellite telemetry

When the objective is to track an animal over long distances and recapture is unlikely, satellite telemetry may be necessary. By relaying the data remotely from any location, these devices can generate impressive quantities of data for animals moving through inaccessible environments (oceans, large rivers, along coasts) and over periods of time that can extend more than a year. Under ideal circumstances, these devices can pinpoint an animal's location to within an error radius of only a few metres.

Relaying a message to a satellite requires a substantial amount of power, and this necessitates larger batteries. Even the smallest satellite telemetry devices generally weigh >5 g; thus, most small species or juveniles of larger species cannot be tracked using this method. Satellite telemetry devices are also the most expensive transmitters; even the cheapest devices cost more than US$1000, with most substantially more than that.

There are two major types of satellite telemetry devices, that is, those that use the Argos satellite system or the GPS. The Argos satellite system computes the location of a transmitter on the basis of the Doppler effect and the repetition period (the time between two consecutive messages sent by the transmitter). To calculate a location, a satellite must receive multiple 'hits' (successful receipts of signals) from a transmitter; the more 'hits' each satellite receives, the better the accuracy of the location estimate. GPS works in a slightly different manner. To compute the location of a GPS transmitter, the device determines the distance between itself and a minimum of three satellites. The device then computes its own location and relays that information directly to the satellite array.

Locations generated by the Argos satellite systems are generally accurate to between a few hundred metres and tens of kilometres. This means that Argos transmitters are generally suited for describing long-distance movements. GPS devices are often accurate to within a few metres, yet generating a location often requires many seconds, even minutes, of uninterrupted satellite connectivity. As a result, GPS devices are not suited to study the movements of animals that are often submerged under water or canopy cover. A compromise exists in the form of Fastloc® GPS devices, which take a 'snapshot' of the GPS satellites overhead, calculate the pseudo-ranges, and then store these data for processing (the GPS unit is not required to be in connection with the satellite). The location of the transmitter is later relayed as an Argos message that is often accurate to within a hundred metres.

9.4 Statistical techniques for analysing telemetry data

Telemetry allows individuals to be tracked over substantial periods of time, but the observations are repeated measures of the same individual. Although informative about an individual's movements, subsequent positions should not be treated as though they were independent observations. Thus, repeated observations on an individual are unavoidably temporally autocorrelated, since where an animal goes next is in part determined by its current location. If we endeavour to assess habitat selection, then what we view as 'available' to an animal may change over time as it moves through the landscape. Furthermore, the area occupied by an individual is not going to stabilize until a full migratory cycle has transpired. Species such as timber rattlesnakes provide an example, wherein individuals often 'loop' through the landscape over the course of a year, not returning to their point of origin until overwintering (Gibson, 2003; Figure 9.1(a)).

A partial solution to this challenge is to monitor an animal for long enough that it revisits, perhaps even repeatedly, the same areas. This tendency varies widely among

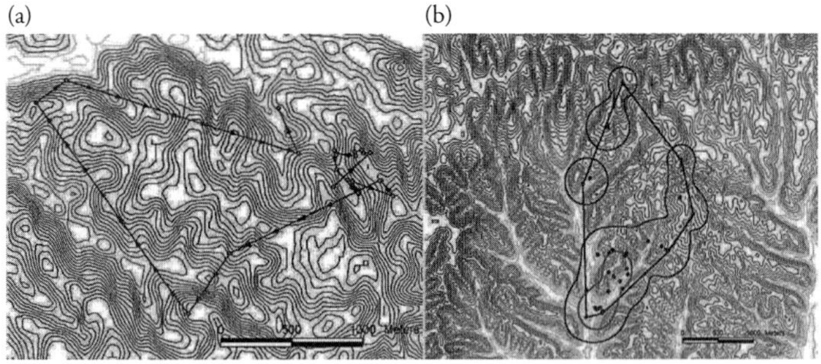

Figure 9.1 Visualizing Timber Rattlesnake relocations from telemetry. In both cases, the background used is a digital elevation model of terrain. (a) Looping path of male over one activity season (lighter shading of path are later in season). (b) Relocations and spatial representations using minimum convex polygon and kernel density analysis (95 and 50% isopleths shown). Adapted from S. Gibson and B. Kingsbury (unpublished data).

species. If it does, tracked animals have an entire area to choose from in terms of habitat, and the area occupied no longer increases, at least within an activity season. Keep in mind that the estimated seasonal range will often increase as more observations are collected; thus, there may be a risk of underestimating the area used if insufficient observations are documented. Researchers must decide how frequently to track individuals. Assuming that the subject is not disturbed, frequent tracking may provide details about localized movements and timing, although observations may be redundant and uninformative and thus a waste of labour that might better have been devoted to tracking other animals over more extended survey periods.

Telemetry can be useful when exploring questions about how much area an animal uses or how far it travels during a particular period. Researchers might then infer how much area an individual or perhaps even a population requires, though such questions are much more challenging to address. We do not know that an animal has elected to use 'just enough' area to satisfy its needs; all we know is where we found it. Although the concept of home range (the area used by an animal to conduct its normal activities over time; Burt, 1943) is simple, there is ample evidence (e.g. Burke et al., 1995) that even with substantial overlap, the area used by many reptiles is not the same every year and that an accurate assessment of home range requires a cumulative examination over at least several years. Researchers should specifically define terms used to maximize comparative benefit among studies. What may be estimated as a 'seasonal range' (cool–warm in temperate zones, wet–dry in the tropics) may in fact not accurately represent a cumulative home range, but only how an animal moved over one field season.

Seasonal range analyses fall into two approaches: drawing boundaries around some proportion of the outermost (most separated) locations, and probabilistic approaches that calculate the area expected to contain proportions of all observations based on the locations observed. The most common approach using outermost points is the minimum convex polygon (MCP) method (Mohr, 1947; Jehnrich and Turner, 1969); the most common probabilistic approach is kernel density analysis (KDA), introduced by Worton (1987).

The MCP approach is intuitive and immediately informative. It allows straightforward comparisons between species, sexes, life stages, and other categorizations. It captures both movement extremes and centres of activity and corridors used to move about the landscape (Figure 9.1(b)). On the other hand, it is simplistic, overly inclusive in terms of area used, and provides no assessment of the interior distribution of points or centres of activity. For example, a terrestrial reptile using the area around a lake would have a MCP that included the area of the lake even though it never entered the water. Researchers may opt for a refined approach by not including all points, perhaps using a 95% criterion to exclude unusual excursions.

KDA is complementary to MCP. Unlike MCP, the precise localities of observations inform the probabilistic distribution, and so additional insights are provided about areas of greater activity (Figure 9.1(b)). Since it is a two-dimensional distribution, the 'tails' of the distribution include extensive areas outside of the area not likely occupied by the animal. For that reason, the 95% isopleth is often used to reduce such an inclusive error. Another issue is that KDA requires that users assign values to parameters that they may not fully understand (e.g. smoothing factor: *k*), and this may often lead to exaggerated estimates of areas occupied. Users are encouraged to be very familiar with the nature and impact of assumptions of the analysis. Row and Blouin-Demers (2006) discuss some of the challenges of KDA and offer a possible solution to the problem of appropriate *k* selection.

Ultimately, any measure of home range is imperfect. It should be understood that the goal of these analyses is to acquire reasonably accurate assessments of area occupied that may allow comparison between groups (e.g. males versus females) to derive answers to research questions.

9.5 Taxonomic considerations

9.5.1 Terrestrial and freshwater turtles

While the shell of a turtle may make it difficult to implant a transmitter within the body cavity, it provides ample options for external attachment. Placement should be to the rear of the shell, such that it does not protrude out of the contours of the individual and thus interfere with movement (Figure 9.2(a)). Avoiding the midline also

(a) (b) (c) (d)

Figure 9.2 Transmitter placement examples for reptiles. (a) Affixed on Box Turtle carapace with epoxy (photo by J. Gibson). (b) Radiograph of snake with an implanted transmitter (anterior portion of the snake in a glass tube to manage the venomous species during procedure; photo by Grayling Hospital for Animals). (c) Secured on nuchal plates of crocodile (photo by J. Beauchamp). (d) Deployed on Leatherback via tether through pygal process (photo by N.J. Robinson). Photos used with permission.

reduces potential interference with mating. Transmitters are attached to the carapace by drilling holes in the marginal scutes and running bolts through to a flange on or around the transmitter. They can also be glued on with epoxy. Areas to be glued should be clean and dry before application, and the epoxy should not be of the fast-curing ('5 minute') type, since the reaction is exothermic and the turtle can be burned during setting. However, long curing times mean managing the uncooperative turtle for extended periods. Whether bolting or gluing, any placement should be on a single scute (avoiding the growth zones in the intervening sutures) unless the turtle is fully grown. Otherwise, proper shell development may be locally impaired during growth.

Antenna placement is another consideration. While from a signal perspective it might be advantageous to have the antenna extend vertically perpendicular to the surface of the carapace, such a position maximizes contact with objects during crawling or swimming, hinders locomotion, increases wear on the unit and antenna, and potentially entraps the animal among debris. An alternative is to affix the antenna around the edge of the carapace so that it does not extend beyond the shell. Such an approach introduces new challenges. The antenna functions best when straightened at full length; curving effectively shortens it. Furthermore, as the ends of an antenna approach forming a loop, the signal is dramatically reduced. Gluing across sutures limits opportunities for expansion during growth. An elaborate solution to this problem is to guide the antenna wire through tubes that are glued to each scute. This allows for growth because the sutures are not involved. However, it may also enhance snagging on objects at any gap along the antenna. Ultimately, after trying all sorts of variations, we settled on letting the whip trail largely behind the turtle and have not observed turtles getting snagged in the habitat.

A challenge for monitoring semiaquatic turtles using VHF frequencies is that the signal may be weak and provide poor directionality when the turtle is in shallow water; it will not work well in deep (>0.5 m) water. Acoustic telemetry may provide a solution, but acoustic telemetry has the challenge of deploying the receiver off the side of a boat in shallow habitats.

9.5.2 Lizards and snakes

Typical transmitter placement on larger snakes and lizards is internal (Figure 9.2(b)) via surgery (e.g. Weatherhead and Anderka, 1984; Goodman et al., 2009), although various efforts have been made to attach transmitters externally, as with duct tape (Wylie et al., 2011). For lizards, there is the additional option of a 'backpack' approach (see Warner et al., 2006; Goodman et al., 2009). Internal placement is invasive and must be done in a manner that precludes infection and does not interfere with normal behaviours, such as eating and mating. Negative impacts such as increased rates of infection and impairment of movement have been observed (Weatherhead and Blouin-Demers, 2004; Lentini et al., 2011).

Implantation surgery should be conducted by trained personnel who are aware of the many seemingly minor complications that can lead to failure. While a thorough description of the technique is beyond the scope of this chapter (see Reinert and Cundall, 1982), we offer some guidance. Transmitter placement should be intraperitoneal rather than subcutaneous, otherwise the resultant bulge is problematic. Entry to the body cavity is ventrolateral through the skin, and access to the peritoneum is achieved by lifting the body wall out of the way rather than by cutting it. Careful cutting of the thin peritoneum without damaging organs provides a hole through which the transmitter can be nestled among the organs. Make sure it is not binding and can move freely. Some researchers stitch the transmitter to a rib to keep it from moving, although many do not take this step as it potentially causes irritation where sutures pass through the body wall.

For large snakes, the whip antenna can be inserted subcutaneously using a small gauge stainless steel tube inserted ventrally so that it rides just above the scutes near the mid-ventral line. This placement is important; deviation from the midline means that as the snake bends its body, the tip of the antenna wire moves fore and aft. It may then repeatedly poke the interior body cavity and possibly creep backward, thus looping the antenna back around the transmitter. Antenna kinking will also promote looping and may even lead to failed stitching and rupture of the integument. For lizards, a helical transmitter antenna potted internally with the rest of the transmitter precludes complications involving placement of the whip antenna, but at the expense of signal range.

9.5.3 Crocodilians

Considering that adult crocodilians can move distances up to 400 km in search of new mates or foraging areas (Lance et al., 2011; Cherkiss et al., 2014) and that juveniles can migrate distances up to 100 km (Magnusson et al., 1979; Sah and Stuebing, 1996), satellite telemetry studies on crocodilians are likely to provide a wealth of information. Indeed, satellite telemetry studies on Saltwater Crocodiles (*Crocodylus porosus*) have provided important insights into how this species uses surface currents to facilitate long-distance migrations when travelling both within estuaries and in open-ocean habitats (Campbell et al., 2010).

For juveniles, transmitters can be attached either through implantation or using neoprene harnesses, while the preferred method for adults is to anchor the device through the nuchal scutes (Figure 9.2(c); Franklin et al., 2009). A transmitter applied to an adult should be placed anteriorly, as the tail is often a target in conspecific aggressive encounters (Strauss et al., 2008).

9.5.4 Sea turtles

Similar to freshwater turtles, most sea turtles' shells provide an excellent location to attach a transmitter. Transmitters are generally attached using a marine epoxy to glue the device to the carapace. The Leatherback (*Dermochelys coriacea*) and Flatback (*Natator depressus*), whose soft carapaces and skin may not be suited for contact adhesives, are exceptions. Instead, transmitters are attached to leatherback turtles using plastic 'backpacks' that loop around the animal's front flippers or are anchored to (man-made) holes in either the central ridge of the carapace or the pygal process (Figure 9.2(d)) to trail behind the animal.

Unlike terrestrial species, where the main concern is the weight of the transmitter, the main issue for marine species is how the animal-borne devices will affect the hydrodynamics of the animal. Partly due to the recent discovery that the widely used 'harness' method for attaching satellite transmitters to Leatherback Turtles could increase hydrodynamic drag (Fossette et al., 2008; Jones et al., 2013), substantial efforts to mitigate the effects of attaching devices to sea turtles are now underway. Reductions have been largely achieved by reducing the size of the device, giving them a more hydrodynamic shape, and placing them further towards the posterior of the animal.

VHF transmitters have limited utility for tracking sea turtles because the signal is only available when the turtle is at the surface and the transmitters have limited range. Acoustic transmitters have similar range restrictions, but have the benefit of being able to relay a signal underwater. Localized tagging efforts coupled with the deployment of arrays of acoustic receivers can be an effective way to investigate fine-scale patterns of habitat use (Taquet et al., 2006). Acoustic transmitters also have the additional benefit that they can be small; acoustic nano-tags are the only telemetry devices small enough to be deployed on sea turtle hatchlings (Scott et al., 2014). Hopefully, these devices will help to uncover the current mysteries surrounding the initial movements of sea turtle hatchlings as they leave their nesting beaches. For larger life-stages, satellite telemetry is the only currently available technique truly suited to unveiling long-distance sea turtle movements.

9.6 Future directions

The scope of what can be achieved using telemetry is linked with technological and scientific innovation. Over time telemetry devices will become smaller and function over longer periods due to improved battery life. These advances will enable tracking an increasing number of species under an increasing variety of scenarios. In turn, the questions that can be answered by these techniques will grow.

Free or at least more affordable geographic information systems (GIS) technologies are becoming widely available. Very high resolution imagery and spatial datasets are increasingly common. Coupled with more accessible means for statistically analysing data, the future accuracy of locality overlays onto spatial data should be extremely informative. An exciting new area of research is the exploration of movements using individual-based models (Patterson et al., 2008; Proulx et al., 2013). By informing Brownian motion with empirical data, such as length and frequency of movement derived from telemetry, investigators will be able to address issues such as barriers, corridor selection, and impacts of habitat type on overall movement patterns.

We anticipate that some of the greatest insights that the next generation of tracking devices (and researchers) will relate to the juvenile life stages of many species. The miniaturization of GPS packages and data loggers also will free researchers from having to relocate larger animals as frequently, thus reducing labour costs and disturbances to study animals.

We are also hopeful that harmonic radar tags will find broader application (see Gourret et al., 2011). These devices do not require batteries, relying instead on returning a signal that is a harmonic frequency to that received. Consequently, they can be quite small and lightweight (<0.3 g). Challenges to using them include configuration (protruding wires) and common frequency (no unique identity).

References

Burke, V.J., Greene, J.L., and Gibbons, J.W. (1995). The effect of sample size and study duration on metapopulation estimates for slider turtles (*Trachemys scripta*). *Herpetologica*, **51**, 451–6.

Burt, W.H. (1943). Territoriality and home range concepts as applied to mammals. *Journal of Mammalogy*, **24**, 346–52.

Campbell, H.A., Watts, M.E., Sullivan, S., et al. (2010). Estuarine crocodiles ride surface currents to facilitate long-distance travel. *Journal of Animal Ecology*, **79**, 955–64.

Cherkiss, M.S., Mazzotti, F.J., Hord, L., et al. (2014). Remarkable movements of an American crocodile (*Crocodylus acutus*) in Florida. *Southeastern Naturalist*, **13**, N52–6.

Doody, S, Young, J.E., and Georges, A. (2002). Sex differences in activity and movements in the Pig-nosed Turtle, *Carettochelys insculpta*, in the wet-dry tropics of Australia. *Copeia*, **2002**, 93–103.

Ferrara, C.R., Vogt, R.C., and Sousa-Lima, R.S. (2013). Turtle vocalizations as the first evidence of posthatching parental care in chelonians. *Journal of Comparative Psychology*, **127**, 24.

Fossette, S., Corbel, H., Gaspar, P., et al. (2008). An alternative technique for the long-term satellite tracking of leatherback turtles. *Endangered Species Research*, **4**, 33–41.

Franklin, C.E., Read, M.A., Kraft, P.G., et al. (2009). Remote monitoring of crocodilians: implantation, attachment and release methods for transmitters and data-loggers. *Marine and Freshwater Research*, **60**, 284–92.

Gibson, S.E. (2003). *Movements and habitat preference of the timber rattlesnake* (Crotalus horridus) *in southern Indiana*. M.S. Thesis, Purdue University, USA.

Goodman, R.M., Knapp, C.R., Bradley, K.A., et al. (2009). Review of radio transmitter attachment methods for West Indian rock iguanas (genus *Cyclura*). *Applied Herpetology*, **6**, 151–70.

Gourret, A., Alford, R., and Schwarzkopf, L. (2011). Very small, light dipole harmonic tags for tracking small animals. *Herpetological Review*, **42**, 522–5.

Guilhon, A.V., Vogt, R.C., Schneider, L., et al. (2011). Assessment of turtle tracking technologies in the Brazilian Amazon. *Herpetological Review*, **42**, 525–30.

James, M.C., Eckert, S.A., and Myers, R.A. (2005). Migratory and reproductive movements of male leatherback turtles (*Dermochelys coriacea*). *Marine Biology*, **147**, 845–53.

Jehnrich, R.I., and Turner, F.B. (1969). Measurement of non-circular home range. *Journal of Theoretical Biology*, **22**, 227–37.

Jones, T.T., Van Houtan, K.S., Bostrom, B.L., et al. (2013). Calculating the ecological impacts of animal-borne instruments on aquatic organisms. *Methods in Ecology and Evolution*, **4**, 1178–86.

Kay, W.R. (2005). Movements and home ranges of radio-tracked *Crocodylus porosus* in the Cambridge Gulf region of Western Australia. *Wildlife Research*, 31, 495–508.

Kays, R., Tilak, S., Crofoot, M., et al. (2011). Tracking animal location and activity with an automated radio telemetry system in a tropical rainforest. *The Computer Journal*, **54**, 1931–48.

Knapp, C.R., and Owens, A. (2005). Home range and habitat associations of a Bahamian iguana: implications for conservation. *Animal Conservation*, **8**, 269–78.

Lance V.A., Elsey R.M., Trosclair, P.L., III, et al. (2011). Long-distance movement by American alligators in southwest Louisiana. *Southeastern Naturalist*, **12**, 389–98.

Lentini, A.M., Crawshaw, G.J., Licht, L.E., et al. (2011). Pathologic and hematologic responses to surgically implanted transmitters in eastern massasauga rattlesnakes (*Sistrurus catenatus catenatus*). *Journal of Wildlife Diseases*, **47**, 107–25.

Magnusson, W.E. (1979). Dispersal of hatchling crocodiles *(Crocodylus porosus)* (Reptilia, Crocodilidae). *Journal of Herpetology*, **13**, 227–31.

McMahon, C.R., Collier, N., Northfield, J.K., et al. (2011). Taking the time to assess the effects of remote sensing and tracking devices on animals. *Animal Welfare*, **20**, 515–21.

Mohr, C.O. (1947). Table of equivalent populations of North American small mammals. *American Midland Naturalist*, **37**, 223–49.
Patterson, T.A., Thomas, L., Wilcox, C., et al. (2008). State–space models of individual animal movement. *Trends in Ecology & Evolution*, **23**, 87–94.
Proulx, C.L., Proulx, L., and Blouin-Demers, G. (2013). Improving the realism of random walk movement analyses through the incorporation of habitat bias. *Ecological Modelling*, **269**, 18–20.
Reinert, H.K., and Cundall, D. (1982). An improved surgical implantation method for radio-tracking snakes. *Copeia*, **1982**, 702–5.
Row, J.R., and Blouin-Demers, G. (2006). Kernels are not accurate estimators of home-range size for herpetofauna. *Copeia*, **2006**, 797–802.
Sah, S.A.M., and Stuebing, R.B. (1996). Diet, growth and movements of juvenile crocodiles *Crocodylus porosus* Scheider in the Klias River, Sabah, Malaysia. *Journal of Tropical Ecology*, **12**, 651–62.
Scott, R., Biastoch, A., Roder, C., et al. (2014). Nano-tags for neonates and ocean-mediated swimming behaviours linked to rapid dispersal of hatchling sea turtles. *Proceedings of the Royal Society of London B: Biological Sciences*, **281**, 20141209.
Strauss, M., Botha, H., and Van Hoven, W. (2008). Nile crocodile *Crocodylus niloticus* telemetry: observation on transmitter attachment and longevity. *South African Journal of Wildlife Research*, **38**, 189–92.
Taquet C., Taquet, M., Dempster, T., et al. (2006). Foraging of the green sea turtle *Chelonia mydas* on seagrass beds at Mayotte Island (Indian Ocean), determined by acoustic transmitters. *Marine Ecology Progress Series*, **306**, 295–302.
Thums, T., Whiting, S.D., Reisser, J., et al. (2013). Tracking sea turtle hatchlings—a pilot study using acoustic telemetry. *Journal of Experimental Marine Biology and Ecology*, 440, 156–63.
Tucker, C.R., Strickland, J.T., Edmond, D.S., et al. (2015). Activity patterns of Ornate Box Turtles (*Terrapene ornata*) in northwestern Illinois. *Copeia*, **103**, 502–11.
Waldron, J.L., Bennett, S.H., Welch, S.M., et al. (2006). Habitat specificity and home-range size as attributes of species vulnerability to extinction: a case study using sympatric rattlesnakes. *Animal Conservation*, **9**, 414–20.
Ward, M.P., Sperry, J.H., and Weatherhead, P.J. (2013). Evaluation of automated radio telemetry for quantifying movements and home ranges of snakes. *Journal of Herpetology*, **47**, 337–45.
Warner, D.A., Thomas, J., and Shine, R. (2006). A simple and reliable method for attaching transmitters to lizards. *Herpetological Conservation and Biology*, **1**, 129–31.
Weatherhead, P.J., and Anderka, F.W. (1984). An improved radio transmitter and implantation technique for snakes. *Journal of Herpetology*, **18**, 264–9.
Weatherhead, P.J., and Blouin-Demers, G. (2004). Long-term effects of radiotelemetry on black ratsnakes. *Wildlife Society Bulletin*, **32**, 900–6.
Worton, B.J. (1987). A review of models of home range for animal movement. *Ecological Modeling*, **38**, 277–98.
Wylie, G.D., Smith, J.J., Amarello, M., et al. (2011). A taping method for external transmitter attachment on aquatic snakes. *Herpetological Review*, **42**, 187–91.

Part 3
Sampling Reptiles

10
Surface-dwelling reptiles

John D. Willson

10.1 Introduction

Perhaps the most fundamental aspect of conducting a successful field study of reptiles is selecting a capture method, but how do you choose the appropriate method from the numerous options? Many terrestrial or semi-arboreal lizards are common, conspicuous, and easily sampled using a variety of techniques. Their amenability to sampling contributed to the use of lizards as model organisms in community and physiological ecology. At the opposite end of the spectrum, many snakes are notoriously secretive, rare, cryptic, or otherwise difficult to observe or capture. My objective in this chapter is to introduce the reader to proven methods for sampling a wide variety of surface-dwelling reptiles, as well as touch on notable recent modifications or methodological advances. Each of these techniques has been described in great detail elsewhere (e.g. McDiarmid et al., 2012; Graeter et al., 2013); thus, this chapter aims to give a broad overview of available methods, with particular attention to strengths and weaknesses of each method, and considerations that may be important when attempting to select the optimal methods for the study species/system and question of interest. With these goals in mind, I first outline factors that contribute to sampling success and their relevance to common study objectives. The remainder of the chapter describes specific sampling methods, divided into methods that require observation or capture of free-ranging animals by the researcher (active capture techniques) and those that intercept or attract animals and trap them for subsequent collection (passive capture techniques). Each section concludes with a discussion of important considerations and limitations of each method. Most of the methods described in this chapter have been used extensively for terrestrial and semi-aquatic/freshwater squamate reptiles (other guilds are treated separately in Chapters 11–16), but they are likely effective for other taxa, such as tuatara and terrestrial chelonians.

10.2 Selecting a capture method

Selecting the best capture method for a particular study may seem simple—just use whatever technique will yield the most captures, given the availability of equipment and manpower. In reality, however, overall capture rates are only one factor that should be considered. In many cases, a research project will fail to yield meaningful data, despite

Reptile Ecology and Conservation. Edited by C. Kenneth Dodd, Jr. © Oxford University Press 2016.
Published 2016 by Oxford University Press.

high capture rates, if other factors such as repeatability and sources of bias are not taken into account. To illustrate the importance of careful planning, consider the following scenarios:

1. As part of a management programme for a threatened species of lizard, you set out to assess whether lizard abundance is higher in forests than in nearby agricultural habitats. You send one field team to visually survey for lizards in the forest, and they report back that they saw 100 lizards over the course of a morning. Another team surveys the agricultural habitat in the afternoon and reports observing 50 lizards.

   Despite having observed a large number of lizards at each site, is it valid to conclude that lizards are more abundant in the forest? Probably not. Counts were different, but how much of that difference was due to differences in skill among the two teams of observers or the time of day when the sampling occurred? Are differences consistent across forest and agricultural habitats in general, or specific to the single locations where each survey was conducted? Could differences be due to lizards being easier to observe in the more open vegetation of the forested habitat? These are just a subset of the many potential sources of bias that could confound a simple visual survey like the one described here.
2. You are interested in studying the diet of a large snake species. You have read that drift fences with funnel traps are great for capturing snakes, so you spend most of your research funds building a drift fence and hiring a technician to check the traps daily. After one week of sampling, your technician reports that they have captured the target sample size of adult animals and their stomachs all contain frogs, which are also commonly captured in the traps.

   In this case, the target sample size was easily attained, but was it worth all of the investment in an intensive, highly standardized sampling technique? Are the diet items recovered really representative of what the snakes feed on in the wild, or were they all consumed within the traps? If the frogs are a natural diet item, are they particularly common in the area where the drift fence was located, or are they an important prey item across the entire study site? Why were only adult animals captured? Clearly, some careful consideration during the study design phase might have addressed some of these potential confounding questions.

10.2.1 Study goals and preliminary considerations

Before beginning a field study of reptiles, researchers should clearly define the goals and design of their study. Many excellent resources are available that detail approaches and study designs appropriate for answering a variety of research questions about reptiles (e.g. Chapters 2, 17, and 18; Dorcas and Willson, 2009; Fisher and Foster, 2012; McDiarmid et al., 2012; Rodda, 2012). Specifically, the researcher would be wise to consider the following: How important is it that capture rates reflect true abundance of the species? How important is it that sampling is representative of the entire area being studied? Does the study require that all sizes or sexes be sampled equally? How important are large sample sizes and what is the sampling unit (i.e. is larger sample size achieved through higher numbers of animals, more surveys, more study sites, or a combination of these)? Ultimately, these factors will help in weighing the considerations listed in the following subsections and in selecting the capture method that will yield the most meaningful data.

10.2.2 Capture rates

Nearly any study will require capture or observation of a reasonable sample size of the target species. In some cases (e.g. collecting animals for laboratory or genetic studies), the total number of animals collected may be the only important factor. Some species are common and easily captured using a variety of techniques, allowing the researcher considerable freedom to choose between several methods based on other considerations of the study design. However, other species are not effectively sampled using particular methods, ruling out these techniques immediately. For example, free-standing baited traps are highly effective for some aquatic snakes and carnivorous lizards but are generally ineffective for most terrestrial snakes and insectivorous lizards. Finally, some species may be so secretive or rare that they are only encountered incidentally. In such cases, researchers may be limited in their ability to conduct standardized surveys, regardless of the capture technique used.

10.2.3 Cost and effort

Capture rates must be considered relative to cost of materials, equipment, and time required. Some sampling methods (e.g. visual surveys) are very inexpensive, but may yield low capture rates for some species or suffer from strong biases. Traps are often expensive and time consuming to construct, but thereafter usually yield high capture rates per unit effort. Traps may require less sampling effort per capture than active capture methods, but they absolutely must be checked on a regular basis to avoid harming captured animals.

10.2.4 Repeatability

Although some studies simply require capture of a large sample of animals, other studies hinge on sampling in as standardized, repeatable, and unbiased a manner as possible. Repeatability refers to the comparability of sampling events, independent of differences among observers (see discussion of observer bias in Section 10.2.5). Thus, repeatability reflects how sensitive capture rates are to variation in environmental conditions during the survey. Generally, active capture methods that require a searching observer to detect animals engaged in specific behaviours (e.g. basking) are highly sensitive to conditions at the time of the survey. Traps and other passive capture methods are typically more repeatable because they accumulate captures over time, even when the researcher is not present (Dorcas and Willson, 2009). Repeatability issues are perhaps most critical when raw capture rates are compared among locations or over time (relative abundance comparisons). In such studies, capture rates are assumed to reflect true abundances, and lack of repeatability can blur or even mislead conclusions. Thus, when using comparisons of relative abundance, it is critical to use a method that is as repeatable as possible and to conduct sampling under similar conditions (e.g. weather, season). Most modern methods for assessing presence/absence (occupancy) or population characteristics (capture–mark–recapture (CMR)) require replicated, standardized sampling events (Chapters 26 and 27) and benefit from use of methods that are as repeatable as possible.

10.2.5 Bias

In the context of reptile sampling, bias refers to variation that causes the sample to not be representative of the population at large, especially variation in capture rates that cause them to be poor indicators of true abundance. For example, some methods may under-sample certain sexes or size classes, leading to a biased perception of the demography of the population. An excellent example of demographic bias has been revealed through research on invasive Brown Treesnakes (*Boiga irregularis*) in Guam. Although baited traps have been developed that are highly effective for capturing adults, ontogenetic shifts in prey preference make this method totally ineffective for capturing juveniles (Rodda et al., 2007). One particularly important source of bias is variation in capture rates among observers due to skill, experience, or technique (observer bias). Observer bias can be a major source of confounding variation in capture probability for some techniques, especially those that rely on the skill of the observer to seek out and detect or capture animals. Biases can also result from various aspects of the study species (e.g. activity patterns, habitat associations, prey preferences, variation in coloration or behaviour, response to capture) or the sampling method (e.g. aspects of trap design, area or habitats sampled). For example, when using a technique that targets individuals engaged in particular behaviours (thermoregulation, long-distance movement, foraging), capture rates will reflect not only abundance of the species, but also what fraction of the population is engaged in the particular activity during the time of the survey. Although detailed discussion of sampling biases is beyond the scope of this chapter, particularly important sources of bias are mentioned here. Many modern analytical techniques (e.g. CMR and occupancy analyses; Chapters 26 and 27) allow the researcher to minimize bias by modelling variation in detection probability among species, demographic groups, over time, or with site or sampling covariates. Nonetheless, it is advisable to carefully consider how a particular capture method or sampling scheme might bias your sample, prior to beginning any field project. Minimizing bias up-front through thoughtful study design will greatly simplify interpretation of results and lend power to statistical analyses.

10.3 Active capture techniques

10.3.1 Visual encounter surveys

The most basic survey method for surface-dwelling reptiles is the visual encounter survey (VES). In its simplest form, a VES is an opportunistic search of a study area or habitat, usually targeting microhabitats or environmental conditions favoured by the target species. VESs may be conducted diurnally or nocturnally and are usually constrained by area, time, or effort (e.g. thorough search of a 10 m × 10 m plot; hiking a 300 m transect; 1 man-hour of opportunistic searching; flipping 100 rocks or other natural cover objects) to increase repeatability (Chapter 17). Depending on the biology of the target species, VES may involve simple visual observation while walking through habitat, searching under rocks, logs, leaf litter, or other cover objects with a snake hook or rake, or watching for actively moving animals (e.g. fast-moving lizards) from a fixed vantage point. VESs are best conducted by trained observers with a thorough understanding of the natural history, activity patterns, and habitat use of the target species. Simple VESs are inexpensive, require little equipment, and can be effective for species that are conspicuous or reliably use particular microhabitats. For example, the Dog-faced Water Snake (*Cerberus rynchops*) was found to be abundant in mangrove forest with open muddy pools and accounted for 72.7% ($N = 159$) of homalopsid snakes captured during nocturnal VESs in Singapore (Voris, 2002). When VESs require examination of natural cover objects (e.g. rocks, bark, leaf litter), care should be taken to avoid crushing animals, and cover should be carefully replaced to avoid degradation of these sensitive habitat elements (Goode et al., 2005; Pike et al., 2010).

10.3.2 Cover boards

An extension of a cover object-based VES is the use of artificial cover objects, or 'cover boards'. Herpetologists have long appreciated that many reptiles seek out surface cover, including discarded human trash, for refugia and thermoregulation. Thus, it is possible to deploy artificial cover objects, such as sheets of roofing tin or non-chemically treated plywood boards, that can be easily 'flipped' to sample the reptiles they conceal. Cover boards for reptiles are generally fairly large (>0.5 m × 0.5 m) and may be deployed in a variety of habitats, especially in fields, along habitat edges, or the margins of aquatic habitats (Figure 10.1).

Many herpetologists believe that cover boards improve with age once placed, and some suggest clearing litter under boards to prepare the site or placing tin in stacks to provide a more thermally variable habitat. Cover boards are usually deployed in groups (arrays) or along transects consisting of a standardized number of boards. Cover boards are particularly effective for fossorial species during mild weather (spring), when reptiles presumably seek warming surface cover for thermoregulation. Alternatively, during hot weather, diurnal species may use cover boards as nocturnal refugia and may be targeted

Figure 10.1 Examples of cover boards used to sample surface-dwelling reptiles: (a) terrestrial plywood cover board set along a wetland edge; (b) tin cover board used to sample semi-aquatic snakes in a heavily vegetated wetland; (c) Copperhead (*Agkistrodon contortrix*) found under a terrestrial metal cover board. Photos by J.D. Willson.

by censusing boards at night. In addition to being an effective way to capture many reptile species, cover boards have two major advantages over traditional VES. First, they provide a standardized level of survey effort that can be repeated at multiple sites (including those with limited natural cover) or at the same site over time. Second, they greatly reduce observer biases that can be strong in traditional VES. However, it should be noted that because animals are not trapped, capture rates are still strongly affected by the conditions at the time of the survey, and thus it is advisable to census boards under as similar conditions as possible when comparison among sites or over time is desired (Grant et al., 1992). Further, although the effects of cover boards on reptile populations have not been quantified, there is the possibility that addition of artificial refugia could actually improve the quality of the habitat, thereby allowing for larger population sizes than would occur naturally. Finally, cover boards may be a desirable alternative to natural cover searches when searching of natural cover could detrimentally affect sensitive habitats. For example, plywood boards can be used as an alternative to ripping bark off decaying trees or disturbance of rock features that are easily destroyed. However, it is advisable to deploy cover boards in secluded locations at public sites, in order to avoid boards being removed as trash or scrap metal or being discovered by reptile enthusiasts who might disturb or collect animals.

10.3.3 Road surveys

Road surveys are an additional subset of the visual encounter survey that deserves special mention. Driving slowly (~30–60 km/h) along low-traffic roads that bisect suitable habitat and looking for crossing animals is one of the most reliable methods for detecting many reptile species, and it may be the only feasible method to cost-effectively sample some rare or secretive snakes. For example, the only in-depth ecological study of the Southern Hognose Snake (*Heterodon simus*), a rare species restricted to xeric habitats in the south-eastern USA, was based on over 700 individuals captured using diurnal road surveys (Beane et al., 2014). Road surveys have been used to gather large sample sizes of many snake species and to assess seasonal activity patterns and habitat associations (e.g. Dalrymple et al., 1991; Bernardino and Dalrymple, 1992; Mendelson and Jennings, 1992; Enge and Wood, 2002). Nocturnal road surveys are usually most effective for detecting high species richness, but diurnal surveys may be effective for diurnal snakes, some slow-moving lizards (e.g. *Phrynosoma* spp., *Tiliqua* spp., *Helodermia* spp.), and chelonians. As with cover boards, road surveys provide a highly standardized (by time or distance) survey effort that is less prone to observer bias than VES, but capture rates are still strongly influenced by timing and conditions during the survey. Additionally, despite the efficacy of road surveys for some species, their utility for abundance estimation is questionable because researchers are unable to spread samples randomly or systematically across the area of interest. In fact, the road itself may result in an abundance estimate that is not representative of the overall landscape due to depression of populations from cumulative road mortality or attraction of species to edge habitats or high ground along the road. One added benefit of road surveys is that road-killed animals may be collected as voucher or study specimens without additional harm to the population.

10.3.4 Lizard noosing

Although most snakes and many lizards are easily captured by hand or with a snake hook/tongs once located visually, some fast-moving lizards may require additional tools to capture reliably. One tried-and-true technique for capturing wary diurnal lizards is a 'lizard noose', consisting of a small slip-loop of light twine attached to the end of a long rod, such as a telescoping fishing pole (see Chapter 11). Many difficult-to-catch lizards will allow approach within a distance that allows the noose to be gently lowered around its neck or fore-body. A gentle lift of the rod tightens the noose, and the lizard is caught. Noosing works well for mid-sized lizards with enlarged heads, such as Anoles, *Sceloporus*, Agamids, most Teiids, and Lacertids. However, this technique may not be effective for fossorial or smooth-scaled species without enlarged heads (e.g. smaller skinks, legless lizards) and should be used with care for delicate species, such as many geckos.

10.3.5 Considerations and limitations

Active capture methods, including traditional VES, cover boards, and road surveys are among the most effective and inexpensive survey methods for surface-dwelling reptiles, and they may be the only reliable methods for capturing some species that are rare, secretive, or not effectively trapped. Thus, these techniques are desirable when the primary goal is to document species presence, to simply gather samples for focal animal (e.g. demographic, genetic, telemetry, diet) or laboratory studies, or to rapidly conduct surveys across a large number of sites (Chapters 18 and 26). Despite their efficacy, active methods are generally less standardized and more prone to bias than other capture methods, and thus are less likely to yield meaningful data if the goal is to use capture rates as indices of relative abundance or to allow for meaningful CMR analyses. For example, Rodda et al. (2005) found no correlation between relative abundances of Brown Treesnakes during visual surveys and true abundances estimated via CMR.

10.4 Passive capture techniques

10.4.1 Pitfall traps

Passive capture methods intercept or attract animals and trap them for subsequent collection. Thus, passive sampling methods do not require the presence of the researcher to coincide with animal activity, and some methods can accrue many captures over time. Nearly all traps used for surface-dwelling reptiles fall into two general categories, pitfalls and funnel traps (but see sticky traps for lizards; Chapter 11). Pitfall traps consist of some type of smooth-sided container (most commonly a plastic or metal can or bucket, 1–19 l in volume) sunk in the ground such that the rim is at surface level (Figure 10.2).

Animals that fall into the 'pit' are unable to climb out and are trapped. Due to their limited jumping abilities, small terrestrial chelonians, most small to mid-sized lizards, and small snakes are easily contained in most pitfall traps, but larger lizards and snakes can often climb out and are generally under-sampled (Todd et al., 2007). Pitfalls are most commonly used in conjunction with drift fences (see Section 10.4.3) and are not baited. In order to ensure the health of captured animals, pitfalls should be checked

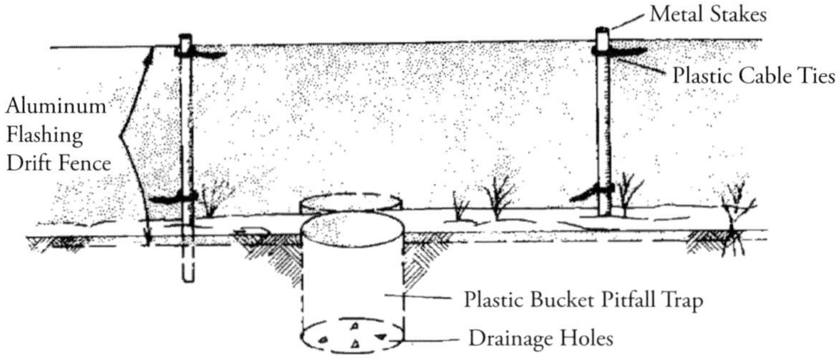

Figure 10.2 Schematic of a terrestrial drift fence with large pitfall traps. Reprinted from Gibbons and Semlitsch (1981).

regularly (usually daily). A large moist sponge or other cover object is usually placed in each pitfall to provide shelter and moisture during dry conditions; a raft can be used if standing water accumulates in the trap. Excess water should be removed from traps whenever it accumulates, and during hot weather, a cover (small board) can be placed on short stilts over the top of the trap or propped against the drift fence to provide shade and to reduce the possibility of predation on captured animals. Drainage holes are often drilled in the bottom of the container, but these may allow escape of small, slender snakes and lizards.

10.4.2 Funnel traps

Funnel traps consist of a funnel-shaped entrance to a larger cage or holding area. Animals are guided through the narrow funnel opening and then are unable to find their way back out. Numerous funnel trap designs have been used to sample surface-dwelling reptiles (Figure 10.3). Commercially available 'minnow traps' are highly effective for sampling many aquatic or semi-aquatic snakes in shallow-water habitats (Keck, 1994; Willson et al., 2011). Similar mesh funnel traps appropriate for terrestrial sampling can be constructed of hardware cloth or window screen (Fitch, 1951). For terrestrial applications and larger species, box-type funnel traps can be constructed of plywood or other materials and outfitted with mesh funnels and latched doors that allow the trap to be checked without being disassembled (Burgdorf et al., 2005). These funnel traps are effective for most snakes and small to mid-sized lizards. However, the maximum and minimum size (diameter) of animals that can be captured are determined by the size of the funnel opening and mesh size, respectively, potentially biasing the size distributions of animals that are captured (Willson et al., 2008). Most chelonians, large lizards, and giant snakes (boas and pythons) are so large that they would be excluded from traditional funnel traps, but some large carnivorous lizards (*Varanus* spp., *Tupinambis* spp.) can be effectively captured in baited 'snap-door' traps typically used for mid-sized mammals (Auliya and Erdelen, 1999). Funnel traps can be effective in a variety of terrestrial, aquatic, and even arboreal habitats, but as with pitfalls, care must be taken to ensure the health of captured animals. This generally means that terrestrial traps should be shaded

Figure 10.3 Several varieties of funnel traps are commonly used to sample reptiles: (a) commercially available plastic 'minnow trap' set in shallow water and allowed to self-bait with fish, amphibian larvae, and invertebrates; (b) screen funnel trap set along one side of a drift fence; (c) box funnel trap set along one side of a drift fence; (d) large box funnel trap located at the centre of a cross-shaped drift-fence array and used to capture secretive upland snakes in Florida, USA. Photos by J.D. Willson (a, c) and Jonathan Mays (b, d).

in hot weather, and aquatic traps must be set shallow enough or floated (e.g. Casazza et al., 2000) such that some part of the trap protrudes above the water. Large box-type traps used in conjunction with drift fences may be large enough that they can be outfitted with cover objects and water sources, allowing for less frequent monitoring, but predation of captured animals on one another may be a problem (Burgdorf et al., 2005).

Funnel traps usually require bait or a drift fence to direct animals into the trap (Figure 10.3). In many cases, traps may be sufficiently 'self-baiting' through bycatch of prey taxa that additional bait is not necessary (Winne, 2005). Alternatively, supplemental bait can be added; Keck (1998) baited stand-alone aquatic minnow traps with dead fish to increase captures of Diamondback Water Snakes (*Nerodia rhombifer*) and Cottonmouths (*Agkistrodon piscivorus*). A variety of stand-alone funnel traps baited with mice, rats, or chicks have been developed to capture nuisance Habu vipers (*Trimeresurus flavoviridis*) in Japan (Hattori, 1999). Years of trap development research has resulted in a highly effective stand-alone funnel trap for invasive Brown Treesnakes on Guam (Rodda et al., 1999). These traps are baited with a live mouse that is maintained within a chamber inside the trap and provisioned with raw potato. It is likely that baited traps are more effective for species that forage actively than for sit-and-wait foragers. Indeed, a baited trap similar to those used for Brown Treesnakes was not effective for invasive Burmese Pythons (*Python molorus*) in Florida, USA (Reed et al., 2011). Unbaited or naturally baited stand-alone funnel traps have generally proven to be ineffective for terrestrial reptiles.

10.4.3 Drift fences

Drift fences (Gibbons and Semlitsch, 1981) are barriers that intercept moving animals and direct them into funnel or pitfall traps. In some cases, landscape features such as rock ledges, logs, or the banks of waterbodies, can function as natural drift fences and stand-alone funnel traps are often most effective when set along such barriers. Alternatively, drift fences can be constructed out of sections of aluminium or galvanized flashing, erosion (silt) fencing, hardware cloth, or other materials, with pitfall or funnel traps set along their length (Figures 10.2 and 10.3). Several studies have compared the efficacy of different trap types and drift fence configurations for sampling reptiles (e.g. Campbell and Christman, 1982; Friend et al., 1989; Hobbs et al., 1994; Crosswhite et al., 1999; Enge, 2001; Todd et al., 2007). Fences are usually buried at least several centimetres into the substrate, but height will depend on the sizes and climbing abilities of the target species and how important it is to prevent trespass over the fence. For example, small terrestrial lizards might only require a fence a few centimetres high, but a fence 1-m tall, or higher, might be needed if the goal is to capture all snakes emerging from a hibernaculum. Likewise, short sections of fence may be adequate for small species, but generating sufficient captures of large or uncommon species, such as many large snakes, may necessitate fences at least 100 m long. Drift-fence arrays can be constructed in a variety of configurations; popular variations include cross-shaped arrays with a large box trap at the centre of the array and/or smaller funnel or pitfall traps at the end of each arm; linear arrays along habitat edges; or circular arrays around isolated features such as hibernacula or wetlands (Figure 10.4).

Frequently, the array is used as the sampling unit in statistical comparisons, and thus several arrays are often set in different 'treatments' (e.g. habitat types, sites). Because drift fences intercept moving animals, they can be particularly useful for assessing activity patterns or shifts in habitat use. For example, several studies have used drift fences encircling wetlands to document movement of animals in and out of wetlands in association with drought (Dodd, 1993; Seigel et al., 1995; Willson et al., 2006). Likewise, seasonal shifts in activity associated with male mate-searching behaviour are often discernible in drift-fence capture data (Todd et al., 2008).

10.4.4 Considerations and limitations

Passive traps generally suffer from fewer sources of bias than do active surveys. For example, because traps can integrate captures over time, they tend to 'average out' the effects of short-term shifts in activity on capture rates. Likewise, when properly constructed and set, traps are relatively insensitive to observer biases. Traps can be effective for capturing some highly secretive species that do not use habitats that are visually searchable. In fact, there are many cases where drift fences or stand-alone traps have captured rare species not known to be present at a site or have revealed that secretive species are much more common than had been previously suspected. Despite these

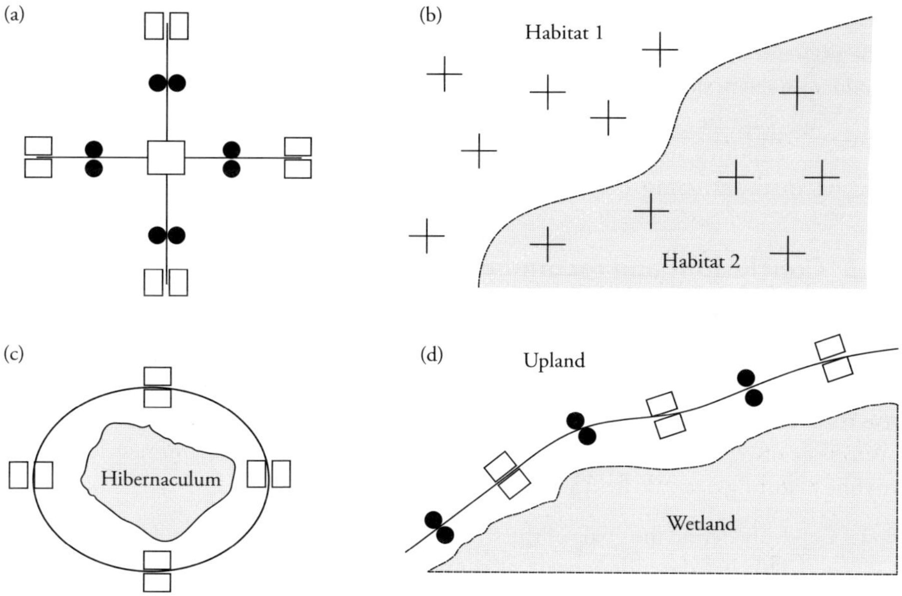

Figure 10.4 Examples of drift fence and passive trap configurations used to sample reptiles. Solid lines represent sections of drift fence, filled circles represent pitfalls, and open rectangles represent funnel traps. (a) Cross-shaped array with central box trap, pitfall, and funnel traps; (b) replicated arrays deployed to compare reptile communities in two habitat types; (c) circular drift fence deployed to capture snakes emerging from a hibernaculum; (d) linear drift fence along a wetland edge to assess movement between wetland and upland habitats.

strengths, the major drawback of passive traps is that they require a substantial initial investment of time and equipment. In forested areas, drift fences also require constant maintenance to repair damage caused by falling timber. Baited stand-alone traps are not as expensive but are only effective for some species. Traps must absolutely be monitored regularly to avoid harm to captured animals, and the potential exists for complications that result in death of numerous animals. Some common complications or sources of mortality include flooding or overheating of traps, predation (especially by insects [ants and beetles] and carnivorous mammals), fire damage, and vandalism (by humans or animals, e.g. raccoons, bears, or crocodilians). Researchers should also appreciate the tendency of passive traps to capture substantial by-catch of non-target species. These may include other reptile and amphibian species, small mammals, and numerous species of invertebrates. Care and safety precautions (e.g. avoiding blindly searching traps with bare hands) must be used when checking traps, as a variety of potentially dangerous taxa (e.g. spiders, scorpions, fire ants, velvet ants, venomous snakes, small mammals) are frequently captured in pitfall and funnel traps. The effort required to monitor traps makes passive methods impractical for sampling in remote locations and can limit the researcher's ability to replicate sampling events over short periods of time. Additionally, passive capture methods usually do not allow the researcher to target specific individuals or demographic groups that may be desired for a particular project. Finally, researchers should appreciate that different trap types sample animals engaged in different behaviours. Thus, capture rates in baited traps are directly related to foraging activity, whereas drift fence captures are reflective of various behaviours that prompt movement (e.g. foraging, mate-searching, dispersal/migration).

10.5 Conclusions and recommendations

This chapter has provided an overview of well-established methods for field sampling of surface-dwelling reptiles. In describing the basic methods, I have attempted to pay particular attention to the strengths and weakness of each method for sampling various species and for yielding capture data that will be most appropriate for addressing common goals of reptile field studies. I will close with several final recommendations for anyone planning a reptile field study.

1. Carefully define the goals of the study and consult resources (e.g. Chapters 2, 17, and 18) that describe aspects of study and sampling design that are critical for successful analysis of the data that will be collected.
2. Consider how important large sample sizes (numbers of animals captured) are relative to number and repeatability of surveys and minimization of bias.
3. Thoroughly research the biology and natural history of your study organism(s) and consult authorities, if possible, to reveal important species/system-specific sampling considerations beyond the scope of this chapter.
4. Once you have settled on a sampling method, seek training from someone who has experience with that method in similar field systems and be sure that all individuals involved in sampling receive standardized training.

References

Auliya, M.A., and Erdelen, W. (1999). A field study of the water monitor lizard (*Varanus salvator*) in West Kalimantan, Indonesia: new methods and old problems. *Mertensiella*, **11**, 247–66.

Beane, J.C., Graham, S.P., Thorp, T.J., et al. (2014). Natural history of the southern hognose snake (*Heterodon simus*) in North Carolina, USA. *Copeia*, **2014**, 168–75.

Bernardino, F.S., and Dalrymple, G.H. (1992). Seasonal activity and road mortality of the snakes of the Pa-Hay-Okee wetlands of Everglades National Park, USA. *Biological Conservation*, **62**, 71–5.

Burgdorf, S.J., Rudolph, D.C., Conner, R.N., et al. (2005). A successful trap design for capturing large terrestrial snakes. *Herpetological Review*, **36**, 421–4.

Campbell, H.W., and Christman, S.P. (1982). Field techniques for herpetofaunal community analysis. In N.J. Scott, Jr. (ed) *Herpetological Communities*. Wildlife Research Report 13. Washington, DC:U.S. Department of Interior, Fish and Wildlife Service, pp. 192–200.

Casazza, M.L., Wyle, G.D., and Gregory, C.J. (2000). A funnel trap modification for surface collection of aquatic amphibians and reptiles. *Herpetological Review*, **31**, 91–2.

Crosswhite, D.L., Fox, S.F., and Thill, R.E. (1999). Comparison of methods for monitoring reptiles and amphibians in upland forests of the Ouachita Mountains. *Proceedings of the Oklahoma Academy of Science*, **79**, 45–50.

Dalrymple, G.H., Steiner, T.M., Nodell, R.J., et al. (1991). Seasonal activity of the snakes of Long Pine Key, Everglades National Park. *Copeia*, **1991**, 294–302.

Dodd, C.K., Jr. (1993). Population structure, body mass, activity, and orientation of an aquatic snake (*Seminatrix pygaea*) during a drought. *Canadian Journal of Zoology*, **71**, 1281–8.

Dorcas, M.E., and Willson, J.D. (2009). Innovative methods for studies of snake ecology and conservation. In S.J. Mullin and R.A. Seigel (eds) *Snakes: Applied Ecology and Conservation*. Ithaca, NY:Cornell University Press, pp. 5–37.

Enge, K.M. (2001). The pitfalls of pitfall traps. *Journal of Herpetology*, **35**, 467–78.

Enge, K.M., and Wood, K.N. (2002). A pedestrian road survey of an upland snake community in Florida. *Southeastern Naturalist*, **1**, 365–80.

Fisher, R.N., and Foster, M.S. (2012). Study design and sampling. In R.W. McDiarmid, M.S. Foster, C. Guyer, et al. (eds) *Reptile Biodiversity: Standard Methods for Inventory and Monitoring*. Berkeley, CA:University of California Press, pp. 27–50.

Fitch, H.S. (1951). A simplified type of funnel trap for reptiles. *Herpetologica*, **7**, 77–80.

Friend, G.R., Smith, G.T., Mitchell, D.S., et al. (1989). Influence of pitfall and drift fence design on capture rates of small vertebrates in semi-arid habitats of Western Australia. *Wildlife Research*, **16**, 1–10.

Gibbons, J.W., and Semlitsch, R.D. (1981). Terrestrial drift fences with pitfall traps: an effective technique for quantitative sampling of animal populations. *Brimleyana*, 7, 1–16.

Goode, M.J., Horrace, W.C., Sredl, M.J., et al. (2005). Habitat destruction by collectors associated with decreased abundance of rock-dwelling lizards. *Biological Conservation*, **125**, 47–54.

Graeter, G.J., Buhlmann, K.A., Wilkinson, L.R., et al. (eds) (2013). *Inventory and Monitoring: Recommended Techniques for Reptiles and Amphibians*. Technical Publication IM-1. Birmingham, AL: Partners in Amphibian and Reptile Conservation.

Grant, B.W., Tucker, A.D., Lovich, J.E., et al. (1992). The use of coverboards in estimating patterns of reptile and amphibian biodiversity. In D.R. McCullough and R.H. Barrett (eds) *Wildlife 2001*. London: Elsevier Science, pp. 379–403.

Hattori, S. (1999). Development of the box trap for Habu. In G.H. Rodda, Y. Sawai, D. Chiszar, et al. (eds) *Problem Snake Management: The Habu and the Brown Treesnake*. Ithaca, NY:Cornell University Press, pp. 257–63.

Hobbs, T.J., Morton, S.R., Masters, P., et al. (1994). Influence of pit-trap design on sampling of reptiles in arid spinifex grasslands. *Wildlife Research*, **21**, 483–9.

Keck, M.B. (1994). A new technique for sampling semi-aquatic snake populations. *Herpetological Natural History*, **2**, 101–3.

Keck, M.B. (1998). Habitat use by semi-aquatic snakes at ponds on a reclaimed strip mine. *Southwestern Naturalist*, **43**, 13–19.

McDiarmid, R.W., Foster, M.S., Guyer, C., et al. (eds) (2012). *Reptile Biodiversity: Standard Methods for Inventory and Monitoring*. Berkeley, CA:University of California Press.

Mendelson, J.R., and Jennings, W.B. (1992). Shifts in the relative abundance of snakes in a desert grassland. *Journal of Herpetology*, **26**, 38–45.

Pike, D.A., Croak, B.M., Webb, J.K., et al. (2010). Subtle—but easily reversible—anthropogenic disturbance seriously degrades habitat quality for rock-dwelling reptiles. *Animal Conservation*, **13**, 411–18.

Reed, R.N., Hart, K.M., Rodda, G., et al. (2011). A field test of attractant traps for invasive Burmese pythons (*Python molurus bivittatus*) in southern Florida. *Wildlife Research*, **38**, 114–21.

Rodda, G. (2012). Population size and demographics. In R.W. McDiarmid, M.S. Foster, C. Guyer, et al. (eds) *Reptile Biodiversity: Standard Methods for Inventory and Monitoring*. Berkeley, CA:University of California Press, pp. 283–322.

Rodda, G.H., Campbell, E.W.I., Fritts, T.H., et al. (2005). The predictive power of visual searching. *Herpetological Review*, **36**, 259–64.
Rodda, G.H., Fritts, T.H., Clark, C.S., et al. (1999). A state-of-the-art trap for the brown treesnake. In G.H. Rodda, Y. Sawai, D. Chiszar, et al. (eds) *Problem Snake Management: The Habu and the Brown Treesnake*. Ithaca, NY:Cornell University Press, pp. 285–305.
Rodda, G.H., Savidge, J.A., Tyrrell, C.L., et al. (2007). Size bias in visual searches and trapping of brown treesnakes on Guam. *Journal of Wildlife Management*, **71**, 656–61.
Seigel, R.A., Gibbons, J.W., and Lynch, T.K. (1995). Temporal changes in reptile populations: effects of a severe drought on aquatic snakes. *Herpetologica*, **51**, 424–34.
Todd, B.D., Willson, J.D., Winne, C.T., et al. (2008). Ecology of the southeastern crowned snake (*Tantilla coronata*). *Copeia*, **2008**, 388–94.
Todd, B.D., Winne, C.T., Willson, J.D., et al. (2007). Getting the drift: examining the effects of timing, trap type, and taxon on herpetofaunal drift fence surveys. *American Midland Naturalist*, **158**, 292–305.
Voris, H.K. (2002). Ecology of oriental-Australian rear-fanged water snakes (Colubridae: Homalopsinae) in the Pasir Ris Park mangrove forest, Singapore. *The Raffles Bulletin of Zoology*, **50**, 487–98.
Willson, J.D., Todd, B.D., and Winne, C.T. (2011). Ecological and methodological factors affecting detectability and population estimation in elusive species. *Journal of Wildlife Management*, **75**, 36–45.
Willson, J.D., Winne, C.T., Dorcas, M.E., et al. (2006). Post-drought responses of semi-aquatic snakes inhabiting an isolated wetland: insights on different strategies for persistence in a dynamic habitat. *Wetlands*, **26**, 1071–8.
Willson, J.D., Winne, C.T., and Keck, M.B. (2008). Empirical tests of biased body size distributions in aquatic snake captures. *Copeia*, **2008**, 401–8.
Winne, C.T. (2005). Increases in capture rates of an aquatic snake (*Seminatrix pygaea*) using naturally baited minnow traps: evidence for aquatic funnel trapping as a measure of foraging activity. *Herpetological Review*, **36**, 411–13.

11

Arboreal and fossorial reptiles

Robert W. Henderson, Robert Powell, Jose Martín, and Pilar Lopez

11.1 Arboreal reptiles

11.1.1 Introduction

Researchers working with, surveying, and/or collecting arboreal reptiles (largely limited to lizards and snakes) are faced with challenges that are mostly absent from projects that focus on ground-dwelling species. Whereas researchers rarely exceed 2.0 m in height, arboreal habitat frequently exceeds 30 m. Furthermore, although some arboreal species may have a narrow (2–3 m) vertical range of activity over a 24-hour period, others might spend the day resting at 10–20 m above ground, but forage at 0.5–3 m at night. In addition to the vertical challenge, the three-dimensional nature of tree crowns and forest canopies provides various obstacles, including a potentially unstable substrate and observations impeded by leafy vegetation. Many techniques have been developed since the days when the only means of observing or sampling arboreal reptiles was to look up while standing on the ground, lifting or knocking a reptile out of the tree with a long pole or branch, climbing into the tree (with or without a ladder), cutting down the tree or branch with a saw, or shooting the desired animal.

Before undertaking any project in which data are to be quantitatively analysed, the research protocol must assure that data are collected uniformly and clearly understood. This applies whether the project is community based or focuses on a single species. Arboreal habitats provide an extremely wide variety of microhabitats (e.g. tree trunks and branches of various diameters and surface textures; use of retreats under peeling bark, in holes in trunks or branches, and in epiphytes) at various perch heights and under varying amounts of canopy cover. All of these variables must be taken into consideration when developing protocols.

Although some methods might apply to more than one situation, chosen techniques will vary depending on whether the protocol involves random sampling or targets a particular species and whether it calls for the collection of specimens or live animals. Regardless, in addition to the information that follows, biologists considering research with arboreal reptiles are encouraged to review Das (2012) for more details regarding various techniques.

11.1.2 General methods

These methods deal primarily with gaining access to greater heights in order to observe or capture animals and must be combined with other methods to collect specimens.

Climbing trees

Although climbing can be a simple means of sampling arboreal habitat, it is often fraught with hazards, and the dangers involved might offset the benefits. Besides problems involved in actually climbing a tree (e.g. rotten branches), one is likely to encounter vegetation with spines and thorns, biting and stinging insects, or other unpleasant surprises. Tree-climbing biologists must plan carefully, dress appropriately, and do whatever is necessary to prevent accidents. Only experienced climbers should attempt ascending tall trees. However, if accomplished successfully (safely), tree-climbing can provide access to a reptilian fauna not seen at or near ground level, or garner data on species that divide their time between the canopy and heights closer to the ground.

Ladders

Wooden or aluminium ladders can be used for sampling microhabitats that are inaccessible when working at ground level. They facilitate, for example, the examination of epiphytes (e.g. bromeliads), tree holes, peeling bark, and birds' nests. In certain habitats, a ladder may reach into the canopy and expedite collecting by various means.

Tree towers, canopy walkways, canopy cranes, and canopy rafts

Das (2012) described permanent towers of wood or metal that stand in a few areas, primarily for botanical or meteorological studies, that can provide access to an arboreal community. He also described aerial walkways or catwalks that are useful for vertical sampling, but ideal for visual surveys at a given elevation. Construction cranes with a suspended personnel basket can provide a flexible three-dimensional system capable of monitoring various heights in a single visit, and rafts of helium balloons with air-inflated beams and interspersed netting connected by ropes permit observers to extend their stay in the canopy. Although all of these methods have the potential for making new observations or collecting otherwise rarely seen species, access to situations where these methods can be employed are rare and the cost of implementing them is high. Consequently, they are of limited utility for most biologists.

Tree cutting

Trees being felled provide an opportunity to sample canopy reptiles that might otherwise be impossible to survey or collect. Areas where commercial logging is in progress can be especially productive. On a smaller scale, a small folding hand saw can be useful for cutting saplings or branches that are known to harbour desired animals.

11.1.3 Collecting methods

Specific methods will vary depending on whether specimens are to be collected or animals are to be released after marking. The following are arranged so those that are most likely to be lethal to the animal are listed last.

Hand capture

With the obvious exceptions of dangerously venomous species, grabbing arboreal snakes by hand is very efficient (although bites might ensue). Lizards are often more difficult to capture by hand, but some simple modifications can increase the efficiency of hand capture for certain taxa. Nordberg and Schwarzkopf (2015) used tree bark as artificial cover boards to increase hand capture of several lizard species in Australia. In Costa Rica, Fitch (1973) used a straight, smooth stick about 1.5 m in length to 'freeze' *Anolis* lizards on tree trunks and branches. The stick is slowly extended towards the lizard and within several centimetres in front of its head. It is then drawn slowly back and forth in a horizontal plane, in theory simulating a foraging snake. When the lizard's attention becomes focused on the moving stick, capture by hand is facilitated. Horn and Hanula (2006) described the collection of several lizard species under burlap bands used for monitoring insects in mature pine forests. Paterson (1998) used bridal veil fabric to facilitate capture of *Anolis distichus*. The anoles were reluctant to pass a barrier of the fabric wrapped several times around the trunk (providing an overhang; Figure 11.1).

Similarly, wrapping an arm around a tree trunk often restricts lizards, loath to climb over an obstruction, to heights where hand capture is possible. Hamilton et al. (2007) used mosquito netting flush around a tree and extended out from the trunk to catch lizards climbing down the trunk, thrown down by persons climbing the tree, or knocked from arboreal perches with a blowgun. Once lizards hit the mesh, they could quickly be encapsulated and captured. For males of territorial species, a tethered male can be introduced low on the trunk of a tree inhabited by a resident male; the latter will often descend from high in the crown to drive away the intruder (e.g. Trivers, 1976). Similarly, tethered females can be used to entice males to descend to within reach (e.g. Garcia et al., 1994). Fitch and Henderson (1987) placed *Anolis bahorucoensis* in a clear, inflated, plastic bag, set the bag in the crook of a tree, and observed the bag to quickly attract male *A. cybotes* that would attack the bag, trying to get at its contents. Das (2012) described the use of laser pointers to attract arboreal reptiles such as gekkonids. Perhaps mistaking the projected light as food or a potential predator, lizards can be lured to a

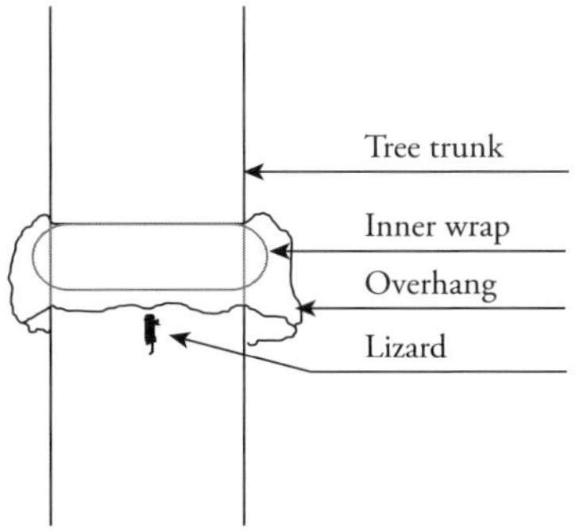

Figure 11.1 Paterson (1998) found that anoles were reluctant to pass a barrier of bridal veil fabric wrapped several times around a tree trunk and providing an overhang (used with permission of the author).

lower height and captured by a co-worker. Care must be taken to not point the laser beam directly at the eyes of co-workers or study animals.

Nooses

Noosing techniques can be simple and effective. Most involve a noose on some sort of pole. Children around the world apparently have been capturing lizards with nooses made from stems of grass for some time. The statue Apollo Sauroctonus from 350 to 340 BC appears to depict a youth trying to noose a lizard (Eimer, 1882), and Wood (1863) mentioned children catching anoles with nooses. Today, biologists collecting arboreal lizards often favour extensible fishing poles; many are less than 45 cm in length when collapsed, but capable of telescoping to over 4 m when extended. These, however, lack the strength necessary for larger lizards. Vaughan et al. (2007) used a 2.0 m aluminium pole to capture the corytophanid lizard *Basiliscus plumifrons*, and catch-poles, more frequently associated with dogs or crocodilians, may be used to noose iguanas or varanids. Preferred noose materials range from grass stems, coconut frond fibres, dental floss, and various types of fishing line to small-diameter rope and nylon-covered wire for larger species. Caution is advised when using fine lines for collecting even moderately sized animals, as small-diameter monofilament, for example, can easily inflict injuries on captured lizards.

Bertram and Cogger (1971) constructed a 'noosing gun' in which a retractor band tightens the noose, but which can be used safely even for very small lizards. Bennett et al. (2001) described a more elaborate technique developed specifically for capturing *Varanus olivaceus*. It includes placing a barrier around a tree trunk and then constructing an elaborate (and clever) concoction of triggers, a tensioned branch, and, of course, a noose.

Extensible poles

Poles can be used to lift, noose, or knock a lizard or snake from a tree. Poles can simply be convenient branches (e.g. bamboo) or commercially produced products. Extensible (telescoping) poles manufactured for working on utility lines are exceptionally useful; when collapsed the unit is about 1.2 m long, but can be extended to about 6.0 m. They work especially well when collecting snakes with prehensile tails (e.g. *Corallus* spp.; Henderson, 2015; Figure 11.2). Poles can be modified with hooks or nooses, depending on what species is being sampled. Durtsche (1996) fastened a portion of glue board to a pole and was able to collect a variety of small lizards. On a smaller scale, clamp sticks, often used for handling venomous reptiles, or modified bolt retrievers (Witz, 1996) can be used for species encountered closer to the ground.

Drift fences

These usually consist of vertical surfaces (wood, plastic, aluminium) extending along the ground and more or less directing small animals along a path until they drop into a pitfall trap (e.g. a plastic bucket with smooth sides). Vogt (1987) demonstrated that drift fences could be set in a forest canopy and he was able to collect a variety of lizards and snakes (as well as amphibians) with this method (Figure 11.3(c)). Using drift

Figure 11.2 An extensible (telescoping) pole is here used to remove a tree boa (*Corallus grenadensis*) from a tree at a site in central Grenada. These poles are especially efficient for capturing snakes with prehensile tails (photo by R.A. Sajdak).

fence material of flashing (lubricated with oil or Vaseline) in concert with cricket-baited funnel traps, Davis et al. (2008) were successful in collecting arboreal gekkonids and scincids in Australia.

Baited traps

Traps have been developed to capture a variety of arboreal reptiles, but often to target particular species rather than random sampling. Zani and Vitt (1995) used modified minnow traps to collect treehole-dwelling *Uracentron flaviceps* in Ecuador. In Australia, Davis et al. (2008) used crickets as bait in funnel traps in conjunction with drift fences to collect gekkonids and scincids. On Guam, Vice et al. (2005) compared three trap designs (baited with live mice) based on modified minnow or crayfish traps for the capture of Brown Treesnakes (*Boiga irregularis*).

Fishing

Krysko (2000) used a fishing rod and a small 'barbless' hook baited with dead dragonflies to successfully (nearly 95% success rate) 'fish' giant anoles (*Anolis equestris*) from trees. Benefits include low expense, not physically demanding, efficient, all size classes can be sampled, and habitats are not disturbed.

Figure 11.3 (a) Attaching a sticky trap to a limb to catch arboreal lizards on Guam. (b) Sticky trap in place. (c) Arboreal drift fence used to sample canopy reptiles in an evergreen rain forest in southern Veracruz, Mexico ((a, b) photos by Bjorn Lardner; (c) photo by Richard Vogt).

Canopy fogging

Arboreal invertebrates have been collected by fogging the canopy of a tree with biodegradable pesticides of the pyrethrin group. Incidental to the collection of invertebrates, canopy-fogging in northern Borneo resulted in the collection of two genera of arboreal skinks (*Sphenomorphus* and *Lipinia*) (Das, 2012). Further studies on the applicability of this technique are needed, as well as data on the effects of the chemicals (which are arthropod-specific) on reptiles.

Adhesive traps

Adhesive traps (also known as glue traps, sticky traps, or mouse glue traps) are especially useful for collecting species that avoid capture by hand or noosing (Figure 11.3(a, b)). Bauer and Sadlier (1992) were able to collect specimens of an arboreal gecko on a small island in the southwestern Pacific that they had been unable to collect by hand.

Although useful in tropical rainforest habitats (Ribeiro-Júnior et al., 2006), glue traps lose their efficiency after a hard rain (Zani and Vitt, 1995). Glor et al. (2000) provided a quantitative assessment of glue-trap sampling, including cost effectiveness, based on a study in the Dominican Republic. The traps were especially successful trapping *Anolis* lizards with traps placed on tree trunks at three heights, but also collected several snake species including juveniles of the boa *Chilabothrus striatus*. On Guam, Rodda et al. (1993) captured lizards and snakes with masses up to about 100 g. Drawbacks of adhesive traps include relatively high mortality due to exposure or predation, inability to release animals unless carefully cleaned, and collecting unwanted species (including birds and mammals). Vargas et al. (2000) compared mortality rates in three methods of collecting lizards (*Anolis carolinensis*). Of the three, adhesive traps had by far the highest mortality rate (47.6%), followed by rubber bands (25%) and noosing (0%).

Rubber bands and sling shots

Large rubber bands (size 107, ca. 12.5 × 1.5 cm), especially when two are looped together to create a more efficient (i.e. powerful) projectile, can be used to dislodge lizards from elevated perches at distances to about 6 m. Commercially available or homemade rubber-band guns generally use smaller bands suitable only for small lizards at shorter distances. The impact often stuns the animal, making it easy to grab after it falls. However, lizards can be injured or even killed (Vargas et al., 2000), so this method is not suitable for catch-and-release studies. Likewise, homemade or commercially produced slingshots have the ability to stun or kill small lizards, depending on species and type of projectile used.

Blowguns

Aboriginals have long used blowguns to capture small animals (Yost and Kelley, 1983). Cherokee boys, for example, killed 'mice, fence lizards, and other small critters which would be added to the soup kettle' (Freeman-Witthoft, 1992). Tinkle and Lawrence (1956) noted the effectiveness of blowguns for collecting 'swift terrestrial and arboreal lizards', especially when large samples are needed. With experience, 'an accuracy of 80–90% is common'. They also indicated that blowguns are as effective as shot pistols, cheaper to obtain and operate, and not objectionable to landowners who do not allow the use of firearms. Blowguns and darts are commercially available, but blowguns can be inexpensively made from lightweight materials such as aluminium or electrical conduit. They can, with practice, be an efficient means of capturing reptiles, especially lizards. Blunt projectiles can stun or kill with the force of impact, and pointed projectiles are designed to be lethal. Blowguns are 'most effective at distances of 5 to 10 m' (Fitzgerald 2012). Keeley and Keeley (2012) used blunt, sticky blowgun darts to collect genetic samples from mammals (a technique potentially transferable to reptiles), and noted that blowgun use is difficult during heavy rain or high winds.

Firearms

A 0.22-calibre revolver or rifle loaded with cartridges of dust shot is effective for collecting lizards and snakes (e.g. Barbour, 1946). According to Fitzgerald (2012), #12 shot can kill 'large snakes at distances of about 6 m. Shooting small lizards and snakes with a revolver at distances of 3 to 6 m causes surprisingly little external damage, small specimens shot at close range are visibly damaged'. Local regulations may preclude the use of firearms.

11.2 Fossorial reptiles

11.2.1 Introduction

As many as 28% of the world's squamates, that is, more than 2000 species including many skinks, legless lizards, blind snakes, and amphisbaenians, are fossorial or semi-fossorial and spend all or most of their lives underground (Measey, 2006). The biology, ecology, and conservation status of fossorial species is much less well understood than that of their epigeal relatives because of the apparent low densities of some fossorial reptiles and the difficulty in finding and sampling individuals (Böhm et al., 2013). Most encounters with fossorial reptiles occur opportunistically when researchers are looking for other species; not surprisingly, studies focused on fossorial reptiles are rare. The study of fossorial reptiles is important and rewarding, however, because these species face different ecological challenges than epigeal reptiles, and such challenges are often solved by very peculiar morphological, functional, and behavioural adaptations to living underground. Moreover, fossorial species may be at particular risk from anthropogenic disturbances affecting soils and landscapes, with the result that local extinction of fossorial reptile populations may be occurring unnoticed.

Sampling techniques for fossorial reptiles can be divided into active searching by digging or by flipping rocks or artificially provided cover boards, or by passive below-ground trapping. The choice of a method depends on the characteristics of the area's habitat and the objectives of the study. Methods may vary depending on whether researchers are interested solely in confirming the presence of one or more species in an area, whether they seek to make quantitative estimates of abundance, or whether they are performing more in-depth ecological studies. The latter may require finding large number of individuals in a local population at different times, or even marking and recapturing the same individuals on different occasions.

11.2.2 Active searching

Fossorial reptiles are often found by flipping rocks or fallen logs and other surface objects, including debris and material of domestic or industrial origin in anthropic areas, or artificial cover objects (e.g. cover boards) that have been placed in advance (see Chapter 10). A second more laborious method is to actively dig in the open soil or deep leaf litter within favourable locations. These techniques can be combined during intensive searches of an area, and require finding the animals at the sampled location at the same moment that researchers are looking for them. Thus, the effectiveness of these sampling techniques depends on a researcher's ability and prior knowledge in order to select the microhabitats and times of the day when animals might be found at specific locations (McCoy et al., 1999).

Flipping surface objects

In areas with abundant cover of small to medium-sized rocks, slowly and cautiously lifting and turning them or other surface objects can be an effective method to find amphisbaenians or fossorial snakes and lizards, such as skinks, during general surveys. Unfortunately, researchers usually need to turn many rocks or objects to find only a few individuals. Fossorial reptiles are often found directly under the rock or semi-buried in the substrate under a rock, so additional gentle 'probing' with the fingers or a stick in the loose substrate may reveal the presence of hidden individuals (Civantos et al., 2013). Some species remain more or less immobile for long periods, but others tend to flee quickly underground or away from the rock immediately after it has been turned. Thus, researchers should be alert to quickly catch the animals by hand, something that can be difficult when simultaneously holding the cover object. Many times hand-capture is feasible only when reptiles are cold early in the morning, as capture becomes increasingly difficult as the animals attain optimal body temperatures and are able to escape rapidly. Two persons working together when sampling cover objects can be advantageous—one to lift and one to catch. Researchers should be alert to the possible presence of dangerous animals under rocks, such as spiders, scorpions, or venomous snakes. Special care should be taken to replace all rocks in their original position, including covering the edges with earth or leaf litter, because microclimatic conditions under rocks or other surface objects might be otherwise altered, thus negatively affecting many soil animals and possibly resulting in later avoidance of the rock by reptiles (Schlesinger and Shine, 1994). Naturally, avoidance would introduce a future sampling bias in mark–recapture studies. Rocks should be replaced carefully to avoid crushing other animal life under them.

Fossorial reptiles are often found under rocks because they preferentially use rocks to attain an optimal body temperature through thigmothermy (i.e. direct rock-to-body heat transfer) (López et al., 1998, 2002). Fossorial reptiles also may go under rocks to forage, as many invertebrates are found under rocks but are less abundant in other open-soil areas (López et al., 1991). The daily and seasonal activity periods of fossorial reptiles should be considered with respect to rock use (Díaz-Paniagua et al., 1995; López et al., 1998), not only to avoid investing needless time and effort in flipping rocks, but also to use this technique accurately to assess the presence or densities of animals. During hot and dry periods, for example, fossorial animals often move deeper underground or to cooler or more moist soil under bushes or other vegetation and are not found under rocks. For species whose life histories are better known, specific patterns of microhabitat and rock selection (Martín et al., 1991, 2013) should be considered when choosing which areas to sample and which rocks to flip. The number of individuals found in relation to search effort (time invested in the survey or number of rocks flipped) can be used as a simple semi-quantitative index of abundance with the results presented as *catch per unit effort*.

Cover boards

In areas where rock cover is scarce, some researchers have positioned artificial cover boards, such as small pieces of untreated plywood, metal, plastic, or tiles (Grant et al.,

1992; Díaz-Paniagua et al., 1995; Sutton et al., 1999), throughout a well-defined grid pattern (see also Chapter 10). The size of cover boards has been variable among studies (e.g. 30 × 30 cm^2, 60 × 60 cm^2, 66 × 133 cm^2). Cover boards may create special microhabitat conditions that attract reptiles, just as if they were natural rocks or logs, and usually are more effective the longer they remain in place. After a number of days (e.g. one week), the cover boards can be flipped and checked regularly thereafter as if they were natural surface objects. Researchers periodically look for animals underneath the boards or buried in the substrate; it may be necessary sometimes to dig under the cover board to some moderate depth (e.g. 15–20 cm) to capture hidden reptiles. Using artificial cover objects may be inexpensive and easy to put out, but for some species or in some areas they are not as effective as other methods for capturing fossorial reptiles (Kuhnz et al., 2005). As a result, using artificial cover objects might fail to establish the presence of a given fossorial species having small population sizes, especially when used during short-term studies or with a low number of cover boards. Even if animals are not observed directly, however, the presence or absence of typical tracks or slither patterns of fossorial reptiles under cover boards has been used to estimate abundance, activity patterns, and microhabitat use (Sutton et al., 1999). When cover boards are effective in attracting a species, their placement (e.g. distributed along different microhabitats in similar or proportional numbers) may allow for more comprehensive spatial sampling than when relying upon detection under natural rocks or other cover objects.

Digging

Some amphisbaenians and other fossorial reptiles can be found by digging the soil with hoes (e.g. a forged metal blade set at right angles to a wooden handle) to approximately 30 cm depth in favourable areas (Gomes et al., 2009). Deeper excavations may not be necessary because most fossorial reptiles usually live within the leaf litter and first few centimetres of soil. This method can, however, require much labour and time to find only a few animals, and can have large detrimental impacts on the animal and the environment if the area disturbed cannot be restored to its original state. In sandy areas such as dunes, digging carefully and probing gently with the hands under the loose substrate can reveal the presence of, for example, small skinks. However, this method requires quickness and agility to catch the animals after they are detected, as many sand-swimming reptiles disappear rapidly into the substrate.

A standardized method for quantitatively comparing surveys of fossorial herpetofauna may employ digging up large sample quadrats and searching intensively within them (Measey et al., 2003; Measey, 2006). The method consists of digging the soil with a hoe in a large area (5 × 5 m^2) to a superficial depth (e.g. 5 cm), and intensively searching the area without disturbing large vegetation (trees and bushes). Then, a small 1 × 1 m^2 random quadrat can be excavated to a more substantial depth (e.g. 30–40 cm) within this area in order to look for subterranean species (Measey, 2006). Alternatively, a standard method may employ a larger 10 × 10 m^2 survey grid marked at 1 m intervals. Five 1 × 1 m^2 small quadrats are randomly selected within the larger area; these small quadrats are then dug to a depth of approximately 30 cm. Each survey would then consist of three of these large grids positioned either side by side or at intervals of 10 m, resulting in a total of 15 m^2 being dug per survey (Measey et al., 2003).

Digging and flipping rocks or other objects can be combined to obtain semi-quantitative indexes of abundance (see Chapter 21). Researchers have used time constrained surveys, where several observers intensively search a given plot during a time-limited survey (e.g. 30 min) using low-impact (i.e. searching by hand for fossorial reptiles under dry vegetation or objects to a depth of 5–7 cm below the surface, while minimizing disturbance to vegetation and restoring the disturbed materials as close as possible to pre-sampling conditions) or moderate-impact methods (i.e. with more extensive disturbance of vegetation and duff layers while using hand tools and looking for animals to a depth of 15 cm below the surface, removing patches of annual vegetation, and by pushing aside but not uprooting larger perennial plants) (Kuhnz et al., 2005). These methods usually underestimate true abundance, based on studies where mark–recapture procedures were used to determine actual population size (Kuhnz et al., 2005).

Opportunistic sampling

Researchers have taken advantage of large excavations during construction projects and soil removal by earth-moving equipment (e.g. bulldozers) to collect and rescue fossorial reptiles (Esteves et al., 2008; De Souza e Lima et al., 2014). Opportunistic sampling may provide large numbers of individuals for study, but should not be relied upon when planning research projects because they are not repeatable or predictable.

11.2.3 Below-ground trapping

To trap semi-fossorial reptiles, researchers can use passive methods such as pitfall traps and drift fences, although capturing truly fossorial reptiles may be difficult and require specially designed traps. These methods have the advantage of accumulating captures without requiring the presence of the researcher, who only needs to check the traps regularly.

Pitfall traps and drift fences

The use of trap lines consisting of series of pitfall traps with or without drift fences has been successfully used to capture some types of semi-fossorial and 'sand swimming' reptiles, such as skinks and fossorial snakes (How and Shine, 1999; Sutton et al., 1999; Goodyear and Pianka, 2008), although these traps are not always as useful as for surface-dwelling reptiles (see Chapter 10). Pitfall drift fences are effective in capturing individuals because animals are diverted by the fence until they can fall into the traps. Pitfall arrays are relatively expensive to construct and maintain. Drift fences of variable length (from a few to many metres) are made of mesh, plastic, or flashing aluminium. For fossorial reptiles, the fences do not need to be very high, but must be buried 5–30 cm into the substrate to ensure that individuals do not pass under them. Pitfall traps are placed at the ends and at different length intervals along the fence. Pitfall traps are open containers that are usually made of plastic or PVC and buried in the ground such that the tops of the containers are level with the ground. The bottom is perforated or

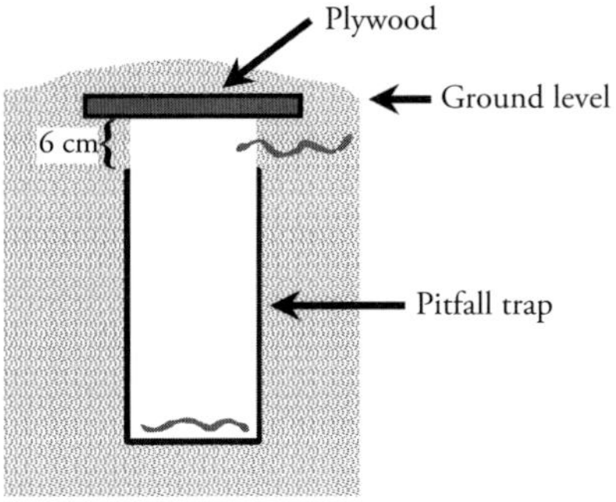

Figure 11.4 Schematic of a pitfall trap specifically designed to capture fossorial amphisbaenians. Figure reprinted with permission from Román and Ruiz (2003).

open and provided with a mesh to promote drainage. The trap can contain sand or leaf litter in the bottom to allow captured animals to burrow and seek refuge inside. The traps are buried at variable depths depending on the target species (e.g. 3–8 cm for small skinks, or 60 cm for fossorial snakes) and the top is covered to provide shade and prevent predation. Traps can also be provided with a funnel at the top to prevent escape. Trap lines can be arranged randomly or form specific array designs (e.g. crosses). Traps must be checked daily or every few days to prevent mortality. Traps should be closed during periods when sampling is not possible to avoid unwanted captures.

In order to capture strictly fossorial reptiles such as amphisbaenians, modified pitfall traps made of PVC can be buried such that the superior edge of the trap also remains buried 6–7 cm under the substrate. The trap is covered with a plywood sheet at ground level, and everything is then covered with sand (Figure 11.4). Animals moving under the soil surface fall into the trap and are retained there. These traps have been useful in capturing the amphisbaenian *Blanus cinereus* in sandy habitats (Román and Ruiz, 2003).

References

Barbour, T. (1946). *A Naturalist's Scrapbook*. Cambridge, MA:Harvard University Press.

Bauer, A.M., and Sadlier, R.A. (1992). The use of mouse glue traps to capture lizards. *Herpetological Review*, **23**, 112–13.

Bennett, D., Hampson, K., and Yngente, V. (2001). A noose method for catching a large arboreal lizard, *Varanus olivaceus*. *Herpetological Review*, **32**, 167–8.

Bertram, B.P., and Cogger, H.G. (1971). A noosing gun for live captures of small lizards. *Copeia*, **1971**, 371–3.

Böhm, M., Collen, B., Baillie, J.E.M., et al. (2013). The conservation status of the world's reptiles. *Biological Conservation*, **157**, 372–85.

Civantos, E., Ortega, J., López, P., et al. (2013). Diet selection by the threatened Chafarinas skink *Chalcides parallelus* in North Africa. *African Journal of Herpetology*, **69**, 78–89.

Das, I. (2012). Arboreal reptiles: tree-trunk and canopy-dwelling species. In R.W. McDiarmid, M.S. Foster, C. Guyer, et al. (eds) *Reptile Biodiversity: Standard Methods for Inventory and Monitoring*. Berkeley, CA:University of California Press, pp. 175–8.

Davis, C., Fleming, P., Craig, M., et al. (2008). A funnel trap for capture of small arboreal reptiles. *Amphibia-Reptilia*, **29**, 413–23.

De Souza e Lima, F.A.N., Gasparini, J.L., De Padua Almeida, A., et al. (2014). Sexual dimorphism in *Amphisbaena nigricauda* (Reptilia, Squamata, Amphisbaenidae) from southeastern Brazil. *Iheringia, Série Zoologia*, **104**, 299–307.

Díaz-Paniagua, C., Blázquez, M.C., Séller, C., et al. (1995). Observations on seasonal and diel surface activity of the amphisbaenian *Blanus cinereus* in southwestern Spain. *Herpetological Journal*, **5**, 217–20.

Durtsche, R.D. (1996). A capture technique for small smooth-scaled lizards. *Herpetological Review*, **27**, 12–13.

Eimer, T. (1882). Mode of capture of lizards in southern Europe. *The Annals and Magazine of Natural History, including Zoology, Botany, and Geology*, **9**, 138–40.

Esteves, F.A., Brandáo, C.R.F., and Viegas, K. (2008). Subterranean ants (Hymenoptera, Formicidae) as prey of fossorial reptiles (Reptilia, Squamata: Amphisbaenidae) in central Brazil. *Papéis Avulsos de Zoologia*, **48**, 329–34.

Fitch, H.S. (1973). A field study of Costa Rican lizards. *University of Kansas Science Bulletin*, **50**, 39–126.

Fitch, H.S., and Henderson, R.W. (1987). Ecological and ethological parameters in *Anolis bahorucoensis*, a species having rudimentary development of the dewlap. *Amphibia-Reptilia*, **8**, 69–80.

Fitzgerald, L.A. (2012). Detecting, counting, and sampling squamates. In R.W. McDiarmid, M.S. Foster, C. Guyer, et al. (eds) *Reptile Biodiversity: Standard Methods for Inventory and Monitoring*. Berkeley, CA:University of California Press, pp. 78–80.

Freeman-Whithoft, B. (1992). Changing Cherokee games, amusements, and economy as an index of acculturation. *Keystone Folklore*, **4**(2), 25–40.

Garcia, R., Queral, A., Powell, R., et al. (1994). Evidence of hybridization among green anoles (Lacertilia: Polychrotidae) from Hispaniola. *Caribbean Journal of Science*, **30**, 279–81.

Glor, R.E., Townsend, T.M., Benard, M.F., et al. (2000). Sampling reptile diversity in the West Indies with mouse glue-traps. *Herpetological Review*, **31**, 88–90.

Gomes, J.O., Maciel, A.O., Costa, J.C.L., et al. (2009). Diet composition of two sympatric amphisbaenian species (*Amphisbaena ibijara* and *Leposternon polystegum*) from the Brazilian cerrado. *Journal of Herpetology*, **43**, 377–84.

Goodyear, E.E., and Pianka, E.R. (2008). Sympatric ecology of five species of fossorial snakes (Elapidae) in Western Australia. *Journal of Herpetology*, **42**, 279–85.

Grant, B.W., Tucker, A.D., Lovich, J.E., et al. (1992). The use of coverboards in estimating patterns of reptile and amphibian biodiversity. In D.R. McCullough and R.H. Barret (eds) *Wildlife 2001: Populations*. New York: Elsevier, pp. 379–403.

Hamilton, A.M., Klein, E.R., Eckstut, M.E., et al. (2007). A simple, inexpensive method to capture arboreal lizards. *Herpetological Conservation and Biology*, **2**,164–7.

Henderson, R.W. (2015). *Natural History of Neotropical Treeboas* (Genus Corallus). Frankfurt-am-Main, Germany:Chimaira.

Horn, S., and Hanula, J.L. (2006). Burlap bands as a sampling technique for green anoles (*Anolis carolinensis*) and other reptiles commonly found on tree boles. *Herpetological Review*, **37**, 427–8.

How, R.A., and Shine, R. (1999). Ecological traits and conservation biology of five fossorial 'sand-swimming' snake species (*Simoselaps*: Elapidae) in south-western Australia. *Journal of Zoology*, **249**, 269–82.

Keeley, B.W., and Keeley, A.T.H. (2012). Using a specialized blowgun dart to obtain genetic samples from mammals. *Wildlife Society Bulletin*, **36**, 185–8.

Krysko, K.L. (2000). A fishing technique for collecting the introduced knight anole (*Anolis equestris*) in southern Florida. *Caribbean Journal of Science*, **36**, 162.
Kuhnz, L.A., Burton, R.K., Slattery, P.N., et al. (2005). Microhabitats and population densities of California legless lizards, with comments on effectiveness of various techniques for estimating numbers of fossorial reptiles. *Journal of Herpetology*, **39**, 395–402.
López, P., Civantos, E., and Martín, J. (2002). Body temperature regulation in the amphisbaenian *Trogonophis wiegmanni*. *Canadian Journal of Zoology*, **80**, 42–7.
López, P., Martín, J., and Salvador, A. (1991). Diet selection by the amphisbaenian *Blanus cinereus*. *Herpetologica*, **47**, 210–18.
López, P., Salvador, A., and Martín, J. (1998). Soil temperatures, rock selection and the thermal ecology of the amphisbaenian reptile *Blanus cinereus*. *Canadian Journal of Zoology*, **76**, 673–9.
Martín, J., López, P., and García, L.V. (2013). Soil characteristics determine microhabitat selection of the fossorial amphisbaenian *Trogonophis wiegmanni*. *Journal of Zoology*, **290**, 265–72.
Martín, J., López, P., and Salvador, A. (1991). Microhabitat selection of the amphisbaenian *Blanus cinereus*. *Copeia*, **1991**, 1142–6.
McCoy, E.D., Sutton, P.E., and Mushinsky, H.R. (1999). The role of guesswork in conserving the threatened sand skink. *Conservation Biology*, **13**, 190–4.
Measey, G.J. (2006). Surveying biodiversity of soil herpetofauna: towards a standard quantitative methodology. *European Journal of Soil Biology*, **42**, S103–10.
Measey, G.J., Gower, D.J., Oommen, O.V., et al. (2003). Quantitative surveying of endogeic limbless vertebrates—a case study of *Gegeneophis ramaswamii* (Amphibia: Gymnophiona: Caeciliidae) in southern India. *Applied Soil Ecology*, **23**, 43–53.
Nordberg, E.J., and Schwarzkopf, L. 2015. Arboreal cover boards: using artificial bark to sample cryptic arboreal lizards. *Herpetologica*, **71**, 268–73.
Paterson, A. (1998). A new capture technique for arboreal lizards. *Herpetological Review*, **29**, 159.
Ribeiro-Júnior, M.A., Gardner, T.A., and Ávila-Pires, T.C.S. (2006). The effectiveness of glue traps to sample lizards in a tropical rainforest. *South American Journal of Herpetology*, **1**, 131–7.
Rodda, G.H., McCoid, M.J., and Fritts, T.H. (1993). Adhesive trapping II. *Herpetological Review*, **24**, 99–100.
Román, J., and Ruiz, G. (2003). Un modelo de trampa para la captura en vivo de culebrillas ciegas (*Blanus cinereus*). *Boletín de la Asociación Herpetológica Española*, **14**, 55–7.
Schlesinger, C.A., and Shine, R. (1994). Choosing a rock: perspectives of a bush-rock collector and a saxicolous lizard. *Biological Conservation*, **67**, 49–56.
Sutton, P.E., Mushinsky, H.R., and McCoy, E.D. (1999). Comparing the use of pitfall drift fences and cover boards for sampling the threatened sand skink (*Neoseps reynoldsi*). *Herpetological Review*, **30**, 149–51.
Tinkle, D.W., and Lawrence, W.C. (1956). Blowguns for reptile sampling. *The Southwestern Naturalist*, **1**, 133–4.
Trivers, R.L. (1976). Sexual selection and resource-accruing abilities in *Anolis garmani*. *Evolution*, **30**, 253–69.
Vargas, G.A., Krakauer, K.L., Egremy-Hernandez, J.L., et al. (2000). Sticky trapping and lizard survivorship. *Herpetological Review*, **31**, 23.
Vaughan, C., Ramirez, O., Herrera, G., et al. (2007). Home range and habitat use of *Basiliscus plumifrons* (Squamata: Corytophanidae) in an active Costa Rican cacao farm. *Applied Herpetology*, **4**, 217–26.
Vice, D.S., Engeman, R.M., and Vice, D.L. (2005). A comparison of three trap designs for capturing brown treesnakes on Guam. *Wildlife Research*, **32**, 355–9.
Vogt, R.C. (1987). You *can* set drift fences in the canopy! *Herpetological Review*, **18**, 13–14.

Witz, B.W. (1996). A new device for capturing small and medium-sized lizards by hand: the lizard grabber. *Herpetological Review*, **27**, 130–1.

Wood, J.G. (1863). *The Illustrated Natural History*. London: Routledge, Warne, and Routledge.

Yost, J.A., and Kelley, P. (1983). Shotguns, blowguns and spears: the analysis of technological efficiency. In R.B. Hames and W.T. Vickers (eds) *Adaptive Responses of Native Amazonians*. New York: Academic Press, pp. 189–224.

Zani, P.A., and Vitt, L.J. (1995). Techniques for capturing arboreal lizards. *Herpetological Review*, **26**, 136–7.

12
Sea snakes

Xavier Bonnet, Arne R. Rasmussen, and François Brischoux

12.1 Introduction

Tropical reptiles have been less intensively studied than temperate species, and marine organisms are less accessible to observation than terrestrial species. Consequently, relatively few field studies have been performed on sea snakes, especially regarding their ecology and natural history. During the last century, pioneer researchers laid the basic foundation for current knowledge (e.g. Dunson, 1975; Heatwole, 1999), and several long-term field studies have provided additional data (Burns and Heatwole, 2000; Guinea, 2006, 2007; Lukoschek et al., 2007; Bonnet, 2012; Goiran and Shine, 2013; Sanders et al., 2013a, 2013b). Although fragmented, sufficient information is available to examine the applicability of the techniques developed for terrestrial species to sea snakes. The marine environment imposes strong constraints to observers and equipment, whereas adaptations to marine life deeply modified life-history traits of sea snakes. As a result, studying sea snakes requires specific techniques.

We first present information on how to locate and observe sea snakes and how to collect morphological data, and then progressively shift towards a description of more demanding techniques. Many sea snakes are fragile when handled, especially when they are pulled out of the water. Sea snakes are also potentially dangerous, although most species are docile unless seized. Several species are reluctant to bite, whereas some others may bite when threatened. Sea snakes are relatively slow-moving and possess short fangs, but the inherent dangers associated with working with these snakes should not be underestimated (e.g. minor incidents can be complicated during scuba diving). Dusky Sea Snakes (*Aipysurus fuscus*) retaliate vigorously during capture and can chase an observer (Arne R. Rasmussen, personal observation); *Aipysurus laevis* males become aggressive towards humans during the mating season. Many *Hydrophis* species bite defensively when handled, especially when manipulated out of water. In contrast, sea kraits (*Laticauda* spp.) are remarkably tame and reluctant to bite. Nonetheless, some individuals might bite and cause moderate symptoms of envenomation (e.g. eyelid ptosis, stiffness). Thus, caution is required for almost all species of sea snakes, turtle head sea snakes (*Emydocephalus* spp.) being the only harmless species. Simple precautions (e.g. wearing a swimming suit and thick neoprene gloves) usually are sufficient to limit the risk of envenomation. Overall, an important issue when studying sea snakes is to not harm the snakes. This chapter focuses on practical field issues; therefore, much

Reptile Ecology and Conservation. Edited by C. Kenneth Dodd, Jr. © Oxford University Press 2016.
Published 2016 by Oxford University Press.

essential information regarding taxonomy or general biology is omitted. Unfortunately, scientific references are often not available.

12.2 Locating, catching, and identifying sea snakes

12.2.1 Locating and catching snakes

For inventories, presence/absence information is usually sufficient. At a broad geographic scale, environmental DNA bar-coding is very efficient (Bohmann et al., 2014). However, a DNA-library and important logistics are required (Cristescu, 2014). Rapid developments considerably relax these constraints, and environmental DNA sampling may soon be the key method. Anecdotal observations are subject to errors of identification, especially when performed under water, during skin sloughing or pre-sloughing phases, or in melanistic individuals. Photographs and videos should be examined by people who are well trained in snake identification (Brischoux et al., 2013).

For population studies, capturing individuals is necessary. We first present information regarding amphibious sea snakes on land, followed by a consideration of amphibious and truly marine sea snakes at sea.

12.2.2 Amphibious sea snakes on land

Sea kraits (*Laticauda* spp.) return periodically to land, notably to lay their eggs. All species can be observed on the shore, although the most terrestrial species (*Laticauda colubrina* complex; Heatwole et al., 2005) are the easiest to detect. They rarely venture far inland, but some species are excellent climbers and can be found on the tops of cliffs. Sea kraits select the easiest way to travel between the sea and their terrestrial refuges; they prefer downwind shores, sheltered harbours, and rocky jetties (see Liu et al., 2012). These sites should be surveyed carefully. Lifting leaves (e.g. palm fronds), logs, beach rocks, or artificial refuges, and exploring cavities and crevices are often fruitful (Bonnet et al., 2009). Beach rocks must be manipulated with caution to not harm animals underneath (e.g. snakes, snails), and must be replaced carefully.

During hot weather, sea kraits remain sheltered. Detectability increases in winter or under cool conditions when snakes bask in the sun. Rainfall after drought allows finding many individuals and offers an opportunity to assess population structure; natural and artificial sources of freshwater are attractive (Lillywhite et al., 2008). Males are more visible during the mating season (late spring, early summer). Sea kraits travel between land and ocean from early dusk to the first hours of full night; dawn is also favourable. To a lesser extent, the tide can be important because snakes tend to avoid crossing large open areas on land; thus, fewer sea kraits are visible during low tides. Sand tracks can remain visible for prolonged periods, are easily identified, and represent an excellent clue as to the presence of snakes. Sloughed skins can be found under beach rocks, in logs and crevices, and within thick herbaceous layers of vegetation. Capturing sea kraits by hand is generally easy, both on land and at sea. It is recommended to gently grab individuals around mid-body and to maintain them horizontally; indeed, strong escape attempts are triggered when the snakes are sized and lifted by the tip of the tail. Using two hands provides a better support of the snake, and thus permits a smoother capture.

12.2.3 Sea snakes at sea

Snakes regularly surface to breathe and can be observed from a boat. In coral reefs and lagoons, snakes can be located by navigating slowly (~4 knots) along transects using small boats, preferably avoiding windy conditions. Most species can be identified at depths up to 3 m in clear water. At low tide, surveys can be performed by walking on the reef flats along transects, notably to search for individuals resting under coral or rocky structures (e.g. *Aipysurus* spp., *Hydrophis (Astrotia) stokesii*). It is important to not disturb fragile ecosystems and to not destroy living corals; inspection of reef flats is thus not always possible.

Manta board surveys, a standard tool of the Australian Institute of Marine Science, are often used to assess snake abundance (Michael Guinea, Arne R. Rasmussen, personal observations). The board is attached with a 25–30 m rope to the stern of a boat, and the observer is dragged at 2–3 knots to look for sea snakes through a mask, usually during 15–30 minute sessions. The trip is recorded with a GPS, and the perpendicular distance of the animal from the transect line is estimated to calculate densities. The observer's identity, time, sea snake species, and other information (e.g. age, condition) should be noted. Video recordings can be used to improve a record's accuracy. For security, one assistant must continuously monitor the observer(s), and hand signals should be used in accordance with international diving rules. Two parallel manta boards can be used to improve survey efficiency and to estimate the heterogeneity of the data associated with each observer. Manta board surveys are very efficient in shallow waters (~10 m depth) during the day and at low tide. This technique also permits collecting sea snakes; the observer indicates to the assistant when to release the board in order to capture the snakes with a net.

Hydrophis (Pelamis) platurus is an exception among sea snakes, insofar as it is the only pelagic species. It ranges from the coastal waters of eastern Africa along the southern Asian coasts to Japan, southward and eastward to Australia and islands of the western Pacific, and eastward to the Americas. This species is found in tight association with oceanic slicks (Dunson and Ehlert, 1971), which are smooth glassy streaks forming drift lines in the ocean while accumulating foam, floating parts of plants, and other debris. Slicks are typically small, ephemeral, mobile oceanic structures resulting from Langmuir circulations, internal waves or convergent currents. Locating slicks and snakes requires experienced skippers. Slicks containing a high density of debris, high solar irradiance, and searched in the early morning (06:00–11:00 h) offer better conditions (Brischoux and Lillywhite, 2011). Captures are relatively easy with a scooping net, but great care is required; the circulatory system of sea snakes is not adapted to handle gradients of gravitational (or hydrostatic) pressure when held vertically outside of water (Lillywhite, 2014). In addition, *H. platurus* can readily bite.

Some species can be easily located during snorkelling, for example, *Emydocephalus* snakes foraging intensively in the shallow reefs near the shore in search of fish eggs deposited on rocky substrates. Other species tend to dive at greater depths (Cook and Brischoux, 2014; Cook et al., 2015). *Hydrophis (Acalyptophis) peronii*, for example, often forages on soft bottoms between 10 and 20 m depth, and surveys for them generally require scuba diving. Sea snakes are more easily located when they forage during the day. In general, most species can be approached by divers without exhibiting signs of disturbance. Some individuals are curious and approach divers, whereas others slowly move away, yet rapidly escape during capture attempts. Sea snakes have been regularly pictured or filmed when hunting. Continuous records during several hours suggest that sea kraits tend to ignore the observer (Xavier Bonnet, personal observation). This tolerance during long sessions of close observation is far greater compared to most terrestrial species, facilitating behavioural projects under natural conditions. Monitoring, photographing, and videotaping sea snakes are relatively simple in shallow waters (5–20 m). Unfortunately, observations become far more complicated in deep waters (40–100 m) and as a consequence records by divers can be anecdotal. On average, sea snakes are not easy to spot under water at night, even using powerful lights and selecting appropriate timing.

Snakes swimming at the surface can be captured using a dip net (Lillywhite et al., 2015). This technique is particularly successful at night when snakes remain relatively motionless (Arne R. Rasmussen, personal observation). Swimming snakes can be captured using a cylindrical net 300–400 mm diameter, 1 m long, with 10 mm mesh. If researchers wear neoprene gloves and a swimming suit, snakes can be gently grasped behind the head and the mid-body simultaneously. After capture, snakes are stressed; they try to escape and they require frequent breaths. During prolonged dives (>15 minutes), calico bags can be partly filled with air bubbles (e.g. blowing air into the bag), thus avoiding the risk of drowning snakes (Xavier Bonnet, personal observation); nets should not be used in such cases. After capture, snakes must be kept in appropriate conditions on board or in captivity. During transport or short-term captivity, sea kraits can be kept in calico bags or nets placed in shaded, moist, and ventilated places. Water can be poured on the snakes. Out of water, net-bags or dry calico bags must be employed because snakes can suffocate in wet bags (high moisture makes the fabrics airproof). Truly marine snakes must be kept in sea water but they must not be crowded inside plastic holding barrels. The snakes tend to wrap around each other, and this creates pressure on the snakes' bodies and prevents voluntary surfacing for air breathing. Plastic holding barrels can be provided with pieces of dead coral or other objects on which snakes can anchor themselves (Harvey Lillywhite, personal communication). Temperature must be maintained below 30°C, preferably around 20–25°C.

Dead snakes should be collected, frozen, or photographed. At some localities, very large numbers of snakes are accidentally or voluntarily collected. Agreements with fishermen enable researchers to gather important information, yet it is important to not support snake harvesting that can be highly destructive (Van Cao et al., 2014).

12.3 Identifying sea snakes

Several genera and species are relatively easy to identify using appropriate guides. Note that many snakes pictured on the internet are not correctly identified, and reference books should be preferably used. We point out several additional difficulties in Section 12.4. Identifying sea snakes to species is not always an easy task (Shine et al., 2002); the genus *Hydrophis* shows particularly great interspecific and intraspecific variation (Sanders et al., 2013b). External characters are crucial, although internal characters are sometimes important. Many species can be identified using a combination of head shields, counts of scale rows around the neck and the body, and the number of ventral scales (multiple counts are required to obtain mean, minimum, and maximum values; Smith, 1926). The shape and size of the head, position of the maxillary bone, number of maxillary teeth, and the colour pattern are useful, but not easy to collect in living specimens. To examine maxillary teeth, a small rod is used to open the mouth of the snake and to gently push impression material (e.g. clay) upwards to a level above the maxillary bone. Two rods can be used to keep the impression material in the desired position against the roof of the mouth. Imprinted marks of the dentition can then be examined. In the genus *Hydrophis,* it is sometime necessary to use additional characters: e.g. vertebral counts (using radiography) of the body and of the tail (from the first pair of forked ribs in the cloacal region). In general, the colour pattern should be recorded, including the number of bands on the body and on the tail.

12.4 Measuring and describing sea snakes

Body size and sex represent important information to collect; other morphological traits are usually less important. Body size is difficult to measure in snakes. Snout–vent length (SVL) and total length (TL) are generally recorded by a single person by gently stretching the snake on a flexible ruler (Figure 12.1). This measurement requires training to avoid overstretching (revealed by vertebrae creaking), to limit the influence of contraction, and thus to get the measurement exactly when the snake relaxes between contractions. Long individuals (>1 m) can be measured in two steps (e.g. anterior and posterior parts). Repeated measurements are required to determine imprecision and observer variation. Photographing a snake while it rests on a substrate next to a metric ruler is a simple and safe alternative method (Harvey Lillywhite, personal communication). Subsequently, precise measurements of length can be obtained using computer programs that scale the length of the snake with the ruler for calibration (e.g. ImageJ; http://imagej.nih.gov/ij/features.html). Measurements of body mass simply require an appropriate device (e.g. ±1.0 g or ±0.1 g electronic spring). However, palpation is important to take into account clutch or litter mass and food items in addition to the actual body mass of the snake. Palpation often permits estimating clutch or litter mass and allows determination of whether prey were ingested head or tail first. Palpation must be quick and gentle and thus requires some training.

Hemipenes are well developed in male sea snakes. Examining the shape of the tail is often sufficient to determine the sex of individuals, although a close comparison of the tail of a female and a male is necessary to appreciate sexual dimorphism. Males have longer and wider tails, and, ventrally, the medium part of the tail is markedly wider in males. In juveniles or in problematic cases, the hemipenes can be everted through a quick manual squeezing of the back portion of the tail, thereby pushing out the hemipenes towards the cloaca (Figure 12.1). The use of probes to determine sex is possible but delicate in sea snakes because openings (cloaca, nostrils, hemipenis orifices) are partly obstructed by twisted valves.

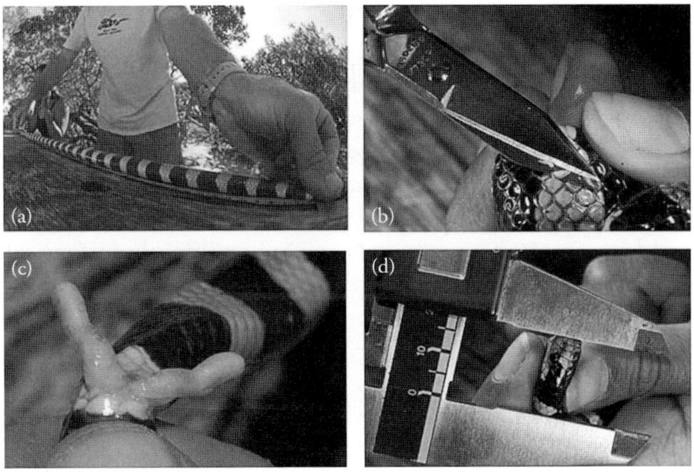

Figure 12.1 Illustration of several field techniques used in sea kraits: (a) measuring snout–vent length (SVL); (b) taking marginal parts of scales; (c) hemipenis eversion; (d) measuring jaw length.

Head dimensions are important traits. Jaw length, from the quadrato-articular joint to the tip of the snout, can be measured with a calliper. The flat part of the external jaws of the calliper preferably is used. It is important to take measurements when the jaws of the calliper touch the joint and the tip of the snout (Figure 12.1).

Many sea snakes exhibit banded patterns with dark bands or rings, i.e. dorsal bands or complete rings depending upon the species or individual (Shine et al., 2010). The number of rings on the body (head included) and on the tail can be counted. These counts can help to determine the sex of individuals (females tend to have more rings on the body and fewer on the tail). Colour background, unusual patterns (e.g. divided rings), injuries, and scars (shape, colour, location) are important traits (e.g. skin rugosity that often varies with sex and with season; Avolio et al., 2006) that enhance the description of individuals and are useful for recognizing recaptures.

As with many living and inert objects, sea snakes are subjected to epibiosis. The description and quantification of external epibionts and parasites (e.g. ticks in sea kraits) can provide useful information (Pfaller et al., 2012).

12.5 Photographing sea snakes

Technical aspects of photographing marine animals are available in specialized books and magazines. This section focuses on subjects. Both terrestrial and marine habitats where the snakes are observed are important to photograph. A comprehensive description of the habitats and microhabitats used by the different snake species, different age classes, and seasonal variation is lacking for almost all species. Photographs help to fill in this lack of information. Additional measurements should be recorded where possible—notably air, substrate, or water temperature and body temperatures as well as oxygen levels, and other parameters such as salinity and pH—that are of potentially high in value to investigate climate change scenarios.

Photographs are useful to minimize identification errors, particularly to avoid confusion between species because of colour changes under water or attributable to skin sloughing. In addition to classical photographs of the head and of the entire individual, injuries, scars, unusual patterns, and external parasites can be photographed. Regurgitated prey can often be identified with photos; however ingested items must be first cleaned with water to remove the coating resulting from ingestion. Macrophotography of the dentition of the prey (also cleaned with water) can be very useful.

12.6 Recapture studies

Gathering accurate ecological information is essential to address questions that are fundamental to ecology and conservation. Under natural conditions, there is no surrogate to mark–recapture studies. Although demanding, this is the best approach to monitor fluctuations and trends in populations, assess inter-individual variability, and to reveal unexpected and important patterns of population functioning (e.g. Voris, 1985; Bonnet et al., 2015).

12.6.1 Marking snakes

A modified version of the classical scale-clipping method works well in sea kraits. Light iron burning of the targeted scale rows (to code units, tens, hundreds) entails a permanent colour change (Bonnet, 2012; see Chapter 4). Electric or gas tools can be employed, although medical cauterizers powered by batteries must be used on small individuals (Winne et al., 2006). The tip of the iron must be hot enough to induce a fast superficial burn. When paired with a comprehensive description of individuals (sex, ring numbers, scars), this technique is very efficient. In our sea-krait studies, failure to recognize an individual at recapture remained below 0.5% on more than 15,000 individuals marked over 10 years (Bonnet, 2012). Cauterizing is cheaper and more reliable than passive integrated transponder (PIT) tags (i.e. radiofrequency identification) because this method is not subjected to tag loss and does not necessitate an electronic reader. In addition, very small snakes, including neonates, can be marked whereas pit tags are too large. However, cauterizing requires training to ensure that the snakes are not injured during marking. Further, setting up specific codes to circumvent the blurring effect of abnormal scales or injuries on those scales that must be counted represent a complication. It also takes a certain amount of time to mark each snake.

In truly marine sea snakes, scale-clipping does not work sufficiently well; the marks tend to disappear over time, and sea snakes do not tolerate prolonged handling. Therefore, PIT tags are preferred (Voris, 1985; Goiran and Shine, 2013). This method is accurate, very fast and it does not require much specialized training. In certain species, numbers can be directly 'painted' on the side of the snakes using liquid nitrogen (Heatwole, 1999). Snakes must be rapidly released at the place of capture, i.e. a few hours after capture for most species and preferably within 24 hours for sea kraits.

Capture sessions are usually organized annually. A typical annual field session should last two weeks, although one month (or more) is preferable. When several sites are surveyed, alternating between them in order to survey each site two times (or more) is profitable, especially in sea kraits. Indeed, foraging bouts, skin sloughing, or digestion that often requires more than one week may influence individual catchability and can heavily bias estimates during short field sessions. Intensive recapture studies on individuals and on populations were shown to have no noticeable impact on survival in sea kraits (Fauvel et al., 2012).

12.6.2 Organizing data

Because disturbance associated with transect surveys and recapture studies is limited and because different observers can be involved simultaneously, these approaches are compatible with repeated/multiple surveys. Repeated surveys allow researchers to take into account important sources of heterogeneity in their analyses. Matrices can be built with time sessions in columns and species or individuals in rows, thus filling cells with observation codes (e.g. 0: absence, 1: male, 2: female) and co-variables (e.g. water temperature). They open avenues for robust analyses, notably to examine important effects of time, season, year, and observer ability to spot and accurately identify species (providing that observers do not influence each other during observations). Implementing repeated surveys (e.g. weeks) within broader sessions (e.g. years) enables researchers to use robust designs to calculate various demographic parameters and abundance or richness estimates, or to perform site occupancy analyses (MacKenzie et al., 2003; Chapter 26). Species richness, abundance, survival, and population trends, for example, can be estimated accurately (Chapters 21 and 27). Overall, mark recapture studies should be encouraged in sea snakes.

12.7 Blood and other tissue sampling

Blood is one of the main connective tissues of an animal and is conveniently studied. Many assays can be performed to address various biological and conservation questions, notably following centrifugation to separate plasma from cell platelets. Withdrawing blood from the heart (cardiocentesis) of snakes is an acceptable and usually satisfactory method of collecting blood, but cardiac blood withdrawal can be more delicate to perform in sea snakes compared to terrestrial species. The laterally flattened body shape of truly marine snakes poses difficulties, and thus training is required. The heart is often easily located via gentle palpation. The difficulty is to target the tip of the cardiac ventricle without harming delicate tissues situated just anterior. Using very small needles is important (27 G–30 G), connected to a 1 ml syringe. Very small amounts of heparin are required and the tip of the needle must remain sharp (e.g. not passed through the rubber lid of a vial). In adults (body mass >150 g), 200 µl to 1 ml of blood can be taken. This quantity must be adjusted in smaller individuals (~100 µl/100 g of body mass when body mass <150 g). We never attempted to take blood from neonates or very small individuals (<50 g). Sinuses and lymphatic vessels (e.g. sometimes impaled during tail puncture) should be avoided as they deliver unknown proportions of blood mixed with lymph that can influence assays.

For eco-physiological investigations, centrifuging the blood immediately in the field is preferable (using a battery powered centrifuge). The plasma can be stored in liquid nitrogen (cryotubes are required) or frozen at −25°C, especially to assay hormones, metabolites, and organic contaminants. Cell platelets can be frozen and/or preserved in alcohol for genetic analyses or to assay stable isotopes and trace metals. A blood volume of 100 µl is sufficient for many analyses (e.g. several steroids, genetic material, and stable isotopes) and 1 ml is sufficient for almost all routine analyses (e.g. including trace metals, various metabolites, and organic contaminants). For RNA preservation, as for other elements that degrade rapidly (e.g. enzymes), immediate preservation of tissues in liquid nitrogen or RNAlater can be critical.

For DNA, isotope, and several trace metal analyses, the tip of scales can be collected (notably when marking snakes; Figure 12.1) and stored in alcohol within screw-cap vials. Sloughed skins that offer larger amounts of material should be collected opportunistically.

Stomach contents obtained from spontaneous or forced regurgitation must be weighed, identified (photographed) and stored in alcohol for DNA, trace metal and stable isotope analyses, or preferably frozen. Similarly, ticks and other parasites can be collected and preserved in alcohol.

Snakes found freshly dead offer excellent opportunities to obtain different tissues in relatively large quantities and for donations to local museums. Individuals can be dissected, and all the organs and gut contents can be carefully measured, weighed, and examined. If only a few tissues can be obtained, fat bodies and liver should be collected and frozen in priority because these organs accumulate contaminants and provide important ecological indexes (e.g. fat content and/or liver mass correlate with body condition). Muscles, skin, and gonads also are important to collect and freeze. Drying whole specimens or tissues represents an alternative that should be preferred over formalin. Formalin hampers eco-toxicological and genetic investigations, is carcinogenic, and is difficult to transport. For museum collections, formalin offers important benefits, such as better long-term storage, more natural looking specimens, ability to study gut contents, and morphometrics. A simple solution is to take tissue samples of the specimen for eco-toxicological and genetic studies and then fix the specimen in formalin.

12.8 Bio-logging

Current knowledge about diving, displacement, or home range is extremely meagre in sea snakes (e.g. Burns and Heatwole, 1998). Bio-logging is appropriate to monitor free-ranging individuals (Cook et al., 2015). However, the flexible morphology and small body size of sea snakes usually preclude using external devices (but see Rubinoff et al., 1986, 1988). Nonetheless, externally attached loggers can be used in controlled conditions (Brischoux et al., 2010). Sea snakes display remarkable capacities to recover from wounds (e.g. caused by aggressive prey) and from surgery; therefore, implantation of internal devices is possible. Anaesthesia and tolerance to implanted devices must be carefully considered, and our field experience has revealed unexplained differences (i.e. recovery from anaesthesia) between closely related species (e.g. *L. colubrina* vs. *L. saintgironsi*). Despite similar dosages, intra-peritoneal injections (e.g. ketamine) or inhalation (e.g. isoflurane) were either successful (i.e. rapid anaesthesia and rapid recovery) or posed major complications (prolonged recovery requiring breathing assistance, or sometimes death). We also observed that combining local analgesics (e.g. lidocaine) and anaesthesia delayed recovery. Truly marine sea snakes are generally more fragile than sea kraits, but information regarding anaesthesia, surgery, and bio-logging is fragmentary. Overall, bio-logging methodology for sea snakes is in its very early stages of development. Consequently, we strongly suggest adopting a practical step-by-step approach to select the best method, e.g. by gradually increasing minimal dosage of a drug. Miniaturization now makes bio-logging investigation possible with limited perturbation, although sea snakes appear to be more sensitive to the presence of internal devices compared to their terrestrial relatives. We further suggest using the smallest devices available, to avoid attachment to the ribs, and not to combine devices. For instance, implanting depth time recorders but not transmitters is recommended, even in individuals weighing more than 300 g when the total load of both devices may remain below 15 g. Indeed, transmitters have a long and problematic antenna that must be inserted below the fragile skin of sea snakes (since skin contributes to respiration). Selecting a study system where recapture probabilities are elevated to offset the lack of transmitters is important. Further technical improvements (e.g. shape of devices, coating material) would be welcome for sea snakes.

12.9 Captivity

Sea kraits are relatively robust compared to other sea snakes. They accept various prey (e.g. freshwater eels) and they can be maintained in both marine and fresh water tanks. Emergent shelters and freshwater bowls must be provided. Other species are less tolerant to captivity; *Emydocephalus* are particularly fragile and unexplained mortality has been observed during brief captivity episodes (only a few hours). Captivity should be limited to short-term experiments (hours, days) to record locomotor, physiological, and behavioural traits, for example, and should be performed with great care in the most sensitive species. Our limited experience suggests that few aquariums display sea snakes, individuals are not often fed, and they are regularly replaced. However, encouraging exceptions exist. For example, olive sea snakes have been maintained and fed in the Reef HQ Aquarium of Townsville (Australia). It is noteworthy that both sea kraits and sea snakes require regular access to freshwater to survive (Lillywhite et al., 2008, 2015). In general, improvements are required to study and breed sea snakes in captivity, particularly for threatened species, and not simply to maintain individuals alive.

12.10 Conclusions

The ecology and general biology of sea snakes are insufficiently documented, although appropriate study techniques are now available for most species. Because scientific curiosity and knowledge represent the foundation for education and conservation, investigations are urgently needed to improve our information about their distribution and population status, to address fundamental scientific questions, and eventually to implement conservation and management programmes. In the following paragraphs, we list several suggestions to focus practical action.

In addition to the widespread pelagic *Hydrophis (Pelamis) platurus*, several sea snake species have a broad distribution in both the Indian and Pacific Oceans and were/are locally abundant, e.g. *Hydrophis (Lapemis) curtus, H. cyanocinctus, H. ornatus, H. (Acalyptophis) peronii, H. (Astrotia) stokesii, Laticauda colubrina* complex, and *L. laticaudata*. Investigations concerning geographic variation and connectivity across populations would likely be profitable. Other species have been collected in restricted areas and might well be vulnerable. Only five *Hydrophis parviceps* have been observed in a small area in the southern part of Vietnamese waters of the South China Sea (Rasmussen et al., 2012). The recently described species *Hydrophis sibauensis* is known only from three individuals collected more than 1000 km upstream in rivers of Borneo. Two individuals of *Hydrophis laboutei* were collected in the Chesterfield Reefs in New Caledonia. Other examples of species with a limited distribution are *Aipysurus apraefrontalis, A. foliosquama*, and *A. fuscus* from Western Australia. *Hydrophis semperi* occurs only in Lake Taal in the Philippines, and *Laticauda crockeri* occurs only in Lake Te-Nggano, Rennell Island. In these species, both field population studies and captive breeding programmes are needed.

For almost a century, intensive commercial exploitation of skin, various organs, blood, and meat severely impacted sea snakes in the Philippines, Indonesia, Japan, Taiwan, Thailand, and Vietnam, as well as in Australia, New Caledonia, and China. This includes unjustified massive killing for venom research, where thousands of individuals were decapitated, although alternatives already existed (e.g. Tu, 1976). Several species (e.g. *Laticauda* spp. and some *Hydrophis* spp.) have been exploited to the point of pushing populations to extinction. Vietnamese squid fishers currently take more than 220,000 sea snakes annually from the Gulf of Thailand (Van Cao et al., 2014). Large numbers of other species are likely taken by fisheries. Unfortunately, sea snakes are not protected by CITES and only rarely by national law or regulation. For example, in New Caledonia none of the 14–15 sea snake species is protected. In a few instances, local protection of sea snakes has avoided extinction, e.g. in the Philippines where a catastrophic decline of *Laticauda* species occurred in the 1970s. Populations never recovered, however, likely because a minimum population threshold was exceeded from which the population cannot recover. Over large geographic areas, dramatic declines over the last 10–15 years remain unexplained (e.g. Ashmore Reef; data collected by Mick Guinea; Lukoschek et al., 2013), although habitat destruction (e.g. coral bleaching) may be the main cause. Changing rainfall patterns associated with global climate change may also be an important factor (Brischoux et al., 2012; Lillywhite et al., 2014). Overall, many species previously abundant and/or with previously dense populations are now relictual (e.g. *Hydrophis torquatus* and *H. klossi*) or locally extinct (e.g. *Aipysurus apraefrontalis* and *A. foliosquama* in Ashmore Reef). Fishermen mention that the mean size of harvested sea snakes has decreased over time (Van Cao et al., 2014). Specific effort is thus needed to improve the worldwide protection of all sea snakes.

To protect sea snakes and to determine sustainable harvest levels, scientific information is required. Foraging and breeding areas for most species of sea snakes are unknown; thus, the impact of habitat destruction (e.g. mangrove clearings) or oil exploration (e.g. sonar blasting) has not been evaluated yet. Future studies should focus on breeding cycles, mortality associated with fisheries, population abundance, sexual maturity, taxonomy, and diet, especially in Asia (Voris, 2015).

Finally, sea snakes are useful bio-indicators to monitor the health status of coral reefs (Reed et al., 2002; Brischoux et al., 2009). Catching and measuring approximately 30 individuals per year in a given site is sufficient for a broad monitoring programme over time and for site comparisons (e.g. polluted versus non-polluted areas). In addition, prey, blood, sloughed skin, and dead individuals offer excellent opportunities for contaminant investigation (Bonnet et al., 2014).

References

Avolio, C., Shine, R., and Pile, A.J. (2006). The adaptive significance of sexually dimorphic scale rugosity in sea snakes. *American Naturalist*, **167**, 728–38.

Bohmann, K., Evans, A., Gilbert, M.T.P., et al. (2014). Environmental DNA for wildlife biology and biodiversity monitoring. *Trends in Ecology & Evolution*, **29**, 358–67.

Bonnet, X. (2012). Long term field study of sea kraits in New Caledonia: fundamental and conservation issues. *Integrative and Comparative Biology*, **52**, 281–95.

Bonnet, X., Briand, M.J., Brischoux, F., et al. (2014). Anguilliform fish reveal large scale contamination by mine trace elements in the coral reefs of New Caledonia. *Science of the Total Environment*, **470**, 876–82.

Bonnet, X., Brischoux, F., Pearson, D., et al. (2009). Beach-rock as a keystone habitat for sea kraits. *Environmental Conservation*, **36**, 62–70.

Bonnet, X., Brischoux, F., Pinaud, D., et al. (2015). Spatial variation in age structure among colonies of a marine snake: the influence of ectothermy. *Journal of Animal Ecology*, **84**, 925–33.

Brischoux, F., and Lillywhite, H.B. (2011). Light- and flotsam-dependent 'float-and-wait' foraging in pelagic sea snakes (*Pelamis platurus*). *Marine Biology*, **158**, 2343–7.

Brischoux, F., Bonnet, X., and Legagneux, P. (2009). Are sea snakes pertinent bio-indicators for coral reefs? A comparison between species and sites. *Marine Biology*, **156**, 1985–92.

Brischoux, F., Kato, A., Ropert-Coudert, Y., et al. (2010). Swimming speed variation in amphibious seasnakes (Laticaudinae): a search for underlying mechanisms. *Journal of Experimental Marine Biology and Ecology*, **394**, 116–22.

Brischoux, F., Peacock, S., and Bonnet, X. (2013). *Laticauda* spp. (sea krait). Avian predation. *Herpetological Review*, **44**, 331–2.

Brischoux, F., Tingley, R., Shine, R., et al. (2012). Salinity influences the distribution of marine snakes: implications for evolutionary transitions to marine life. *Ecography*, **35**, 994–1003.

Burns, G., and Heatwole, H. (1998). Home range and habitat use of the Olive Sea Snake, *Aipysurus laevis*, on the Great Barrier Reef, Australia. *Journal of Herpetology*, **32**, 350–8.

Burns, G., and Heatwole, H. (2000). Growth, sexual dimorphism, and population biology of the olive sea snake, *Aipysurus laevis*, on the Great Barrier Reef of Australia. *Amphibia-Reptilia*, **21**, 289–300.

Cook, T.R., and Brischoux, F. (2014). Why does the only 'planktonic tetrapod' dive? Determinants of diving behaviour in a marine ectotherm. *Animal Behaviour*, **98**, 113–23.

Cook, T.R., Bonnet, X., Fauvel, T., et al. (2015). Foraging behaviour and energy budgets of sea snakes: insights from implanted data-loggers. *Journal of Zoology*, in press.

Cristescu, M.E. (2014). From barcoding single individuals to metabarcoding biological communities. *Trends in Ecology & Evolution*, **29**, 566–71.
Dunson, W.A. (1975). *The Biology of Sea Snakes*. Baltimore, MD: University Park Press.
Dunson, W.A., and Ehlert, G.W. (1971). Effects of temperature, salinity, and surface water flow on distribution of the sea snake *Pelamis*. *Limnology and Oceanography*, **16**, 845–53.
Fauvel, T., Brischoux, F., Briand, M.J., et al. (2012). Do researchers impact their study populations? Assessing the effect of field procedures in a long term population monitoring of sea kraits. *Amphibia-Reptilia*, **33**, 365–72.
Goiran, C., and Shine, R. (2013). Decline in sea snake abundance on a protected coral reef system in the New Caledonian Lagoon. *Coral Reefs*, **32**, 281–4.
Guinea, M.L. (2006). *Sea Snakes of Ashmore Reef, Hibernia Reef and Cartier Island*. Final Report Survey 2005, Department of the Environment, Water, Heritage and the Arts (DEWHA), Canberra, ACT, Australia.
Guinea, M.L. (2007). *Sea Snakes of Ashmore Reef, Hibernia Reef and Cartier Island with comments on Scott Reef*. Final Report Survey 2007, Department of the Environment, Water, Heritage and the Arts (DEWHA), Canberra, ACT, Australia.
Heatwole, H. (1999). *Sea Snakes*. 2nd ed. Australian Natural History Series. Sydney: University of New South Wales Press.
Heatwole, H., Busack, S., and Cogger, H. (2005). Geographic variation in sea kraits of the *Laticauda colubrina* complex (Serpentes: Elapidae: Hydrophiinae: Laticaudini). *Herpetological Monographs*, **19**, 1–136.
Lillywhite, H.B. (2014). *How Snakes Work: Structure, Function, and Behavior of the World's Snakes*. New York: Oxford University Press.
Lillywhite, H.B., Babonis, L.S., Sheehy, C.M., III, et al. (2008). Sea snakes (*Laticauda* spp.) require fresh drinking water: implication for the distribution and persistence of populations. *Physiological and Biochemical Zoology*, **81**, 785–96.
Lillywhite, H.B., Heatwole, H., and Sheehy, C.M. (2015). Dehydration and drinking behavior in true sea snakes (Elapidae: Hydrophiinae: Hydrophiini). *Journal of Zoology*, in press.
Lillywhite, H.B., Sheehy, C.M., III, Brischoux, F., et al. (2014). Pelagic sea snakes dehydrate at sea. *Proceedings of the Royal Society of London B: Biological Sciences*, **281**, 20140119.
Liu, Y.-L., Chen, Y.-H., Lillywhite, H.B., et al. (2012). Habitat selection by sea kraits (*Laticauda* spp.) at coastal sites of Orchid Island, Taiwan. *Integrative and Comparative Biology*, **52**, 274–80.
Lukoschek, V., Beger, M., Ceccarelli, D., et al. (2013). Enigmatic declines of Australia's sea snakes from a biodiversity hotspot. *Biological Conservation*, **166**, 191–202.
Lukoschek, V., Heatwole, H., Grech, A., et al. (2007). Distribution of two species of sea snakes, *Aipysurus laevis* and *Emydocephalus annulatus*, in the southern Great Barrier Reef: metapopulation dynamics, marine protected areas and conservation. *Coral Reefs*, **26**, 291–307.
MacKenzie, D.I., Nichols, J.D., Hines, J.E., et al. (2003). Estimating site occupancy, colonization, and local extinction when a species is detected imperfectly. *Ecology*, **84**, 2200–7.
Pfaller, J.B., Frick, M.G., Brischoux, F., et al. (2012). Marine snake epibiosis: a review and first report of decapods associated with *Pelamis platurus*. *Integrative and Comparative Biology*, **52**, 296–310.
Rasmussen, A.R., Elmberg, J., Sanders, K.L., et al. (2012). Rediscovery of the rare sea snake *Hydrophis parviceps* Smith 1935: identification and conservation status. *Copeia*, **2012**, 276–82.
Reed, R.N., Shine, R., and Shetty, S. (2002). Sea kraits (Squamata: *Laticauda* spp.) as a useful bioassay for assessing local diversity of eels (Muraenidae, Congridae) in the western Pacific Ocean. *Copeia*, **2002**, 1098–101.
Rubinoff, I., Graham, J.B., and Motta, J. (1986). Diving of the sea snake *Pelamis platurus* in the Gulf of Panama I. Dive depth and duration. *Marine Biology*, **91**, 181–91.

Rubinoff, I., Graham, J.B., and Motta, J. (1988). Diving of the sea snake *Pelamis platurus* in the Gulf of Panama II. Horizontal movement patterns. *Marine Biology*, **97**, 157–63.

Sanders, K.L., Lee, M.S.Y., Bertozzi, T., et al. (2013a). Multilocus phylogeny and recent rapid radiation of the viviparous sea snakes (Elapidae: Hydrophiinae). *Molecular Phylogenetics and Evolution*, **66**, 575–91.

Sanders, K.L., Rasmussen, A.R., Elmberg, J., et al. (2013b). Recent rapid speciation and ecomorph divergence in Indo-Australian sea snakes. *Molecular Ecology*, **22**, 2742–59.

Shine, R., Brischoux, F., and Pile, A.J. (2010). A seasnake's colour affects its susceptibility to algal fouling. *Proceedings of the Royal Society of London B: Biological Sciences*, **27**, 2459–64.

Shine, R., Reed, R.N., Shetty, S., et al. (2002). Reproductive isolating mechanisms between two sympatric sibling species of sea snakes. *Evolution*, **56**, 1655–62.

Smith, M. (1926). *Monograph of the Sea-Snakes*. London: Oxford University Press.

Tu, A.T. (1976). Investigation of the sea snake, *Pelamis platurus* (Reptilia, Serpentes, Hydrophiidae), on the Pacific coast of Costa Rica, Central America. *Journal of Herpetology*, **10**, 13–18.

Van Cao, N., Thien Tao, N., Moore, A., et al. (2014). Sea snake harvest in the Gulf of Thailand. *Conservation Biology*, **28**, 1677–87.

Voris, H.K. (1985). Population size estimates for a marine snake (*Enhydrina schistosa*) in Malaysia. *Copeia*, **1985**, 955–61.

Voris, H.K. (2015). Marine snake diversity in the mouth of the Muar River, Malaysia. *Tropical Natural History*, **15**, 1–21.

Winne, C.T., Willson, J.D., Andrews, K.M., et al. (2006). Efficacy of marking snakes with disposable medical cautery units. *Herpetological Review*, **37**, 52–4.

13

Freshwater turtles

Richard C. Vogt

13.1 Aquatic turtles on land

13.1.1 Miscellaneous techniques

The most basic methods for capturing aquatic turtles are by hand or dip net, probing for them in the leaf litter with a 10-penny nail on the end of a 2 m stick, or capturing them as they are crossing roads. Many aquatic species cross roads during nesting migrations, spring dispersal, fall migration to hibernating sites, migration from drying ponds, or as hatchlings after emerging from nests. Most of these types of captures are serendipitous, but they can be valuable references for locality data. Evidence of thriving populations often can be documented by the quantity of roadkills.

Trained dogs are more efficient than people. More than 90% of the 3832 box turtles collected by Schwartz and Schwartz (1974) were found by their dogs. I have used a mongrel bitch to help find White-lipped Mud Turtles (*Kinosternon leucostomum*) when they were aestivating in the forest (Morales-Verdera and Vogt, 1997). She would look for them in the leaf litter or under boulders and fallen trees. She helped to locate turtles with transmitters once we were within a 3 m radius. The breed of dog is important. It is better to train dogs that are not hunting dogs so that they are not distracted by other animal scents while working. Today, I have a Belgian shepherd bitch that was professionally trained to find four different species of Brazilian rainforest pond turtles by their scent. When she finds the scent of one of the species, she walks around the odour in a circle and sits down and waits for the handler to come and find the turtle or the turtle nest. She exhibits no interest in touching the turtles or chasing any kind of animal.

Drift fences with pitfall traps and funnel traps have been used with success in sampling populations of freshwater turtles when they are migrating to or from nesting sites, or emigrating when ponds are drying. Sexton (1959) used low 25 cm barriers that were 100 m long to block the nesting migration route of Painted Turtles (*Chrysemys picta*). Gibbons (1970) used modified drift fences with pitfall traps completely encircling ponds to collect turtles in their nesting or other terrestrial migrations. Drift fences must be monitored on a regular basis depending on the number of turtles moving through the area. During hatchling dispersal, 19 l buckets should be fitted with funnels that will allow the hatchlings to pass through the entrance but prevent predators from removing them (Vogt and Hine, 1982; see Chapter 10).

Reptile Ecology and Conservation. Edited by C. Kenneth Dodd, Jr. © Oxford University Press 2016.
Published 2016 by Oxford University Press.

The best material to use for drift fences is determined by a combination of factors: efficiency in catching turtles, ease in installation and maintenance, permanence, and cost. The permanent complete enclosure method is most effective (Gibbons and Semlitsch, 1981), especially if galvanized metal is used; once the fences are installed, the researcher has a relatively maintenance-free system. Industrial plastic tarp is low cost and easy to carry. When not in use, the plastic can be dropped to the ground, rolled up, and covered with leaves and then erected again when needed. Plastic has high maintenance; wind, large animals, and tree branches may tear or destroy the plastic. Replacement time for plastic is much shorter than for metal material that can remain in place for years. Plastic or aluminium window screening also can be used. If the study subjects are pond turtles, a researcher can set up sections of at least 30 m of drift fences with pitfalls bisecting the overland routes of aquatic turtles as they move from the water to nesting sites.

13.1.2 Nest surveys

Without getting into the water or buying traps or even seeing turtles, one can document the presence of even rare species by making surveys at potential nesting sites. Surveys can be run most successfully during the nesting season when tracks of the females can be seen leading to nest sites. Most species can be identified by tracks or by their eggs, at least to genus. It was sometimes difficult to identify the eggs and nests of the three sympatric species of map turtles (*Graptemys*) I studied on the Mississippi River (Vogt, 1980a). However, nesting seasons are relatively short and even later in the season when the eggs have hatched or the nests have been predated, egg fragments can lead to proper species identification. Dichotomous keys have been developed recently by Raymond Saumure and Joel Bonin (personal communication) that identify egg shell fragments to species. These types of non-traditional survey methods combined with other methods can be used to enhance the quality of any inventory or survey for rare or endangered species without being overly costly.

13.2 Aquatic turtles in water

13.2.1 Surprise, snorkelling, muddling, and polling

Basking turtles can be rushed at full speed and often plucked off a log or swept into a dip net. Fred Cagle used this technique day and night, as many species often slept on submerged branches in clear water (Chaney and Smith, 1950) where they could be captured by deft hands. Fred would drive an outboard full speed ahead with his students at the bow to capture the turtles as he crashed the boat into the tree branches (James Dobie, personal communication). Airboats have been used to rush Florida Red-bellied Turtles (*Pseudemys nelsoni*) in north Florida lakes. Two observers laid flat on the bow as the driver ran the boat up on mats of floating vegetation where sleeping and basking turtles were caught by hand (Ken Dodd, personal communication). In the rivers of Chiapas, Mexico, we could see Central American river turtles (*Dermatemys*) swimming underwater in clear water streams, but they were too fast to catch if we were snorkelling. We could get the additional speed by diving off the boat to capture them, and we used this technique by day and at night with a spotlight. John Legler also caught most of the turtles in his Australian studies by diving for them. I have camouflaged my head with a mat of aquatic vegetation, leaving only my eyes above water, and approached map turtles (*Graptemys*) while they were surface feeding in Mississippi River backwaters. When they submerged their heads to feed, I moved forward rapidly and, when they emerged to look around, I froze. Because it would take 20–30 minutes to catch a turtle, this method is not very efficient.

When waters are crystal clear, biologists have surveyed turtles by snorkelling (Buhlmann et al., 2013). Usually two or three persons swim parallel to the shore and with the current (in streams) or along transects in clear water lakes. When turtles are observed, swimmers record the species and their behaviour, and may attempt to catch them. Turtles may either be taken to shore for processing or researchers may follow along in a canoe or small boat where measurements and other data are taken. Water (clarity using a Secchi disk, depth, current velocity, temperature) and habitat (substrate, extent of rocky or woody debris) conditions should be recorded, in addition to time of day, weather, season, biota, and other variables of possible interest. Snorkelling surveys have been used quite successfully for inventorying streams and for long-term monitoring of turtle communities (Huestis and Meylan, 2004; Chapin and Meylan, 2011; Johnston et al., 2011).

Muddling for turtles in shallow water involves putting hands below water and feeling in the mud, below logs, snags, rocks, and under overhanging banks for turtles. These types of incidental capture may suffice for documentation purposes, but these are hardly systematic methods to use in qualitative or quantitative sampling. In the Brazilian Amazon, I have caught more than 40 Yellow-spotted Amazon Sideneck Turtles (*Podocnemis unifilis*) within 2 hours in small backwater ponds during the dry season. Mike Ewert spent most of his life collecting thousands of turtles this way throughout the United States. An extension of muddling is sounding or polling, where a wooden pole is plunked down into the mud in springs or areas where bubbles are surfacing. The sound heard/felt when the carapace of a turtle is hit is a hollow 'plunk', as opposed to the solid feel and sound when a rock or log is hit. When a turtle is hit, reach down underwater and grasp for it. Professional turtle trappers in Wisconsin could diminish Snapping Turtle (*Chelydra serpentina*) and Wood Turtle (*Glyptemys insculpta*) populations effectively using this technique when the turtles were aggregated in winter. This technique also is used in Mexico during the dry season to find Giant Musk Turtles (*Staurotypus triporcatus*).

In the Brazilian Amazon, an extension of this technique is to have a spear point on the other end of the pole so that if a turtle is hit, the pole is inverted and the turtle speared. The spear head is actually quite small, a rectangular point of steel 2–3 mm in diameter and 10 mm long. The head is attached to a pole by a string such that when a turtle is struck, the head will be released from the pole yet the point will remain in the turtle's carapace. A similar technique is used in Mexico for terrestrial turtles, particularly Tabasco Mud Turtles (*Kinosternon acutum*) and Mexican Wood Turtles (*Rhinoclemys areolata*). A blunt two-penny nail is driven into a 25 mm wooden pole and horizontally thrust through the leaf litter at the base of shrub clumps. When the nail hits a turtle, it makes a plunking sound, differing from that of a rock or tree branch. We were able to find several dozen turtles using this technique on a regular basis. The technique is especially valuable for locating turtles during the dry season when they are not naturally moving around. Carpenter (1955) using a similar technique and located five species this way.

13.2.2 Basking traps

Basking traps work for many species; plastic or metal baskets can be attached to basking logs and turtles scared into them while they are basking (Carr, 1952). The collector must first survey the area to be sampled with binoculars or spotting scope to find out where the preferred basking sites are located. Baskets should be attached below areas where turtles are accustomed to falling into the water; the lips of the baskets should be flush with the water surface so that when the baskets are not attended the turtles can escape. This type of basking trap is an active trap, that is, the collector must rush the basking log from the opposite side from which the baskets are attached and pull out any turtles that have fallen into the traps. Horne et al. (2003) used this technique successfully while studying Yellow-blotched Sawbacks (*Graptemys flavimaculata*) on the Pascagoula River in southern Mississippi. The advantage of this technique is that once the turtle is in hand, fresh stomach contents can be collected unimpeded by trap bait. Such turtles have not been stressed by trap confinement so hormone levels in the blood can be sampled without fear of biased results. The main disadvantage is that it is an active, time-consuming collecting technique that requires ideal basking sites.

More complicated basking traps have been in use for decades. Floating basking traps have a square or rectangular wooden frame with a wire mesh or nylon bag attached to the centre. Planks are angled from the edge of the water towards the centre of the trap, allowing the turtles to come out of the water to bask but not to escape out of the basket once they have fallen into the water. These traps are good for long-term population studies, but are not adequate for routine sampling. They require that the turtles become accustomed to the newly offered basking area; the traps are bulky to carry; and many species do not bask. Various more complicated basking traps with multiple treadles have been invented; these are summarized in Plummer (1979).

Another active basking trap has been taken from the birds of prey collecting manual, the bal-chatri. Hundreds of snares of monofilament fishing line are tied to 0.5 m square pieces of galvanized chicken wire or by tying them to cloth ropes that can be tied around basking sites at the water's surface; this latter technique is far less cumbersome and allows the rope to be rolled to get the loops into upright positions easily. These then can be moulded to fit onto any basking surface and on logs fastened with nails, tied onto rock surfaces, or staked onto sand beaches. Like any basking trap, the trap has to be set where the turtles are basking. After the trap is in place, traps must be checked every few hours or the turtles may drown, suffer from sun exposure, or be attacked by birds of prey (Braid, 1974).

13.2.3 Basking surveys

Basking surveys are conducted using high powered binoculars or spotting scopes. Obviously, these surveys are appropriate only for species that bask at the appropriate time of day and season. C.J. McCoy and I used this technique most adequately for assessing population densities of sawbacks (*Graptemys flavimaculata, G. nigrinoda*, and *G. oculifera*) throughout their ranges. Our data were substantiated by simultaneous trapping with fyke nets. We floated downstream in a boat or canoe, one person in the stern steering and the other in the bow with a 30× spotting scope. To standardize our results, we floated rivers on sunny days for 2 hour stretches from 09:00 to 11:00 when turtles would most likely be basking. With this technique, we were able to identify and determine the sex of adults of eight species of turtles as well as to identify hatchlings to species; we were able to identify but not determine sex for two additional species. Lindeman (1998, 1999) also used this technique for assessing map turtle (*Graptemys*) populations in the southern USA, as did Kornilev et al. (2010, 2012) in the Santa Fe River in north Florida. By combining daily surveys with painting numbers on turtle carapaces, the latter authors tracked movement patterns and habitat occupancy, and estimated the abundance of Suwannee Cooters (*Pseudemys suwanniensis*) in a black-water river where subsurface visibility was limited. Basking surveys of Giant South American River Turtles (*Podocnemis expansa*) at the nesting beaches can be conducted by airplane surveys since during their characteristic short nesting season they spend most of the daylight hours basking. Since this species is large, newly designed land-sat radar techniques should allow for counting basking individuals at aggregations of this species throughout the Amazon without the rigors of long-distance travel.

13.2.4 Trapping

Legler and Iverson traps

The standard for mud turtle (*Kinonsternon*) and slider (*Trachemys*) biologists is a variation on the ingenious design of John M. Legler (1960), hence the name Legler trap (Figure 13.1). The basic design has been modified by various authors, for example, by using chicken wire, metal hoops, fibreglass hoops, wooden hoops, or the box traps of Australians (Kennett, 1992). The principle of the hoop trap is about 6000 years old (Singer, 1954, in Legler, 1960). However, Legler was the first to use a simplified, standardized construction of lightweight durable materials. For more than 40 years, this has been the basic turtle trap used by researchers sampling and studying turtles that are attracted to baited traps. The hoops are made of 1 cm aluminium tubing for 50 cm diameter hoops or 1.5 cm for 1 m diameter hoops; for hoops of larger diameter, galvanized steel or fibreglass is used. Four hoops are covered with a rectangle of 2 cm nylon fish netting. The loose ends of the netting are inverted into hoops to form the throats of the trap. The traps are kept rigid by a pair of dowel stiffeners with metal screw hooks placed on each side of the trap. These stiffeners are important in that they keep the form of the trap and, in particular, maintain the throats taut such that turtles must push their way into the traps and then have a hard time finding their way out. Sardines in oil, chicken parts, fresh fish, shrimp, or other smelly bait can be placed in a bait container made of window screen; I have found aluminium cans to be easier to find. The bait is placed in the can and the can is crimped and punctured profusely so that the scent escapes but the turtle cannot devour the bait. Thus, turtles are more likely to stay in the trap trying to get at the bait. Also, the natural stomach contents will not be confused with the bait.

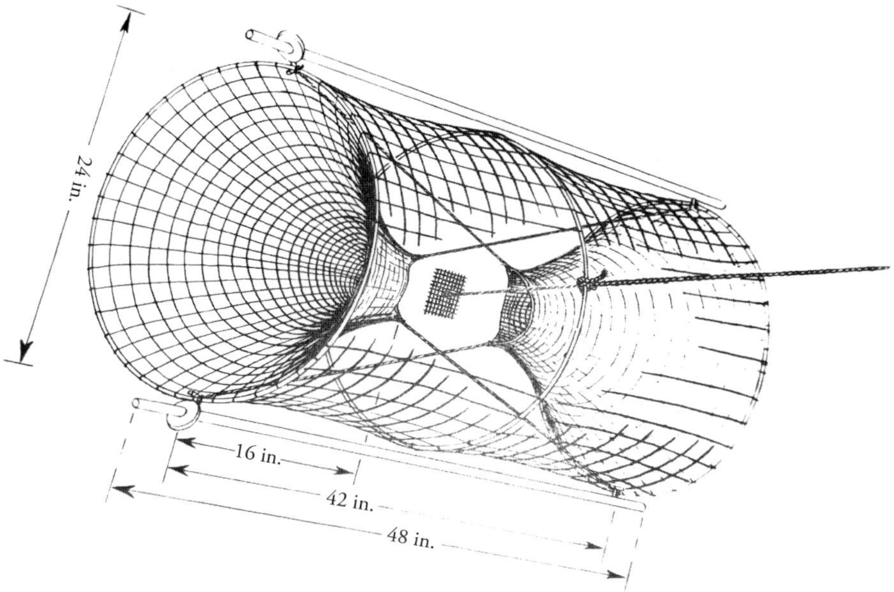

Figure 13.1 Legler trap. Original 1960 line drawing by John Legler (courtesy of Richard Vogt).

The bait cans are hung from the centre of the trap. Checking traps at 4–8-hour intervals usually produces a greater catch than if the traps are checked once every 24 hours. Legler found that traps of 48 cm diameter, 83 cm long, with a throat of 30 cm deep, were most effective. These traps are easy to use, inexpensive to build, and easy to transport.

A 2–4 m nylon line is attached to one of the end hoops and the traps can be hurled into the water with the grace of a fly fisherman, landing next to a log or eddy or other probable capture site. The loose end is then tied to a tree, stake or rock such that if the trap fills with turtles they do not move away with it; other animals, such as crocodilians, also cannot move it great distances. With this line, the trap can be positioned above the water so that turtles do not drown. One advantage of the Legler trap design is their convenience, since researchers do not have to get into the water to set or check them. The traps are also lightweight, of small size, easy to transport and store, and relatively inexpensive. Iverson (1979) modified the trap and made them out of galvanized chicken wire; his design was considerably less expensive. The traps can be collapsed and later reshaped when used, but they may not hold up well over a long (3–4 month) period of sampling. Since they are made of metal, there are fewer problems with otters, piranhas, and crocodiles making holes in the mesh, although I have found these traps to be less effective than the standard Legler trap. This standard trap design can also be modified to trap turtles of larger sizes.

There are two very important considerations when using these funnel traps: the length and diameter of the throats. The throats must be kept taut and be long enough, and the opening small enough, such that it is not just as easy to swim out as it is to swim in. Long throats keep the turtles either above the opening or below it trying to find a way out in the corners. The classic Leger trap had throats that were 7.5 cm high and 25 cm wide.

Legler and Iverson traps are highly useful for carnivorous or omnivorous species; more herbivorous species are difficult to attract into hoop traps without leads. These traps work on the principle that the turtles are active and hungry or sexually active; the turtles must be active, otherwise they will not enter traps. I found that traps baited with live Painted Turtles (*Chrysemys picta*) attracted other painted turtles into the traps in the early spring when they were copulating, but not yet feeding (Vogt, 1979). A decade later, Frazer et al. (1990) independently discovered this phenomenon again for the same species. However, they did not place their turtles in escape proof containers in the traps and noted that a number of their 'bait turtles' escaped; this was their first notion that turtles were readily going in and out of their traps, an obvious bias in interpreting any results.

Getting out of the trap is one of the major problems with these types of traps, especially if the throats are not long enough and the openings are too large. Kennett (1992) described a further more complicated modification in this trap design to make the traps virtually escape-proof. His basic design of the single throated turtle trap is the same, with an entry section composed of a funnel entrance to reach the bait and a holding section from which the turtles cannot escape. The two parts are joined by a rectangular tunnel of wire mesh with a one-way plastic mesh door that allows turtles to enter the holding compartment but not leave. This solves the problems of turtles going in and out of the traps, and also allows for the traps to be more easily positioned such that the holding pen is at least partially out of water. In this manner, traps can be left unattended for several days without drowning or losing the catch.

Hoop and fyke nets

The commercial fishing industry uses double-throated hoop nets of the same general design as the basic turtle trap, but uses seven hoops of varying sizes from 40 to 60 cm diameters. Long-fingered, crowfoot-style throats tied to the second and fourth hoops let the turtles swim in but make it very difficult for them to swim out. These traps must be set either by boat and or by walking into the water. A further modification of this trap is to put a lead on the front of it and another trap on the other end of the lead creating an aquatic drift fence (Vogt, 1980b). These can be constructed of varying sizes, depending on the depth of the water. The lead should be at least as deep as the water so that when a turtle hits the lead, it cannot get above it or below it and so swims along the lead; either way it swims, it ends up in a trap. Turtles are not attracted to these traps by bait; they are merely guided into them as they are trying to get by the leads. Care should be taken to clean the traps of leaves and debris should they interfere with the entrance to the trap.

There are a number of potential capture biases when using hoop nets. Nets set in front of nesting beaches are obviously going to be biased for females; males of some species swim around more searching for females and so are more likely to be captured. I have used this technique with great success in ponds, lakes, streams, and fast-moving rivers, and for all types of turtles and habitats, from the Mississippi River in Wisconsin to the Amazon Basin in Brazil. Hoop nets are the best way to catch a large number and variety of turtles from whatever aquatic habitat; all species regardless of feeding preference are caught in these traps.

A further modification produced an even more efficient trap, the fyke net (Figure 13.2). It has a rectangular net box attached to the front of the hoop net. Besides having the lead that runs from the centre of each fyke net, wings can be attached to the sides of each net extending 10–20 m. The wings interrupt turtle migration routes or intercept movements to and from basking sites, and direct the turtles into the traps. This further enhances the efficiency of capturing herbivorous turtles that are not attracted to bait. It is also the most cumbersome and expensive of all turtle traps. Donald Tinkle and Justin Congdon, in their long-term turtle studies on the E.S. George Reserve in Michigan (USA), made stationary leads and merely put the traps into the water attached to the leads whenever they were ready to sample. This likely enhanced their capture effort over the years in that the turtles were accustomed to moving along the leads when the traps were not there.

Figure 13.2 Fyke net. (a) Fyke net being set near San Roque, Oaxaca, Mexico, in 2014: this trap caught 38 *Kinosternon oaxaca* in one night; (b) close-up of throat of fyke net; (c) baiting a fyke net with a female turtle (courtesy of Richard Vogt).

Trap placement

The use of all of these traps requires at least some basic knowledge of where to put the traps. All species of aquatic turtles I have studied can be trapped effectively with turtle traps with leads in all bodies of water. We sometimes had eight species of turtles in a single trap in our studies on the Pearl River in Mississippi (McCoy and Vogt, unpublished data). Smaller bodies of water require smaller diameter hoops and shorter leads, or only the use of wings. A variety of different habitats need to be sampled in any heterogeneous body of water in order to capture a representative sample of the resident turtles. Traps should be set with the leads of the trap parallel to basking logs and between basking logs and deep water. Softshell turtles (Trionychidae) and many other species bask or nest on sandbars, so nets set parallel to sandbars or in front of nesting areas in 1–2 m of water are often effective.

Map turtles (*Graptemys*) can be trapped in the fast moving portions of rivers. The nets must be set parallel to and attached to the shoreline in 1–2 m of water. Traps must often be tied in several places to keep at least part of the trap out of the water, or with floats placed in the end chamber to allow the turtles to get their heads above water to breath and to keep the current from rotating the traps, making the leads ineffective. Turtles usually feed and forage within 5–10 m along the shoreline in water 0.5–2 m deep. Traps should be set parallel to the shoreline as turtles often move along the shoreline in search of prey or forage plants. Cursory observations before setting traps are important. Central American River Turtles (*Dermatemys mawi*) leave tell-tale signs where they have recently been feeding along the shoreline by the distinctive bite marks left on the leaves of shoreline vegetation. In shallow 2–3 m deep lakes where there is aquatic vegetation throughout the lake, turtles can often be trapped everywhere. I had success collecting Alabama Red-bellied Turtles (*Pseudemys alabamensis*) by stringing a series of 10 turtle traps with leads, together spanning 300 m across a large bed of submerged aquatic vegetation, in Mobile Bay, Alabama (USA).

Trammel nets

There are times when turtles are not active or are in deep water where fyke nets or turtle traps will not work; that is when trammel nets become important (Vogt, 1980b). Trammel nets are made of three tiers of nylon netting hung on common float and lead lines (Figure 13.3). The two outside nets (called walling) are usually made of thicker, #9, nylon twine, and have a square mesh size of 30–75 cm, depending on the size of the turtles to be captured. The finer inside mesh is made of lightweight multifilament gill netting of 3–12 cm square mesh size. The inside net hangs loosely between the two outside walls and is 30% deeper than the outside walls. When the turtle pushes into the net, it pushes the lightweight netting though the outside walls forming a pocket of netting surrounding it, causing the turtle to be held in the net. Using trammel nets will catch a much wider size range of turtles than standard gill nets. These nets usually are used in sections of 100 m, and range in depth from 2 to 6 m. Several of these nets can be strung along the shore parallel to a nesting beach, across bays and inlets to lakes, parallel to shorelines in rivers, and with great success in water 4–6 m deep where herbivorous turtles are foraging. These nets must be checked regularly (at least every 4 hours) to ensure that the turtles do not drown. Monofilament trammel nets are also available; they are

Figure 13.3 Trammel net being set in Rio Trombetas, Amazonas, Brazil (courtesy of Richard Vogt).

less expensive, but harder to repair. Trammel nets alone only work if turtles are moving. However, inactive turtles can be stimulated to move and be captured in trammel nets by driving them with a carp horn (Vogt, 1980b), even when they are in winter dormancy.

The carp horn was designed by commercial fishermen on the Mississippi River and was sold commercially in hardware stores in Lansing, Iowa, specifically for driving carp and other schooling fish down the river channels into gill nets, particularly during the colder months of the year. It is a funnel constructed out of heavy-gauge galvanized sheet metal 16 cm in diameter, 21 cm high, and tapering into a sleeve 4 cm in diameter. A 4 cm diameter, 3 m long wooden pole is bolted onto the sleeve. It is important to use heavy gauge sheet metal or the funnel will collapse upon being pounded forcefully into the water. Any device that makes a lot of disturbance in the water is perhaps effective. Henry Bates (1863) reported Amazonian Indians beating sticks and branches on the surface of the water to drive sideneck turtles (*Podocnemis* spp.) into their nets. Amazonian fishermen today often beat the surface of the water with canoe paddles to drive fish and turtles out of their hiding places and into their nets.

Trammel nets can be set completely across ponds, lakes, backwaters, inlets, river backwaters, between basking logs and deep water, adjacent to nesting beaches, or below hibernacula. When turtles are seen on a basking log, the idea is to get the trammel net between the turtles and their escape route as fast as possible. This is done by driving the boat as fast as possible to the shore near the basking log. Upon nearing the shore, the motor is put in reverse and the person on the bow of the boat throws out the anchor attached to a trammel net and proceeds to feed the net into the water as the boat makes a semicircle around the basking site; the other end of the trammel net is hurled into the water along with an anchor. Once the net is in place, move between the shore and the basking log going back and forth parallel to the net plunging the carp horn rapidly and forcefully into the water making a loud popping sound. I have never tested how the sound carries underwater, but turtles move rapidly into the net and in a matter of minutes 50–100 turtles have been caught. This works successfully for capturing turtles even when they are dormant or aestivating (Vogt, 1980b).

Locations of turtle aggregations must be known in advance to use this technique. In the Mississippi River in Wisconsin, three species of map turtles (*Graptemys* spp.) aggregate behind wing dams to overwinter. By setting trammel nets 30 m below these wing dams and pounding the water with a carp horn above the hibernaculum, turtles are induced to move out and into the net. The nets should be bottom weighted such that they sink to the bottom to be most effective at hibernacula or aestivating sites. We used this technique at aestivating sites for Central American river turtles (*Dermatemys*) in the deep whirlpool areas of the Rio Lacantun in Chiapas, Mexico, and also for Yellow-spotted Sidenecks (*Podocnemis unifilis*) in the Brazilian Amazon. The technique is also effective for rapid sampling of small ponds 100–300 m in diameter. Once the net is set in the centre of the pond, pound the carp horn from the shore to the net on both sides of the net.

13.3 Capture biases

All methods are perhaps biased towards a particular sex, size class, life stage, or species of turtle. Ream and Ream (1966) discussed some of the problems with some of the methods. However, biases can be recognized and factored into data interpretation (by developing a correction factor) during long-term mark–recapture studies. One of the problems with some of the concerns about sampling biases is the misconception that turtle populations are supposed to have a 1:1 sex ratio, so that if there is any drastic deviation from this ratio, there must be a sampling bias. However, researchers have documented unequivocally that sex is determined by incubation temperature in many species of turtles (Bull and Vogt, 1979), that hatchling sex ratios of some species in any given year can be highly biased in favour of one sex (Vogt and Bull, 1984), and that some species of turtles have a highly biased adult sex ratio that can fluctuate through time, no matter which way they were collected (e.g. 4:1 females to male; Vogt, 1980a). It may be that some sampling techniques are not biased at all, especially during long-term studies!

The basic thing to remember about sampling for turtles is that traps or nets must be set where the turtles are likely to be found; do not randomly throw traps out into lakes and rivers in order to satisfy rigid sampling requirements for analysing data afterwards (e.g. completely randomized sampling), or there will not be sufficient data to analyse!

Acknowledgements

I thank all of my field assistants and collaborators from the USA, Mexico, and Brazil who helped me perfect these trapping methods; special thanks to Mike Ewert, Jack McCoy, and John Legler.

References

Bates, H.W. (1863). *The Naturalist on the River Amazon*. London: John Murray.

Braid, M.R. (1974). A bal-chatri trap for basking turtles. *Copeia*, **1974**, 539–40.

Buhlmann, K.A., Jones, J.L., and Humphries, W.J. (2013). Snorkeling surveys. In G.J. Graeter, K.A. Buhlmann, L.R. Wilkinson, et al. (eds) *Inventory and Monitoring: Recommended*

Techniques for Reptiles and Amphibians. Technical Publication IM-1. Birmingham, AL:Partners in Amphibian and Reptile Conservation, pp. 83–6.
Bull, J.J., and Vogt, R.C. (1979). Temperature controlled sex determination in turtles. *Science*, **206**, 1186–8.
Carpenter, C.C. (1955). Sounding turtles: a field locating technique. *Herpetologica*, **11**, 120.
Carr, A.F. (1952). *Handbook of Turtles*. Ithaca, NY:Comstock Publishing.
Chaney, A., and Smith, C.I. (1950). Methods for collecting map turtles. *Copeia*, **1950**, 323–4.
Chapin, K.J., and Meylan, P.A. (2011). Turtle populations at a heavily used recreational site: Ichetucknee Springs State Park, Columbia County, Florida. *Herpetological Conservation and Biology*, **6**, 51–60.
Frazer, N.B., Gibbons, J.W., and Owens, T.J. (1990). Turtle trapping: preliminary tests of conventional wisdom. *Copeia*, **1990**, 1150–2.
Gibbons, J.W. (1970). Terrestrial activity and the population dynamics of aquatic turtles. *American Midland Naturalist*, **83**, 404–14.
Gibbons, J.W., and Semlitsch, R.D. (1981). Terrestrial drift fences with pitfall traps—an effective technique for quantitative sampling of animal populations. *Brimleyana*, 7, 1–16.
Horne, B.D., Brauman, R.J., Moore, M.J.C., et al. (2003). Reproductive and nesting ecology of the yellow-blotched map turtle, *Graptemys flavimaculata*: implications for conservation and management. *Copeia*, **2003**, 729–38.
Huestis, D.L., and Meylan, P.A. (2004). The turtles of Rainbow Run (Marion County, Florida): observations on the genus *Pseudemys*. *Southeastern Naturalist*, **3**, 595–612.
Iverson, J.B. (1979). Another inexpensive turtle trap. *Herpetological Review*, **10**, 55.
Johnston, G.R., Lau, A., and Kornilev, Y.K. (2011). Composition of a turtle assemblage in a northern Florida blackwater stream. *Florida Scientist*, **74**, 126–33.
Kennett, R. (1992). A new trap design for catching freshwater turtles. *Wildlife Research*, **19**, 443–5.
Kornilev, Y.V., Dodd, C.K., Jr., and Johnston, G.R. (2010). Linear home range, movement, and spatial distribution of the Suwannee Cooter (*Pseudemys concinna suwanniensis*) in a blackwater stream. *Chelonian Conservation and Biology*, **9**, 196–204.
Kornilev, Y.V., Dodd, C.K., Jr., and Johnston, G.R. (2012). Retention of paint marking for individual identification of free-ranging basking aquatic turtles (Suwannee Cooters; *Pseudemys concinna suwanniensis*). *Herpetological Review*, **43**, 61–4.
Legler, J.M. (1960). A simple and inexpensive device for trapping aquatic turtles. *Utah Academy of Science Proceedings*, **37**, 63–6.
Lindeman, P.V. (1998). Of deadwood and map turtles (*Graptemys*): using replicated spotting scope counts of basking turtles. *Chelonian Conservation and Biology*, **3**, 137–41.
Lindeman, P.V. (1999). Surveys of basking map turtles *Graptemys* spp. in three river drainages and the importance of deadwood abundance. *Biological Conservation*, **88**, 33–42.
Morales-Verdera, S.A., and Vogt, R.C. (1997). Terrestrial movements and activity patterns of *Kinosternon leucostomum*. *Copeia*, **1997**, 123–30.
Plummer, M.V. (1979). Collecting and marking. In M. Harless and H. Morlock (eds) *Turtles Perspectives and Research*. New York: John Wiley and Sons, pp. 45–61.
Ream, C., and Ream, R. (1966). The influence of sampling methods on the estimation of population structure in painted turtles. *American Midland Naturalist*, **75**, 325–38.
Schwartz, C.W., and Schwartz, E.R. (1974). The three-toed box turtle in central Missouri: its population, home range, and movements. *Missouri Department of Conservation, Terrestrial Series*, **5**, 1–28.
Sexton, O.J. (1959). Spatial and temporal movements of a population of the painted turtle, *Chrysemys picta marginata* (Agassiz). *Ecological Monographs*, **29**, 113–40.

Vogt, R.C. (1979). Spring aggregating behavior of painted turtles, *Chrysemys picta* (Reptilia, Testudines). *Journal of Herpetology*, **13**, 363–5.

Vogt, R.C. (1980a). Natural history of *Graptemys pseudogeographica* and *Graptemys ouachitensis* in Wisconsin. *Tulane Studies in Zoology*, **22**, 17–48.

Vogt, R.C. (1980b). New methods for trapping aquatic turtles. *Copeia*, **1980**, 368–71.

Vogt, R.C., and Bull, J.J. (1984). Ecology of hatchling sex ratio in map turtles. *Ecology*, **65**, 582–7.

Vogt, R.C., and Hine, R.L. (1982). Evaluation of techniques for assessment of amphibian and reptile populations in Wisconsin. In N.J. Scott (ed) *Herpetological Communities*. Wildlife Research Report 13. U.S. Department of the Interior, Fish and Wildlife Service, pp. 201–17.

14

Terrestrial turtles and tortoises

Margaretha D. Hofmeyr and Brian T. Henen

14.1 Introduction

Terrestrial chelonians include all species in the family Testudinidae (tortoises) and some species of the families Emydidae and Geoemydidae. Chelonians are among the most endangered vertebrate groups with more than 60% of modern species considered threatened or already extinct (Turtle Taxonomy Working Group, 2014). Increased threats to the conservation status of terrestrial chelonians necessitate surveys to demarcate areas of occupancy and monitor changes in the status of species.

It is vital to define the goal of a study clearly and set objectives to attain this goal before initiating surveys (Tucker et al., 2005; see Chapter 2). Study objectives can vary in complexity from the presence or absence of a species to changes in population abundance over time and space (Table 14.1). Numerous methods exist to survey terrestrial chelonians, and researchers should consider all relevant factors before selecting a survey method and study design. Of primary importance is that results from the technique should accomplish the study objectives (see Table 14.1). Other factors to consider are species traits (e.g. rare or common; small- or large-bodied), habitat characteristics (e.g. rocky outcrops or plains; dense or sparse vegetation), and the temporal and/or geographic scope of the study (Foster, 2012). Because many of these factors influence cost, one should consider in advance if there are sufficient funds, personnel, and equipment for the study.

This chapter introduces concepts that need consideration when planning surveys, and outlines assumptions, biases, and outcomes of methods suitable for terrestrial chelonians. Further reading on methodology and data analyses is necessary before initiating a study.

14.2 Concepts in survey design

A complete count (census) of individuals in a population or study area is frequently not realistic because the population or study area is too large. An alternative is to count individuals in replicate samples, representing a subgroup of the study area or population, and extrapolate the sample estimate to the whole population (Greene, 2012). Counts are seldom complete because observers do not detect all individuals during surveys. It is possible to estimate absolute abundance from counts if the samples were representative

Reptile Ecology and Conservation. Edited by C. Kenneth Dodd, Jr. © Oxford University Press 2016.
Published 2016 by Oxford University Press.

Table 14.1 *Common methods to survey terrestrial turtles and tortoises with primary outcomes and limitations; many methods can support additional data collection, such as activity level and location. Outcomes include PA (presence–absence), RA (relative abundance), AA (absolute abundance), B (behaviour), HP (habitat preference), PT (predation threats), and PSS (priority survey sites).*

Survey technique	Outcomes	Limitations
Mark–recapture	AA	Costly, time and staff intensive
Visual encounters		
Haphazard trails	PA	Statistical inferences limited or impossible
Randomized walks	PA; RA	Low accuracy for RA
Transect surveys	RA; AA	Costly, time and staff intensive
Stratified surveys	RA; AA; HP	Costly, time and staff intensive
Distance sampling	RA; AA	Less suitable for cryptic, rare, small, and rupicolous species
Surrogates	PA; RA	Low accuracy except for burrowing species
Dogs	PA; RA	Costly, dogs vulnerable to extreme environmental conditions
Motion cameras	B; PT	Limited application
Road driving	PA	No statistical inferences possible
Pole checking	PA	No statistical inferences possible
Niche modelling	PSS	Field verification needed for localities where species is not known to occur

of the population and the counts were adjusted for undetected animals (Greenwood and Robinson, 2006a, 2006b). When the counts are incomplete or biased, but correlate to the actual total number of animals, they represent an index and not an estimate (Greenwood and Robinson, 2006a, 2006b). The usefulness of indices depends on knowledge of the type of relationship between the index and actual abundance, and on the condition that the relationship remains constant over space and time (Greene, 2012).

The selection of sample locations can be random (probability based) or non-random (judgemental or convenience based). Researchers often use convenience sampling for exploratory purposes and choose sampling sites because they are easily accessible. In judgemental sampling, there is prior knowledge of the species' ecology and researchers use professional judgement to select sampling areas considered suitable for the species. Although non-random sampling is generally not advisable, judgemental sampling may be the best option for presence–absence assessments or when studying rare species (Tucker et al., 2005). A random sampling strategy allows one to calculate sampling error and infer population size. Researchers can use simple random sampling when the habitat is homogenous, but it is better to use stratified sampling in heterogeneous habitats to account for possible differences in abundance among strata (Greenwood and Robinson, 2006a). The number of sampling units from each stratum should reflect its contribution to the total study area (Greene, 2012).

The reliability of population estimates depends on accuracy and precision. Accuracy indicates how close the estimate is to its true value, that is, how small the bias of the estimate is. Precision refers to the level of variation among repeated estimates (Greenwood and Robinson, 2006a). Precision increases with the number of sampling units but sample number should not compromise unit size (Greenwood and Robinson, 2006a), which needs to be relatively large to find terrestrial chelonians. Effort and costs will rise with an increase in sample number; consequently, it is advisable to calculate in advance how many samples are needed to obtain an acceptable percentage relative precision (PRP; Greenwood and Robinson, 2006a). These authors provide equations to calculate sample size at a pre-determined PRP, the PRP for specific sample sizes, and the precision attainable for fixed costs.

A biased estimate deviates from the true population value and lowers accuracy. It is possible to reduce bias by applying good field techniques and a standardized protocol that would limit differences in environmental conditions, observer skills, effort, and equipment over time and space (Greenwood and Robinson, 2006a). Bias also results from non-compliance to assumptions inherent to a survey method. Assumptions common to many abundance estimation procedures are that (1) the population is closed; (2) observers record each individual only once during a survey; (3) there is no observer-related bias; (4) observers detect all individuals; and (5) all individuals are equally detectable.

Closure means that population size remains constant during the sampling period, that is, there are no births, deaths, immigration, or emigration (Greene, 2012). Due to specific traits of terrestrial chelonians (see Couturier et al., 2013), closure is often satisfied for short sampling periods, for example, within season estimates, but not for longer periods. It is easy to meet the second assumption if researchers mark individuals (see Chapter 4) properly for later identification. The third assumption is more difficult to meet. Observers differ in their ability to detect individuals and experience or training does not eliminate observer bias (see Gardner et al., 1999). One solution is to use telemetered animals or models to establish detection abilities of observers and adjust for this bias in population estimates.

Detectability is important when making inferences about population abundance from survey results (Anderson et al., 2001; Refsnider et al., 2011; see Chapter 27). Inactive terrestrial turtles and tortoises often hide under dense vegetation or in crevices and burrows. These individuals are not detectable unless observers search the habitat systematically and thoroughly. Individuals in the open are detectable but observers do not always see and count them all (Greene, 2012). Because time of day, season, and specific weather conditions influence chelonian activity, these factors affect detectability, and it is best practice to schedule surveys for peak activity periods. Body size also influences detectability; small-bodied species are less visible than are large-bodied species, and juveniles are typically less detectable than are adults (Gardner et al., 1999). Habitat also plays a role; the detectability of individuals in densely vegetated or rocky habitats is lower than in sparsely vegetated terrain (Gardner et al., 1999). Consequently, if males and females use different microhabitats, detectability may differ between sexes.

Several procedures exist to estimate detection probabilities, which differ among survey methods and species. A useful procedure is to compare the chosen survey technique (e.g. visual encounter survey) with the detection of telemetered individuals (see Chapter 9) spaced across the study site. Researchers can estimate detectability by comparing the number of telemetered animals encountered during surveys to the total number known within the survey area (see Refsnider et al., 2011, and references therein). Similarly, researchers can place a known number of tortoise models within the survey area to estimate detectability (Gardner et al., 1999), although the models' lack of authenticity may lessen their detection. Another option is to express counts for a quick survey method relative to counts of a complete census for subsamples of the habitat surveyed (see Bart and Earnst, 2002). Whichever method is used, it is preferable to estimate and report detectability for each cohort, habitat type, season, and year.

Non-detection of a species does not prove that it is absent; its density might simply be low. To prove absence at an acceptable significance level (α) requires a minimal number of sampling trials (N_{min}) for a species with known detectability (p) in a particular environment and conditions (Refsnider et al., 2011; Kapfer et al., 2012). Equation 14.1 allows one to calculate N_{min} to state absence at a specific confidence level, for example, 95%:

$$N_{min} = \log\alpha \,/\, \log(1-p) \qquad (14.1)$$

Conversely, equation 14.2 allows one to calculate α for absence after conducting a specific number of surveys (Refsnider et al., 2011; Kapfer et al., 2012):

$$\alpha = (1-p)^{N_{min}} \qquad (14.2)$$

For example, application of equation 14.1 shows that managers need to do three surveys ($\log 0.05/\log(1-0.64) = 2.93$) to have 95% confidence that a species with detectability of 0.64 is absent. A lower detectability will require even more surveys, for example, eight surveys would be necessary if detectability were 0.32.

14.3 Review of survey methods

14.3.1 Mark–recapture

Many chelonian researchers consider mark–recapture (see Chapter 27) the most reliable method to estimate species abundance (Buică et al., 2013; Couturier et al., 2013). The technique requires that researchers capture and mark a subsample of the population, release the marked animals, and recapture some of them again in follow-up surveys (Fasham and Mustoe, 2005). Terrestrial chelonians are easy to mark (see Chapter 4), either permanently through scute filing, or through painting temporary spots or identification numbers on the shell. The basic assumption underlying this technique is that marked individuals disperse through the population so that the ratio of marked to unmarked individuals in a second capture effort represents the ratio of marked animals to the whole population (Fasham and Mustoe, 2005; Greenwood and Robinson, 2006b). To test if full mixing occurred, researchers can use a contingency table approach to evaluate second capture data on subsections of the study area (see Greenwood and Robinson, 2006b, for calculations).

Other fundamental assumptions are that capture probability is equal for all individuals in the population, and in some applications, that the population is closed (Fasham and Mustoe, 2005). For terrestrial chelonians, juveniles are more difficult to detect than are adults, and the activity rate for males and females may differ. Consequently, it may be necessary to exclude individuals below a certain size from the calculations or to evaluate groups separately (Greenwood and Robinson, 2006b; Buică et al., 2013). It is reasonable to assume closure in short-term studies because terrestrial chelonians have limited mobility, high adult survivorship, and low detection of new-borns (see Couturier et al., 2013, and references therein). One can test for closure over several recaptures, and if the population is not closed, one can use an open population method with fewer assumptions, although these estimates are less precise (Greenwood and Robinson, 2006b).

The simplest application of mark–recapture requires only one recapture session, but long-term monitoring programmes require regular sampling, sometimes over years. To limit bias in long-term monitoring, it is good practice to keep trapping effort constant among surveys (Greenwood and Robinson, 2006b). Data analysis is relatively straightforward for an estimate that requires only two-samples (e.g. Lincoln–Peterson index of density), but can be complex for estimates that require multiple samples, for which there are several software programs (e.g. Capture, Jolly, and Mark) available to assist with analysis.

14.3.2 Visual encounter surveys

Visual encounter surveys (VESs) are probably the most widely used technique to inventory or monitor terrestrial turtles and tortoises. The technique entails non-random or random searches by one or more observer(s) for individuals of one or more species within a selected area. Researchers or managers can modify the methodology easily to answer simple or more complex questions, such as to confirm presence or absence of a target species, or to estimate abundance and trends in abundance over time or space (Foster, 2012). By collecting more than locality data during surveys (see Table 14.1), the method can also provide information on population characteristics (e.g. sex ratios and population structure), the species' habitat preferences, and its activity pattern under different weather conditions and at different times of the day and year.

VESs to estimate relative abundance can be time-, distance-, or area-constrained, and the results can be expressed as either individuals per person-hour, per distance, or per unit area (Fasham and Mustoe, 2005; Foster, 2012). For example, if four observers participate in a five-hour survey and find 12 turtles, sampling effort would amount to 20 person-hours and relative abundance to 0.6 individuals per person-hour. Time-constrained surveys should identify a pre-defined search time and exclude time required to process captured individuals (Eekhout, 2010). The search method can be haphazard or randomized walks, using computer software to generate the latter as a random sequence of compass headings (Vonesh et al., 2009). Time-constrained surveys are suitable for multiple searches of the same study site or a comparison of areas with similar habitats, but do not allow realistic comparisons if terrain differs among sites. In such instances, a difference in catch per unit effort may be due to habitat differences rather than differences in the abundance of the species. In most instances, area-constrained searches within transects are more productive, although often more costly.

Transect sampling is an area-constrained VES technique (Figure 14.1(a)) that has been used extensively for terrestrial chelonians (Gardner et al., 1999; Foster, 2012, and references therein). When surveying transects, a number of observers usually walk

Figure 14.1 (a) Using visual encounter surveys after a fire to detect critically endangered Geometric Tortoises (*Psammobates geometricus*) in South Africa. Observers walk parallel to one another along a designated transect. Increasing the number of observers helps to increase detection probabilities. (b) Dogs have a keen sense of smell and have been used to find terrestrial turtles in many regions of the world. Here, a Belgian shepherd has found an angulate and tent tortoise in one study area. (c) Dogs can find terrestrial turtles and tortoises in terrain where dense groundcover obscures hard to find species. (d) An example of an environmental niche model for the Greater Padloper (*Homopus femoralis*) in South Africa. The greyscale map indicates locality points as circles with habitat suitability declining from excellent (white) to hardly suitable (grey). Photos by Bernard Wooding (a), Margaretha Hofmeyr (b), and Theunis Hofmeyr (c), and map by Pierre Fouche (d).

a fixed distance apart in parallel lines along the transect length. Surveys that involve intensive searches for individuals in the open and in cover potentially realize complete detection, and allow estimates of population abundance (see Chapter 17). Most often, however, search effort is less intensive and detection is incomplete. As a result, transect surveys often underestimate animal abundance (Foster, 2012) and simply provide an index of population size. It is possible to convert an index of abundance to an estimate by incorporating the species' detectability (if known) into the equation. It is also possible to enhance detection by adjusting the distance between observers according to characteristics of the species and habitat. Although observers may detect most or all individuals when spaced 5–10 m apart in desert habitat, this spacing would impair detection in dense vegetation, where a smaller distance between observers would be more appropriate.

It is best practice to standardize transect size to maximize precision (Greenwood and Robinson, 2006a), but there are advantages to not fixing dimensions for a specified size. Such flexibility allows one to modify transect width according to the number of available surveyors and the distance they are spaced. Furthermore, researchers can combine two strips (away from and back to the starting point) into one transect width to maximize efficiency. After width has been determined, transect length is calculated according to the required transect size. For example, if transect size is set at 5 ha or 0.05 km^2 and five surveyors walk 5 m apart away and back to the starting point, transect width would be 50 m and length 1000 m or 1 km.

14.3.3 Line distance sampling

Distance sampling has been used frequently to estimate population size of terrestrial chelonians (Anderson et al., 2001; Leuteritz et al., 2005; Inman et al., 2009) but line distance sampling can underestimate density and abundance compared to other survey techniques (Walker, 2012; Couturier et al., 2013). Distance sampling differs from transect sampling in that the method fits a detection function to the distances of individuals from the line to compensate for decreasing detectability with distance (Buckland et al., 2001; Fasham and Mustoe, 2005; Thomas et al., 2010). The survey protocol for distance sampling can be through lines or points. Line distance sampling (LDS) is more suitable than point sampling for terrestrial chelonians because the area covered by point sampling is too small to find enough individuals. Important assumptions of LDS are that (1) target objects are distributed randomly with reference to the line, (2) all objects on the line are detected, (3) the object does not move in response to the observer, and (4) the distance to each object is measured accurately (Fasham and Mustoe, 2005).

The first and fourth assumptions are relatively easy to meet, respectively, by using a systematic or random sampling design and by training observers to measure distances accurately or use a laser range finder (Fasham and Mustoe, 2005). The second assumption is often difficult to meet, particularly for species that exhibit seasonal or daily periods of inactivity (Inman et al., 2009; Couturier et al., 2013), or when body size is small (e.g. subadults; Anderson et al., 2001). When detectability on the line (g_0) is not 100%, it is necessary to independently survey to determine g_0 (Buckland et al., 2001), increasing the effort and cost of the survey. The third assumption should not be problematic for terrestrial chelonians because they often halt when disturbed. However, if they move away, they may go undetected or be counted twice if not marked. Fasham and Mustoe (2005) recommend that surveyors standardize methodology to avoid additional bias; for example, they should maintain a constant walking speed and not survey when weather conditions are unfavourable.

For LDS to be effective, surveyors need to detect at least 60–80 individuals, although 40 individuals might provide sufficient precision (Buckland et al., 2001). It is advisable to perform an exploratory study to assess if LDS is suitable for the species and terrain before starting surveys. The exploratory procedure entails walking a number of line-transects until approximately 10–15 observations are made and then use the distance of the pilot transects, in combination with the number of observations, to calculate encounter rate. Researchers then use encounter rate to calculate the total distance they need to sample for a density estimate with a coefficient of variation (CV) between 10 and 20% (Buckland et al., 2001; Smith et al., 2009). If the pilot study found 10 individuals over a 5 km distance, one needs to sample 67 km for a 15% CV. For LDS, one or two surveyors walk along a line and measure the perpendicular distance of the object from the line when detecting the target species. If the object is not perpendicular to the line when detected by the surveyor, surveyors can use a compass to measure the angle between the object and the line to calculate perpendicular distance. Instead of measuring the exact distance, it is acceptable to record individuals as being within a band of specific width from the transect line, for example, 0–2 m, 2–5 m, 5–10 m, and >10 m (Smith et al., 2009; Foster, 2012). The free software 'Distance' (http://distancesampling.org/) can evaluate results of the pilot study, plan the final study and analyse the data to obtain estimates of density and abundance with confidence intervals (Smith et al., 2009; Thomas et al., 2010).

LDS is used widely to survey adult *Gopherus polyphemus* and *G. agassizii* (Smith et al., 2009; U.S. Fish and Wildlife Service, 2015). When estimating abundance, researchers consider not only the probability that a tortoise is above ground, but also the detectability of individuals above ground. LDS is often not feasible for small or cryptic species, particularly if population density is low, because it would not be cost-effective to survey hundreds of kilometres to find sufficient individuals for a reliable estimate. Walker (2012) found that time-constrained VESs were more effective than LDS in detecting the small, cryptic tortoise *Pyxis arachnoides*. For another cryptic species, *Testudo hermanni* (Couturier et al., 2013), mark–recapture provided better estimates of abundance than did either LDS or N-mixture models. These authors highlighted the difficulties of implementing monitoring designs for species with widely fluctuating activity rates that influence detectability.

14.3.4 Surrogates

Surrogates are species-specific sign that scientists use to quantify relative abundance or indicate species presence. For terrestrial turtles and tortoises, sign may include carcasses or partial remains (e.g. bones and scutes), refugia (e.g. burrows or pallets), scat, tracks, nests, eggshell fragments, and courtship rings (Woodman et al., 1986; Keith et al., 2008). Carcasses, scat, burrows, and eggshells may persist for months or years, providing a time-line for activity periods or time-since- and cause-of-death assessments (e.g. predators, drought, or disease; Keith et al., 2008; Berry et al., 2014, and references therein). Some sign provides information on sex, age, or habitat use. Furthermore, managers can use tortoise sign to identify suitable sites for monitoring surveys in areas where tortoise density is low (Keith et al., 2008).

Combinations of burrows, scat, or other sign can indicate the relative density of *G. agassizii* (see Woodman et al., 1986), providing sign tallies, categories, indices, or ranks of tortoise density. Nevertheless, burrows are the only surrogate that managers have used widely to estimate population parameters and monitor populations, particularly of *G. polyphemus*. Earlier assessments equated burrow counts to tortoise density or abundance, but various studies have shown that neither burrow density nor activity status is a reliable index of *G. polyphemus* density (see Smith et al., 2009).

Furthermore, incomplete detection applies equally to burrows as to individual animals (see Section 14.1). Surveyors detect fewer burrows in dense than in sparse habitats (Nomani et al., 2008); consequently, it is good policy to include detectability in calculations.

Gopherus polyphemus digs multiple burrows and it is necessary to inspect every burrow for occupancy instead of using a standard occupancy estimate to convert burrow number to tortoise number (Nomani et al., 2008; Smith et al., 2009, and references therein). Burrow inspection involves the use of a burrow camera system, but it is not always possible to determine if a tortoise occupies a burrow that may be too long for the scope, or when it is impossible to manoeuvre the scope around obstructions in the burrow (Smith et al., 2009). Juveniles use burrows of other wildlife, and their numbers fluctuate considerably among seasons and years, making burrow correlations impossible for juveniles. Factor and multiple regression analyses may demonstrate whether surrogates can successfully estimate the density of terrestrial turtles and tortoises.

14.3.5 Wildlife detector dog surveys

Researchers have effectively employed wildlife detector dogs (WDD) to survey terrestrial chelonians (Cablk and Heaton, 2006; Nussear et al., 2008; Kapfer et al., 2012; Chapter 13). Dogs use visual and olfactory cues to detect target animals, giving them an advantage over human observers. In addition, dogs can cover distances much quicker than humans can (Nussear et al., 2008; Dahlgren et al., 2012). Cablk and Heaton (2006) quantified the efficacy and reliability of dogs in surveys of *G. agassizii* and found that the technique was effective over a range of environmental conditions. The usefulness of this technique in the conservation and management of threatened chelonians is currently being validated in Asia (McCormack, 2010) on Indochinese Box Turtles and in South Africa on Geometric Tortoises (Figure 14.1(b, c); Hudson, 2013).

Domestic dogs must undergo intensive training to make them efficient in wildlife detection. The dogs are trained to find one or more target species (or their sign, e.g. scat) and to indicate their find by a trained alert (e.g. by sitting down) without harming the animal (Cablk and Heaton, 2006). The training programme should be followed by a validation process (certification tests; Cablk and Harmon, 2011) to evaluate safety (no harm to target species or handler), and assess the dog's ability to find the target (accuracy or efficiency) and display a trained alert (reliability; Cablk and Heaton, 2006).

When comparing the efficacy of WDD against humans under natural field conditions, Nussear et al. (2008) found that detectability of *G. agassizii* (ca. 70%) did not differ between dog and human teams, but that dogs found a higher proportion of tortoises in vegetation than did humans. Dog teams required less time to finish the surveys than human teams, but the costs to contract professional dog teams were higher than for human teams. They also found that dogs are more prone to overheating than humans, which potentially restricts the use of dog teams to certain seasons, time of day, or habitats. When comparing the efficacy of dogs versus humans to find Eastern Box Turtles (*Terrapene carolina carolina*), Kapfer et al. (2012) found that dogs were significantly better in finding the target species than humans; dogs detected a similar number of box turtles over 9 dog-hours compared to 316.5 person-hours required by humans. WDD were also more effective in finding juvenile *T. c. carolina*. The authors concluded that WDD might be of particular value for terrestrial species preferring dense vegetation.

Managers can use WDD in surveys to establish presence or absence for distribution maps, to count animals, to collect demographic information, in mark–recapture studies, or even for density estimates if the study design accounts for biases. Potential biases in using WDD in surveys include variability in the efficacy of dogs to find the target, differences in dog handler skills, and environmental effects on olfaction (Cablk and Heaton, 2006; Nussear et al., 2008). It is possible to reduce biases by using the same dog-handler team throughout a study, by restricting searches to a certain period of the day, and by conducting the training and trials under similar environmental conditions (Dahlgren et al., 2012). In view of the potential advantages of WDD surveys in chelonian conservation, a standardized protocol for WDD surveys, as proposed by Kapfer et al. (2012), is highly desirable to improve comparability of population estimates over space and time.

14.3.6 Other methods

Camera traps are useful tools to record the presence of wildlife and to study their activity (Foster, 2012). In contrast to freshwater turtles (see Chapter 13), this method has limited application for terrestrial tortoises and turtles. Nevertheless, cameras are suitable to study species that use a home base (e.g. burrowing tortoises), as has been shown by Boglioli et al. (2003) who used cameras to study mating opportunities of female gopher tortoises. Another application is to use cameras to assess the presence of tortoise predators, such as coyotes (Boarman, 2014).

Herpetologists often survey along roads to gain information on the diversity, geographic distributions, population structure, and relative abundance of reptiles, including chelonians (see references in Steen and Smith, 2006; Foster, 2012). The technique also provides information on road mortality, which can be high in some regions, and used to identify areas of tortoise concentration (Loehr, 2012). Observers drive at a standardized speed to identify and count individuals of one or more species of interest, and express the counts per unit distance or time. The technique has many shortcomings that may bias outcomes, particularly measures of abundance. Biases include that the high temperatures on, and diverse vegetation alongside, the road attract some species, cohorts, or individuals more than others; the population near the road may have declined due to high road mortality; and individuals may have learnt to avoid the road (Steen and Smith, 2006). Specific advantages, however, are that the technique is not labour-intensive and that observers can cover larger areas in shorter time compared to other survey techniques. For this reason, daytime road driving in South Africa is often more effective than VES when assessing chelonian geographic distributions (M.D. Hofmeyr, personal observation), or obtaining study animals for species with low densities (Keswick, 2012). Furthermore, road driving combined with checking for carcasses under telephone poles or property fences have been a useful strategy in South Africa to collect material for genetic evaluations (M.D. Hofmeyr, personal observation).

Environmental niche modelling has important applications in chelonian conservation. The technique relates locality data of a species with environmental variables at these locations to create maps of habitat suitability. Such information helps estimate geographic distributions of species and their genetic lineages, improve survey efficacy for rare species, identify priority conservation areas, and identify areas suitable for species reintroduction (Figure 14.1(d); Anderson, 2012; Marcer et al., 2013). Factors that influence the quality and reliability of niche models include, among other, the accuracy of occurrence data, the level of sampling bias, and an appropriate selection of study region and quantification of model performance (Anderson, 2012). This field of research made significant methodological advances in recent years and has huge potential to identify hotspots and assist in the conservation of rare and threatened species (Marcer et al., 2013).

14.4 Conclusions

The effects of habitat destruction and climate change on terrestrial tortoise and turtle species necessitate studies that enhance the understanding of geographic distributions and size of populations (see Chapter 29). Niche modelling can help design survey strategies to define distribution limits, and identify candidate locations for rare species surveys. When choosing among the many available survey techniques, it is essential to consider budgetary constraints, and to ensure that the selected method will provide the required information. The low activity levels of some terrestrial chelonians make them difficult to study, particularly if they are small and occur in densely vegetated or rocky habitats. In such instances, mark–recapture appears to be the best method to estimate abundance. It is essential to get as much information as possible about the target species and its habitat in advance, and carefully consider which method will work for the species. Also, consider whether it is essential to obtain absolute estimates, which require great effort and cost, or if indices will provide sufficient information to meet the study objective. Whichever study method is selected, researchers should standardize sampling design over time and space, and meet all assumptions as best as possible to limit bias and variance.

References

Anderson, D.R., Burnham, K.P., Lubow, B.C., et al. (2001). Field trials of line transect methods applied to estimation of desert tortoise abundance. *Journal of Wildlife Management*, **65**, 583–97.

Anderson, R.P. (2012). Harnessing the world's biodiversity data: promise and peril in ecological niche modelling of species distributions. *Annals of the New York Academy of Sciences*, **1260**, 66–80.

Bart, J., and Earnst, S. (2002). Double sampling to estimate density and population trends in birds. *The Auk*, **119**, 36–45.

Berry, K.H., Lyren, L.M., Yee, J.L., et al. (2014). Protection benefits desert tortoise (*Gopherus agassizii*) abundance: the influence of three management strategies on a threatened species. *Herpetological Monographs*, **28**, 66–92.

Boarman, W.I. (2014). *Measuring Raven and Coyote Predation of Desert Tortoises: Phase 1.* Unpublished report to Clark County Desert Conservation Program, Las Vegas, NV.

Boglioli, M.D., Guyer, C., and Michener, W.K. (2003). Mating opportunities of female gopher tortoises, *Gopherus polyphemus*, in relation to spatial isolation of females and their burrows. *Copeia*, **2003**, 846–50.

Buckland, S.T., Anderson, D.R., Burnham, K.P., et al. (2001). *Introduction to Distance Sampling. Estimating Abundance of Biological Populations.* New York: Oxford University Press.

Buică, G., Iosif, R., and Cogălniceanu, D. (2013). Demography and conservation of an isolated spur-thighed tortoise *Testudo graeca* population in Dobrogea (Romania). *Ecologia Balkanica*, **5**, 97–106.

Cablk, M.E., and Harmon, R. (2011). Final report: validation and development of a certification program for using K9s to survey desert tortoises, RC-200609, ESTCP project. Available at: http://www.serdp.org/Program-Areas/Resource-Conservation-and-Climate-Change/Natural-Resources/Arid-Lands-Ecology-and-Management/RC-200609.

Cablk, M.E., and Heaton, J.S. (2006). Accuracy and reliability of dogs in surveying for desert tortoise (*Gopherus agassizii*). *Ecological Applications*, **16**, 1926–35.

Couturier, T., Cheylan, M., Bertolero, A., et al. (2013). Estimating abundance and population trends when detection is low and highly variable: a comparison of three methods for the Hermann's tortoise. *The Journal of Wildlife Management*, 77, 454–62.

Dahlgren, D.K., Elmore, R.D., Smith, D.A., et al. (2012). Use of dogs in wildlife research and management. In N.Silvy (ed) *Wildlife Techniques Manual, Volume 1*. 7th ed. Washington, DC:The Wildlife Society Inc., pp. 140–53.

Eekhout, X. (2010). Sampling amphibians and reptiles. In J.Eymann, J.Degreef, C.L.Häuser, et al. (eds) *Manual on Field Recording Techniques and Protocols for All Taxa Biodiversity Inventories, Volume 8*. Belgium:Abc Taxa, pp. 530–57.

Fasham, M., and Mustoe, S. (2005). General principles and methods for species. In D.Hill, M.Fasham, G.Tucker, et al. (eds) *Handbook of Biodiversity Methods: Survey, Evaluation and Monitoring*. Cambridge:Cambridge University Press, pp. 255–70.

Foster, M.S. (2012). Standard techniques for inventory and monitoring. In R.W.McDiarmid, M.S.Foster, C.Guyer, et al. (eds) *Reptile Biodiversity: Standard Methods for Inventory and Monitoring*. Berkeley, CA:University of California Press, pp. 205–71.

Gardner, S., Baard, E.H.W., and le Roux, N.J. (1999). Estimating the detection probability of the geometric tortoise. *South African Journal of Wildlife Research*, **29**, 62–71.

Greene, T. (2012). A guideline to monitoring populations. Version 1.0. In T.Greene and K.McNutt (eds) *Biodiversity Inventory and Monitoring Toolbox*. Wellington, New Zealand:Department of Conservation. Available at: http://www.doc.govt.nz/biodiversitymonitoring/.

Greenwood, J.J.D., and Robinson, R.A. (2006a). Principles of sampling. In W.J.Sutherland (ed) *Ecological Census Techniques: A Handbook*, 2nd ed. Cambridge: Cambridge University Press, pp. 11–86.

Greenwood, J.J.D., and Robinson, R.A. (2006b). General census methods. In W.J.Sutherland (ed) *Ecological Census Techniques: A Handbook*, 2nd ed. Cambridge: Cambridge University Press, pp. 87–185.

Hudson, V. (2013). The development of a conservation detection dog team as an additional tool to survey geometric tortoises. In *Fynbos Forum*. Cape Town, South Africa:Kirstenbosch.

Inman, R.D., Nussear, K.E., and Tracy, C.R. (2009). Detecting trends in desert tortoise population growth: elusive behavior inflates variance in estimates of population density. *Endangered Species Research*, **10**, 295–304.

Kapfer, J.M., Muñoz, D.J., and Tomasek, T. (2012). Use of wildlife detector dogs to study eastern box turtle (*Terrapene carolina carolina*) populations. *Herpetological Conservation and Biology*, 7, 169–75.

Keith, K., Berry, K.H., and Weigand, J.F. (2008). When desert tortoises are rare: testing a new protocol for assessing status. *California Fish and Game*, **94**, 75–97.

Keswick, T. (2012). *Ecology and morphology of the Kalahari tent tortoises,* Psammobates oculifer, *in a semi-arid environment*. Ph.D. Thesis, University of the Western Cape, Bellville, South Africa.

Leuteritz, T.E.J., Lamb, T., and Limberaza, J.C. (2005). Distribution, status, and conservation of radiated tortoises (*Geochelone radiata*) in Madagascar. *Biological Conservation*, **124**, 451–61.

Loehr, V.J.T. (2012). Road mortality in the greater padloper *Homopus femoralis* (Testudinidae). *Chelonian Conservation and Biology*, **11**, 226–9.

Marcer, A., Sáez, L., Molowny-Horas, R., et al. (2013). Using species distributions for rare species conservation. *Biological Conservation*, **166**, 221–30.

McCormack, T. (2010). Training detection dogs for research and conservation of tortoises and freshwater turtles in southeast Asia. *Turtle Survival Alliance Magazine*, **2010**, 70–2.

Nomani, S.Z., Carthy, R.R., and Oli, M.K. (2008). Comparison of methods for estimating abundance of gopher tortoises. *Applied Herpetology*, **5**, 13–31.

Nussear, K.E., Esque, T.C., Heaton, J.S., et al. (2008). Are wildlife detector dogs or people better at finding desert tortoises (*Gopherus agassizii*)? *Herpetological Conservation and Biology*, **3**, 103–15.

Refsnider, J.M., Mitchell, T.S., Streby, H.M., et al. (2011). A generalized method to determine detectability of rare and cryptic species using the ornate box turtle as a model. *Wildlife Society Bulletin*, **35**, 93–100.

Smith, L.S., Stober, J., Balbach, H.E., et al. (2009). *Gopher Tortoise Survey Handbook*. ERDC/CERL TR-09–7. Final report for U.S. Army Corps of Engineers, Washington, DC, 20314–1000.

Steen, D.A., and Smith, L.L. (2006). Road surveys for turtles: consideration of possible sampling biases. *Herpetological Conservation and Biology*, **1**, 9–15.

Thomas, L., Buckland, S.T., Rexstad, E.A., et al. (2010). Distance software: design and analysis of distance sampling surveys for estimating population size. *Journal of Applied Ecology*, **47**, 5–14.

Tucker, G., Hill, D., and Fasham, M. (2005). Planning a programme. In D.Hill, M.Fasham, G.Tucker, et al. (eds) *Handbook of Biodiversity Methods: Survey, Evaluation and Monitoring*. Cambridge: Cambridge University Press, pp. 6–64.

Turtle Taxonomy Working Group: Van Dijk, P.P., Iverson, J.B., Rhodin, A.G.J., et al. (2014). Turtles of the world, 7th edition: annotated checklist of taxonomy, synonymy, distribution of maps, and conservation status. In A.G.J.Rhodin, P.C.H.Pritchard, P.P.Van Dijk, et al. (eds) *Conservation Biology of Freshwater Turtles and Tortoises: A Compilation Project of the IUCN/SSC Tortoise and Freshwater Turtle Specialist Group*. IUCN/SSC.

U.S. Fish and Wildlife Service (2015). *2015 Desert Tortoise Monitoring Handbook*. Reno, NV: Desert Tortoise Recovery Office, U.S. Fish and Wildlife Service.

Vonesh, J.R., Mitchell, J.C., Howell, K., et al. (2009). Rapid assessments of amphibian diversity. In C.K.Dodd, Jr. (ed) *Amphibian Ecology and Conservation: A Handbook of Techniques*. Oxford: Oxford University Press, pp. 263–80.

Walker, R.C.J. (2012). A critical evaluation of field survey methods for establishing the range of a small, cryptic tortoise (*Pyxis arachnoides*). *Herpetological Journal*, **22**, 7–12.

Woodman, A.P., Juarez, S.M., Humphreys, E.D., et al. (1986). Estimated density and distribution of the desert tortoise at Fort Irwin, National Training Center and Goldstone Space Communications Complex. *Proceedings of the Symposium of the Desert Tortoise Council*, **1986**, 81–99.

15
Sea turtles

Seth Stapleton and Karen L. Eckert

15.1 Introduction

Sea turtles have plied the world's oceans for more than one hundred million years (Hirayama, 1998), and zooarchaeological research has uncovered evidence of human–sea turtle interactions that span millennia (Frazier, 2003). Once plentiful stocks provided important sustenance, commodities, and cultural resources to coastal communities and were described as limitless and inexhaustible by early naturalists and conquistadors (Frazier, 2003). Today the vast flotillas of sea turtles described in these early accounts are greatly depleted (Bjorndal and Bolten, 2003; Bell et al., 2007). The current population of Green Turtles (*Chelonia mydas*) in the Caribbean, for example, is estimated to represent 3–7% of the pre-human population (Jackson et al., 2001).

Contemporary threats to sea turtle populations include the direct take of turtles and eggs, incidental capture in commercial and artisanal fisheries, and habitat degradation, including alterations associated with anthropogenic climate change. Remnant populations persist in many places but generally comprise fractions of historical stocks, resulting in six of the seven extant species (Figure 15.1)—the Loggerhead (*Caretta caretta*), Green (*C. mydas*), Leatherback (*Dermochelys coriacea*), Hawksbill (*Eretmochelys imbricata*), Kemp's Ridley (*Lepidochelys kempii*), and Olive Ridley (*L. olivacea*)—persisting for some four decades on *The IUCN Red List of Threatened Species* (IUCN, 2015). Only the Flatback Sea Turtle (*Natator depressus*), endemic to Australian waters, is not included on the Red List, being classified as 'Data Deficient'. Laudable are conservation successes (e.g. Chaloupka et al., 2008; Stewart et al., 2014) that have moved some regional populations to an IUCN classification of 'Least Concern', including the Hawaiian Green Turtle (Pilcher et al., 2012) and the North Atlantic Leatherback (Tiwari et al., 2013).

Complex life histories complicate management and recovery efforts. Because sea turtles are highly migratory at all life stages, relying on tropical and subtropical nesting beaches, as well as coastal and high seas feeding grounds and migratory corridors (Lohmann et al., 1997; Musick and Limpus, 1997), researchers must integrate diverse methods suitable for specific life history stages in order to obtain information relevant to the development, implementation, and evaluation of management and conservation programmes.

This chapter outlines methods most commonly used to gather management-relevant data from the study of nesting beaches; specifically, monitoring (e.g. surveys, species

Reptile Ecology and Conservation. Edited by C. Kenneth Dodd, Jr. © Oxford University Press 2016.
Published 2016 by Oxford University Press.

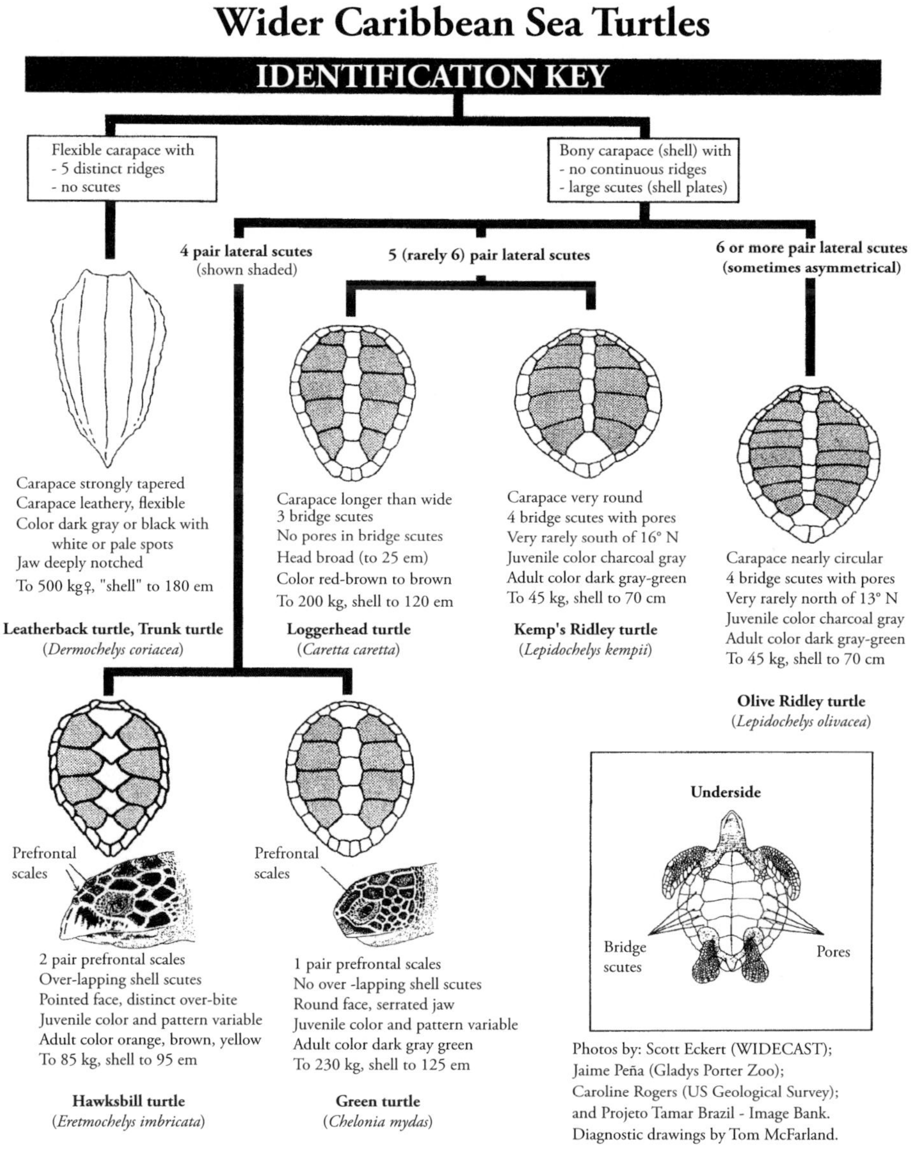

Figure 15.1 Seven extant species of sea turtles are recognized largely by carapacial scute patterns in combination with interocular scale patterns. Australia's flatback turtle (*Natator depressus*) is not shown.

Source: Wider Caribbean Sea Turtle Conservation Network (WIDECAST), used with permission.

identification, nest location, egg counts), tagging, and local interviews. These and other techniques are extensively detailed in manuals developed by Eckert et al. (1999), NMFS Southeast Fisheries Science Center (2008), and Eckert and Eckert (2012), as well as the National Research Council's (2010) publication on population and demographic assessment methods and the State of the World's Sea Turtles Scientific Advisory Board's (SWOT SAB, 2011) resources outlining standards for study design, implementation, and analysis. Regional expert networks, including, *inter alia*, the Wider Caribbean Sea Turtle Conservation Network (http://www.widecast.org), Indian Ocean-Southeast Asia Marine Turtle Memorandum of Understanding (http://ioseaturtles.org/), and the Mediterranean Association to Save the Sea Turtles (http://www.medasset.org/en/), also provide valuable resources for technical guidance.

15.2 Monitoring nesting beaches

Because sea turtles are most accessible on nesting beaches, much of what we know has been gleaned from or facilitated by studies of gravid females. Programmes that monitor nesting beaches are implemented worldwide for all species at variable intensities and under a variety of environmental and social circumstances. Given such diversity, standardized protocols can be important in encouraging the adoption of methodologies that are consistent and replicable. Such consistency helps to ensure that data are compatible when assessing local, regional, and global population trends. Standardized protocols are also useful in providing fledgling projects with guidelines for efficient and effective data collection.

With this in mind, SWOT SAB (2011) outlines recommended monitoring strategies for designing programmes to achieve particular objectives. Eckert and Eckert (2012) provide additional detail on how to determine the temporal and spatial scale of the annual reproductive effort, deciding what to monitor, what to count, and how often to count, validating index sites, and analysing data for estimates of abundance. Project objectives, available resources, and site-specific and biological constraints are among the factors that inform selection of specific protocols (see Chapter 2).

Whether using ground and/or aerial methodologies to monitor nesting beaches, determining if old crawls can be distinguished from fresh crawls, whether successful nesting attempts can be differentiated from unsuccessful attempts, and how to account for factors that might compromise the ability of observers to identify nesting crawls, including wind, rain, and observer experience, are important initial decisions (Schroeder and Murphy, 1999). Regardless of the methodology used, ensuring that observers are well-trained is essential for the collection of accurate data. Table 15.1 introduces terms commonly associated with nesting that will be used throughout this chapter.

15.2.1 Ground-based methods

Ground-based methods are widely implemented where beaches are relatively accessible and/or not conducive to aerial surveys due to their structure, substrate, or target species' behaviours (e.g. Hawksbills nesting in or very near littoral vegetation) (Schroeder and Murphy, 1999). Regular patrol of a nesting beach by foot or vehicle facilitates the

Table 15.1 *Common terms associated with nesting sea turtles*

Term	Definition
Body pit	Depression excavated by sea turtle during the nesting process. Primary body pits are excavated prior to digging an egg chamber. Secondary body pits are excavated after deposition of eggs, effectively obscuring the location of the egg chamber and camouflaging the nesting site. Leatherbacks and Green Turtles create deeper and more extensive body pits than other species.
Clutch	Eggs deposited by a turtle in a single nest; clutch size refers to the number of eggs in a nest; clutch frequency refers to the number of nests oviposited by an individual over the course of a reproductive season.
Crawl	A turtle's tracks, associated with an emergence from the ocean onto a sandy beach in an attempt to lay eggs.
Egg chamber	Cavity that a sea turtle excavates with her rear flippers into which eggs are oviposited, typically flask-shaped (i.e. wider base than neck).
False crawl	Unsuccessful nesting attempt; an emergence from the sea, onto the nesting beach, and a return to the sea without ovipositing eggs.
Index beach	A nesting beach (or series of nesting beaches) where the consistent application of standardized population monitoring protocols ensures that data collected are suitable for long-term analyses of population abundance and/or trends.
Internesting interval	Number of days between a successful nesting and a subsequent attempt (varies between species, but generally 8–18 days; longer for *Lepidochelys*).
Nest	Successful nesting attempt resulting in the deposition of eggs.

tagging of turtles, monitoring of nests, and collection of data and generally yields more detailed and accurate demographic information—and at lower cost—than does an aerial survey.

Ground-based patrols span a diversity of intensities, from periodic morning crawl counts to intensive protocols that assess distribution, composition, and abundance using night-time patrols (most sea turtles are nocturnal nesters). Night-time patrols afford investigators the opportunity to record and uniquely mark nesting turtles, facilitating estimation of demographic parameters such as reproductive output, population size and trend, and survival rates. Ideally, an index site is integrated into a long-term monitoring programme. The index beach, generally an accessible primary nesting site, is intensively covered and the more detailed demographic information collected at this site is assumed to reflect nesting activities on peripheral beaches that are less rigorously monitored (Beggs et al., 2007; Eckert and Eckert, 2012).

For nesting beaches that host synchronized mass nesting events (*arribadas*), traditional survey methods may be inappropriate given the sheer volume of crawls. Gates et al. (1996) and Valverde and Gates (1999) outline the strip transect in time method for these situations. With this technique, transects are established at intervals along the nesting beach and egg-laying females within these transects are counted to estimate total nesting volume.

Ground-based patrols may be conducted by foot or vehicle. Dragging a foot or driving a vehicle across a crawl marks it as counted, informing observers that the crawl was previously recorded (Eckert and Eckert, 2012). High tide marks—specifically identifying those nesting crawls that extend below the high tide mark—provide another diagnostic cue on crawl freshness (see Schroeder and Murphy, 1999).

15.2.2 Aerial survey methods

By offering rapid access to remote and expansive habitats, aerial surveys provide a means to study wildlife occurring at low densities over large areas. However, such surveys are costly and the resulting information relatively coarse (e.g. limited biometric and fine-scale habitat use data can be collected, species or crawl signs may be missed or misidentified from the air). Aerial surveys of nesting grounds are most useful when the territory of interest is too large to survey with ground-based patrols (Schroeder and Murphy, 1999). Calibrating aerial survey data with ground-based counts and establishing correction factors for, *inter alia,* crawl (species) identification, observer bias, and the rate of successful versus unsuccessful nesting attempts are critical for accuracy and interpretation (Witt et al., 2009).

From the air, crawl signs must be closely yet quickly scrutinized to identify freshness, generally using the high tide mark as a guide (see Schroeder and Murphy, 1999, for details) and, if appropriate, to categorize the likely outcome of the turtle's emergence from the sea. Selecting an appropriate flight speed and altitude is important to ensure that observers have sufficient opportunity to scan the beach and categorize crawls. To this end, Schroeder and Murphy (1999) provide general flight guidelines and discuss the importance of identifying optimal flight parameters for specific sites and conditions. They note that early morning is ideal for aerial surveys, since low-angled sunlight enhances the visibility of crawl signs. Moreover, early morning surveys help to ensure that crawl signs have not been degraded by wind, drying sand, rain, foot traffic, or changing tides. Other considerations include selecting a survey window during greatest tidal variation to improve accuracy of counts.

Although a detailed review of in-water research is beyond the scope of this chapter, we note that aerial surveys of foraging sites can be used to obtain or improve estimates of sea turtle abundance and to identify study areas for more intensive in-water research (e.g. observational snorkel transects and hand captures; Henwood and Epperly, 1999). Such aerial surveys are typically conducted with line transect sampling (e.g. Seminoff et al., 2014) and generate site-specific snapshots of abundance and distribution. Important technical considerations must be addressed during aerial surveys of foraging sites. Raw or standardized count data (e.g. captures per unit effort of time or distance) are often used to index abundance and evaluate distribution, but un-calibrated indices are widely criticized and often of limited value; techniques such as distance sampling and multiple observer platforms can be used to overcome some limitations. These methods, however, cannot account for turtles that are completely unavailable for observation from an aircraft because they are beneath the surface of the water. This challenge, known as availability bias, results in underestimates of density and abundance. Availability can be quantified through documentation of dive-surfacing behaviours by outfitting turtles with depth recorders. However, Thomson et al. (2012, 2013) report that surfacing behaviours vary spatially and temporally, both within and among species, significantly influencing bias and associated correction factors for density and abundance. Fuentes et al. (2015) also discuss the importance of adjusting aerial survey counts for both availability and perception bias.

15.2.3 Nesting crawl identifications

Identifying species from nesting crawl signs is useful in characterizing species composition, seasonal variation in nesting activities, and reproductive output. Distinguishing among species solely on the basis of crawl signs can be difficult, but a variety of diagnostic cues, used together, can increase accuracy. These signs include gait, maximum crawl width, crawl depth, presence and extent of body pitting (nest preparation phase), and presence and depth of tail drag (Table 15.2). A sea turtle's gait describes the movement of the fore flippers: symmetrical tracks result from flippers moving in synchrony, such that the flipper marks are parallel to one another, whereas an

Table 15.2 *Diagnostic characteristics of sea turtle nesting crawls. Source: adapted from Pritchard and Mortimer (1999).*

Species	Gait	Nesting crawl depth	Nesting crawl width (cm)	Body pit	Tail drag	Common nesting beach characteristics
Leatherback	symmetrical	deep	150–230	deep	usually deep	large, high energy tropical beaches with deep water access
Green	symmetrical	deep	100–130	deep	present, solid or broken line	variably sized beaches, often with open (non-reef) access
Loggerhead	alternating	moderate–deep	70–90	moderate	none	large beaches with moderate to steep profile
Hawksbill	alternating	shallow	70–85	shallow	often present	often small, pocket beaches with offshore reefs; typically nest in or near vegetation
Kemp's Ridley	alternating	very shallow	70–80	shallow	none or minimal	large beaches of mainland and barrier islands with scrubby vegetation
Olive Ridley	alternating	very shallow	70–80	shallow	none or minimal	tropical mainland and barrier island beaches
Flatback	alternating or symmetrical	shallow to moderate	90	shallow	none	endemic to Australia and surrounding waters

alternating gait indicates a left-right 'walking' crawl with asymmetrical flipper marks (Figure 15.2(a–c)).

Basic biological information, such as preferred nesting beach characteristics and latitudes, can inform diagnosis. For example, Leatherbacks (*Dermochelys coriacea*) typically do not nest on beaches fringed by reef since they require deep-water access to avoid injury; Kemp's and Olive Ridleys display identical crawl signs but there is no overlap in their latitudinal nesting distributions; Green and Loggerhead Turtles have similar crawl widths and may nest alongside one another, but their gait and body pit signs are very different (Table 15.2). Pritchard and Mortimer (1999) provide detailed information on species identification using morphology and crawl signs.

There is no substitute for gaining practical experience in identifying field signs associated with nesting sea turtle crawls. Witnessing a turtle as she emerges from the surf, selects her nesting site, prepares the nesting site ('body pitting'), excavates an egg chamber, lays her eggs, backfills the chamber, and camouflages the site before returning to the sea (see Miller, 1997, for a description of nesting stages) provides an opportunity to observe how the turtle moves throughout the nesting process and therefore to better understand the distinctive features of her crawl.

15.2.4 Locating nests

Sea turtle nesting attempts are not always successful. Being able to distinguish successful from unsuccessful attempts (the latter often referred to as a 'false crawl'; Table 15.1), particularly for research methodologies (e.g. morning crawl counts) in which the nesting turtle is not seen, is a critical step for assessing reproductive success. Watching eggs being laid is the best way to confirm the presence and location of a nest. This may not always be feasible, however, so learning how to 'read' a turtle crawl is an invaluable skill. Key features to identify include the emergence (entrance) crawl, the body-pitting site, and the return (exit) crawl. As a turtle crawls, sand is pushed backwards by the flippers, resulting in distinct patterns that indicate the direction of travel (Figure 15.2(d)).

The structure of false crawls varies widely, from an emergence that does not extend beyond the high water mark to extensive meanderings that can include multiple unsuccessful attempts to excavate an egg chamber. Crawls with no evidence of body-pitting or with empty, partially excavated egg chambers and no signs of additional body-pitting are easily recognized as 'false crawls'. In contrast, a crawl high onto the beach platform and extensive disturbance at the location of body-pitting (Figure 15.2(e)) are cues that a turtle may have successfully nested. In the absence of a clear disturbance associated with a body pit (Figure 15.2(f)), sand thrown over the emergence crawl (e.g. for Loggerheads) or onto the leaves of nearby vegetation more than a few cm off the ground (e.g. for Hawksbills) suggest that the turtle was digging an egg chamber. Such sites also should be marked and monitored for signs of hatching. When investigating crawls signs, always verify the presence of an exit crawl—if you suspect that the turtle is still present, quietly and slowly follow the crawl with lights off, listening for cues as to her location and activity.

False crawls can occur for a number of reasons, including insufficient sand depth, inadequate (e.g. lack of vegetation for Hawksbills) or inaccessible (e.g. due to artificial

Figure 15.2 Sea turtle nesting crawls can be identified to species based on width and symmetry. (a) Asymmetrical Loggerhead (*Caretta caretta*) crawl reflects an alternating gait. (b) Green Turtle (*Chelonia mydas*) flippers move in synchrony, leaving flipper marks parallel to one another and symmetrical across the midline. (c) The tail drag is clearly visible in a Leatherback's (*Dermochelys coriacea*) symmetrical emergence crawl and (d) crawl signs are useful in determining the direction of travel. (e) A successful Green Turtle (*C. mydas*) nesting event clearly shows an emergence crawl, body-pit (concealing the egg chamber), and a return crawl; the deep body pit is uniquely characteristic of this species. (f) In contrast, when a Gravid Hawksbill (*Eretmochelys imbricata*) enters littoral vegetation, evidence of a body-pit is much less clear. For more detail, see Schroeder and Murphy (1999) and the Florida Fish and Wildlife Conservation Commission's crawl identification guide (online at http://myfwc.com/media/3055670/crawlidentificationguidelines.pdf). Photos: (a) Kelly Stewart (North Carolina, USA); (b, e) Turtugaruba Foundation (Aruba, West Indies); (c) Ocean Spirits (Grenada, West Indies); (d) Sonia Gautreau (Playa Gandoca, Costa Rica); (f) Rosalie Sea Turtle Initiative (Dominica, West Indies).

walls or erosion-induced embankments) habitat, light pollution, or human disturbance. Understanding the frequency with which false crawls occur relative to nests thus can provide information about the suitability of the nesting habitat. However, individual turtles also can exhibit high variability in their tendency to false crawl: some are efficient and typically lay their eggs on the first nesting attempt, whereas others may false crawl multiple times before successfully laying. An individual's tendency to false crawl may be linked to a physical condition or injury that does not allow her to complete an egg chamber.

In addition to learning to read crawl signs, spending time on the nesting beach and observing the nesting process is helpful in understanding how to diagnose species-specific crawl signs, distinguish between false crawls and successful nests, and identify where the egg chamber is likely to be located.

15.2.5 Egg counts and nest excavations

Ground- and aerial-based patrols of nesting beaches yield valuable data about species presence, including seasonality and abundance. Patrols that simply tally nesting activities, however, do not provide a comprehensive picture of reproductive productivity. Quantifying the number of eggs per nest (clutch size), the number of hatched turtles (hatch success), and the number of hatchlings emerging from the nest onto the surface of the beach (emergence success) using egg counts and post-hatch nest excavations can improve our understanding of reproductive output and success.

Viable sea turtle eggs are white, spherical in shape, and vary in diameter (ca. 40 mm to >50 mm) depending on species (Miller, 1997). At times, eggs may be oddly shaped and vary greatly in size, from very small to very large. Very large abnormal eggs can appear as chains of eggs linked together, often containing multiple yolks, and can develop into viable embryos; very small eggs (less than half the size of normal eggs; particularly common with Leatherbacks) are yolkless and will not develop (Miller, 1997, 1999). Egg counts document the number of viable (yolked) eggs. The presence and number of yolkless eggs are reported separately.

Determining clutch size can be accomplished at deposition or, if eggs are moved, during reburial. During deposition, eggs can be counted by an observer lying directly behind the turtle. Removing a small amount of surface sand may be necessary to gain unobstructed visual access to the egg chamber. The observer allows eggs to pass through a clean or gloved hand placed beneath the ovipositor while tallying the number of eggs with a click-counter in the other hand. If nests are deposited in vulnerable sites (e.g. in close proximity to the high-water line or in areas with high rates of depredation or poaching), eggs may be relocated to zones of lower risk and counted during reburial. Boulon (1999) and Mortimer (1999) provide details and best practices associated with *in situ* relocation and hatchery burials, respectively.

A typical emergence of hatchlings is characterized by small tracks radiating from the nest depression in a funnel shape towards the sea. The emergence may take place over several evenings. Once the hatchlings have left the nest, contents can be exhumed and categorized. Post-hatch nest excavations provide a third opportunity to obtain egg counts and estimate other metrics of reproductive success. Excavations should be completed within 24–48 hours post-emergence to prevent decay and deterioration of contents. Inert protective gloves should be worn during excavations to prevent exposure to bacteria or other infectious agents. Sand is gently scooped from the nest chamber, and nest contents are carefully sorted and set aside. Because the nest chamber may adjoin other nests, observers should take care to ensure that incubating nests are not disturbed.

The exhumed contents are categorized as empty eggshells (i.e. hatched eggs), undeveloped eggs, embryos (unhatched eggs may be subdivided into embryonic stages), and depredated eggs (broken eggs with residue remaining; cf. Miller, 1999). It is not uncommon during an excavation to encounter live or dead hatchlings residual within the nest cavity; these are recorded as separate categories. Healthy residual hatchlings should not be held captive, but rather allowed to crawl unaided to the sea.

Caution should be used while exhuming nest contents to minimize eggshell fragmentation. As a general rule and to avoid overestimating clutch size, an eggshell is tallied when ≥50% of the shell is intact. Comparing egg counts obtained during deposition or reburial with post-hatch excavation tallies provides an opportunity to estimate error rates.

In addition to facilitating egg counts, nest excavations also enable estimation of hatching and emergence success rates. Hatching success, defined as the proportion of hatchlings that are completely out of (hatched from) the eggshell, is estimated by dividing the number of empty eggshells by the total number of eggs laid (excluding yolkless eggs; Miller, 1999). Emergence success, on the other hand, refers to the proportion of hatchlings that successfully reach the sand surface. Emergence success is calculated as the number of empty eggshells in the nest, minus any live and dead hatchlings remaining in the nest, divided by the total number of eggs in the nest (Miller, 1999). By definition, hatching success must be equal to or greater than emergence success. Estimating these metrics can inform management by identifying environmental or biological factors that impact hatching success, such as beach profile, organic content, distance to the high-water line, site disturbance, and climatic events (e.g. beach erosion, water table rise, or temperature fluctuation due to severe storms).

15.3 Tagging

Information obtained from individually marking sea turtles for subsequent identification forms the foundation for much of our current understanding of sea turtle ecology, including population demography, reproductive output, internesting intervals, habitat use, growth rates, and movements. Although accessing turtles for marking can be labour-intensive, these capture–mark–recapture programmes are recognized as the 'gold standard' of sea turtle monitoring because of the detailed information they provide (SWOT SAB, 2011). Balazs (1999), Eckert and Beggs (2006), NMFS Southeast Fisheries Science Center (2008), and Bluvias and Eckert (2010) describe considerations, outline best practices, and detail step-by-step protocols for tagging sea turtles.

The most common marking technique involves the application of plastic or metallic tags to the trailing edges of front and/or rear flippers, often coupled with the subcutaneous insertion of a passive integrated transponder (PIT) tag (see Chapter 4). Shell notching, living tags (i.e. tissue plug transplants; see NMFS Southeast Fisheries Science Center, 2008), photo-identification (McDonald and Dutton, 1996; Reisser et al., 2008; Chapter 5), and the advent of genetic markers (Chapter 25) provide other mechanisms for identification. We focus on flipper tagging and PIT tagging here, which are generally restricted to individuals larger than 30 cm straight-line carapace length (see NMFS Southeast Fisheries Science Center, 2008, for details on measuring sea turtles).

External tags pierce the flipper, are clasped or locked in place, and are inscribed with a unique alphanumeric identifier on one side and a return address on the other. With hard-shelled sea turtles, tags are commonly applied directly into or between the large scales along the posterior edges of the front flippers (Figure 15.3), and should be placed in a proximal location to reduce drag and improve retention. For Leatherbacks, flipper tags are often applied in the thinnest section of skin between the rear flippers and the tail (Figure 15.3).

PIT tags are about the size of a grain of rice, have unique 10-digit alphanumeric identifiers, and are injected under the skin and into the muscle. Scanners read PIT tags

Figure 15.3 Flipper tags are typically applied to hard-shelled sea turtles either between (a) or directly into (b) the large scales along the posterior edges of the front flippers. (c) It's useful to experiment with metal and plastic tags to determine the best tag retention under local conditions. (d) Metal tags are typically inserted in the inguinal area of the rear flippers of Leatherbacks (*Dermochelys coriacea*). Photos: (a) Erik Martin (Florida, USA); (b) Stacy Kubis (Florida, USA); (c) Sea Turtle Conservation Bonaire (Bonaire, West Indies); (d) KIDO Foundation (Carriacou, Grenada).

at deployment and upon subsequent sighting. PIT tags are commonly injected into the musculature of a Leatherback's front shoulder, whereas in hard-shelled turtles, injection sites generally target connective tissue and muscle to improve retention and reduce potential tag migration (NMFS Southeast Fisheries Science Center, 2008; Wyneken et al., 2010). PIT tags often show higher rates of retention than do flipper tags. However, PIT tags and associated scanners are significantly more expensive than flipper tags, and tag migration due to improper injection techniques can preclude successful identification of a PIT-tagged individual. Additionally, scanners may not be available to an individual subsequently coming into contact with the turtle, thus reducing opportunities for collecting tag recovery data. Epperly et al. (2015) provide a comprehensive review of PIT tag technology and scanner performance.

Before applying any tags, thoroughly inspect the turtle for previous tags. All four flippers should be examined for flipper tags or signs of tag scars or tears. Similarly, if a PIT tag reader is available, the turtle should be thoroughly scanned (especially flippers and shoulder muscles) because there are no global standards for the placement of PIT tags in sea turtles; moreover, the tag may have migrated. In general, tags are applied during the latter stages of egg-laying or immediately after egg-laying is completed, which reduces stress and the probability of nest abandonment.

Tag loss can result from tearing, tissue necrosis, or poor positioning, and tags differ in their retention or failure rates (e.g. Van Dam and Diez, 1999; Reisser et al., 2008). To minimize the likelihood that a previously marked turtle is rendered unidentifiable due to tag loss, individuals should be marked with at least two tags. PIT tagging can increase the likelihood of a turtle retaining a unique identity, and a tissue sample can also facilitate the identification of turtles with subsequent genetic analyses, thereby providing a truly permanent 'tag'. In demographic modelling based on capture–mark–recapture data, permanency of marks (relative to the study period) is a fundamental assumption, and unmodelled tag loss will result in an overestimation of abundance and underestimation of survival (see Chapter 27).

15.4 Local interviews

To this point, we have focused on the collection of data through ecological field methods. However, inhabitants of coastal communities, including commercial and artisanal fishers, are often well-positioned to provide broad insights into the ecology, management, and historic and contemporary uses of sea turtles. Indeed, interdisciplinary approaches that incorporate methods from both the natural and social sciences are recognized as effective means to inform and address conservation and management challenges (Ban et al., 2013). Interviews, used here as a catch-all term for gathering information from the public via direct conversations, questionnaires, or other methods, are widely implemented to study sea turtles and complement other means of data collection. For sea turtles, interviews are used, *inter alia*, to assess the magnitude and impacts of by-catch associated with fisheries (Moore et al., 2010; Mancini et al., 2011), the prevalence of legal and illegal subsistence use of sea turtles (Weir et al., 2007), and human attitudes towards and perceptions of turtle conservation and harvest (Hart et al.,

2013). Interviews can be a cost-effective mechanism to obtain general information, but there are limitations in reliability and accuracy (Moore et al., 2010).

Tambiah (1999) outlines three general strategies and associated caveats related to conducting interviews:

1. Questionnaires allow for the rapid collection of large volumes of quantitative information, but are limited by a lack of open questions and an inability to foster discussion.
2. Lead questions, used to obtain data on specific topics, are less structured and constrained than questionnaires and can yield both quantitative and qualitative data.
3. Open-ended discussions, which are the least structured, are useful in communities in developing countries and help to understand and address complex issues; however, they require cultural and linguistic familiarity and can be more challenging to interpret and analyse, given the qualitative nature of the data.

Whether one of these methods or a hybrid or alternative approach is selected, careful consideration must be given to the structure and implementation of interviews. In particular, researchers must be aware of the ethical considerations of information-sharing through interviews, particularly regarding confidentiality about sensitive topics (e.g. harvest of protected species), how the interview will be conducted, and how information gleaned from an interview will be used (Tambiah, 1999; Moore et al., 2010). Understanding local sensitivities and political context can help to promote trust in information sharing and enhance participation in research (Carruthers and Neis, 2011). Interviews should be structured to generate unbiased responses and not disclose or interject the beliefs of the interviewer (Marcovaldi and Marcovaldi, 1999; Tambiah, 1999). Obtaining the same information multiple times through differently structured questions can safeguard against misunderstandings and inconsistencies in responses. Finally, pre-interview site visits help to establish familiarity and can identify potential collaborations, and understanding cultural norms, listening attentively, and demonstrating respect are essential elements of an effective interview (Tambiah, 1999).

A wealth of information can potentially be gleaned from interviews including, *inter alia*, species composition, abundance, and seasonality of nesting and foraging populations, changes in population levels, uses of sea turtles and their prevalence, existing regulations and effectiveness of enforcement, and by-catch in fisheries. Moore et al. (2010) provide specific recommendations for by-catch related interviews, including sampling design, questionnaire formatting, interviewer training, and improving reliability of data.

15.5 Summary

The complex life histories of sea turtles necessitate integrating land-based and in-water research techniques to fully inform conservation and management. Nesting beach studies are implemented worldwide and provide a foundation from which to estimate reproductive output and population size and trend. Although monitoring nesting beaches via aerial surveys provides rapid access to remote and expansive beaches, ground-based patrols generally provide more reliable information at lower cost and without reliance on advanced technologies.

Among the key elements of a nesting beach monitoring programme are accurate species identification based on visual observation of the turtle or an accurate interpretation of crawl signs, location and confirmation of successful versus unsuccessful nesting attempts, and collection of metrics (e.g. clutch size, interesting interval, clutch frequency, hatch success, emergence success) related to reproductive output. Tagging initiatives strengthen inferences and data analyses by facilitating the identification of individuals over time.

Residents and stakeholders can provide valuable information about sea turtle populations, including both historic and contemporary patterns of nesting, exploitation, and commerce. This information can improve management, as well as involve local populations in sea turtle research and conservation programming. Interdisciplinary approaches that incorporate natural and social science methodologies and integrate local people in the planning, implementation, and evaluation of conservation initiatives are viewed as critical to the recovery of depleted sea turtle populations worldwide.

References

Balazs, G.H. (1999). Factors to consider in the tagging of sea turtles. In K.L. Eckert, K.A. Bjorndal, F.A. Abreu-Grobois, et al. (eds) *Research and Management Techniques for the Conservation of Sea Turtles*. IUCN/SSC Marine Turtle Specialist Group Publication No. 4. Washington, DC: IUCN/SSC, pp. 101–9.

Ban, N.C., Mills, M., Tam, J., et al. (2013). A social-ecological approach to conservation planning: embedding social considerations. *Frontiers in Ecology and the Environment*, **11**, 194–202.

Beggs, J.A., Horrocks, J.A., and Krueger, B.H. (2007). Increase in hawksbill sea turtle *Eretmochelys imbricata* nesting in Barbados, West Indies. *Endangered Species Research*, **3**, 159–68.

Bell, C.D., Solomon, J.L., Blumenthal, J.M., et al. (2007). Monitoring and conservation of critically reduced marine turtle nesting populations: lessons from the Cayman Islands. *Animal Conservation*, **10**, 39–47.

Bjorndal, K.A., and Bolten, A.B. (2003). From ghosts to key species: restoring sea turtle populations to fulfill their ecological roles. *Marine Turtle Newsletter*, **100**, 16–21.

Bluvias, J.E., and Eckert, K.L. (2010). *Marine Turtle Trauma Response Procedures: A Husbandry Manual*. WIDECAST Technical Report No. 10. Ballwin, MO: Wider Caribbean Sea Turtle Conservation Network.

Boulon, R.H. (1999). Reducing threats to eggs and hatchlings: *in situ* protection. In K.L. Eckert, K.A. Bjorndal, F.A. Abreu-Grobois, et al. (eds) *Research and Management Techniques for the Conservation of Sea Turtles*. IUCN/SSC Marine Turtle Specialist Group Publication No. 4. Washington, DC: IUCN/SSC, pp. 169–74.

Carruthers, E.H., and Neis, B. (2011). Bycatch mitigation in context: using qualitative interview data to improve assessment and mitigation in a data-rich fishery. *Biological Conservation*, **144**, 2289–99.

Chaloupka, M., Work, T.M., Balazs, G.H., et al. (2008). Cause-specific temporal and spatial trends in green sea turtle strandings in the Hawaiian Archipelago (1982–2003). *Marine Biology*, **154**, 887–98.

Eckert, K.L., and Beggs, J. (2006). *Marine Turtle Tagging: A Manual of Recommended Practices*. WIDECAST Technical Report No. 2. Revised ed. Beaufort, NC: Wider Caribbean Sea Turtle Conservation Network.

Eckert, K.L., and Eckert, S.A. (2012). *Designing Surveys of Abundance at Sea Turtle Nesting Beaches.* WIDECAST Technical Report No. 15. Ballwin, MO: Wider Caribbean Sea Turtle Conservation Network.

Eckert, K.L., Bjorndal, K.A., Abreu-Grobois, F.A., et al. (eds). (1999). *Research and Management Techniques for the Conservation of Sea Turtles.* IUCN/SSC Marine Turtle Specialist Group Publication No. 4. Washington, DC: IUCN/SSC.

Epperly, S.P., Stokes, L.W., and Belskis, L.C. (2015). Radio frequency identification technology and marine turtles: investigation of passive integrated transponder (PIT) tags and readers. *Marine Turtle Newsletter,* **145**, 4–15.

Frazier, J. (2003). Prehistoric and ancient historic interactions between humans and marine turtles. In P.L. Lutz, J.A. Musick, and J. Wyneken (eds) *The Biology of Sea Turtles, Volume II.* Boca Raton, FL:CRC Press, pp. 1–39.

Fuentes, M.M.P.B., Bell, I., Hagihara, R., et al. (2015). Improving in-water estimates of marine turtle abundance by adjusting aerial survey counts for perception and availability biases. *Journal of Experimental Marine Biology and Ecology,* **471**, 77–83.

Gates, C.E., Valverde, R.A., Mo, C.L., et al. (1996). Estimating arribada size using a modified instantaneous count procedure. *Journal of Agricultural, Biological, and Environmental Statistics,* **1**, 275–87.

Hart, K.A., Gray, T., and Stead, S.M. (2013). Consumptive versus non-consumptive use of sea turtles? Stakeholder perceptions about sustainable use in three communities near Cahuita National Park, Costa Rica. *Marine Policy,* **42**, 236–44.

Henwood, T.A., and Epperly, S.P. (1999). Aerial surveys in foraging habitats. In K.L. Eckert, K.A. Bjorndal, F.A. Abreu-Grobois, et al. (eds) *Research and Management Techniques for the Conservation of Sea Turtles.* IUCN/SSC Marine Turtle Specialist Group Publication No. 4. Washington, DC: IUCN/SSC, pp. 65–6.

Hirayama, R. (1998). Oldest known sea turtle. *Nature,* **392**, 705–8.

IUCN. (2015). *The IUCN Red List of Threatened Species, Version 2014.3.* Available at: http://www.iucnredlist.org (accessed 13 January 2015).

Jackson, J.B.C., Kirby, M.X., Berger, W.H., et al. (2001). Historical overfishing and the recent collapse of coastal ecosystems. *Science,* **293**, 629–38.

Lohmann, K.J., Witherington, B.E., Lohmann, C.M.F., et al. (1997). Orientation, navigation, and natal beach homing in sea turtles. In P.L. Lutz and J.A. Musick (eds) *The Biology of Sea Turtles.* Boca Raton, FL:CRC Press, pp. 107–36.

Mancini, A., Koch, V., Seminoff, J.A., et al. (2011). Small-scale gill-net fisheries cause massive green turtle *Chelonia mydas* mortality in Baja California Sur, Mexico. *Oryx,* **46**, 69–77.

Marcovaldi, M.Â., and Marcovaldi, G.G. (1999). Marine turtles of Brazil: the history and structure of Projeto TAMAR-IBAMA. *Biological Conservation,* **91**, 35–41.

McDonald, D.L., and Dutton, P.H. (1996). Use of PIT tags and photo identification to revise remigration estimates of leatherback turtles (*Dermochelys coriacea*) nesting in St. Croix, US Virgin Islands, 1979–85. *Chelonian Conservation and Biology,* **2**, 148–52.

Miller, J.D. (1997). Reproduction in sea turtles. In P.L. Lutz and J.A. Musick (eds) *The Biology of Sea Turtles.* Boca Raton, FL:CRC Press, pp. 51–81.

Miller, J.D. (1999). Determining clutch size and hatching success. In K.L. Eckert, K.A. Bjorndal, F.A. Abreu-Grobois, et al. (eds) *Research and Management Techniques for the Conservation of Sea Turtles.* IUCN/SSC Marine Turtle Specialist Group Publication No. 4. Washington, DC: IUCN/SSC, pp. 124–9.

Moore, J.E., Cox, T.M., Lewison, R.L., et al. (2010). An interview-based approach to assess marine mammal and sea turtle captures in artisanal fisheries. *Biological Conservation,* **143**, 795–805.

Mortimer, J.A. (1999). Reducing threats to eggs and hatchlings: hatcheries. In K.L. Eckert, K.A. Bjorndal, F.A. Abreu-Grobois, et al. (eds) *Research and Management Techniques for the Conservation of Sea Turtles*. IUCN/SSC Marine Turtle Specialist Group Publication No. 4. Washington, DC: IUCN/SSC, pp. 175–8.

Musick, J., and Limpus, C. (1997). Habitat utilization and migration in juvenile sea turtles. In P.L. Lutz and J.A. Musick (eds) *The Biology of Sea Turtles*. Boca Raton, FL:CRC Press, pp. 137–64.

NMFS Southeast Fisheries Science Center. (2008). *Sea Turtle Research Techniques Manual*. NOAA/NMFS Technical Memorandum NMFS-SEFSC-579. Miami, FL: National Marine Fisheries Service.

National Research Council. (2010). *Assessment of Sea Turtle Status and Trends: Integrating Demography and Abundance*. Washington, DC:National Academies Press.

Pilcher, N.J., Chaloupka, M.Y., and Woods, E. (2012). *Chelonia mydas* (Hawaiian subpopulation). In *The IUCN Red List of Threatened Species, Version 2015.2*. Available at: http://www.iucnredlist.org (accessed 17 July 2015).

Pritchard, P.C.H., and Mortimer, J.A. (1999). Taxonomy, external morphology, and species identification. In K.L. Eckert, K.A. Bjorndal, F.A. Abreu-Grobois, et al. (eds) *Research and Management Techniques for the Conservation of Sea Turtles*. IUCN/SSC Marine Turtle Specialist Group Publication No. 4. Washington, DC: IUCN/SSC, pp. 21–40.

Reisser, J., Proietti, M., Kinas, P., et al. (2008). Photographic identification of sea turtles: method description and validation, with an estimation of tag loss. *Endangered Species Research*, **5**, 73–82.

SWOT SAB. (2011). *The State of the World's Sea Turtles (SWOT) Minimum Data Standards for Nesting Beach Monitoring*. Arlington, VA:SWOT Scientific Advisory Board. Available at: http://www.seaturtlestatus.org/data/standards.

Schroeder, B., and Murphy, S. (1999). Population surveys (ground and aerial) on nesting beaches. In K.L. Eckert, K.A. Bjorndal, F.A. Abreu-Grobois, et al. (eds) *Research and Management Techniques for the Conservation of Sea Turtles*. IUCN/SSC Marine Turtle Specialist Group Publication No. 4. Washington, DC: IUCN/SSC, pp. 45–55.

Seminoff, J.A., Eguchi, T., Carretta, J., et al. (2014). Loggerhead sea turtle abundance at a foraging hotspot in the eastern Pacific Ocean: implications for at-sea conservation. *Endangered Species Research*, **24**, 207–20.

Stewart, K.R., Martin, K.J., Johnson, C., et al. (2014). Increased nesting, good survival and variable site fidelity for leatherback turtles in Florida, USA. *Biological Conservation*, **176**, 117–25.

Tambiah, C. (1999). Interviews and market surveys. In K.L. Eckert, K.A. Bjorndal, F.A. Abreu-Grobois, et al. (eds) *Research and Management Techniques for the Conservation of Sea Turtles*. IUCN/SSC Marine Turtle Specialist Group Publication No. 4. Washington, DC: IUCN/SSC, pp. 156–64.

Thomson, J.A., Cooper, A.B., Burkholder, D.A., et al. (2012). Heterogeneous patterns of availability for detection during visual surveys: spatiotemporal variation in sea turtle dive–surfacing behaviour on a feeding ground. *Methods in Ecology and Evolution*, **3**, 378–87.

Thomson, J.A., Cooper, A.B., Burkholder, D.A., et al. (2013). Correcting for heterogeneous availability bias in surveys of long-diving marine turtles. *Biological Conservation*, **165**, 154–61.

Tiwari, M., Wallace, B.P., and Girondot, M. (2013). *Dermochelys coriacea* (Northwest Atlantic Ocean subpopulation). In *The IUCN Red List of Threatened Species, Version 2015.2*. Available at: http://www.iucnredlist.org (accessed 17 July 2015).

Valverde, R.A., and Gates, C.E. (1999). Population surveys on mass nesting beaches. In K.L. Eckert, K.A. Bjorndal, F.A. Abreu-Grobois, et al. (eds) *Research and Management Techniques for the Conservation of Sea Turtles*. IUCN/SSC Marine Turtle Specialist Group Publication No. 4. Washington, DC: IUCN/SSC, pp. 56–60.

Van Dam, R.P., and Diez, C.E. (1999). Differential tag retention in Caribbean hawksbill turtles. *Chelonian Conservation and Biology*, **3**, 225–9.

Weir, C.R., Ron, T., Morais, M., et al. (2007). Nesting and at-sea distribution of marine turtles in Angola, West Africa, 2000–2006: occurrence, threats and conservation implications. *Oryx*, **41**, 224–31.

Witt, M.J., Baert, B., Broderick, A.C., et al. (2009). Aerial surveying of the world's largest leatherback turtle rookery: a more effective methodology for large-scale monitoring. *Biological Conservation*, **142**, 1719–27.

Wyneken, J., Epperly, S.P., Higgins, B., et al. (2010). PIT tag migration in sea turtle flippers. *Herpetological Review*, **41**, 448–54.

16
Crocodilians

Matthew Brien and Charlie Manolis

16.1 Introduction

There are a wide variety of techniques used to monitor, capture, and handle crocodilians throughout the world. The purpose of this chapter is to provide an overview of how those techniques are used, for what purposes, and in what contexts. More detailed descriptions of the various techniques can be found in the references provided. People working with crocodilians should always ensure they comply with any relevant guidelines, codes, and legislative requirements.

16.2 Surveying

Surveys of crocodilian populations are undertaken to answer specific management questions related to monitoring trends in population distribution, size, and size structure over time, quantifying impacts of harvesting strategies or of feral animals and introduced species, and locating problem animals for public safety and research. The type of survey method is dictated by the management or research 'question' that is being addressed, and in part by the status (e.g. density, habitat, wariness) of the population being examined.

Population surveys to quantify trends over time are typically undertaken in a systematic manner, at the same time of year, and under similar environmental conditions each time, to reduce the biases that may be associated with these factors. Information from such surveys is essential for informing conservation, management, or harvest programmes. Surveys that target specific individuals in a population or that are carried out for research purposes (e.g. capture–recapture studies) may not need to be carried out in a systematic fashion.

16.2.1 Spotlight surveys

Spotlight count surveys are the most commonly used method for crocodilians. The technique takes advantage of the bright red reflection of the crocodilian eye that occurs when a light is shone into it at night (Figure 16.1(a); Richardson et al., 2002). Spotlight surveys are typically undertaken by boat or other watercraft (e.g. canoe, kayak) in open waterways such as creeks, rivers, lakes, or along coastlines (Messel et al., 1981), but they can be undertaken from land banks overlooking water bodies.

Reptile Ecology and Conservation. Edited by C. Kenneth Dodd, Jr. © Oxford University Press 2016.
Published 2016 by Oxford University Press.

Figure 16.1 Crocodilians are commonly surveyed through (a) spotlight surveys at night that rely on detecting individuals by their red 'eye shine' (*Crocodylus porosus*; Photograph by C. Manolis), (b) day time aerial counts (*Alligator mississippiensis*; Photo by J. Brien), and nest counts, which are undertaken for (c) mound nesting (*C. porosus*; Photo by C. Manolis) and (d) hole-nesting species (*C. acutus*; Photo by J. Brien).

The strength of the spotlight that is most effective depends in part on the type of habitat. In large rivers, strong light (e.g. 100 W) allows crocodilian eye shines to be detected some distance away (>200 m) before animals have an opportunity to go underwater. However, such strong lights may not be as effective in small narrow meandering waterways, where the amount of light may 'flood' the area and eye shines are not discernible. Under such conditions, a dimmer light source (e.g. head lamp) may be more effective.

In a vessel, the spotter typically sits at the front and constantly scans ahead and to the sides of the boat for eye shines in a reasonably rapid, horizontal, sweeping motion so the beam goes back and forth, covering the water surface, water's edge, bank, and vegetation. Scanning well ahead of the boat detects distant eye shines, noted mentally, that are still recorded if the animals submerge before the location is reached.

The speed of the vessel during a survey should be suitable for the waterway and density of crocodilians. In large open systems with low densities of crocodilians (1–2 individuals/km) and good visibility of banks, speeds of 15–25 km/h may be possible, but in narrower systems with thicker vegetation or higher densities of crocodilians, slower speeds (<8–10 km/h) may be required to optimize detection (Webb et al., 1989; Fukuda et al., 2013). Survey distance should be recorded as the mid-stream distance, not the actual Global Positioning System (GPS) track of the vessel, as that will include all movements of the boat approaching animals. For spotlight surveys from the land, the spotter normally sits still in the dark for a few minutes, quickly scans the water surface and banks that can be reached by the light, often repeats the exercise, and then moves to another vantage point for the next section of the waterway.

Water level and temperature (air and water) impact the ability to sight crocodilians, and this varies between region (tropical vs. temperate) and species. Spotlight surveys are typically undertaken during drier times of the year when water levels are low and crocodilians are more likely to be sighted. In tropical areas, this tends to coincide with the winter months (northern Australia). In tidal habitats, regardless of time of year, surveys should coincide with low tides, when banks are exposed.

Once detected, crocodilians are normally approached to identify species (where more than one may occur) and estimate size. If species and/or size cannot be ascertained, sightings are recorded as eye shine. Other animals that live in and around waterways also produce an eye shine (various colours including red) when shone with a light (e.g. spiders, frogs, birds, and mammals), but spotters become adept at differentiating them from crocodilians.

Other information that may be recorded for each crocodilian sighting includes: position (e.g. on bank, shallow water on edge, midstream); distance approached before submerging (wariness); and, location (e.g. river kilometres, GPS reading). Other data that relate to a survey include start and finishing points, date, time, personnel and role, vegetation density, and environmental conditions (air and water temperatures, moon phase, water levels, tidal periods, rainfall, wind speed and direction, wave height, fog, boat traffic) (Woodward and Marion, 1978; King et al., 1994).

In most cases, estimates of total length (TL) are based on observing only part of the animal, usually the head. For crocodilians, head length represents around one-seventh of total length. The manner in which size is estimated and recorded varies among researchers, species, and locations, as do the size class categories used (e.g. *C. porosus*: 30 cm increments; *C. acutus*: 50 cm increments). The ability to accurately estimate size also varies greatly between observers due to experience. Training involving the capture and measurement of a number of individuals following size estimation can greatly improve accuracy (Choquenot and Webb, 1987).

Spotlight surveys are less effective in heavily vegetated habitats where crocodilians can hide easily, or in areas where animals are wary and submerge before being sighted (Webb and Messel, 1979). Eye shines are less visible in fog, mist, rain, or smoke and it is often necessary to slow down, survey each bank separately, or even stop the survey for the night. While it has been argued that spotlights with coloured filters (e.g. red) may allow the observer to approach an animal more closely, this is not supported by any published studies (Woodward and Marion, 1978).

The proportion of a crocodilian population that is sighted during a spotlight survey is highly dependent upon a variety of factors, including water levels, temperature, season, habitat, food abundance, reproductive cycle, and level of wariness. During the wet season, crocodilians may disperse away from the main waterways onto flood plains. Some species may spend winter in dens or burrows out of sight. The level of wariness can vary between populations depending on the level of human disturbance (past or present) and generally increases with individual size. These animals are rarely approached close enough to determine size (e.g. eyes only).

16.2.2 Aerial count surveys

Aerial surveys during the day are a cost-effective method of surveying large areas of waterway rapidly (Bayliss, 1987). They are generally undertaken by light aircraft (Games, 1994), ultra-lights (Mourão et al., 1994), or helicopters (Webb et al., 1990) during cooler months, when crocodilians are basking on land (Figure 16.1(b)) or in shallow water for longer periods of the day. Time of day is important, as once crocodilians reach an optimum temperature, they typically move back into the water and may submerge. The height above water and speed at which the survey is undertaken may vary according to the type of aircraft, nature of the waterway (e.g. sinuosity, inlets, bays, extent of fringing vegetation), and density of crocodilians, but these should be consistent over time if surveys are used for trend analysis.

However, aerial surveys are generally biased towards detecting the larger individuals in the population (adults), which may be too wary to be sighted during spotlight surveys (Webb et al. 1990). For larger species, size may be estimated in categories (e.g. small (0.6–1.2 m), medium (1.2–2.1 m), large (2.1–3.3 m), extra-large (>3.3 m) for *C. porosus*; Webb et al., 1990), or for smaller species numbers may simply be recorded. While spotlight surveys are often a more precise survey method, as they provide more detailed information on size structure, aerial count surveys enable a greater area to be surveyed that would not otherwise be possible by boat due to cost, duration, and habitat constraints.

16.2.3 Day count surveys

For some crocodilians (e.g. *Gavialis gangeticus, C. palustris*), day count surveys have been the standard method for assessing population status. The reason for this preference over spotlight count surveys relates to behavioural differences between populations and species, and also to issues of security for personnel at night, from wildlife (e.g. big cats, hippopotamus) and other humans. Surveys are carried out in the cooler times of the year when crocodilians are more likely to be basking during the day, and can be carried out on foot from the bank, or from boat/canoe (O'Brien, 1990). Day count surveys typically provide an index of the larger animals in the population, similar to aerial count surveys.

16.2.4 Nest counts

Nest counts are often the only available index of crocodilian populations, particularly in densely vegetated habitats. Aerial nest counts provide an index of the adult female segment of the population, and can be corrected for the proportion of females nesting each year, and for the contribution they make to the population as a whole.

The technique is generally more effective for species that nest within a short, defined pulse, compared with those that nest over a longer period (Webb et al., 1990), and is more applicable to mound-nesting crocodilians (Figure 16.1(c)), although it has also been successfully used for some hole-nesting species (e.g. *C. acutus*; Figure 16.1(d)). With species such as the American Alligator (*Alligator mississippiensis*) in Louisiana (USA), surveys are carried out each year along predetermined transects. With other species, such as the estuarine crocodile in Australia and Papua New Guinea, the nature of nesting habitat means that nests are surveyed completely.

A known bias affecting nest counts is that the peak of nesting may change annually due to environmental factors. Thus, a number of surveys through the nesting season may provide a more reliable index than a single survey. The proportion of all nests present at the time of survey varies between species (hole vs. mound nests) and habitats (dense vegetation vs. exposed habitats).

16.2.5 Other survey methods

A variety of other methods can be utilized to survey crocodilians, or in the case of species that are rare and/or present in low densities, simply confirming presence/absence may be the desired goal.

In the case of foot prints, slides, and patterns of belly scales in mud, sand, or other substrates, morphometric relationships can be used to estimate the size of individuals (Webb and Messel, 1978; Platt et al., 1990), while scat analysis can also provide information on species presence and diet (Daltry et al., 2003).

Burrow counts and camera surveillance have been used to confirm the presence of crocodilians (Channa et al., 2010; Rathnasiri et al., 2013). The use of camera surveillance is becoming more popular and can also reveal important aspects of behaviour (Merchant et al., 2013).

16.3 Capture

Crocodilians are large, dangerous animals that are often captured for purposes such as public safety, research, relocation, or consumption. The method of capture depends on the objective (live or dead), situation, habitat, species, size (hatchling to adult), tradition, level of wariness, season, experience and skill of the people involved, and how urgently a crocodilian needs to be captured.

16.3.1 Hand capture or tongs

Small crocodilians can be captured either by grabbing the animal around the neck (<0.8 m TL) or with the use of snake tongs (<1.2 m TL; Figure 16.2(a)), which are commonly used for snakes. The advantage of snake tongs is that they enable crocodilians to be captured from a greater distance and without direct physical contact with humans.

16.3.2 Noosing

Crocodilians (>1 m TL) can be captured effectively using a noose secured at one end to the boat, or to a float for larger animals. The noose can be nylon rope with a running-loop on one end, or a self-locking wire cable (Cherkiss et al., 2012). The open noose is fixed to a long pole with tape (Figure 16.2(a)) so that it breaks free when the noose is

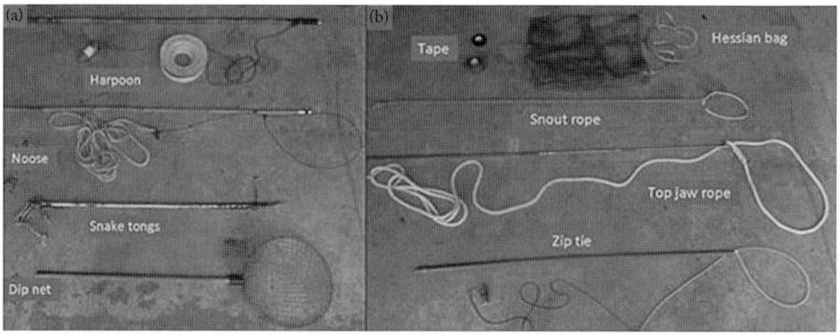

Figure 16.2 A variety of tools and equipment are used to (a) capture (Photo by M. Brien) and (b) restrain (Photo by M. Brien) crocodilians, the selection of which depends on the size, location, and situation.

pulled tight around the neck of the crocodilian. Catch poles should be thin, light, and rigid and can be made from wood, bamboo, or plastic.

16.3.3 Skin harpoon

Harpoons with small detachable heads are a standard method for capturing crocodilians >1.5 m TL (Webb and Messel, 1977). Skin harpoons pierce the relatively thick skin but minimize unnecessary penetration into muscle (Webb and Messel, 1977; Walsh, 1987), and are thus ideal for live capture. The same equipment can be used to capture a wide range of different sized crocodilians (1.5–7.0 m TL). The method relies on the ability to approach the animal close, and it is of limited utility for very wary individuals.

For live capture, crocodilians are approached closely before the harpoon head is jabbed into the skin of the neck region (fewer osteoderms) or the tail (for animals 1.5–2.0 m TL). The harpoon head is mounted on a pole with a retrieval line (e.g. 3–4 mm nylon cord) on a reel, or attached to the vessel or reflective float depending on the size of the animal (Figure 16.2(a)). The barbs of the harpoon head hook into the skin and the animal can be pulled up to the vessel where it can be secured. For species such as *C. porosus* that have little ossification in the skin, the harpoon head has 2–3 barbs <10 mm in length.

Various types of harpoon head fired from a rifle or bow have been tested, but optimal penetration occurs at a precise range, which is hard to achieve in practice. Entanglement of harpoon lines in submerged branches and tree roots needs to be avoided.

16.3.4 Traps

The main types of trap used to capture crocodilians are cylindrical or square tubes, made of metal, rope, or netting, in which a bait at the back entices crocodiles to enter, and a door at the front closes and stops them from escaping. Metal traps made of aluminium or stainless steel, if mounted on pontoons, become floating traps, which can move up and down with the water level (Walsh, 1987). They are fixed in position with anchors or by tethering to the bank, and they usually have shade cloth on the roof to restrict exposure to the sun. The bait is attached to a trigger mechanism, which releases a steel gate that slides down within a metal frame at the front.

Metal and rope traps can both be set up on the bank at the water's edge, if required. Rope traps appear to be effective if crocodiles are wary of entering metal traps (Webb and Messel, 1977). Rope traps are made from 8–12 mm rope, with a 100–150 mm mesh size, and can be fitted to a conventional sliding steel gate frame, or closed with a noose and weight that releases when the bait is pulled.

In tidal areas, traps on the bank need to be set above the high tide mark, in full shade, and may need visual 'wing' barriers (logs and branches) to direct animals into the mouth (rather than sides or back of the trap). Small pieces of meat ('starter baits') hung at the entrance to the trap may encourage animals to enter.

16.3.5 Snatch hook

Weighted treble hooks attached to a fishing rod or hand line with thick nylon or braid line can be used effectively to capture crocodilians 1.5–3.5 m TL (Cherkiss et al., 2012). The hook is cast out over the top of the animal in the water during the day or night in order to snag it in the neck, body, or tail during retrieval. The hook should be barbless so that as soon as pressure is released from the line the hook will fall out, eliminating the risk of remaining embedded if the line snaps. Once the animal is snagged it can be hauled in close to the shore or boat where a noose or harpoon can be used to secure it. Hand lines are effective for capturing animals that have submerged in water with a soft substrate, while rods enable capture of more wary individuals from greater distances.

16.3.6 Fixed-position snares

The use of fixed-position snares is an effective technique in the capture of crocodilians (Kofron, 1989; Murphy et al., 1990). The snare requires a self-locking mechanism and is best constructed of nylon rope or steel wire cable attached to a rope that is adequately tethered to prevent the animal escaping. Ideally, bungee or thick rubber cord should be incorporated into the setup to act as shock absorbers. Snares should be set in a shaded area, left over night, and then checked again before light the next morning.

Walk-through snares are usually suspended (e.g. from a tree) over land, along a known path used by the target crocodilian (Cherkiss et al., 2012). Barricades or shields can be set on either side to direct the animal into the snare. Once it walks through the snare, it tightens on the neck or torso. Baited snares are usually suspended (e.g. from a tree, floating platform) over water with the bait placed on the lower side of the cable to restrain the top jaw of the crocodilian when it grabs the bait. However, this should not be attempted with species that have a thin snout (e.g. *C. johnstoni*).

16.3.7 Nets

Crocodilians inhabiting small waterholes or creeks/rivers may be captured by nets (Webb and Messel, 1977). Dip nets can be used to capture smaller individuals (Figure 16.2(a)). Nets used to block the animal as it swims typically have a large mesh size (20–30 cm mesh size), and are designed to allow the head of a moving crocodilian through and entangle the body. Fine nets (5–10 cm mesh size) are designed to entangle the teeth and head when the crocodilian strikes the net. Nets are usually set in one location and monitored regularly for signs (bobbing floats) that a crocodilian is entangled.

Drag nets can be used to target the capture of specific crocodilians in small water holes or impoundments. The dragging of nets may expose the operator to potential interaction with the crocodilian and should only be implemented if infrastructure exists to eliminate this risk.

16.3.8 Baited hooks

Large baited hooks set over a body of water and anchored to a tree with line, wire cable, or rope are effective for catching crocodilians to be killed (e.g. harvest programmes). The distance between the bait and the water surface can be set to target crocodilians over a certain size.

16.3.9 Baited digestible 'hooks'

Baited digestible hooks can be used for live capture of crocodilians, particularly wary individuals. A digestible baited hook is made by running a rope through a smooth-edged bone and surrounding the bone with the bait (e.g. chicken, meat). The rope can be tied in such a way that the knot will not be inside the animal. When the bait is swallowed, the bone becomes lodged in the animal's stomach, effectively 'hooking' it (Walsh, 1987; Cherkiss et al., 2012). The crocodilian usually settles on the bottom, and can be raised to the surface and noosed or harpooned. The rope can be secured to a bungee cord on a suitable anchor point (e.g. tree), or a float, to assist in locating the baited crocodilian.

16.4 Handling

After capture, the main objective is to close and secure the mouth, and the approach taken to achieve this can vary with size. When pulled from the water, reducing visual stimuli with tape over the eyes, or a wet bag (e.g. hessian, burlap) over the head, will tend to slow or stop struggling, as will locating the crocodile in a cool shaded area versus exposed to the sun.

Because crocodilians struggle using anaerobic metabolism, which leads to metabolic acidosis in the blood that can lead to death in very large individuals (Seymour et al., 1987), every effort should be made to keep restrained animals calm and to minimize noises that may stimulate struggling.

16.4.1 Controlling the head

The first step is to gain control of the head, which can be done by placing a suitably sized rope with a spliced loop fashioned into a noose (Figure 16.2(b)) around the top jaw or neck of the animal using a pole of suitable length. The jaws can then be securely closed prior to handling to avoid injury to handlers. Prior to securing the jaws, covering the animal's eyes with a wet towel or sack to reduce visual stimuli, usually calms the animal.

16.4.2 Securing the jaws

Depending on size, the jaws can be tied together using a rubber band (hatchlings and small juveniles), tape, cord, or heavy rubber straps (larger individuals), around the anterior end of the snout behind the nostrils (Figure 16.2(b)). The nostrils should not be covered. Plastic zip ties can also be applied with a pole or stick to temporarily close the jaws prior to the application of the tape or cord (Figure 16.2(b)).

Once the jaws are secured, a cover over the eyes can be taped to hold it in place. Even with the jaws closed, a crocodilian can cause serious damage by swinging its head to the side. Handlers should remain out of the arc (area around the head) at all times and approach the animal from the front if possible.

16.4.3 Restraining the limbs

It may be important to secure the limbs of a crocodilian to prevent it from gaining traction and either moving off or rolling during handling. This can be achieved by tying the legs back against the side of the body and tail, preferably with wide tape to prevent reducing circulation. Securing the legs off the ground reduces excessive struggling.

16.4.4 Temporary holding and transport

If crocodilians need to be held for any length of time, they should be kept cool and shaded, avoiding direct exposure to hot conditions that may lead to overheating. If there are concerns that animals may regurgitate fluids that may flow back into the lungs, the head should be raised higher than the body. Placing a piece of wood, rubber or plastic tube between the jaws to keep them slightly open prior to securing them may also be useful.

Crocodilians that are bound can be held for short periods by tethering the top jaw rope to a solid object. Another method involves strapping the animal to a carry board (e.g. with ratchet tie down straps; Figure 16.3(a)). The head must be securely tied off to the bottom of the board to prevent struggling. For smaller animals that are to be held in ventilated boxes or pipes with removable caps, it may not be necessary to have the animal restrained.

16.4.5 Immobilizing agents

Chemical immobilization is an effective method of restraining a crocodilian and minimizing risk of harm to both animal and handler. By stopping all struggling, the build-up of lactic acid ceases, blood pH stabilizes, and risk of death by acidosis declines (Seymour et al., 1987). Depending on the situation, chemical agents may be administered intramuscularly using a syringe, pole-syringe, or dart gun. Injection between scales on the tail, front leg, or top or side of the neck appears to result in minimal discomfort for the animal and is safe for handlers.

Commonly used immobilizing agents for crocodilians are gallamine triethiodide (e.g. Flaxedil®, Loveridge and Blake, 1987) and pancuronium bromide (e.g. Pavulon®, NRMMC, 2009), with the effects of both being reversed by neostigmine methyl sulphate. Diazepam-based sedatives (e.g. Valium®) may sometimes be used as a cocktail with neuromuscular blockers such as gallamine triethiodide and pancuronium bromide

Figure 16.3 (a) Large crocodilians can be securely fastened to a board during transport (Photo by M. Brien). Crocodilians can be individually marked (b) by removing a series of raised scutes from the tail (Photo by C. Manolis) or (c) by attaching a numbered metal tag between the webbing of the back feet (Photo by C. Manolis). (d) Transmitters are often attached to the nuchal rosette on the base of the neck in species with pronounced scutes, such as *C. porosus* (Photo by C. Manolis).

(NRMMC, 2009). Recently, Olsson and Phalen (2012a, 2012b) used medetomidine (Zalopine®) on *C. porosus* and *C. johnstoni*, but effective immobilization was dependent on the site of injection, and not all sizes of animal were tested.

The effect of different immobilizing agents (and anaesthetics) varies with species and individuals, and may be influenced by the extent of struggling and temperature. Due to the potential for a build-up of lactic acid in the bloodstream, drugs should not be used with crocodilians that have struggled to the point of exhaustion. Upon release, crocodilians should be placed near, but not in the water, so they can return once the effect of the immobilizing agent has worn off sufficiently.

16.5 Tagging

16.5.1 Scute-clipping

For the purposes of identification, crocodilians are commonly marked in field and captive situations by removing a combination of the raised, vertical scutes on the tail (single and double rows) in accordance with a predetermined numbering system (e.g. Richardson et al., 2002; Figure 16.3(b)). Scutes should be removed at the base using a sharp scalpel, knife, or scissors (for hatchlings) to avoid regrowth. This process is relatively quick, causes minimal discomfort, and can be performed without local anaesthetic.

16.5.2 Webbing tags

Uniquely numbered metal tags placed in the webbing between the toes of the back feet (Saalfeld et al., 2008; Figure 16.3(c)) are commonly used with crocodilians. The tags are attached using a crimping tool, which causes minimal discomfort. A problem with this approach is tag loss, so sometimes two tags (same number) are attached (one on each foot). The size of tag required for a hatchling (<75 g) is different to one that may be required for an adult (<1000 kg), creating problems as the animal grows.

16.5.3 PIT tags

In crocodilians, passive integrated transponder (PIT) tags are injected subcutaneously with an applicator (modified syringe and needle) at the base of the tail (dorsal) or a location where the PIT tag is unlikely to 'migrate' as the animal grows. Information in the PIT tags is read using an electronic reader placed near the site of injection. PIT tags have a high retention rate but require the animal to be close by in order to read the PIT tag.

16.5.4 Ear tags

Numbered plastic livestock (e.g. cattle tags) ear tags, visible from a distance, can be attached to the osteoderms of the neck or the vertical tail scutes (Cooper-Preston, 1991). Tags are attached using an applicator specific for that tag, or using wire inserted through holes drilled into the osteoderm or scute with marine epoxy moulded over the top. However, this method can be time consuming and tag loss is common.

16.5.5 Anchor fish tags

Plastic T-bar anchor fish tags (uniquely numbered and coloured) have also been used with crocodilians and are injected subcutaneously with an applicator (modified syringe and needle) at the base of the skull (Eversole et al., 2014). These fluorescent tags extend out from under the skin (several cm) and are visible at night when shone with a spotlight. Retention rates are higher than livestock tags and web tags but not as high as for PIT tags.

16.5.6 Electronic tags

The robust and heavily armoured nature of a crocodilian, with its many bony deposits (osteoderms, scutes), provides a variety of options for transmitter attachment. Electronic tags are commonly attached directly to the head, mid-dorsal neck, or tail using pins, wires, or nylon line (Franklin et al., 2009).

When attaching devices it is important to ensure the tag will achieve maximum signal attenuation, will be retained long enough to gather the required data, and are small enough not to impede movement, damage underlying tissue, or restrict growth at the attachment site. This is especially important given that most species of crocodilians are often difficult to recapture after being tagged.

The nuchal rosette consists of several raised osteoderms on the mid-dorsal surface of the neck, and provides an ideal platform for attaching electronic tags (Franklin et al., 2009). The tag is placed in the centre of the nuchal rosette, before wires (often plastic coated) are run through holes drilled into the sides of the osteoderms in species in which these are pronounced (e.g. *C. porosus*; Figure 16.3(d)), or subcutaneously under the nuchal rosette using a large needle (Brien et al., 2010), in species with less pronounced osteoderms (e.g. *C. johnstoni*).

Subcutaneous injections of a local anaesthetic around the nuchal rosette are effective for minimizing local pain and prevent the animal moving during the attachment process. The wires are then secured to anchor points on the tag, locking it in place, and a moulding material (e.g. marine epoxy) is often used to form a base underneath the tag and extended over the top of the tag to secure and protect the tag and form a smooth surface that prevents snagging (Figure 16.3(d)). Tags can also be attached in a similar way to the dorsal surface of the tail where the two rows of vertical caudal scutes merge into a single row (Muñoz and Thorbjarnarson, 2000).

Other methods of attachment include back harnesses (Kushlan and Mazzotti, 1989), neck collars (Taylor, 1984), ingestion (Magnusson and Lima, 1991), and surgical implantation (Franklin et al., 2009). Back harnesses and neck collars similar to those used for other animal species have been used sparingly due to the risk of entanglement, fouling, and potential effect on (and from) growth. Ingestion of tags is possible due to the ability of crocodilians to retain hard objects in their stomachs (e.g. gastroliths), but signal attenuation is limited.

Acoustic tags can be surgically implanted into the abdomen through a small incision made with a scalpel on the flank, just posterior to a hindlimb (Franklin et al., 2009). Once implanted, the incision is closed with silk sutures and the area sprayed with an antiseptic spray. This process requires the animal to be anaesthetized during implantation.

References

Bayliss, P. (1987). Survey methods and monitoring within crocodile management programmes. In G J.W. Webb, S.C. Manolis, and P.J. Whitehead (eds) *Wildlife Management: Crocodiles and Alligators*. Sydney:Surrey Beatty and Sons, pp. 157–75.

Brien, M., Webb, G., Manolis, C., et al. (2010). A method for attaching tracking devices to crocodilians. *Herpetological Review*, **41**, 305–8.

Channa, P., Sovanna, P., and Gray, T.N.E. (2010). Recent camera trap records of globally threatened species from the Eastern Plains Landscape, Mondulkiri. *Cambodian Journal of Natural History*, **2**, 87–8.

Cherkiss, M.S., Fling, H.E., Mazzotti, F.J., et al. (2012). *Counting and Capturing Crocodilians*. IFAS Publication CIR1451. University of Florida.

Choquenot, D., and Webb, G.J.W. (1987). A photographic technique for estimating the size of crocodiles seen in spotlight surveys and for quantifying observer bias. In G.J.W. Webb, S.C. Manolis, and P.J. Whitehead (eds) *Wildlife Management: Crocodiles and Alligators*. Sydney:Surrey Beatty and Sons, pp. 217–24.

Cooper-Preston, H. (1991). *Geographic variation in the population dynamics of* Crocodylus johnstoni *(Kreft) in three rivers in the Northern Territory, Australia*. PhD Thesis, University of New England, Armidale, Australia.

Daltry, J.C., Dany, C., Phal, E., et al. (2003). *Status of the Siamese Crocodile in the Central and Southern Cardamom Mountains, Cambodia*. Technical Report: Flora & Fauna International: Cambodia Programme. Phnom Penh:Department of Forestry and Wildlife.

Eversole, C.B., Henke, S.E., Ballard, B.M., et al. (2014). Duration of marking tags on American alligators (*Alligator mississippiensis*). *Herpetological Review*, **45**, 223–6.

Franklin, C.E., Read, M.A., Kraft, P.G., et al. (2009). Remote monitoring of crocodilians: implantation, attachment and release methods for transmitters and data-loggers. *Marine and Freshwater Research*, **60**, 284–92.

Fukuda, Y., Saalfeld, K., Webb G., et al. (2013). Standardised method of spotlight surveys for crocodiles in the tidal rivers of the Northern Territory, Australia. *Northern Territory Naturalist*, **24**, 14–32.

Games, I. (1994). Aerial surveys for monitoring trends and estimating population size of *Crocodylus niloticus* or the theory and practice of aerial survey in Africa. In *Proceedings of the 12th Working Meeting of the IUCN-SSC Crocodile Specialist Group*. Gland, Switzerland:IUCN, pp. 245–55.

King, F.W., Hutton, J., Manolis, C., et al. (1994). Guidelines on monitoring crocodilian populations. Results of workshop D: monitoring—how do we implement practical programs? In *Proceedings of the 2nd Regional Meeting (Eastern Asia, Oceania, Australasia) of the IUCN-SSC Crocodile Specialist Group*. Gland, Switzerland: IUCN.

Kofron, C.P. (1989). A simple method for capturing large Nile crocodiles. *African Journal of Ecology*, **27**, 183–9.

Kushlan, J.A., and Mazzotti, F.J. (1989). Historic and present distribution of the American crocodile in Florida. *Journal of Herpetology*, **23**, 1–7.

Loveridge, J.P., and Blake, D.K. (1987). Crocodile immobilization and anaesthesia. In G.J.W. Webb, S.C. Manolis, and P.J. Whitehead (eds) *Wildlife Management: Crocodiles and Alligators*. Sydney:Surrey Beatty and Sons, pp. 259–67.

Magnusson, W.E., and Lima, A.P. (1991). The ecology of a cryptic predator, *Paleosuchus trigonatus*, in a tropical rainforest. *Journal of Herpetology*, **25**, 41–8.

Merchant, M., Savage, D., Cooper, A., et al. (2013). Assessment of nest attendance of the American alligator (*Alligator mississippiensis*) using a modified motion-sensitive camera trap. In *Proceedings of the 22nd Working Meeting of the IUCN-SSC Crocodile Specialist Group*. Gland, Switzerland:IUCN, p. 184.

Messel, H., Vorlicek, G.C., Wells, A.G., et al. (1981). *Surveys of Tidal River Systems in the Northern Territory of Australia and their Crocodile Populations*. Monograph 1. Sydney:Pergamon Press.

Mourão, M.G., Bayliss, P., Coutinho, E.M., et al. (1994). Test of an aerial survey for caiman and other wildlife in the Pantanal, Brazil. *Wildlife Society Bulletin*, **22**, 50–6.

Muñoz, M.D.C., and Thorbjarnarson, J. (2000). Movement of captive-released Orinoco crocodiles (*Crocodylus intermedius*) in the Capanaparo River, Venezuela. *Journal of Herpetology*, **34**, 397–403.

Murphy, T., Wilkinson, P., Coker, J., et al. (1990). *The Alligator Trip Snare. A Live Capture Method: Construction and Use*. Columbia, SC:South Carolina Wildlife and Marine Resources Department.

NRMMC. (2009). *Code of Practice for the Humane Treatment of Wild and Farmed Australian Crocodiles*. Canberra: Natural Resource Management Ministerial Council.

O'Brien, T. (1990). A comparison of 3 survey methods for estimating relative abundance of rare crocodilians. In *Proceedings of the 10th Working Meeting of the IUCN-SSC Crocodile Specialist Group*. Gland, Switzerland:IUCN, pp. 91–108.

Olsson, A.R., and Phalen, D.N. (2012a). Preliminary studies of chemical immobilization of captive juvenile estuarine (*Crocodylus porosus*) and Australian freshwater (*C. johnstoni*) crocodiles with medetomidine and reversal with atipamezole. *Veterinary Anaesthesia and Analgesia*, **39**, 345–56.

Olsson, A.R., and Phalen, D.N. (2012b). Medetomidine immobilization and atipamezole reversal in large estuarine crocodiles, *Crocodylus porosus*, using metabolically scaled doses. *Australian Veterinary Journal*, **90**, 200–44.

Platt, S.G., Brantley, C.G., Cropanzano, R.S., et al. (1990). A method for determining the size of nesting female alligators. *Wildlife Society Bulletin*, **18**, 296–8.

Rathnasiri, G.W.R.P., De Silva, A., Gabrial, A., et al. (2013). Mugger burrows: preliminary investigations into the unique tunnels excavated by *Crocodylus palustris* in Sri Lanka. In *Proceedings of the 22nd Working Meeting of the IUCN-SSC Crocodile Specialist Group*. Gland, Switzerland:IUCN, pp. 246–50.

Richardson, K.C., Webb, G.J.W., and Manolis, S.C. (2002). *Crocodiles: Inside and Out*. Sydney:Surrey Beatty and Sons.

Saalfeld, D.T., Webb, K.K., Conway, W.C., et al. (2008). Growth and condition of American alligators (*Alligator mississippiensis*) in an inland wetland of east Texas. *Southeastern Naturalist*, **7**, 541–50.

Seymour, R.S., Webb, G.J.W., Bennett, A.F., et al. (1987). Effect of capture on the physiology of *Crocodylus porosus*. In G.J.W. Webb, S.C. Manolis, and P.J. Whitehead (eds) *Wildlife Management: Crocodiles and Alligators*. Sydney:Surrey Beatty and Sons, pp. 253–7.

Taylor, D. (1984). Management implications of an adult female alligator telemetry study. *Proceedings of the Annual Southeastern Association of Fish and Wildlife Agencies*, **38**, 222–7.

Walsh, B. (1987). Crocodile capture methods used in the Northern Territory of Australia. In G.J.W. Webb, S.C. Manolis, and P.J. Whitehead (eds) *Wildlife Management: Crocodiles and Alligators*. Sydney:Surrey Beatty and Sons, pp. 249–52.

Webb, G.J.W., and Messel, H. (1977). Crocodile capture techniques. *Journal of Wildlife Management*, **41**, 572–5.

Webb, G.J.W., and Messel, H. (1978). Morphometric analysis of *Crocodylus porosus* from the north coast of Arnhem Land, northern Australia. *Australian Journal of Zoology*, **26**, 1–27.

Webb, G.J.W., and Messel, H. (1979). Wariness in *Crocodylus porosus*. *Australian Wildlife Research*, **6**, 227–34.

Webb, G.J.W., Bayliss, P.G., and Manolis, S.C. (1989). Population research on crocodiles in the Northern Territory, 1984–86. In *Proceedings of the 8th Working Meeting of the IUCN-SSC Crocodile Specialist Group*. Gland, Switzerland:IUCN, pp. 22–59.

Webb, G.J.W., Dillon, M.L., McLean, G.E., et al. (1990). Monitoring the recovery of the saltwater crocodile (*Crocodylus porosus*) population in the Northern Territory of Australia. In *Proceedings of the 9th Working Meeting of the IUCN-SSC Crocodile Specialist Group*. Gland, Switzerland:IUCN, pp. 329–80.

Woodward, A.R., and Marion, W.R. (1978). An evaluation of factors affecting night-light counts of alligators. *Proceedings of the Annual Southeastern Association of Fish and Wildlife Agencies*, **32**, 291–302.

Part 4
Reptiles in the Community

17

Plot and transect censuses

Tiffany M. Doan

17.1 Introduction

Plot and transect censuses comprise a wide variety of related techniques, all of which utilize a standardized area, distance, or time, making data gathered using these techniques repeatable and comparable to other studies. The primary data collected from plot and transect census surveys may be used to determine abundance, density, and species richness. This family of techniques ranges from visual encounter surveys, which have low intensity of effort but may cover a larger habitat area, to total removal plot methods, which require high intensity over a smaller area. In addition to being standard techniques for surveying reptiles, plot and transect censuses have been used for a multitude of other species, including mammals, insects, amphibians, birds, and plants.

Plot and transect censuses are excellent choices for studies conducting standardized inventories of sites, comparing different habitats or areas, monitoring populations over time, or comparing experimentally altered habitat to control habitat (Marsh and Haywood, 2010). These surveys are appropriate for nearly every reptile species, with the exception of high canopy species or fully aquatic deep-water species. These techniques may be implemented in any terrestrial habitat type, including deserts, rainforests, shorelines, and mountaintops. Some transect methods may even be modified for sampling aquatic species, as has been accomplished with sea snakes (Lillywhite et al., 2015). Both plot and transect techniques may be performed during the day or night, depending on the species being studied. The exact implementation may be tailored to the objectives of the study, habitat, and target reptile species. Especially in areas of high reptile abundance, plot and transect censuses are preferable to other methods such as passive sampling techniques, including pitfall arrays, cover boards, and other trapping methods (see Chapter 10), because they are inexpensive to implement and use very little equipment, though potentially require a larger investment of survey time. A comparison between transect surveys and intensive mark–recapture surveys for population estimation in the endangered tuatara found that transect methods were more accurate, were more cost effective, and caused less stress to the study organisms (Cassey and Ussher, 1999), making transects the technique of choice for monitoring that species.

Reptile Ecology and Conservation. Edited by C. Kenneth Dodd, Jr. © Oxford University Press 2016.
Published 2016 by Oxford University Press.

In this chapter, I describe the differences between plot and transect techniques in general, including requirements for robust plot and transect surveys. I then provide examples of the application of various types of plot and transect techniques. Subsequently, I describe the habitat variables that may be collected during censuses, and finally, how to select among the various plot and transect techniques. Although this chapter does not describe every detail of designing a study using these techniques nor present an exhaustive literature review on the subject, students and herpetologists wishing to use plot and transect censuses in research projects will be armed with enough knowledge to begin to implement these methods in reptile field studies.

17.2 Trade-off between intensity and area

Plot and transect census surveys may be considered along two simultaneous and contrasting gradients: intensity of sampling effort, that is, how much time and personnel are needed to complete a particular method, and area of habitat covered (Figure 17.1). These two variables are inversely related and are important considerations when planning a research project because of the trade-off in experimental design. High intensity methods require a larger investment of time and personnel compared to low-intensity methods and are particularly useful in areas with high abundance of target species, facilitating the capture of a larger number of individuals in each sampling area. However, in addition to requiring a large investment of personnel, high intensity methods can only cover a relatively small amount of habitat area. On the other end of the spectrum, low-intensity methods do not require a large research team and can cover a wider search area. In low-abundance habitats or when researching species that are widely dispersed, low-intensity methods may be preferable because they may capture a higher diversity of mobile species, although they are unlikely to give a complete inventory of reptile diversity and abundance. When planning a research project that will use plot and transect methods, factors such as time, personnel, sample size, and target reptile species are important considerations.

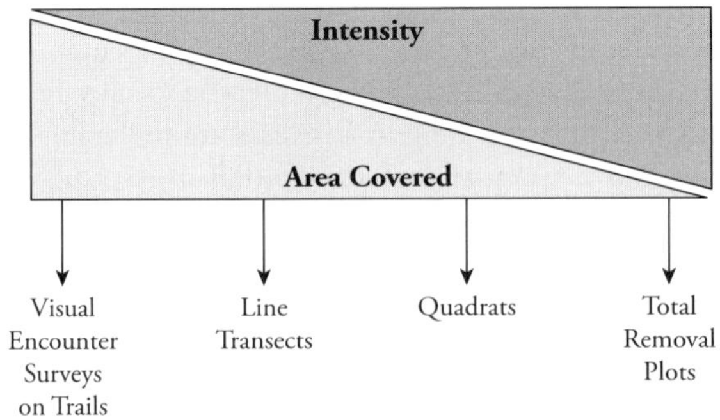

Figure 17.1 Intensity versus area gradient representing the trade-off for common plot and transect census techniques. As intensity increases, area covered decreases and vice versa.

17.3 Plots versus transects

Plots, often called quadrats, are usually square or rectangular areas placed randomly or systematically in habitat to be surveyed for reptiles. Plots are usually searched with medium to high intensity. A project may survey a low number of large plots, or more commonly, a high number of smaller quadrats. Quadrat size should be selected based on habitat, species biology, and the goals of the study. An 8 m × 8 m square (64 m^2) is the recommended standard size that has been successfully used for reptiles in many habitats (Heatwole, 2012; Figure 17.2(a)). Utilizing this standard size facilitates comparison among studies and it is small enough to complete many quadrats in a single day or night.

Transects, on the other hand, are long and relatively narrow, often linear survey routes. While sometimes searched at a relatively high intensity, transects are most commonly implemented at a low intensity. The placement of transects may simply follow an existing trail or may be created specifically for a study. For example, transects are often oriented along or perpendicular to an environmental gradient, depending on the research questions to be addressed (Figure 17.2(b)). Transects may also be arranged

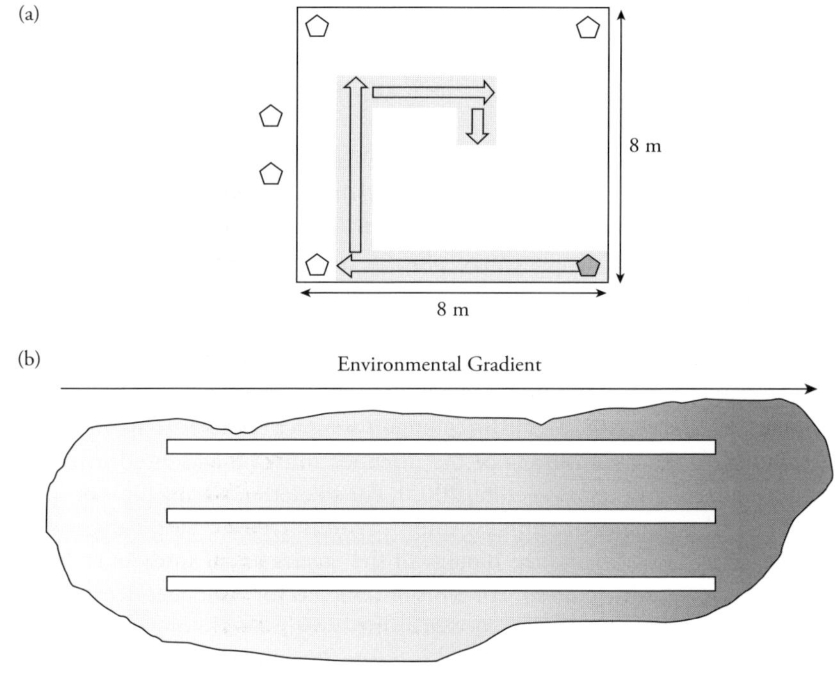

Figure 17.2 Diagrams of two examples of common plot and transect practices. (a) Quadrat pattern with a six-person team. Personnel are indicated by pentagons. The person indicated by the filled-in pentagon demonstrates the pattern with the shaded area and arrows. All four of the quadrat workers follow the same clockwise pattern simultaneously. (b) Transects running parallel along an environmental gradient. The two figures differ in scale.

within larger parcels in an attempt to cover a defined area. A common arrangement is 100 m-long transects arranged parallel to each other within a 1 ha parcel (Doan and Arizábal Arriaga, 2002).

17.4 Valid implementation of plot and transect techniques

Before beginning a project that utilizes plot or transect methods, it is crucial to choose the target organisms, formulate hypotheses, perfect techniques, define the study area, and determine timing, personnel, number of replicates, and analysis to ensure a well-executed study. Additionally, it is critical that all personnel are trained in the techniques and species identification. Making a practice run of the methods prior to sampling will ensure that all observers know their tasks and that the techniques will be performed efficiently. Because experience in herpetofaunal sampling enhances the search image of investigators and makes detection of all target reptiles more likely (Glaudas, 2013; Lardner et al., 2015), inter-observer error is a bias that must be avoided as much as possible. Lardner et al. (2015) found that different observers varied in their lizard detection ability by a factor of six, with variables such as type of headlamp being used, amount of wind on a given sample night, and even the phase of the moon having strong effects on observer detection. To reduce potential inter-observer error, the same personnel should survey each of the transects or quadrats with the same intensity of effort.

Some herpetologists have recommended (or required) that plots and transects be implemented randomly (Anonymous, 1998; Marsh and Haywood, 2010; Heatwole, 2012). That is, each transect or quadrat be placed randomly within the habitat and not in a location determined by other variables. While random placement of plots and transects is a viable and statistically powerful method, it is not a requirement of the methodology (Milner-Gulland and Rowcliffe, 2007). Many researchers have successfully implemented systematically placed plots or transects to examine populations or assemblages of reptiles (Pianka, 1966; Cassey and Ussher, 1999; Schlaepfer and Gavin, 2001; Doan, 2003; Rödel and Ernst, 2004; Garden et al., 2007; Folt and Reider, 2013) and other taxa such as amphibians (Doan, 2004; Veith et al., 2004), insects (Samways et al., 2010), and birds (Gregory et al., 2004).

In order for the data collected using plot and transect censuses to be rigorous and statistically valid, the assumptions of the methods must be addressed in the study design (Milner-Gulland and Rowcliffe, 2007). For all methods using plots or transects as replicates, independence of sampling units is critical. Determining independence is not always straightforward and the biology of the species under study must be somewhat understood in order to do so. For plots or transects to be independent, they must be far enough away from each other so that animals registered in one quadrat or transect will not subsequently be registered in another during the same sampling period. One way to minimize this risk is to have multiple survey teams conduct adjacent quadrats or transects simultaneously (Doan, 2003). Most small leaf litter-dwelling species that are commonly surveyed in quadrats do not travel more than a few metres per day and likely less than a hundred metres in a lifetime, but other species may travel great distances. If quadrats or transects will be resampled over time, permanently marking the captured individuals (see Chapter 4) will ensure that the same organism is not double counted.

In addition to understanding the assumptions, it is important to comprehend what data the various plot and transect census techniques will and will not provide. These techniques can give estimates of population abundance, species richness, diversity, and density, depending on the exact methodology and its implementation. Even though the standard assumption of the quadrat method that all reptiles present will be recorded may be broken, abundances among the quadrats will be comparable as long as the probability of detection of reptiles is the same among the compared quadrats (Marsh and Haywood, 2010). Perhaps with the exception of total removal plots, plot and transect techniques do not provide absolute counts of populations or species richness. However, with appropriate statistical techniques, such numbers may be estimated and compared among estimates of other areas using similar assumptions and methodologies.

Plots or transects should be placed with as little bias as possible, unless patch sampling is the selected technique (see Section 17.5.5). Whether the method being used to place the plots is random or systematic, the habitat in which the sampling is conducted needs to be representative of the total habitat of the target reptile species (Milner-Gulland and Rowcliffe, 2007). In addition to these concerns, the number of plots or transects should be sufficiently large to encompass the diversity of the reptile assemblage (in a community study) or be large enough to get a reliable estimate of population levels (for a single species study). The balance between minimizing the time, funds, and effort needed to complete the study objectives and the gathering of enough data to have valid estimates of populations or diversity is always an important consideration (Newton, 2007).

If the objectives of the study are to compare abundance or species richness among two or more areas, it is critical that the areas of plot or transect censuses be matched for habitat type, which may include distance to bodies of water, forest type, level of disturbance, and other factors, depending on the goals of the study. Because weather may affect the behaviour and visibility of reptiles, matched plots or transects should be sampled in pairs. The same number of each treatment should be sampled on each day, as opposed to sampling all of one treatment during one time period and the other at a later date. Differences in populations and diversity of reptiles among treatment and control areas of experimental burns (Trainor and Woinarski, 1994), tourist disturbance (Doan and Arizábal Arriaga, 2000), pesticide application (Lambert, 2005), habitat disturbance (Folt and Reider, 2013), and fragmentation effect (Schlaepfer and Gavin, 2001) are examples of studies that have used matched parcels of quadrats or transects successfully.

Although most herpetologists who utilize plots and transects sample by capturing the target reptiles, these techniques may also be performed through simple observation, without capture. However, if using strictly observation, extra care must be taken to properly identify species and to avoid double counting individuals. Plot and transect techniques may also include the recording of indirect reptile signs, such as nests, burrows, footprints, tail drags, scat, and shed skins, that may increase the number of taxa sampled.

These plot and transect techniques may be implemented during the day or night, and although many researchers choose census times by convenience, the timing of these surveys should be determined by the biology of the target reptile species. To investigate poorly studied species for which the activity patterns are not understood, sampling throughout the 24-hour period is prudent. For some diurnal species such as *Anolis* lizards or large chameleons, night-time surveying is often more effective because the species can be easily found while sleeping. Alternatively, nocturnal species such as *Bothriopsis* snakes are more easily captured during the day while asleep. If plot or transect censuses are conducted at night, powerful flashlights are needed for maximum visibility (Lardner et al., 2015) and headlamps may be preferred over handheld flashlights to keep the hands free for reptile capture.

A multitude of studies that utilize plots and transects survey the entire herpetofauna, as opposed to only reptiles (Schlaepfer and Gavin, 2001; Doan and Arizábal Arriaga, 2002; Almeida-Gomes et al., 2008; Folt and Reider, 2013). If the objective of a field project is to examine species richness or diversity, this combined approach is recommended because the effort expended to capture amphibians in addition to reptiles is minimal and yields more information about the local community than concentrating solely on the reptile assemblage.

17.5 Standard plot and transect techniques

The techniques presented in the following sections represent the most common plot and transect census types for surveying reptiles. They are ordered from lowest intensity to highest intensity. Researchers should modify these techniques based on the objectives of their projects.

17.5.1 Visual encounter surveys on trails

For this low-intensity method, existing trails are used as transects. The investigator walks the trail slowly and observes both sides of the trail without disturbing the vegetation or substrate. The investigator sets a width a priori, such as 1 or 2 m from the centre of the trail to either side, selecting the distance based on the visibility within the habitat. A vertical distance limit for observing arboreal species is also set. By using an existing trail, no pre-census set up is required. Due to its low intensity, this visual encounter survey (VES) method allows an investigator to cover a relatively large distance in a short amount of time. For more rigor, overall length of the transect should be measured so that the area (or volume) surveyed may be established. This method is particularly useful for large sedentary species that sit in the open, but less useful for secretive species. I have been successful using this method at night in Peru for sleeping diurnal geckos, anoles, arboreal snakes such as *Xenoxybelis*, and other species that sleep in the relative open. Active, yet relatively slow-moving snakes such as terrestrial *Atractus* or arboreal *Imantodes* are also sampled effectively with this method. A visual encounter transect study in Borneo yielded 32 snake species of a variety of ecotypes (Van Rooijen, 2009). Trails that follow water bodies could be used to census aquatic species such as *Nerodia*, some turtles, and crocodilians. Although this method is primarily visual, investigators should also be attuned to audible indicators to the presence of reptiles such as scurrying on the ground or on vegetation or for sounds made directly by species such as *Crotalus* or nocturnal geckos.

A higher intensity variation of this technique uses the same method but incorporates active searching by moving vegetation, flipping rocks and logs, and raking leaf litter on the sides of an existing trail (Wilgers et al., 2006; Glaudas, 2013).

17.5.2 Line transects

In this variation of the transect method, transects are installed in straight lines through a study area in advance of sampling. Depending on the habitat, only beginning and end markers of each transect may be sufficient. In dense habitat such as tropical rainforest, markers should be more frequent (e.g. every 10 m) and light cutting of vegetation may be necessary for surveyors to be able to follow the path. Such set up disturbs the habitat and potentially reptile behaviour, so waiting several days after installation before sampling may be prudent. In dense habitat, it is often helpful to have two observers, one who looks to the right side of the transect and the other to the left. Thus, fast, active reptiles moving through the area will be less likely to escape detection.

Transect position and direction may be placed either randomly in the habitat (Lovich et al., 2012) or selected systematically within a large square or rectangular parcel. One example of this technique involves 1 ha parcels with 24 parallel transects each 4 m apart (Doan and Arizábal Arriaga, 2002). Once all transects have been completed, it is possible to estimate abundance and density of the parcel. Some researchers designate a large parcel and sample it by using a randomized walk design within the area (Rahman et al., 2014). This strategy is a valid technique, but does not allow accurate estimations of density because the entire parcel will not be sampled.

As with trail VES, the search intensity for line transects may be determined by the investigators, depending on their objectives (Ribeiro-Júnior et al., 2008; Sung et al., 2011), ranging from visual searching only to higher intensity searching. In addition to capturing animals, some researchers have used transects to count burrows (e.g. *Gopherus*; Stober and Smith, 2010) for a relative population estimate.

17.5.3 Quadrats

As stated previously, quadrats (or 'leaf litter plots', Marsh and Haywood, 2010) may vary in size depending on habitat and target species, but the standard size for reptiles is 8×8 m^2. The quadrat method assumption is 'get it all', that is, that all reptiles within the quadrat will be recorded (Milner-Gulland and Rowcliffe, 2007), but it is unlikely that this assumption is always met (Rodda et al., 2001; Heatwole, 2012). The entirety of each quadrat is searched intensively by flipping rocks and logs, raking leaf litter, and parting vegetation up to a defined height for every reptile individual of interest. Care must be taken to prevent the disturbance of the quadrat area prior to sampling and to prevent the escape of target reptiles during sampling. Natural cover objects should be returned to their original locations after sampling is complete to reduce disturbance to the local community. Some authors recommend clearing a strip of litter around the periphery of the quadrat so that any escaping reptiles will be observed and captured (Heatwole, 2012).

A common way of performing quadrat sampling uses a team of 5–6 individuals (Figure 17.2(a)). Four people begin at each of the four corners and work in a clockwise fashion from the outside in. Each person searches a 1 m-wide strip from the outer quadrat boundary until reaching the next side. The searcher then turns in to complete the next inner 1 m-wide strip of the subsequent side. With the coordinated movement of all searchers, the entire quadrat is searched and any potentially escaping reptiles run to the inside to be captured by another searcher. The fifth and sixth members of the team remain outside the quadrat to receive animals, register them, measure, mark them (if applicable), and record environmental variables. If individuals are being released, the outer team members keep the reptiles in containers until the quadrat is finished and then release the animals back into the quadrat. One application of the quadrat technique had multiple 70 × 80 m^2 parcels containing 56 8 × 8 m^2 quadrats separated 2 m from adjacent quadrats and 1 m from the edges of the parcel searched using a six-person team (Doan and Arizábal Arriaga, 2002; Doan, 2003).

Although quadrats require a relatively high amount of effort, time, and personnel, they are particularly good for catching small leaf litter lizards and snakes. Many quadrats may be completed in a day or night, providing for a more robust study through the use of replicates. Care must be taken to space the quadrats far enough apart so that the individual reptiles from one quadrat are not likely to travel into another. Because this method disturbs the habitat, quadrats should not be repeated in identical locations within a short period of time (months may be needed for the habitat to recover). In a long-term monitoring study where animals are not being permanently removed, a parcel of quadrats may be sampled repeatedly, with a portion of the quadrats per parcel sampled each time and selected at random (e.g. half of quadrats sampled during each site visit; Doan, 2003).

17.5.4 Total removal plots

Also known as the fenced quadrat technique (Heatwole, 2012), this method is similar to the quadrat method but it requires the highest level of intensity. To implement this method, the boundary of each plot is cleared so that the plot is isolated. An extreme variation of this technique involves creating a 1.5 m-wide clear outer boundary from the ground surface up through the canopy so that arboreal species will not be able to escape (Rodda et al., 2001). The plot is fenced with aluminium flashing, nylon mosquito net, plastic, or other material. The fence is then sunk into the ground, rising approximately 50–100 cm above ground to prevent escape of any animals (Rodda et al., 2001; Almeida-Gomes et al., 2008). Plot workers move in the same pattern as unfenced quadrats to sample the fenced plot. Not only is all the vegetation searched, everything within the quadrat is removed down to the soil (Heatwole, 2012), which may include cutting down the trees with chainsaws (Rodda et al., 2001), ensuring every reptile present is collected. Heatwole (2012) additionally recommends that the fence remains in place for two days after sampling so that searchers may return to catch any reptiles that eluded capture during initial sampling. Because of the significant disturbance to the habitat and to reptiles in the area, total removal plots should be set up in advance of implementing the technique, such as morning set-up, with sampling at night (Almeida-Gomes et al., 2008). Unlike standard quadrats, herpetologists have not settled on a standard size. Rodda et al. (2001) and Heatwole (2012) recommend 10 × 10 m^2, whereas other researchers have used 5 × 5 m^2 (Almeida-Gomes et al., 2008) and other sizes.

Total removal plots are highly destructive to the habitat and have the potential to negatively affect the population of reptiles under study. The use of total removal plots would be contraindicated in sensitive habitat or for studying endangered species. However, this method has been shown to be superior to haphazard searching, transects, and quadrats for estimating reptile species richness (Heatwole, 2012) and may be used safely for abundant species in non-critical habitats. However, effort, time, and financial costs are also higher for total removal plots, so these factors must be weighed when deciding which technique to implement in a field study.

17.5.5 Other plot and transect techniques

All of the aforementioned techniques may be considered area-constrained search (ACS) methods (Todd, 2013a), with the surface area (or volume) as the standard unit to be compared among areas or treatments. Such ACS methods continue for as long as is necessary to complete the plot or transect, which varies depending on the habitat, time of day or night, and number of reptiles found. An alternative technique that uses similar methodology is a time-constrained search (TCS). In a TCS, an amount of time is decided a priori, which could be as little as five minutes for highly abundant small lizards or snakes, to multiple hours for more dispersed species. A study of lizard activity in Mexico used 1 km transects walked at a pace of 2 km/h to investigate five lizard species (Gienger et al., 2002). Each TCS may be compared to other TCS of the same length of time in comparable habitats (Ribeiro-Júnior et al., 2008; Todd, 2013b). To make TCS more comparable, the activity pattern and time of day each TCS is conducted must be matched. For seasonal species, time of year must also be considered. Time-constrained searches may be especially prone to observer bias, so it is critical that the same investigators conduct each TCS that is being compared.

In patch sampling, also called natural-cover surveys (Marsh and Haywood, 2010), species that are known to occur only in particular patches, such as boulders, fallen logs, and certain tree species, may be surveyed by sampling only the habitat patches where they are likely to be found (Fedewa, 2013). Patches must be clearly definable for this method to be rigorous, and all patches in an area must be sampled. Patch sampling has been effective for small lizards that are confined to life under rocks such as *Proctoporus* (Doan, 2008) and for skinks that seek refuge under bushes (Kerr et al., 2003). Larger lizards that spend most of their active time basking on large rocks, such as *Sauromalus* and some species of *Sceloporus,* would also be sampled efficiently with this method.

Adaptive cluster sampling (Thompson, 1991) may be implemented with a randomly placed primary quadrat. If a target reptile is detected in the primary quadrat, secondary quadrats are installed surrounding the primary quadrat. Ishwar et al. (2001) successfully used adaptive cluster sampling to estimate the number of primary quadrats with animals, cluster size, species richness in clusters, and density. They found that less than 15% of primary quadrats contained reptiles, which demonstrates that adaptive cluster sampling conserves considerable time in areas of patchy reptile distributions.

Point sampling is a census method for which a point is determined in a habitat and the investigator randomly selects distances and directions from that point to conduct a small quadrat sample (Fedewa, 2013). This method is effective for very densely distributed small species that may be captured in a small quadrat. A pre-formed quadrat frame made of PVC or similar material quickly laid on the ground will facilitate sampling. This method may be preferable to line transects for sampling small cryptic species in dense habitat (Gregory et al., 2004).

Distance transects are performed by two observers along a transect. One observer walks in the centre of the transect while a second observer measures the distance from the observed reptile to the transect line (Stober and Smith, 2010; Lovich et al., 2012). This method allows the simultaneous estimation of detection probability along with abundance or density of the target species (Milner-Gulland and Rowcliffe, 2007). In this case, the transect does not have a fixed width and, for some species, may require less time and transect length to obtain comparable results to a fixed width transect (Lovich et al., 2012).

17.6 Individual and habitat variables

When reptiles are captured during plot and transect censuses, it is useful to record data on each individual, such as species, sex, body size (snout–vent length or carapace length), and tail length. In addition, recording the activity of each animal, the substrate upon which it was found, perch diameter (for arboreal species), and the height (cm) of each capture may contribute to the knowledge about the species.

Although not required for a population census, recording habitat variables for each of the plots, transects, or parcels evaluated requires relatively little time and could add important explanatory information to a project (see Chapter 19). Common habitat variables that could be recorded at each quadrat or transect are: date, start and end times, weather, temperature, relative humidity, cloud cover, elevation, slope, aspect, and canopy cover percentage. Recording additional variables such as tree diameters at breast height within each sampling unit, leaf litter depth or mass, shrub cover, coarse woody debris cover, and rock cover may be useful because they have been shown to correlate with the abundance of some reptile species (House and Spellerberg, 1983; Kerr et al., 2003; Babbitt et al., 2010).

17.7 Selecting the appropriate plot and transect techniques

When determining which plot or transect technique to select for a particular study, many factors should be considered including time, personnel, habitat, and species biology (see Figure 17.3). Durations for each technique vary with habitat type, number of personnel, abundance of reptiles, and other factors. For example, during a study in Amazonian rainforest, a single 100 m parcel transect with two observers lasted an average of 26 minutes during the day and 40 minutes at night, whereas 8×8 m^2 quadrats with a team of four searchers conducted during the day lasted an average of 11 minutes and 15 minutes at night (Doan, 2003). Nocturnal techniques had longer duration because of the narrower field of view of searchers and the greater abundance of reptiles (Doan, 2003). For another study in the Atlantic forest of Brazil, 5×5 m^2 total removal plots conducted by five searchers at night lasted approximately 30 minutes (Almeida-Gomes et al., 2008).

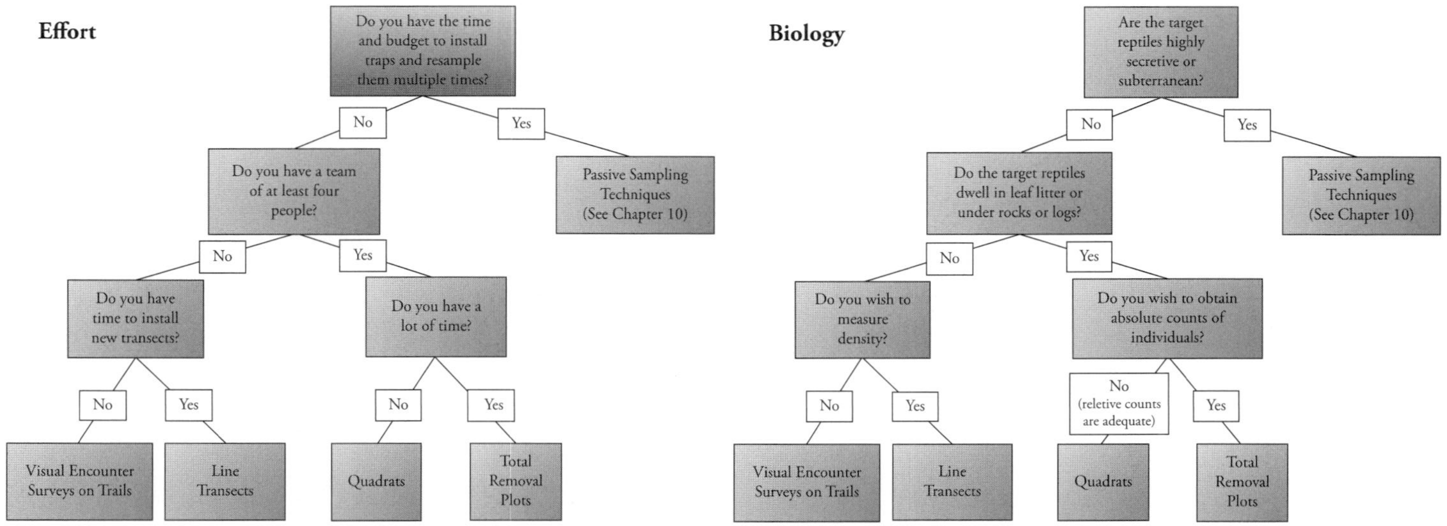

Figure 17.3 Decision trees for common plot and transect census techniques based on the effort the team is willing to expend and the biology of the target reptile species.

When comparing VES parcel transects and quadrat methods done in the same habitat types in the Amazonian rainforest, I found that overall VES parcel transects captured more individuals and more species, but for particular species and ecotypes quadrats were more effective (Doan, 2003). Transects may be preferable in a patchy habitat because they cover more distance (Marsh and Haywood, 2010), whereas randomly or systematically placed quadrats may miss such patches of target reptiles. Transects may also be placed along an environmental gradient, which would not be appropriate for quadrats. An additional advantage of visual transects is that the habitat and fauna are minimally disturbed, which allows resampling the same transect within a short period of time. Such resampling would not be possible when using quadrats or total removal plots.

Quadrat techniques have been shown to be superior for leaf litter-dwelling species or otherwise cryptic species that would be missed in visual transects (Doan, 2003; Marsh and Haywood, 2010). When the sampling team is large, individual quadrats may be completed in much less time than individual transects. In conjunction with a mark–recapture study (Chapter 27), quadrats are preferable because the movement of reptiles is easily missed in a linear transect (Marsh and Haywood, 2010), but adequately sampled in a parcel of quadrats.

Transects and quadrats have been demonstrated to be complementary methods for attempting to survey the entire reptile assemblage. These methods may also be combined with additional passive sampling techniques (see Chapter 10) in a single study (Ribeiro-Júnior et al., 2008). Time and resource permitting, quadrats, transects, and other techniques can be very efficient if combined in the same field project to more fully describe the reptile assemblage of an area (Doan, 2003; Almeida-Gomes et al., 2008).

Acknowledgements

I thank J. Strickland, A. Mason, M. Lawrance, R. Rautsaw, A. Osorio, and the University of Central Florida herpetology class for their helpful comments which improved the manuscript.

References

Almeida-Gomes, M., Vrcibradic, D., Siqueira, C.C., et al. (2008). Herpetofauna of an Atlantic rainforest area (Morro São João) in Rio de Janeiro State, Brazil. *Anais da Academia Brasileira de Ciências*, **80**, 291–300.

Anonymous. (1998). *Inventory Methods for Snakes*. Standards for Components of British Columbia's Biodiversity No. 38. Vancouver, BC:Resources Inventory Committee.

Babbitt, K.J., Veysey, J.S., and Tanner, G.W. (2010). Measuring habitat. In C.K. Dodd, Jr. (ed) *Amphibian Ecology and Conservation: A Handbook of Techniques*. Oxford: Oxford University Press, pp. 299–317.

Cassey, P., and Ussher, G.T. (1999). Estimating abundance of tuatara. *Biological Conservation*, **88**, 361–6.

Doan, T.M. (2003). Which methods are most effective for surveying rain forest herpetofauna? *Journal of Herpetology*, **37**, 72–81.

Doan, T.M. (2004). Extreme weather events and the vertical microhabitat of rain forest anurans. *Journal of Herpetology*, **38**, 422–5.

Doan, T.M. (2008). Dietary variation within the Andean lizard clade *Proctoporus* (Squamata: Gymnophthalmidae). *Journal of Herpetology*, **42**, 16–21.

Doan, T.M., and Arizábal Arriaga, W. (2000). The impact of tourism on the herpetofauna of Tambopata, south-eastern Peru. In C. Kirkby, T.M. Doan, H. Lloyd, et al. (eds) *Tourism Development and the Status of Neotropical Lowland Wildlife in Tambopata, South-eastern Peru: Recommendations for Tourism and Conservation*. London:Tambopata Reserve Society, pp. 63–76.

Doan, T.M., and Arizábal Arriaga, W. (2002). Microgeographic variation in species composition of the herpetofaunal communities of Tambopata Region, Peru. *Biotropica*, **34**, 101–17.

Fedewa, L.A. (2013). Quadrat sampling. In G.J. Graeter, K.A. Buhlmann, LR. Wilkinson, et al. (eds) *Inventory and Monitoring: Recommended Techniques for Amphibians and Reptiles*. Birmingham, AL:Partners in Amphibian and Reptile Conservation, pp. 66–8.

Folt, B., and Reider, K.E. (2013). Leaf-litter herpetofaunal richness, abundance, and community assembly in mono-dominant plantations and primary forest of northeastern Costa Rica. *Biodiversity and Conservation*, **22**, 2057–70.

Garden, J., McAlpine, C., Jones, D., et al. (2007). Using multiple survey methods to detect terrestrial reptiles and mammals: what are the most successful and cost-efficient combinations? *Wildlife Research*, **34**, 218–27.

Gienger, C.M., Beck, D.D., Sabari, N.C., et al. (2002). Dry season habitat use by lizards in a tropical deciduous forest of western Mexico. *Journal of Herpetology*, **36**, 487–90.

Glaudas, X.A. (2013). Visual encounter surveys. In G.J. Graeter, K.A. Buhlmann, LR. Wilkinson, et al. (eds) *Inventory and Monitoring: Recommended Techniques for Amphibians and Reptiles*. Birmingham, AL:Partners in Amphibian and Reptile Conservation, pp. 73–4.

Gregory, R.D., Gibbons, D.W., and Donald, P.F. (2004). Bird census and survey techniques. In W.J. Sutherland, I. Newton, and R.E. Green (eds) *Bird Ecology and Conservation: A Handbook of Techniques*. Oxford: Oxford University Press, pp. 17–55.

Heatwole, H.F. (2012). Quadrat sampling. In R.W. McDiarmid, M.S. Foster, C. Guyer, et al. (eds) *Reptile Biodiversity: Standard Methods for Inventory and Monitoring*. Berkeley, CA:University of California Press, pp. 220–6.

House, S.M., and Spellerberg, I.F. (1983). Ecology and conservation of the sand lizard (*Lacerta agilis* L.) habitat in southern England. *Journal of Applied Ecology*, **20**, 417–37.

Ishwar, N.M., Chellam, R., and Kumar, A. (2001). Distribution of forest floor reptiles in the rainforest of Kalakad-Mundanthurai, Tiger Reserve, South India. *Current Science*, **80**, 413–18.

Kerr, G.D., Bull, C.M., and Burzacott, D. (2003). Refuge sites used by the scincid lizard *Tiliqua rugosa*. *Austral Ecology*, **28**, 152–60.

Lambert, M.R.K. (2005). Lizards used as bioindicators to monitor pesticide contamination in sub-Saharan Africa: a review. *Applied Herpetology*, **2**, 99–107.

Lardner, B., Rodda, G.H., Yackel Adams, A.A., et al. (2015). Detection rates of geckos in visual surveys: turning confounding variables into useful knowledge. *Journal of Herpetology*, **49**, 522–32.

Lillywhite, H.B., Sheehy, C.M., Brischoux, F., et al. (2015). On the abundance of a pelagic sea snake. *Journal of Herpetology*, **49**, 184–9.

Lovich, R.E., Hayes, W.K., Mushinsky, H., et al. (2012). Transect surveys, including line distance. In R.W. McDiarmid, M.S. Foster, C. Guyer, et al. (eds) *Reptile Biodiversity: Standard Methods for Inventory and Monitoring*. Berkeley, CA:University of California Press, pp. 227–34.

Marsh, D.M., and Haywood, L.M.B. (2010). Area-based surveys. In C.K. Dodd, Jr. (ed) *Amphibian Ecology and Conservation: A Handbook of Techniques*. Oxford:Oxford University Press, pp. 247–62.

Milner-Gulland, E.J., and Rowcliffe, J.M. (2007). *Conservation and Sustainable Use: A Handbook of Techniques*. Oxford: Oxford University Press.

Newton, A.C. (2007). *Forest Ecology and Conservation: A Handbook of Techniques*. Oxford: Oxford University Press.

Pianka, E.R. (1966). Convexity, desert lizards, and spatial heterogeneity. *Ecology*, **47**, 1055–9.

Rahman, S.C., Reza, A.A., Datta, R., et al. (2014). Niche partitioning and population structure of sympatric mud snakes (Homalopsidae) from Bangladesh. *Herpetological Journal*, **24**, 123–8.

Ribeiro-Júnior, M.A., Gardner, T.A., and Ávila-Pires, T.C.S. (2008). Evaluating the effectiveness of herpetofaunal sampling techniques across a gradient of habitat change in a tropical forest landscape. *Journal of Herpetology*, **42**, 733–49.

Rodda, G.H., Campbell, E.W., and Fritts, T.H. (2001). A high validity census technique for herpetofaunal assemblages. *Herpetological Review*, **32**, 24–30.

Rödel, M.-O., and Ernst, R. (2004). Measuring and monitoring amphibian diversity in tropical forests. I. An evaluation of methods with recommendations for standardization. *Ecotropica*, **10**, 1–14.

Samways, M.J., McGeoch, M.A., and New, T.R. (2010). *Insect Conservation: A Handbook of Approaches and Methods*. Oxford: Oxford University Press.

Schlaepfer, M.A., and Gavin, T.A. (2001). Edge effects on lizards and frogs in tropical forest fragments. *Conservation Biology*, **15**, 1079–90.

Stober, J.M., and Smith, L.L. (2010). Total counts versus line transects for estimating abundance of small gopher tortoise populations. *Journal of Wildlife Management*, **74**, 1595–600.

Sung, Y.-H., Karraker, N.E., and Hau, B.C.H. (2011). Evaluation of the effectiveness of three survey methods for sampling terrestrial herpetofauna in South China. *Herpetological Conservation and Biology*, **6**, 479–89.

Thompson, S.K. (1991). Adaptive cluster sampling: designs with primary and secondary units. *Biometrics*, **47**, 1103–15.

Todd, B.D. (2013a). Area-constrained searches. In G.J. Graeter, K.A. Buhlmann, LR. Wilkinson, et al. (eds) *Inventory and Monitoring: Recommended Techniques for Amphibians and Reptiles*. Birmingham, AL:Partners in Amphibian and Reptile Conservation, pp. 65–6.

Todd, B.D. (2013b). Time-constrained searches. In G.J. Graeter, K.A. Buhlmann, LR. Wilkinson, et al. (eds) *Inventory and Monitoring: Recommended Techniques for Amphibians and Reptiles*. Birmingham, AL:Partners in Amphibian and Reptile Conservation, pp. 63–5.

Trainor, C.R., and Woinarski, J.C.Z. (1994). Responses of lizards to three experimental fires in the savanna forests of Kakadu National Park. *Wildlife Research*, **21**, 131–48.

Van Rooijen, J. (2009). Estimating the snake species richness of the Santubong Peninsula (Borneo) in two different ways. *Contributions to Zoology*, **78**, 141–7.

Veith, M., Lötters, S., Andreone, F., et al. (2004). Measuring and monitoring amphibian diversity in tropical forest. II. Estimating species richness from standardized transect censing. *Ecotropica*, **10**, 85–99.

Wilgers, D.J., Horne, E.A., Sandercock, B.K., et al. (2006). Effects of rangeland management on community dynamics of the herpetofauna of the tallgrass prairie. *Herpetologica*, **62**, 378–88.

18

Rapid assessments of reptile diversity

Indraneil Das

18.1 Introduction

As of March 2015, a total of 10,178 reptile species have been described (Uetz and Hošek, 2015), at least 165 (>1.62%) of these in the preceding year (2014) alone. The conservation status of a majority of the newly described, or even of relatively familiar species, remains unknown. An earlier assessment of the conservation status of a large representative sample of the world's reptile fauna indicates that a significant proportion is threatened (Böhm et al., 2013). This makes a case for urgency in initiating studies on their conservation requirements, foremost among these being distributional and abundance data, as typically obtained during surveys constrained by time and other resources.

Reptiles play important ecological roles, *inter alia* forming significant animal biomass (Iverson, 1982), constituting important linkages in the ecosystem by providing dispersal mechanism for plants (Hnatiuk, 1978; Fialho, 1990; Olesen and Valido, 2010), contributing to environmental heterogeneity (Kaczor and Hartnett, 1990), having keystone functions in maintaining ecosystem structure (Ashton, 2010), and fostering important symbiotic associations with an array of organisms (Lago, 1991; Witz et al., 1991). Crocodilians are also known to maintain wet refugia during droughts, which are used by a variety of other organisms from macroinvertebrates and fish to turtles (Mazzotti et al., 2008). Many turtles and several crocodilians are scavengers, helping release nutrients locked up in dead tissue (Burroughs et al., 2014). Reptiles are regularly on the menu of predatory mammals, birds, fish, large invertebrates, including spiders, and even themselves (see Cook, 1987; Bauer, 1990; Martín and Lopez, 1990); they are also predators for a range of invertebrate and vertebrate species. Fossorial snakes may help aerate hard soils, allowing air to access rainforest tree roots (Rajendran, 1977) in a sense behaving as ecosystem engineers (*sensu* Jones et al., 1994) by significantly modifying and maintaining habitats.

Reptiles are important predators of insect (Bhanotar and Bhatnagar, 1976) and rodent (Lim, 1974; Whitaker and Advani, 1983) agricultural pests. Additionally, venom extracted from certain snakes is used for the production of life-saving drugs, including anti-venin serum for snake-bites (McCleary and Kini, 2013; Zouari-Kessentini et al., 2013). A number of large lizards and snakes, and nearly all of the world's turtles, are sought for food, medicine, or the pet trade (Valencia-Aguilar et al., 2013). Because of

Reptile Ecology and Conservation. Edited by C. Kenneth Dodd, Jr. © Oxford University Press 2016.
Published 2016 by Oxford University Press.

their lineage and life-history diversity and pan-global distribution, reptiles are considered model organisms for the study of vertebrate life (Pianka, 1986). Additionally, they feature in many indigenous cultural practices and belief systems, further studies of which may provide valuable insights into conservation strategies and management.

A majority of the world's reptiles inhabit the tropics and subtropics, while temperate regions have far lower richness of reptiles. Nonetheless, many sites outside tropical areas remain to be surveyed, while a significant portion of the world's tropics have never been adequately inventoried for their reptile faunas.

18.2 What is an RA?

Rapid assessments (RAs) have been described as 'synoptic assessments that are often undertaken as a matter of urgency in the shortest timeframe possible to produce reliable and applicable results for its designed purpose' (Anonymous, 2006). RAs thus comprise short bouts of field data collection for estimation of species richness. Developed primarily in temperate regions, RAs collect data in a standardized (and therefore, comparable across time and space) and cost-effective manner, and are widely used for the preparation of environmental impact assessments (EIAs); they also may be useful for 'BioBlitz' surveys (or intensive surveys by competing groups to record biodiversity within a fixed period of time; Graham and Timpe, 2007; Robinson et al., 2013). RA techniques are favoured especially in cases of constraints on time, personnel, and other resources, and are themselves constrained by species' biology. The length of the sampling periods used to gather data during field work, purported to be 'rapid', appears not to be uniform in the herpetological literature, but generally is considered short enough to gather preliminary information, such as two weeks or less (see D'Cruze et al., 2006). The typical time frame for the Rapid Assessment Program (RAP) of Conservation International, a U.S. non-governmental environmental organization involved in international biodiversity conservation, is 4–6 weeks of field surveys, of which 5–7 days are allocated per specific site. In places showing seasonality, the decision of when to sample is therefore critical, and is likely best during the wet season when more species are actively foraging and breeding than during dry periods. However, this may not be possible in some cases for reasons of access to the study site (e.g. D'Cruze et al., 2006), delays in receiving permits, logistic arrangements, or other reasons. Given time limitations, it is important to target rare or threatened species in the survey area, as suggested by previously known distributional ranges and from information received through community questionnaire surveys (Section 18.4.1).

18.3 Planning components of RAs

18.3.1 Assembling literature and other resources

Familiarity with the plant and animal literature (published, the so-called 'grey literature', reports, theses, and online material such as databases and electronic publications) concerned with the site where a RA will be conducted is essential for investigators. Moreover, to enable the activities that follow to be more useful, researchers need to be sensitive to topics of biology or human welfare that may be impacted by subsequent conclusions and recommendations.

Tropical herpetofaunas are complex, and rarely are identification resources comprehensive or optimal for RA activities. Sources include monographs, traditional print editions of field guides, online guides, computer (delta) and traditional keys, and systematic and taxonomic papers to specific taxonomic groups or national faunas. Extensive library research and networking among co-specialists are essential in order to gather all relevant taxonomic and related literature for field and laboratory identification. Nonetheless, RAs also produce new species to science in poorly sampled parts of the world. Examination of critical museum specimens, especially type specimens and other comparative material and associated data, becomes essential, as is discussion and networking with colleagues in the same or related fields. Familiarity with the fauna is also helpful when investigators plan searches for threatened or other species of conservation importance that may be expected. Inclusion of local researchers is highly recommended. Not only do they tend to be familiar with the fauna and study site, they can also assist with logistical challenges onsite, besides helping in the permitting process (Chapter 2).

18.3.2 Permitting

Legislation surrounding resource access can be complex, and in all regions of the world permits from local and national agencies are required before survey teams can access the site. These range from villages and towns in the vicinity of the proposed sampling/survey locality to local councils, district, county, state, and/or national governmental agencies that may be located many kilometres from the study sites. Exporting biological specimens, tissue samples and other biological material is similarly controlled by legislation, nationally or bilaterally enforced or under international regulations. The most important among these is the Convention in International Trade in Endangered Species of Wild Fauna and Flora (CITES), which lists species controlled under various Appendices that require export and import permits. Permitting regimes may be tedious (considering that rapid assessments are measurable in days), and may take many months to years to obtain from some agencies after submission of required documents. It is the responsibility of the field investigator to understand and be willing to comply with regulations before application for permits is made. After permits have been obtained, it is important that courtesy calls be made to local stakeholders, such as large landowners and village headmen, to brief them on research activities and possibly recruit assistants (including field researchers, porters, guides, and cooks) from among the local inhabitants. Researchers should consider providing as much economic assistance to these communities (in terms of homestays and purchase of provisions) as possible, in addition to on-the-job training as field biologists or parataxonomists.

18.3.3 Training of field personnel

Biodiversity surveys often are prone to observer bias, and when the time-frame for sampling becomes constrained, such bias can introduce significant problems in data reliability; in extreme cases, bias can result in fewer than expected sightings or collections (see

Zhang et al., 2014). Observer bias can result from: fatigue, poor weather conditions, visibility issues, heterogeneity in target species availability (i.e. observers tend to see animals that are more frequently available or are available in areas where they are more likely to be detected; Borchers and Samara, 2007), differences in observers' eyesight or capacity for walking transects, technical (especially in handling appropriate field instruments) and analytical skills (i.e. employing appropriate sampling methods as well as techniques for their analyses), field skills (such as the ability to interpret tracks and signs), and even soft skills, such as interpersonal skills and empathy with focus groups during interviews. Familiarity with field protocols will significantly decrease wastage of time, permitting investigators to 'hit the ground running'. Field investigators also need to be aware of dangers and annoyances in the field, such as noxious arthropods, predatory mammals, and from the study subjects themselves, such as venomous snakes and large crocodilians. Knowledge of first-aid and emergency medical plans are therefore essential, including location of the nearest primary health centre or hospital and whether they stock the appropriate anti-venin serum for locally occurring venomous snakes.

18.3.4 Timing

RAs essentially are snapshots of biodiversity over a short time span, and need to be scheduled with care, taking into account the phenology of the group being sampled and pre-existing knowledge of their activities. Behaviour and activity periods, both diel and seasonal, may be different, especially in seasonally dry–wet and/or hot–cold areas, and it becomes imperative to choose periods when the greatest number of species are active (usually, wet and warm periods) to obtain realistic estimates of species richness. A majority of reptiles are nocturnal or crepuscular. Many may be diurnal, however, such as most lizards of the families Agamidae, Cordylidae, Dactyloidae, Iguanidae, Lacertidae, Phrynosomatidae, Scincidae, Teiidae, Varanidae, and some members of the snake families Colubridae and Elapidae. Other members of these latter two snake families, and with few exceptions, the geckos (Gekkonidae), are nocturnal. Most sit-and-wait predatory reptiles, including many members of the snake families Boidae, Pythonidae, and Viperidae, crocodilians, and some turtles, may be encountered at any time of the day or night, whereas activity patterns of the fossorial snake families Dibamidae, Leptotyphlopidae, Typhlopidae, and Uropeltidae are largely unknown, although classified as nocturnal because they tend to be sighted after dark. In areas with strong seasonality, RAs may be conducted during different seasons (spring, summer, autumn, and even winter at some sites). Given the (often) vagaries of receiving permits and solving all logistic challenges, including access to sites in many tropical regions, pre-survey data on the life histories of the local reptiles are essential.

18.4 Field sampling

18.4.1 Community questionnaire surveys

For very short field visits or site reconnaissance prior to actual sampling, structured questionnaire surveys often prove useful for investigators, and arguably are the most cost-efficient techniques to gain new knowledge. Data sheets should be developed and tested for errors prior to being used at a field site on a particular focus group (local human inhabitants, who may be stakeholders of the survey site), and should include a series of questions addressing observed species in an area, seasons of observation, impressions of abundance, exploitation, habitat use, distinctive behaviour, folklore, and taboos. The list of questions should be relatively brief, inasmuch as the quality and accuracy of the responses decrease towards the end of a long questionnaire survey (Bogen, 1996). In some cases, anonymity may be desired by the interviewees.

The knowledge of local residents understandably varies according to age, aptitude, experience, and profession, information that should be recorded in the data sheets. Residents often reveal the existence of rare or cryptic species that may not be discovered during short inventory periods (such as aquatic turtles or snakes) when interviews are conducted with individuals most familiar with the environment, such as fishermen, hunters, and farmers. Other species that may be familiar to local human inhabitants include seasonally active species and those restricted to special habitats. Experience in conducting sociological studies and familiarity with the native language of the interviewee (including vernacular names of target organisms) is an asset for RA personnel collecting such data. Descriptions of morphology and distinctive behaviour of reptiles should be gathered from interviewees. In the absence of established vernacular names, images of species likely to occur in the area can be shown to respondents in order to avoid introducing bias into the reporting. RA personnel will do well to heed local taboos and restrictions, such as the capture of species that are culturally protected. Remuneration (money, other products, such as fishing gear, or even the prospect of future employment) may be appropriate in some cases to information providers. Semi-structured questionnaires, whereby interviewers collect additional (especially important anecdotal) data, are preferred. Attributes such as a genuine interest on the part of interviewees and general empathy enhance the accuracy of response. Data sheets should be sequentially numbered and cross-linked to georeferenced maps that include information on elevation, habitat/vegetation, wetland areas, and land cover. Community questionnaire surveys offer unparalleled opportunity for public education and outreach (Chapter 30), and skills in such areas are highly desirable.

18.4.2 Visual encounter survey

Perhaps the easiest RA technique for reptiles is the visual encounter survey (VES; Crump and Scott, 1994; Guyer and Donnelly, 2012; see Chapter 10). A VES comprises time-constrained searches along pre-established transects for visually or acoustically locating animals from the target group. Techniques include using rakes or sticks to turn over leaf litter and logs, looking inside tree holes and rock crevices, and netting streams and other water bodies (Chapters 10, 11 and 17). Transect location and position are important in the planning phase. If the objective of the RA is to record as many species as possible (approaching a comprehensive species inventory), representative habitats should be covered. For quick comparisons of species richness among sites or habitats, the effort expended in different habitats needs to be equivalent to permit standardization. Every individual is accurately identified to species and georeferenced to the point of observation; each data sheet should contain a record of individuals, description of the habitat, time of survey, and number of field personnel employed in order to estimate survey effort. Sites for assessment need to be selected carefully based on available maps and transects established during daylight hours. As transects are typically established along existing forest trails, investigators need to be aware of potential bias. In such edge habitats, visibility may be significantly greater with different species assemblages than within forested habitats. If multiple transects are used, these should be located sufficiently distant from one another to avoid recounting the same individuals and to exploit the within-habitat diversity. Where faunal turnover with distance is high, such as within tropical sites, the transect technique is quantitatively more effective and easier to use than pitfall trapping or cover boards (Sung et al., 2011).

18.4.3 Species list technique

Originally a technique for rapidly estimating bird species within a small geographical area when constrained by time, the species list technique (SLT) of MacKinnon and Phillipps (1993) has been used for herpetofaunas (Muir and Muir, 2011) and consists of preparing faunal lists of three, five, or 10 species, and once completed, a second list started, and the process repeated during the length of the survey period. The strength of the technique is that multiple observers can pool their data into one large database. The cumulative species richness then is related to the number of observations, rather than to space and time, thus permitting moderate differences in techniques employed and inter-observer skill.

18.4.4 Trapping

A variety of trapping techniques is utilized for sampling reptiles, and while a majority are suitable for long-term studies, a few may be employed during sampling periods that last a fortnight or less (suiting the requirements of a RA). Traps typically sample species that are not encountered during VES, and result in data on presence (but not absence) in addition to valuable life history information and perhaps relative abundance. One trapping technique is pitfall trapping (Chapter 10), which comprises burying an array of buckets flush with the ground surface. Buckets need to be as large as possible (as the depth of the same will determine the size of animal to be trapped) and with smooth sides. In association with a drift fence that directs animals to the mouth of the bucket, captures can be made of a variety of surface-dwelling or subfossorial species of squamates and small turtles (in addition to small mammals, amphibians, and arthropods). Initial capture rates may be high in pitfall trapping. These traps often capture species not otherwise encountered, making the technique appropriate for reptile RAs.

18.4.5 Taxon-specific techniques

Sea turtles

Beach surveys are a useful RA technique for gathering information on sea turtles. When time is limited, sea turtle presence may be indicated through egg shells, crawl tracks, and nesting pits that are species-specific (see Pritchard and Mortimer, 1999, for details of tracks and nests). These can be examined carefully and classified at least as fresh (e.g.

made within the last 24 hours or a few days, and indicative of ongoing nesting) or aged (tracks of some species, such as *Dermochelys coriacea* may persist for weeks or even months under fair weather conditions). Survey platforms include ground and aerial surveys, both having their strengths and weaknesses, and a combination of these two techniques is naturally the most useful (see Chapter 15).

Non-marine turtles

The frequent low densities of tortoises and freshwater turtles make them difficult candidates when applying RA techniques for reptiles. Successful methods are frequently restricted to a few species, including visually locating terrestrial turtles on horseback, using trained dogs, detecting well-used trails and burrows, using blunt-tipped metal rods to probe holes underground ('sounding'), using a rake in vegetation-choked waterbodies, hooks, and electroshocking (Plummer, 1979; Vogt, 2012; Chapters 13 and 14).

Crocodiles

Inventories for crocodilians are facilitated by the fact that the group contains the fewest number of extant species relative to other groups of reptiles, and rarely do more than two species occur in syntopy. Additionally, community questionnaire surveys (Section 18.4.1) can be useful for investigators in a RA for learning much about which species occur at a site and other biological data, including habitat use and even estimates of abundance. Techniques that can be employed for rapid collection of data relative to RA requirements include boat surveys when investigators use spotlights to detect reflective eyeshine of crocodilians at night, especially under conditions of low tide (Bayliss, 1987). Bright lights from headlamps or hand-held flashlights are employed to detect eyeshine during these surveys, although too bright a light (outside the 50,000 and 20,000 candlepower range) may fail to detect crocodilians in the vicinity of the investigators (Mazzotti, 2012). Another RA technique includes the use of either fix-winged aircraft or helicopters to count crocodilians during the day, where a constant height and speed is recommended, although aerial surveys have only been successfully carried out in tidal forests and other relatively open landscapes (Mazzotti, 2012). Other RA techniques for crocodilians are in Chapter 16.

Squamates

Snakes and lizards comprise the bulk of reptiles encountered at most field sites. Fitch (1987) described several methods used by investigators in locating snakes, including the behaviour of prey species (e.g. amphibian distress calls, bird mobbing activities), road-cruising (where sampling is conducted over the same stretch of road 'transect' by car and live and dead snakes and lizards are recorded), and trapping. The latter may involve several techniques, most of which require a relatively large effort and time. One applicable technique for a RA is the use of cover boards (strategic placement of sheet metal or boards that are quick heating), under which thermophilous squamates may shelter, especially in temperate regions. Species' activities, natural history observations, and perhaps relative abundance data can be collected through the use of cover boards. Cover boards have not been used very successfully in tropical areas.

Pitfall trapping is a useful technique for sampling terrestrial and fossorial reptiles (Chapters 10 and 11). Because of the time expended in locating appropriate sites, establishing arrays of pitfalls and associated drift fences, and removing them at the end of the sampling period, it may be an unnecessarily tedious and time-consuming technique for many RAs targeting reptiles, especially relative to VES. Nonetheless, the technique is recommended when sufficient resources exist, as it can rapidly aid in the acquisition of additional reptile taxa that are not encountered using VES, such as fossorial or subsurface active lizards and snakes.

18.4.6 Environmental DNA

Development of techniques in environmental DNA promises a new approach to monitoring reptiles in aquatic environments (Pilliod et al., 2013; Goldberg et al., 2015; see Chapter 25). The concept is based on the fact that aquatic animals leave DNA—either nuclear or mitochondrial DNA, in cellular or extracellular (dissolved DNA) form, through excretion, or shed skin (environmental DNA or eDNA). The presence of target species in biodiversity surveys can thus be potentially detected by analysing water samples for eDNA without the necessity of observing them. In water, eDNA is diluted and transported by currents and other hydrological processes, and may last 7–21 days, depending on environmental conditions (Dejean et al., 2011).

Studies on eDNA have targeted invertebrates, fish, frogs, and small mammals in North America and Europe (Ficetola et al., 2008; Kelly et al., 2014). More recently, studies are being initiated using eDNA for rare or secretive freshwater turtles in North America and China. The costs of detecting species via eDNA has been argued to be lower than field surveys, with a better chance of detection, especially of ecologically cryptic species. The procedure includes collection of eDNA by sampling the water body. The usually miniscule quantity of DNA is amplified via a polymerase chain reaction (PCR). By using species-specific primers that bind to the DNA of a target species, the presence of the species is confirmed via bar-coding (that helps identify species using short DNA sequences from a standard position in the genome). Apart from presence data for focal species in a particular habitat, other potential uses include estimating species abundance, noting the presence of invasive species, and generating species inventories from eDNA samples (Chapter 25).

18.5 Data analyses and limitations

The limitations of time and, typically, collectors' bias have the potential to seriously underestimate biodiversity and, consequently, undermine the scientific value of RAs. Developing a solid theoretical basis for understanding the relationship between (collecting) effort and the number of species obtained, therefore, becomes important. Species identification errors need to be kept minimal, and if collection is not possible, digital images of vouchers (Chapter 5) and non-lethal DNA sampling (such as buccal swabs) may be appropriate (Chapter 25).

Since estimating species diversity and richness is challenging, especially at tropical sites, statistical extrapolation of data gathered may be helpful in many cases. Derived from this is the species accumulation curve (Chapter 21), a widely used predictive tool that has been argued to be without bias of collectors' attention given to uncollected species (Colwell and Coddington, 1994). Species accumulation curves are produced by plotting the cumulative number of new species encountered after each period of sampling against sampling effort (either transect length or hours of observations), other factors being constant (including weather and number of observers). The asymptote of the curve indicates the likely species richness for that habitat. For all models employed, a well-defined asymptote is required for reliable estimates of species richness at a particular geographical area. The general trend is for a rapid accumulation of species encountered (representing the more abundant representatives), reaching a higher 'shoulder', and plateauing off (including at this stage, the relatively more rare species), to show that all species have been sampled at the site. At sites with high reptile species richness, curves rise to a point of clear upper inflection or 'crest', indicative of a need for greater sampling that may be beyond the scope of a RA. Thus, in most tropical sites, curves are crested, and finding a model that can accurately predict species richness may be challenging. Species accumulation curves are influenced by ecological characteristics of the sites (Thompson et al., 2003), and the comparative ecologically cryptic nature of the study organisms, including their rarity, non-trapability, and their transient presence.

The SLT familiar to ornithologists has been recently employed with herpetofaunal groups. A bias observed in the use of the technique for avian groups is towards solitary and terrestrial species (as opposed to monospecific flocking species). Consequently, the SLT may not reflect community structure, as quantified using other techniques (O'Dea et al., 2004). When such information is desirable, a combination of techniques needs to be employed in order to gather additional species names in lists of sites where the RA is employed.

18.6 Summary

A successful RA programme depends much on having realistic objectives, project planning, addressing resources available for the task, and understanding the limitations of the work itself (see also Chapter 2). It is important to recognize that absolute estimates of species diversity and richness are elusive figures even for long-term studies in most parts of the world, and complete species inventories may take up a person's lifetime (Myers and Rand, 1969; Das, 1996). Limitations on assessing richness and diversity increase with the size and vegetation complexity of the geographical region sampled (e.g. tropical rainforests, deserts) or when shy, ecologically cryptic species (such as fossorial squamates or aquatic turtles) are concerned. Current inventory techniques are biased towards reptile faunas of terrestrial and aquatic environments; those that are fossorial and arboreal tend to be poorly sampled because of limited sampling methods (Das, 2012; Chapter 11). Nonetheless, RAs have the potential to reveal the existence of unexpected and undescribed species (see Hawkins et al., 1990) and the identification of areas of high species richness (Graham et al., 2010). RAs are superior to another rapid source of acquiring similar information, that is, compilations based on publications, databases and museum materials that tend to be biased and/or of low temporal or spatial resolution (Elith and Leathwick, 2007).

Indeed, many RAs targeting reptiles are conducted in poorly-known tropical or other species-rich sites and remain the only source of information for years. Thus, rapid surveys of reptile biodiversity need careful planning and execution, thereby eliminating the mismatch between expectations and results. Investigators would do well to incorporate emerging technologies, perhaps in related disciplines, into RAs for reptiles. These include the use of environmental DNA and bar-coding, camera trapping for larger species (sea turtles, squamates, and crocodilians), and new methods of accessing tall trees in rainforest canopies. Finally, every field technique has its strengths and weaknesses, and a combination of field methods may be appropriate for reptile assessments that are of short duration.

Acknowledgements

I thank Ken Dodd for inviting me to prepare this chapter, my colleagues and students over many years for collaborating in field studies, and Jean-Marc Hero for ideas. I am grateful to Genevieve V.A. Gee for commenting on an early draft, and two anonymous reviewers for comments. Manuscript preparation was supported by a grant from the Niche Research Grant Scheme, awarded by the Ministry of Higher Education, Government of Malaysia NRGS/1087/2013(01).

References

Anonymous. (2006). *Guidelines for the Rapid Ecological Assessment of Biodiversity in Inland Water, Coastal and Marine Areas.* CBD Technical Series no. 22, Secretariat of the Convention on Biological Diversity, Montreal, and Ramsar Technical Report no. 1, the Secretariat of the Ramsar Convention, Gland, Switzerland.

Ashton, P.J. (2010). The demise of the Nile crocodile (*Crocodylus niloticus*) as a keystone species for aquatic ecosystem conservation in South Africa: the case of the Olifants River. *Aquatic Conservation: Marine and Freshwater Ecosystems*, **20**, 489–93.

Bauer, A.M. (1990). Gekkonid lizards as prey of invertebrates and predators of vertebrates. *Herpetological Review*, **21**, 83–7.

Bayliss, P. (1987). Survey methods and monitoring within crocodile management programmes. In G.J.W. Webb, S.C. Manolis, and P.J. Whitehead (eds) *Wildlife Management. Crocodiles and Alligators.* Chipping Norton, NSW, Australia:Surrey Beatty and Sons Pty Ltd, pp. 157–75.

Bhanotar, R.K., and Bhatnagar, R.K. (1976). Reptile predators of the desert locust. *Journal of the Bombay Natural History Society*, **73**, 311–13.

Bogen, K. (1996). The effect of questionnaire length on response rates: a review of the literature. *Proceedings of the Section on Survey Research Methods*, **1996**, 1020–5.

Böhm, M., Collen, B., Baillie, J.E.M., et al. (2013). The conservation status of the world's reptiles. *Biological Conservation*, **157**, 372–85.

Borchers, D.L., and Samara, F.I.P. (2007). Accommodating availability bias on line transect surveys using hidden Markov models. *Centre for Research into Ecological and Environmental Modelling Technical Report*, **2007–05**, 1–13.

Burroughs, R.W., Morris, Z.S., and Marsh, A.D. (2014). *Trachemys scripta* (red-eared slider), *Pseudemys texana* (Texas river cooter), *Chelydra serpentina* (snapping turtle). Feeding behavior and scavenging. *Herpetological Review*, **45**, 321–2.

Colwell, R.K., and Coddington, J.A. (1994). Estimating terrestrial biodiversity through extrapolation. *Philosophical Transactions of the Royal Society of London B: Biological Sciences*, **345**, 101–18.

Cook, W.E. (1987). Amphibians and reptiles: predators and prey. *Smithsonian Herpetological Information Service*, (73), 1–15.

Crump, M.A., and Scott, N.J., Jr. (1994). Visual encounter surveys. In W.R. Heyer, M.A. Donnelly, R.W. McDiarmid, et al. (eds) *Measuring and Monitoring Biological Diversity: Standard Methods for Amphibians*. Washington, DC:Smithsonian Institution Press, pp. 84–92.

Das, I. (1996). Spatio-temporal resource utilization by a Bornean rainforest herpetofauna: preliminary results. In D.S. Edwards, W.E. Booth, and S.C. Choy (eds) *Tropical Rainforest Research: Current Issues*. Dordrecht, The Netherlands:Kluwer Academic Publishers, pp. 315–23.

Das, I. (2012). Arboreal reptiles (tree trunk and canopy-dwelling species). In R.W. McDiarmid, M.S. Foster, C. Guyer, et al. (eds) *Reptile Biodiversity. Standard Methods for Inventory and Monitoring*. Berkeley, CA:University of California Press, pp. 175–9.

D'Cruze, N.C., Green, K.E., Robinson, J.E., et al. (2006). A rapid assessment of the amphibians and reptiles of an unprotected area of dry deciduous forest in north Madagascar. *Herpetological Bulletin*, (96), 17–25.

Dejean, T., Valentini, A., Duparc, A., et al. (2011). Persistence of environmental DNA in freshwater ecosystems. *PLoS One*, **6**, e23398.

Elith, J., and Leathwick, J. (2007). Predicting species distributions from museum and herbarium records using multiple response models fitted with multivariate adaptive regression splines. *Diversity and Distributions*, **13**, 265–75.

Fialho, R.F. (1990). Seed dispersal by a lizard and a tree frog: effect of dispersal site on seed survivorship. *Biotropica*, **22**, 423–4.

Ficetola, G.F., Miaud, C., Pompanon, F., et al. (2008). Species detection using environmental DNA from water samples. *Biology Letters*, **4**, 423–5.

Fitch, H.S. (1987). Collecting and life-history techniques. In R.A. Seigel, J.T. Collins, and S.S. Noval (eds) *Snakes: Ecology and Evolutionary Biology*. New York: McGraw-Hill, pp. 143–81.

Goldberg, C.S., Strickler, K.M., and Pilliod, D.S. (2015). Moving environmental DNA methods from concept to practice for monitoring aquatic macroorganisms. *Biological Conservation*, **183**, 1–3.

Graham, S.P., and Timpe, E.K. (2007). A bioblitz competition to assess a distribution gap in Georgia herpetofaunal records. *Herpetological Review*, **38**, 493–4.

Graham, S.P., Steen, D.A., Nelson, K.T., et al. (2010). An overlooked hotspot? Rapid biodiversity assessment reveals a region of exceptional herpetofaunal richness in the southeastern United States. *Southeastern Naturalist*, **9**, 19–34.

Guyer, C., and Donnelly, M.A. (2012). Visual encounter survey. In R.W. McDiarmid, M.S. Foster, C. Guyer, et al. (eds) *Reptile Biodiversity. Standard Methods for Inventory and Monitoring*. Berkeley, CA:University of California Press, pp. 218–20.

Hawkins, A.F.A., Chapman, P., Ganzhorn, J.U., et al. (1990). Vertebrate conservation in Ankarana Special Reserve, northern Madagascar. *Biological Conservation*, **54**, 83–110.

Hnatiuk, S.H. (1978). Plant dispersal by the Aldabran giant tortoise *Geochelone gigantea* (Schweigger). *Oecologia*, **36**, 345–50.

Iverson, J.B. (1982). Biomass of turtle populations: a neglected subject. *Oecologia*, **55**, 69–76.

Jones, C.G., Lawton, J.H., and Shachak, M. (1994). Organisms as ecosystem engineers. *Oikos*, **69**, 373–86.

Kaczor, S.A., and Harnett, D.C. (1990). Gopher tortoise (*Gopherus polyphemus*) effects on soils and vegetation in a Florida Sandhill community. *American Midland Naturalist*, **123**, 100–11.

Kelly, R.P., Port, J.A., Yamahara, K.M., et al. (2014). Using environmental DNA to census marine fish in a large mesocosm. *PLoS One*, **9**, e86175.

Lago, P.K. (1991). A survey of arthropod associated with gopher tortoise burrows in Mississippi. *Entomology News*, **102**, 1–13.

Lim, B.-L. (1974). Snakes as natural predators of rats in an oil palm estate. *Malayan Nature Journal*, **27**, 114–17.

MacKinnon, J., and Phillipps, K. (1993). *A Field Guide to the Birds of Borneo, Sumatra, Java and Bali*. Oxford: Oxford University Press.

Martín, J., and Lopez, P. (1990). Amphibians and reptiles as prey of birds in southwestern Europe. *Smithsonian Herpetological Information Service*, (82), 1–43.

Mazzotti, F.J. (2012). Finding, counting, and catching crocodiles. In R.W. McDiarmid, M.S. Foster, C. Guyer, et al. (eds.) *Reptile Biodiversity. Standard Methods for Inventory and Monitoring*. Berkeley, CA:University of California Press, pp. 83–6.

Mazzotti, F.J., Best, G.R., Brandt, L.A., et al. (2008). Alligators and crocodiles as indicators for restoration of Everglades ecosystems. *Ecological Indicators*, **9**, S137–49.

McCleary, R.J., and Kini, R.M. (2013). Non-enzymatic proteins from snake venoms: a gold mine of pharmacological tools and drug leads. *Toxicon*, **62**, 56–74.

Muir, A.P., and Muir, M.C.A. (2011). A new rapid assessment technique for amphibians: introduction of the species list technique from San José de Payamino, Ecuador. *Herpetological Review*, **42**, 184–7.

Myers, C.W., and Rand, A.S. (1969). Checklist of amphibians and reptiles of Barro Colorado Island, Panama, with comments on faunal change and sampling. *Smithsonian Contributions to Zoology*, **10**, 1–11.

O'Dea, N., Watson, J.E.M., and Whittaker, R.J. (2004). Rapid assessment in conservation research: a critique of avifaunal assessment techniques illustrated by Ecuadorian and Madagascan case study data. *Diversity and Distributions*, **10**, 55–63.

Olesen, J.M., and Valido, A. (2010). Lizards as pollinators and seed dispersers: an island phenomenon. *Trends in Ecology & Evolution*, **18**, 177–81.

Pianka, E.R. (1986). *Ecology and Natural History of Desert Lizards. Analysis of the Ecological Niche and Community Structure*. Princeton, NJ:Princeton University Press.

Pilliod, D.S., Goldberg, C.S., Laramie, M.B., et al. (2013). Application of environmental DNA for inventory and monitoring of aquatic species. *U.S. Geological Survey Fact Sheet*, (2012–3146), 1–4.

Plummer, M.V. (1979). Collecting and marking. In M. Harless and H. Morlock (eds) *Turtles. Perspectives and Research*. New York: John Wiley & Sons, pp. 45–60.

Pritchard, P.C.H., and Mortimer, J.A. (1999). Taxonomy, external morphology, and species identification. In K.L. Eckert, K.A. Bjorndal, F.A. Abreau-Grobois, et al. (eds) *Research and Management Techniques for the Conservation of Sea Turtles*. IUCN/SSC Marine Turtle Specialist Group Publication No. 4. Blanchard, PA: IUCN/SSC, pp. 21–38.

Rajendran, M.V. (1977). A survey of uropeltid snakes. *Journal of Madurai University*, **6**, 68–73.

Robinson, L.D., Tweddle, J.C., Postles, M.C., et al. (2013). *Guide to Running a BioBlitz*. London: Natural History Museum; Bristol: Natural History Consortium; York: Stockholm Environment Institute; Plymouth, UK: Marine Biological Association of the UK.

Sung, Y.-H., Karraker, N.E., and Hau, B.C.H. (2011). Evaluation of the effectiveness of three survey methods for sampling terrestrial herpetofauna in South China. *Herpetological Conservation and Biology*, **6**, 479–89.

Thompson, G.G., Withers, P.C., Pianka, E.R., et al. (2003). Assessing biodiversity with species accumulation curves, inventories of small reptiles by pit-trapping in Western Australia. *Austral Ecology*, **28**, 361–83.

Uetz, P., and Hošek, J. (2015). *The Reptile Database*. Available at: http://www.reptile-database.org (accessed 8 April 2015).

Valencia-Aguilar, A., Cortés-Gómez, A.M., and Ruiz-Agudelo, C.A. (2013). Ecosystem services provided by amphibians and reptiles in Neotropical ecosystems. *International Journal of Biodiversity Science, Ecosystem Services and Management*, **9**, 257–72.

Vogt, R.C. (2012). Detecting and capturing turtles in freshwater habitats. In R.W. McDiarmid, M.S. Foster, C. Guyer, et al. (eds) *Reptile Biodiversity. Standard Methods for Inventory and Monitoring*. Berkeley, CA:University of California Press, pp. 181–7.

Whitaker, R., and Advani, R. (1983). Preliminary field study on snakes as agents of management of rodent populations. *Indian Forester*, **109**, 417–19.

Witz, B.W., Wilson, D.S., and Palmer, M.D. (1991). Distribution of *Gopherus polyphemus* and its vertebrate symbionts in three burrow categories. *American Midland Naturalist*, **126**, 152–8.

Zhang, J., Nielsen, S.E., Grainger, T.N., et al. (2014). Sampling plant diversity and rarity at landscape scales: importance of sampling time in species detectability. *PLoS One*, **16**, e95334.

Zouari-Kessentini, R., Srairi-Abid, N., Bazaa, A., et al. (2013). Antitumoral potential of Tunisian snake venoms secreted phospholipases A2. *BioMed Research International*, **2013**, 39138.

19

Measuring microhabitats used by non-avian reptiles

Henry R. Mushinsky and Earl D. McCoy

19.1 Introduction

As is true for many concepts in ecology, the definitions applied to 'habitat' and 'habitat structure' are controversial. Some persons (e.g. Morrison and Hall, 2002) argue that these concepts should possess precise definitions that all practitioners use, but we suggest that definitions of concepts should be tailored to specific circumstances and that over-precision could retard theorizing (see McCoy and Bell, 1991). The definition of 'habitat' that we adopt here is tailored to the organization of the book: the environmental requirements and constraints that immediately shape the distributions and sizes of populations. We fully recognize the potential influence of processes occurring at larger spatial and temporal scales and other shortcomings of this definition, however. Although the conceptualization of 'habitat' may be imprecise in many ways, the practice of gathering habitat data should not be. One should think long and hard about answers to the questions why, what, where, when, and how environmental requirements and constraints are to be determined, well before setting out to gather any data. We shall discuss briefly the value of addressing these questions. For more detailed explanations, see relevant chapters in Scott et al. (2002), Morrison et al. (2006), and Fox et al. (2015).

Perhaps the most salient example of gathering data without purpose is the collection of so-called 'baseline data' (see Alagona et al., 2012). The motive for collecting such data often is that the information available for a particular species is so sparse that any new information will be useful. However, the resulting baseline data often are not useful or, worse, are dangerously flawed. Gathering less than useful data wastes valuable time and money and potentially steers additional data gathering activities in errant directions. Methodological and observer biases—which we discuss subsequently—deterministic spatial and temporal changes, environmental stochasticity, and many other factors also can influence the usefulness of data adversely. Although most of these influences cannot be eliminated entirely, their effects can be reduced with a little reflection. Reflection not only helps to steer data gathering in a productive direction, but also to place realistic bounds on what one can infer about a pattern or process from the data.

Which environmental variables one chooses to measure tends to be a function of training and experience. An avian ecologist, for example, might find it difficult to decide

Reptile Ecology and Conservation. Edited by C. Kenneth Dodd, Jr. © Oxford University Press 2016.
Published 2016 by Oxford University Press.

which variables to measure for a tortoise population. The best course probably is to allow the behaviour of the organism to dictate the important variables, but behaviour may not be a reliable indicator in all circumstances (see Kristan, 2003). For example, if the location and/or demographics of a population are influenced by the availability of enemy-free space, then the behaviour of the organisms within that space might not indicate the importance of predator-avoidance as an environmental constraint. Furthermore, the decision about which variables to measure should be based on best available evidence; it is often confounded by other considerations, however, such as the availability of equipment and the knowledge of how to use it. Less-than-careful selection of variables in its worst manifestation is the 'shotgun approach' (see Tremblay et al., 2000), all too common in ecology, in which the researcher records something about a population, takes a variety of environmental measurements, runs a bunch of correlations, and then proclaims the environmental variable that does correlate with the population data an important influence on the population.

The environmental variables one chooses to measure will be useful only if they provide information about some aspect of the population (response variable). Just as one must choose the environmental variables carefully, one also must choose the response variable carefully. Among the possible response variables are locations of individuals, population extent, population density, growth rates of individuals, population growth rate, and recruitment. Choosing incorrectly may cause one to fail to find a relationship where one actually exists. For example, adult herbivores may be able to forage widely to obtain proper nutrition, but juveniles may require their food source to occupy a much more limited area. The details of high-nitrogen plant distribution may, therefore, have little to do with the well-being of adults, but much to do with the well-being of juveniles, and thus may be reflected most in recruitment, not in adult distribution.

Where and when environmental variables are measured tends to be a matter of convenience. In most cases, a chosen research site is not necessarily the best place to conduct the research, but rather it is nearby and/or accessible and/or protected and/or has been used historically, for example. Having to use research sites for reasons such as these usually is unavoidable, but it emphasizes the need to answer yet another suite of questions about the relationship between a population and its environmental requirements and constraints, especially if the research results are to be extrapolated or generalized. Is the site 'typical' and how much replication at the site level is required? Is the site nearer the centre of the range, or nearer the periphery? Is the site in the northern part of the range or the southern part? Likewise, the time selected to conduct research is not necessarily the best time, but rather a time when funding has been obtained, when one is released from other obligations, and/or even when it simply matches one's diel cycle, for example. Again, having to conduct research when one is able to do so probably is necessary; however, a series of questions about temporal conditions run parallel to the spatial ones already listed, and the answers to them are important. Is it a particularly cold or warm year? Is it a particularly wet or dry year? Does the research timing match the life history pattern of the species being studied? Although it may seem obvious that questions such as those we have posed are important, they are not always answered. If they were, the many cases in which virtually all of the information about the interactions of a largely crepuscular or nocturnal species with its habitat comes from data gathered during the day would not exist.

The ecological researcher must make a wide variety of choices about which field and analytical techniques to employ and how to employ them. Most researchers are aware of, and compensate for a variety of potential uncertainties and errors in their data, by worrying about power and independence, for example. Researchers often spend less time worrying about exactly which techniques they will employ and how they will employ them, however, often simply following the lead of previous research. Yet, to some degree, the results one achieves depend upon the methods employed. Even deciding precisely where within a study site to take samples can lead to methodological biases. Often, such biases result from the fact that environmental conditions almost universally occur as gradients, whereas ecological techniques typically treat them as discrete (see Colwell and Futuyma, 1971). Among these methodological biases are underestimating the number of habitat categories recognized by the organism of interest (Abrams, 1980) and failing to recognize temporal and spatial 'veiling' (McCoy, 2002). Methodological biases cannot be addressed in the same way as the statistical biases addressed by, for example, random and stratified-random sampling, and tend to be subtle in their effects. Another common type of methodological bias is failure to design studies adequately to incorporate the multiplicity of environmental relationships (Hilborn and Stearns, 1982; Hilborn and Mangel, 1997). Perhaps the most common and severe methodological bias is the use of indirect measurements of environmental variables, such as indices and surrogates, rather than direct measurements. For example, one thinks that canopy cover is an important environmental variable, but it is difficult and time-consuming to measure; so, one measures the density of trees instead, under the assumption that the two variables are strongly related. No matter how strongly related they may be, however, one is not a perfect surrogate for the other. The adverse effect of indirect measurement may be especially severe in correlative studies, in which causation often is inferred based upon precious little evidence.

In addition to methodological biases, ecological researchers must be vigilant to eliminate, or at least to recognize, observer biases. Consider how the importance of habitat features to an organism is assessed. The use–availability study design (see Thomas and Taylor, 1990, 2006) is appropriate for virtually all such assessments The terms 'use' and 'availability' are sometimes employed interchangeably with 'preference' and 'abundance', respectively. Using 'preference' and/or 'abundance', either explicitly or implicitly, implies that the observer has interpreted the motivation of the study organism correctly. Suppose, for example, that the habitat used by an organism in one place simply reflects a shortage of another habitat that is truly preferred. If the 'habitat preference' of the organisms is based upon this single observation, or even a few such observations as is often the case, then the inference may be quite erroneous.

A variety of ways exist to improve inference. For example, correlative evidence may be 'tested' with randomization. Randomization allows one to determine whether a relationship between a population and an environmental variable is different than the relationship expected by chance (see Crowley, 1992). Such randomization should be a fundamental component of the use-availability study design, mentioned previously.

Real tests of hypotheses or predictions in ecology ('ecological field experiments'), although potentially useful, are difficult to design and execute (see Underwood, 2009). When field experiments are manipulative, they tend to engender their own suites of problems, although they can be very instructive given the right conditions. For example, it is possible to use introductions of individuals into new habitats to assess experimentally the environmental conditions that promote successful establishment (e.g. McCoy et al., 2014).

Efforts to characterize habitat variables that predict the environmental requirements and constraints that immediately shape the distributions and sizes of populations may be aided by use of a conceptual framework (Morrison et al., 2006), which essentially is a standardized set of ideas and hypotheses that guide researchers to collect data in a manner that facilitates meaningful comparisons. We agree that having clear reasons for data collection is a good idea. In this sense, a conceptual framework, and its accompanying research strategy, facilitates elucidation of complex patterns of abundance and distributions by identifying both appropriate scales at which to measure habitat variables and a suite of statistical methods needed to analyse the resulting data properly (Huston, 2002). One research strategy, for example, is to focus efforts on those variables known to limit the abundance and distribution of a species (O'Connor, 2002): of course, knowing which variables are limiting in the first place is itself a substantial challenge.

The extensive morphological, behavioural, physiological, and other types of variation characteristic of extant (non-avian) reptiles presents a substantial challenge to any general statements about the why, what, when, and where aspects of data collection. Habitat variables important to a sea turtle differ greatly from those important to a fossorial lizard or an arboreal snake, for example. This variety makes the necessity of having a clear purpose for data collection all the more pressing. For the remainder of the chapter, we characterize and summarize key habitat variables for the various groups of reptiles. We focus mainly on showing the relevance of these variables to the well-being of reptiles (e.g. Whitmore, 1981). We do not repeat detailed descriptions of techniques, which can be obtained from the many sources on sampling techniques that are available.

19.2 Types of habitats and variables

The habitat variables that are typically measured arrange themselves somewhat naturally into groups. Part of the reason is simple covariation. For example, one can measure canopy attributes only in areas with shrubs and/or trees; if one can measure canopy attributes there, then he/she also can measure stem attributes. Another part of the reason for grouping of variables is similarity of purpose. For example, variables typically measured for species inhabiting open-water marine habitats are measured largely because of their potential influence on movement patterns and homing to nesting beaches; variables measured for species inhabiting shallow marine and brackish waters near land are measured largely because of their potential influence on the availability of sheltering, foraging, and nesting sites. Variables measured in terrestrial habitats often reflect the importance of thermoregulatory opportunities.

Table 19.1 *Some habitat variables frequently measured for reptiles, the rationale for measuring them, and a brief indication of how they are measured*

Variable	Relevance to reptiles	Measurement technique
Marine habitats		
Water current	Associated with movement and migration patterns of sea turtles	Current drogue and global position receiver
Water temperature	Thermoregulation	Temperature data loggers
Albedo	Habitat quality, incoming solar radiation	Multifilter radiometer
Distribution of *Sargassum*	Used by sea turtles as a refuge	Imaging spectrometry and ocean colour satellites
Slope of beaches	Nesting habitat for sea turtles	Pocket transit and level
Reflectivity	Habitat quality, solar energy	Black and white pyranometer
Thermal environment of a beach	Environmental quality for nesting	Data loggers and photographic light meter
Salinity/electrical conductivity	Water quality for crocodilians	Digital conductivity meter or optical refractometer
Freshwater habitats		
Water depth	Quality of habitat, refuge	Calibrated weighted string, meter stick
Turbidity	Alters water transparency	Secchi disc, turbidity meter
Water current	Alters use by turtles	Current drogue
Diffuse solar radiation	Micro-meteorological data for habitat quality, solar energy	Black and white pyranometer
Air/water temperatures	Thermal environment, basking sites for turtles/snakes	Thermometer, thermocouple, data loggers
Terrestrial habitats		
Canopy cover	Quality of thermal environment	Spherical densiometer or GRS densitometer
Tree height	Refuge/residence for reptiles	Clinometer
Tree/branch diameter	Refuge/residence for reptiles	Tree callipers
Air temperature	Thermal environment	Thermometer/thermocouple Data logger
Wind speed	Thermal habitat quality	Anemometer
Fossorial habitats		
Soil moisture	Quality of nesting habitats for many egg-laying species	Drying oven
Soil temperature	Quality of thermal environment	Soil thermometer
Soil hardness	Suitability for burrowing	Soil penetrometer
Rocky habitats		
Reflectivity	Quality of thermal environment, solar energy	Black and white pyranometer
Wind speed	Quality of thermal environment	Anemometer
Air temperature	Thermoregulation and habitat quality	Copper constantan (Cu-Cn) thermocouples, thermometer
Incident radiation, canopy cover	Thermal environment	Hemispherical digital camera
Rock temperature	Thermal environment	Data loggers placed above and below rocks
Crevice geometry	Refuge space	Expanding foam

Arboreal habitats		
Branch/tree dimensions	Available microhabitats	Tree callipers
Canopy cover	Thermal habitat quality	Spherical densiometer or GRS densitometer
Tree height	Refuge/residence for reptiles	Clinometer

Standard habitat classification schemes (e.g. the IUCN (2014) list of 16 distinct habitat types for species on the Red Data List) are not particularly suitable for simultaneously grouping habitat variables. To focus on the goal of this chapter, characterizing and summarizing key habitat variables, we collapsed available habitat descriptions into the smallest set of categories that we thought would still represent adequately both the types of areas most frequently occupied by reptiles and the breadth of habitat variables typically measured. The resulting scheme includes six categories: marine, freshwater, terrestrial, fossorial, rocky, and arboreal. Table 19.1 provides a listing of frequently measured variables, the rational for measuring them, and a brief indication of how they are measured, for each of the six habitat categories (Figure 19.1). We elaborate on the information in this table in the following sections, with abstracts of studies that measured one or more of the variables. Because every study should be designed to address specific goals or questions, our representative examples should not be taken as applicable in all similar situations, nor should their inclusion as representative examples necessarily be taken as our endorsement of them.

A suite of techniques applicable across virtually all habitats involves sensing and positioning technologies. Measuring habitat variables now often relies on remote sensing

Figure 19.1 Using GPS to locate random points for structural analysis of vegetation (a). Using GPS to mark positions of microhabitat features near Gopher Tortoise burrows (b). Seining to determine presence-absence of aquatic snakes for later microhabitat analysis (c). Measuring incident radiation to evaluate the thermal environment of freshwater ponds (d).

to monitor both structural and climatic variables, especially in marine habitats. This modern technology provides numerous options to monitor temperature fluctuations, salinity, and tidal patterns, for example. In all habitat types, researchers also have come to rely increasingly on radiotelemetry to track individuals. Knowing the locations and time budgets of focal individuals permits researchers to assess the what, when, and where aspects of habitat use effectively. Many researchers use the Global Positioning System (GPS), often in combination with radiotelemetry, to record animal movement patterns, to map the positions of habitat variables in space, and to relate movement patterns to habitat variables.

19.3 Marine habitats: sea and brackish water turtles, sea snakes, crocodiles, marine iguanas

Habitats used by sea turtles vary ontogenetically, and include beaches during nesting, ocean surface during the epipelagic life stage, and the benthos where adults and juveniles forage. Much evidence exists (Carr, 1987; Witherington, 2002; Mansfield et al., 2014) that the early life stages of epipelagic turtles are spent at ocean fronts or areas of convergence typically associated with *Sargassum*. *Sargassum*, when in its free-floating stage, provides shelter for juvenile turtles as they move with the flotsam during their early developmental years. Studies of sea turtles in the open oceans require extensive remote sensing (Gower et al., 2006; Hu, 2009). Data collection at the foraging grounds includes water depth, temperature, and activity patterns, all of which can be collected with transmitters (Godley et al., 2002). Variables associated with sea turtle nest microhabitat, including temperature, moisture, salinity, and slope, have been studied by Wood et al. (2000). The relationship between sand albedo and thermal environment 75 cm below the sand surface was studied on nesting beaches by Hays et al. (2001). Albedo was measured with a standard photographic light meter under perfectly clear skies on sand that had been smoothed. Five measurements were made on the sand and sequentially on a grey card that had a known albedo to calibrate the sand measurements.

Reptiles inhabiting brackish waters and estuaries experience changes in salinity, water depth, and temperature on daily, monthly and annual cycles. Habitat use by the Diamondback Terrapin (*Malaclemys terrapin*) in estuaries has been studied by following radio-tagged individuals in habitats fitted with data-loggers to monitor environmental variables (Harden et al., 2007). Variables measured include substrate (mud) and water temperatures, salinity, water depth, and the degree of coverage by marsh grass (*Spartina alterniflora*) (Roosenburg et al., 1999). These variables can influence the movement patterns of the turtles, as well as the availability of prey. Salinity can alter habitat use by crocodiles, because of their limited salt tolerance (Ellis, 1981). Crocodiles were observed only in areas of vegetation in canals, along shorelines, ponds, creeks, and freshwater marshes (Cherkiss et al., 2011). Non-hatchlings were found in intermediate salinities, while hatchlings were found in the lowest salinities.

Studying habitat use by sea snakes presents challenges similar to those encountered when studying sea turtles; such studies often depend upon remote sensing to locate free-ranging individuals (Chapter 12). Sea snakes fitted with transmitters were studied by the use of underwater acoustic locator systems (Burns and Heatwole, 1998) to map their usage of coral reefs in Australia. The same study also used freeze-branded individuals to test homing ability. Dunson and Ehlert (1971) measured the effect of temperature, salinity, and surface water flow patterns to monitor sea snake distributions. More recently, Shine et al. (2003) made visual observations while snorkelling and recorded water depth, percentage cover of coral rubble, live coral, and sand in square metre quadrats directly below an individual's head when first observed. These studies of sea snakes were designed to search for correlations among the habitat variables and the distribution and abundance of individuals.

A unique marine lizard, the Galapagos Marine Iguana (*Amblyrhynchus cristatus*), exhibited ontogenetic and diurnal microhabitat shifts (Buttemer and Dawson, 1993). Researchers measured global and reflected radiation, direct solar irradiation, wind speed, wind direction, air temperature, water temperature, surface albedo and used models of marine iguanas to estimate their body temperatures. Body size strongly influenced microhabitat usage. Because this species spends time on rocks that absorb much solar radiation (for thermoregulation) and swim in cold oceanic currents to forage, measuring their microhabitats requires sophisticated technology.

19.4 Freshwater habitats: freshwater turtles, water snakes, alligators, caimans

Some highly aquatic turtle species leave water only to bask and/or nest. The Western Pond Turtle (*Actinemys marmorata*) has been studied in dammed and undammed river tributaries to understand the relations between habitat use and damming (Reese and Welsh, 1998). Individuals were captured underwater by snorkelers who searched the submerged banks and river bottoms manually and visually. They used a floating rectangular quadrat (3×6 m^2, divided into nine 1×2 m^2 sub-quadrats) to describe the habitat where an individual was first encountered. Within each quadrat they measured shoreline vegetation type, flow types (pool, riffle, glide) along a transect of the river, flow type in each sub-quadrat, water velocity, water depth (in each sub-quadrat), presence of basking sites (large and small), availability of basking sites, sand cover, presence of cover objects, degree of bank undercut (maximum and mean of each sub-quadrat), water temperature, and percentage canopy. For each quadrat that was characterized at a point of capture, they also characterized a random quadrat to represent the available habitat. A random number generator was used to determine the placement of the random quadrats along the river. A detailed, shotgun approach, such as this one, to measure microhabitat variables above and below the surface of the water illustrates a fairly common approach to assessing habitat selection of a species when hard data are lacking. This approach must be used cautiously, as we indicated earlier, as it can produce spurious results. As well, oftentimes measuring multiple variables leads to many autocorrelations. The most productive research occurs when variables meaningful to the focal species are identified a priori.

Many turtle species that live in freshwater use the surrounding riparian habitats for more than nesting and basking sites. A notable example is the well-studied Wood Turtle

(*Glyptemys insculpta*), which mates and hibernates in streams and rivers, nests in riparian gravel bars, and forages in wetlands and nearby uplands (Compton et al., 2002). A detailed investigation of habitat selection by the Wood Turtle measured distance to running water, distance to any water, distance to nearest vernal pool, depth of water, amount of sun, canopy density, raspberry in fruit, mushrooms within 3 m, presence of slugs, presence of earthworms (taking five samples with a bulb planter), distance to nearest edge, type of surrounding habitat (forest, marsh meadow), and cover type (alder, bog, fen). Each of the selected variables contributes to the availability of an adequate food supply and basking sites for thermoregulation, and thus overall habitat quality.

Habitats occupied by water snakes have been studied by several investigators. Working from a boat, Hebrard and Mushinsky (1978) measured the slope of the shoreline, the density of vegetation (used as basking sites), and distance an individual was first observed from the shoreline, in a bayou in southern Louisiana. Basking sites are important thermoregulatory features for many reptiles. Madsen (1984) classified snake habitats in southern Sweden as stone fence, blackberry bushes, blackthorn bushes, arable land, water, and forest. Stone fences with stands of blackberry or blackthorn bushes covered less than 1% of the study site but accounted for 87% of all observations. Microhabitats that provide sources of food and shelter are often those that are highly preferred. Tiebout and Gary (1987) considered five different levels at which a snake must 'decide' where it might be located at a given time: (1) activity area (home range); (2) habitat (vegetation type); (3) substrate; (4) perch height; and (5) exposure to sunlight. They also considered (6) activity patterns, or movements, that link snake locations. These researchers used the term habitat to define areas on a relatively large scale such as deciduous woods, open water, fallow field, cultivated field, cattails, brush-hedgerow, grassy field, hydric woods, yard-domestic, and flooded meadow. Within each of these areas they identified a substrate category and noted if the telemetered snake was observed in water or on land. Substrate categories included dead cattails from last year, dead cattails, dead leaves, leaf litter, dead grass, dead forbs, live grass, live forbs, live tree branch, live tree twig, live bush, and others. The researchers did not measure canopy cover but rather estimated what percentage of an observed individual was in direct sunlight. They computed the area occupied by each of the substrates and reported the percentage of time the focal species spent in each.

Alligators and caimans occupy freshwater habitats in a variety of relatively warm climates across the globe. Numerous studies of American Alligators (*Alligator mississippiensis*) have focused on nesting habitats and movement patterns, but few have focused on adult microhabitats, per se. Using radiotelemetry and making measurements of surface water and air temperatures, Goodwin and Marion (1979) followed individuals for one year as they moved around a large lake and surrounding swamps. Moreno-Arias et al. (2013) studied the Spectacled Caiman (*Caiman crocodilus fuscus*) in lotic and lentic environments at ten sites in Colombia. They measured the number of beaches, tree trunks or woody brides, trees on water, and the amount of litter in the surrounding riparian forests at each site. These structures provide basking sites and nesting materials for females.

19.5 Terrestrial habitats: most lizards, most snakes, terrestrial turtles, Tuatara

Terrestrial habitats are exceedingly diverse, but can be arranged along a broad continuum of relatively open to relatively closed, depending on vegetation cover. In relatively open terrestrial habitats, thermoregulation often presents challenges to reptiles both during the day and night, and thermal refugia may strongly influence microhabitat use. Pianka (1973) divided the microhabitats occupied by the species in desert lizard communities on three continents into the broad categories; subterranean, terrestrial, and arboreal; and the terrestrial microhabitats into open sun, grass, sun bush, sun tree, sun other, shade grass, shade bush shade, tree shade. Huey (1991) made detailed measurements of solar radiation and wind speed, and related these measurements to organismal properties, such as body size, shape, and colouration. Paulissen (1988) studied seasonal changes in microhabitat use by locating undisturbed individuals of *Aspidocelis sexlineata*, marking their positions with flags, classifying them as adults or juveniles, and then collecting data on nine habitat variables. These variables were temperature 1 cm below the surface and 15 cm above ground, total solar radiation, distance to nearest open patch, distance to nearest cover (a burrow or a clump of vegetation), height of the nearest plant, percentage of open ground, percentage of vegetation within 0.1 m^2 centred on the marked position, and wind velocity averaged over 5 minutes. The goal of this study was to evaluate various microhabitat features in different seasons (therefore having different exposures to sunlight) to evaluate their relative importance and contribution to the thermal environment of the lizards.

In studies of relatively closed terrestrial habitats, structural variables often become prominent. Physical structure influences thermoregulation, but also strongly affects prey availability. To characterize the structural variables habitats important to two lizard species in the sclerophyll forests of subtropical Australia, Singh et al. (2002) used a modified point-intercept method of Sutherland (1996). Four 2.5 m transects were marked in a cross (+) pattern radiating from the point of capture, and canopy height (>3 m), presence/absence of shrub cover (1–3 m), percentage cover of ground vegetation (<1 m), percentage cover of logs (>0.1 m diameter), and percentage cover of litter were measured at 0.5 m intervals. Litter depth to the underlying soil was measured at the 1 m mark of each transect. The mean values of measurements were compared to mean values derived from 40 random surveys conducted nearby. These researchers focused on the structural components of the habitat that could influence the thermal biology of two closely related species: sex related differences in thermal preferences proved more important than differences between the species.

Using radiotelemetery, Reinert (1984) collected data on microhabitat use by two snake species, the Timber Rattlesnake (*Crotalus horridus*) and the Northern Copperhead (*Agkistrodon contortrix*) in forested land in Pennsylvania. At each observation, microhabitat variables were measured in a 1 m^2 quadrat centred on the located individual. The detailed description of the microhabitat included rock cover, leaf litter cover, vegetation cover, fallen-log cover, woody stem diversity, woody stem height, distance to nearest rock, length of rocks, distance to nearest log, diameter of log, distance to canopy, diameter at breast height of canopy trees, distance to understory, canopy closure, soil temperature, surface temperature, ambient temperature, surface relative humidity, and ambient relative humidity. He accumulated data on about 100 observations of each species and for comparison with the microhabitats selected by those individuals, he collected data on the same variables at 100 randomly selected sites. Both the physiological condition (gravid versus non-gravid) and morphology (melanistic versus non-melanistic) proved to be important determinants for the thermal ecology of the timber rattlesnake and the overall openness of the habitat was most important to the northern copperhead.

Habitat use by the Three-toed Box Turtle (*Terrapene carolina triunguis*) was studied by Reagan (1974) in a grassland site and a forested site in Arkansas. He collected data only on free-ranging individuals found in shallow resting depressions in litter or soil ('form') and not moving for at least 30 min. Data were collected on 33 structural and climatic variables to define selected microhabitats using a circle of 50 cm radius centred on the depression. Many of the measurements were redundant; for example, he included nine measures of the thermal environment at each depression. Ultimately these data were subjected to principle component analysis. The first three components, temperature, canopy, and moisture, proved to be most informative and accounted for more the 50% of the variation.

19.6 Rocky habitats: lizards and snakes

Rock crevices and rocky outcrops provide a unique set of microhabitats used by saxicolous lizards and snakes. Croak et al. (2008) provide a comprehensive summary of the many studies of saxicolous reptiles. As they point out, researchers often focused on attributes of the overlying rocks to measure exposure to solar radiation and have demonstrated that many reptiles prefer sun-exposed rocks that facilitate active thermoregulation (see Huey, 1991). Croak et al. developed a comprehensive description of microhabitats used as shelter by rock-dwelling lizards and snakes. Using a digital camera, they took a hemispherical photograph above each rock to evaluate canopy characteristics enabling them to predict the amount of incident radiation reaching each rock. To measure the thermal profile rocks, they placed thermal data loggers below them. To measure crevice geometry, they made physical casts of the crevices underneath rocks using expanding foam spray that hardens in about 12–24 hours. They measured rock dimensions, including width, length, and thickness of each overlying rock. The researchers then developed methods to interpret the three-dimensional attributes of the casts based on the known requirements of the focal species. Crevice attributes, such as the height of the crevice above the substrate, the area of the crevice suitable for fitting the entire body of the animal, the degree of closure of the crevice and the amount of fragmentation within the crevice affected the suitability of a crevice as a retreat site. They also found that crevice structure directly alters thermal regimes, and may be more important than rock thickness in this respect.

19.7 Fossorial habitats: some lizards, amphisbaenids, some snakes

Studying the important microhabitat variables for burrowing and semi-fossorial species is challenging because of the difficulty in observing individuals. McCoy et al. (2013) constructed 36 20 × 20 m^2 enclosures to confine and facilitate trapping of the sand-swimming Florida Sand Skink (*Plestiodon reynoldsi*). They were interested in how time since fire affected microhabitat use. Within each enclosure, they measured ten microhabitat variables known to be affected by time since fire (McCoy et al., 1999). These variables were mean litter depth, mean incident light, number of palmetto bushes, mean canopy height, mean sand moisture content at 15 cm depth, and percentage of bare ground, live vegetation, dead vegetation, lichen, and woody debris. The values for these variables within the 36 enclosures were then related to abundances of individuals within the same enclosures. Interestingly and unexpectedly, mean litter biomass had the strongest correlation with the density of the lizard.

Greenville and Dickman (2009) completed a well-planned and executed study of the microhabitat characteristics important for the burrowing Southern Sandslider (*Lerista labialis*). This species is confined to the crests of sand dunes in the Simpson Desert, central Australia. The researchers conducted an elaborate set of experiments to determine what confines this species to sand dune crests. They visually measured vegetation cover, used data loggers to measure the thermal environment, a soil penetrometer to measure the strength of the sand crust layer, and sampled termites in trenches. Predictions that the lizard selects dune crests because of their vegetation characteristics, favourable temperatures, or abundances of preferred prey items were not supported by their data. Rather, the dune crests appeared to provide soft and less compacted sand that facilitated movement by *L. labialis*. Dune crests are covered by sand that is softer than that on the sides and swales, permitting the skinks to reduce the energetic costs of locomotion.

Martín et al. (2013) overturned stones along walking transects in North Africa to locate the fossorial amphisbaenian *Trogonophis wiegmanni*. To characterize the species' microhabitat use, they took measurements within a circular area of 2 m diameter around a stone under which an individual was found. They estimated visually percentage cover of each vegetation type at the ground level, cover-dominant large woody bushes, and calculated a mean bush height. They also estimated percentage cover of small stones, as well as medium stones that might be used for shelter. Soil compaction was measured using a hand penetrometer at five random points near the central stone and calculated an average value for each site. Both physical (depth, % gravel, fine sand, course sand, silt, and clay) and chemical (acidity, inorganic carbonates, electrical conductivity, and organic carbon) characteristics were measured from soil samples taken at each location. The researchers measured microhabitat availability and soil compaction and composition by taking random transects through the study area. They stopped every 25 m and took the same measurements around the nearest stone. The focal species selected soils that were sandy, basic, carbonated, and shallow, having a high cover of rocks, and they avoided marine salinized, more acid and deeper soils, and those covered mainly by small rocks. The energetic costs of burrowing and influences of soil chemistry were used to explain the observed pattern.

Many snakes hibernate below ground and some, such as the Pine Snake (*Pituophis melanoleucus*), also use underground summer dens as thermal refugia. Burger et al. (1988) compared the above-ground characteristics of winter hibernacula with those of summer dens of the pine snake. In the year prior to this research, they had marked burrow opening from which individuals had been observed to emerge. They measured distance between burrow openings, slope of the entrance, slope of the tunnel, air temperature, soil temperature, vegetation cover, soil moisture, leaf cover, distance and height of nearest herb, distance to nearest tree, and distance to nearest log. Measurements of the same set of variables were taken at nearby randomly chosen locations. The researchers excavated the burrows in early spring, before any individuals emerged from them. Those excavated burrows that contained one or more individuals were considered hibernacula; those without a snake were considered summer dens. Temperatures were higher and leaf cover was less at random points than at hibernacula or summer dens. Hibernacula had more vegetation cover, more leaf cover over the burrow entrance, and were closer to trees than summer dens. Hibernacula proved to be more structurally complex than summer dens by having more side chambers.

Two families of small to medium snakes, Typhlopidae and Leptotyphlopidae, are comprised of mostly fossorial species that burrow into the substrate. Akani et al. (2008) studied the distribution of the Spotted Blindsnake (*Typhlops punctatus*) within five study plots. They turned over objects encountered on the ground, such as logs, metallic panels, plastics, and stones while walking along transects to locate individuals. Twelve soil samples were collected on each transect by pushing in a corer to a depth of 10 cm where individuals were found. These samples were used to measure soil texture, moisture content, and organic matter content. Soil particle size distribution was measured using a combination of wet sieving and pipette methods. The percentage composition of the mineral particles was calculated and soil pH was measured. These snakes have a strong affinity for loamy soils, with moisture content of 8–18%, high organic matter (1–6%), and slightly acid pH of 5.4–6.9.

19.8 Arboreal habitats: some snakes, some lizards

Studies of habitat use by arboreal snakes must consider the diversity of microhabitats present in trees of different species and dimensions. Plummer (1981) studied microhabitat use by the arboreal snake *Opheodrys aestivus* in a forest surrounding a lake in Arkansas. Working from a boat, he captured and marked individuals, and recorded location, perch height, perch diameter, perch angle, longitudinal position on branch, distance from shoreline, and perch taxon. Observations were made during daylight hours and at night. No random samples of perch dimensions or availability were taken. The study area was marked in 10 m increments to document an exact location on the shoreline. To obtain data in the daytime, individuals were captured in the hours before daybreak, marked with a narrow band of rapid-drying fluorescent orange spray paint, and released at the site of capture. Perch characteristics were recorded for each focal animal every 15 min, if possible. Observations were made through binoculars from behind a blind fitted on a boat which was anchored about 10 m from the shoreline.

The nocturnal and diurnal perches used by *O. aestivus* were narrowly selected with regard to distance from shoreline, perch height, perch diameter, position on perch, and perch angle. Daytime observations on focal animals revealed that most movement was near the edge, but that occasional, brief forays into the forest occurred. Radiotelemetry was used to follow adult individuals of Stephens' Banded Snake (*Hoplocephalus stephensii*) in southern Australia (Fitzgerald et al., 2002). The researchers measured or categorized a large number of microhabitat variables of the trees in which individuals spent time. These variables included height, diameter at breast height, buttress diameter, percentage canopy cover, bole shape (round, oval), bark texture, basal crevice area, hollow-bearing stage (a measure of tree hollow availability based on the successional stage of a tree), growth stage, position in the canopy, interconnectedness, trunk hollow, branch hollow, termitaria, epiphytes, vines, and fire scars. They compared measurements and categorizations for 139 trees used by the focal species and 1437 randomly chosen trees. Tracked individuals were in trees on >80% of observations, generally hidden within hollows and were inactive in trees for three to five months during winter each year. Individuals selected old, large trees with many hollows or extensive vine cover.

Two studies compared the morphology and behavioural traits of *Anolis carolinensis* in two disparate locations and related the morphology and traits to habitat characteristics. One location was an isolated lowland freshwater swamp that supported large cypress trees and bushes (Irschick et al., 2005b), and the other was an urban setting on the campus of Tulane University that was characterized by simpler vegetation including dense clumps of palmetto (*Aspidistra elatior*) with relatively short (<2 m) leaves and few large trees or bushes (Irschick et al., 2005a). Structural habitat availability was quantified by measuring the perches at 0.5, 1, and 2 m at regular intervals along a predetermined transect. For each perch, the researcher measured perch diameter, perch length, the distance to the nearest perch, and the diameter of that nearest perch. The transect was sampled every 5 m resulting in a total of 156 sample points. To determine habitat use by the lizards, the researcher walked along the transect daily when individuals were active (09:00 to 17:00), and captured any individual sighted and marked its position with a coloured flag. At the points of capture, he measured several habitat variables: substrate type (e.g. tree trunk, branch), perch height, diameter, and length, distance to nearest perch and diameter of nearest perch. After habitat data were collected, individuals were returned to their original points of capture. At the less complex site, individuals perched on broad, smooth leaves, while at the more complex site, individuals perched on branches and tree trunks. The populations differed significantly in morphology. Individuals at the less complex site had shorter distal hindlimb elements and longer forelimb elements, were less robust, and had larger toepads and higher clinging abilities than those at the more complex site.

19.9 Summary and recommendations

Marine, freshwater, terrestrial, arboreal, rocky, and fossorial habitats each presents unique challenges both for the organisms inhabiting them and the researchers studying them. As employed here, 'habitat' is the collection of environmental requirements and constraints that immediately shape the distributions and sizes of populations. Gathering data on habitat variables should have a specific purpose to avoid potentially useless or even erroneous conclusions. Selection of the habitat variables to be measured should be based on the biology of the focal species, be relevant to the purpose of the study, and use the best available background information. Just because a focal species inhabits trees or is fossorial, for example, does not mean, necessarily, that easily measured features of those habitats are important requirements or constraints. Where and when selected variables are measured often is a matter of convenience or habit and, in such cases, the results should be interpreted especially carefully. Selection and assessment of response variables (location of individuals, recruitment of juveniles) also should be done carefully. Indirect measurements of habitat variables (indices, surrogates) often are problematic, and should be avoided or, at least used advisedly. Techniques used to measure habitat variables should be chosen for accuracy and precision. The use–availability design, common in studies of habitat selection, should incorporate a randomization procedure to determine whether the results one obtains were simply to be expected by chance. Given this large set of considerations, testing the applicability of 'standard' techniques to specific studies must be the norm. The increasing use of techniques such as radiotelemetry and GPS most certainly has improved researchers' abilities to locate and track the focal species and to measure habitat variables. Nevertheless, the value of such techniques does not obviate the researcher's responsibility to see that they are used correctly relative to the purpose of the study.

The question naturally arises as to what should be the next step in the evolution of ways to measure habitat variables. We suggest that two courses may be particularly productive. The first is to evaluate the outcomes of the use of 'standard' techniques: when have they proved productive and when have they not; when they have not, why not? Such an evaluation requires careful matching of outcomes to purpose, as well as broad testing of assumptions. Meta-analysis may prove useful in this regard. The second course that we suggest is targeting of both the development of techniques and the elucidation of important variables. Several researchers working on the same problem may prove an effective strategy for accelerating progress. One target, for example, might be the habitats occupied by endangered and threatened species. If a certain habitat supports a number of such species that is disproportionate to its extent, then it may warrant emphasis. To make progress along this course requires researchers to search for commonalities among various locations; they may need to adopt a more habitat-centric view, as opposed to a species-centric one.

Acknowledgements

We thank Peter Meylan, Anne Meylan, Robert Hardy, and Emma Harrison for their advice on microhabitat variables for sea turtles.

References

Abrams, P. (1980). Some comments on measuring niche overlap. *Ecology*, **61**, 44–9.

Akani, G.C., Luiselli, L., Eniang, E.A., et al. (2008). Aspects of the ecology of the spotted blind-snake, *Typhlops punctatus punctatus* in Port-Harcourt, Nigeria. *African Journal of Ecology*, **46**, 533–9.

Alagona, P.S., Sandlos, J. and Wiersma, Y.F. (2012). Past imperfect: using historical ecology and baseline data for conservation and restoration projects in North America. *Environmental Philosophy*, **9**, 49–70.

Burger, J., Zappalorti, R.T., Gochfeld, M., et al. (1988). Hibernacula and summer den sites of pine snakes (*Pituophis melanoleucus*) in the New Jersey Pine Barrens. *Journal of Herpetology*, **22**, 425–33.

Burns, G., and Heatwole, H. (1998).Home range and habitat use of the olive sea snake, *Aipyurus laevis*, on the Great Barrier Reef, Australia. *Journal of Herpetology*, **32**, 350–8.

Buttemer, W.A., and Dawson, W.R. (1993). Temporal pattern of foraging and microhabitat use by Galapagos marine iguanas, *Amblyrhynchus cristatus*. *Oecologia*, **96**, 56–64.

Carr, A.F. (1987). New perspectives on the pelagic stage of sea turtle development. *Conservation Biology*, **1**, 101–21.

Cherkiss, M.S, Romanach, S.A., and Mazzotti, F.J. (2011). The American crocodile in Biscayne Bay, Florida. *Estuaries and Coasts*, **34**, 529–35.

Colwell, R.K., and Futuyma, D.J. (1971). On the measurement of niche breadth and overlap. *Ecology*, **52**, 567–76.

Compton, B.W., Rhymer, J.M., and McCollough, M. (2002). Habitat selection by wood turtles (*Clemmys insculpta*): an application of paired logistic regressions. *Ecology*, **33**, 833–43.

Croak, B.M., Pike, D.A., Webb, J.K., et al. (2008). Three-dimensional crevice structure affects retreat site selection by reptiles. *Animal Behaviour*, **76**, 1875–84.

Crowley, P.H. (1992). Resampling methods for computation-intensive data analysis in ecology and evolution. *Annual Review of Ecology and Systematics*, **21**, 405–47.

Dunson, W.A., and Ehlert, G.W. (1971). Effect of temperature, salinity and surface water flow on the distribution of the sea snake, *Pelamis*. *Limnological Oceanography*, **16**, 845–53.

Ellis, T.M. (1981). Tolerance of sea water by the American crocodile *Crocodylis acutus*. *Journal of Herpetology*, **15**, 187–92.

Fitzgerald, M., Shine, R., and Lemckert, F. (2002). Radiotelemetric study of habitat use by the arboreal snake *Hoplocephalus stephensii* (Elapidae) in eastern Australia. *Copeia*, **2002**, 321–32.

Fox, G.A., Negrete-Yankelevich, S., and Sosa, V.J. (2015). *Ecological Statistics: Contemporary Theory and Application*. Oxford: Oxford University Press.

Godley, B.J., Richardson, S., Broderick, A.C., et al. (2002). Long-term satellite telemetry of the movements and habitat utilisation by green turtles in the Mediterranean. *Ecography*, **25**, 352–62.

Goodwin, T.M., and Marion, W.R. (1979). Seasonal activity ranges and habitat preferences of adult alligators in a north-central Florida lake. *Journal of Herpetology*, **13**, 157–63.

Gower, J., Hu, C., Borstad, G., et al. (2006). Ocean color satellites show extensive lines of floating sargassum in the Gulf of Mexico. *Geoscience and Remote Sensing*, **44**, 3619–25.

Greenville, A.A., and Dickman, C.R. (2009). Factors affecting habitat selection in a specialist fossorial skink. *Biological Journal of the Linnean Society*, **97**, 531–44.

Harden, L.A., Diluzio, N.A., Gibbons, J.W., et al. (2007). Spatial and thermal ecology of diamondback terrapins (*Malaclemys terrapin*) in a South Carolina salt marsh. *Journal of the North Carolina Academy of Science*, **123**, 154–62.

Hays, G.C., Ashworth, J.S., Barnsley, M.J., et al. (2001). The importance of sand albedo for the thermal conditions on sea turtle nesting beaches. *Oikos*, **93**, 87–94.

Hebrard, J.J., and Mushinsky, H.R. (1978). Habitat use by five sympatric water snakes in a Louisiana swamp. *Herpetologica*, **34**, 306–11.

Hilborn, R., and Mangel, M. (1997). *The Ecological Detective: Confronting Models with Data*. Princeton, NJ: Princeton University Press.

Hilborn, R., and Stearns, S.C. (1982). On inference in ecology and evolutionary biology: the problem of multiple causes. *Acta Biotheoretica*, **31**, 145–64.

Hu, C. (2009). Remote sensing of environment: a novel ocean color index to detect floating algae in the global oceans. *Remote Sensing of Environment*, **113**, 2118–29.

Huey, R.B. (1991). Physiological consequences of habitat selection. *American Naturalist*, **137**, S91–115.

Huston, M.W. (2002). Introductory essay: critical issues for improving predictions. In J.M. Scott, P.J. Hugland, M.L. Morrison, et al. (eds) *Predicting Species Occurrences: Issues of Accuracy and Scale*. Washington, DC: Island Press, pp. 7–21.

Irschick, D.J., Carlisle, E., Elstrotti, J., et al. (2005a). A comparison of habitat use, morphology, clinging performance and escape behaviour among two divergent green anole lizard (*Anolis carolinensis*) populations. *Biological Journal of the Linnean Society*, **85**, 234–44.

Irschick, D.J., Van Hooydonck, B., Herrel, A., et al. (2005b). Intraspecific correlations among morphology, performance, and habitat use within a green anole lizard (*Anolis carolinensis*) population. *Biological Journal of the Linnean Society*, **85**, 223–34.

IUCN. (2014). *Habitats Classification Scheme (Version 3.1)*. Available at: http://www.iucnredlist.org/technical-documents/classification-schemes/habitats-classification-scheme-ver3

Kristan, W.B. (2003). The role of habitat selection behavior in population dynamics: source-sink systems and ecological traps. *Oikos*, **103**, 457–68.

Madsen, T. (1984).Movements, home range size and habitat use of radio-tracked grass snakes (*Natrix natrix*) in southern Sweden. *Copeia*, **1984**, 707–13.

Mansfield, K.L., Wyneken, J. Porter, W.P., et al. (2014). First satellite tracks of neonate sea turtle redefine the 'lost years' oceanic niche. *Proceedings of the Royal Society of London B: Biological Sciences*, **281**, 2013–39.

Martín, J., López, P., and García, L.V. (2013). Soil characteristics determine microhabitat selection of the fossorial amphisbaenian *Trogonophis wiegmanni*. *Journal of Zoology*, **290**, 265–72.

McCoy, E.D. (2002). The 'veiled gradients' problem in ecology. *Oikos*, **99**, 189–92.

McCoy, E.D., and Bell, S.S. (1991). Habitat structure: the evolution and diversification of a complex topic. In S.S. Bell, E.D. McCoy, and H.R. Mushinsky (eds) *Habitat Structure: The Physical Arrangement of Objects in Space*. London: Chapman and Hall, pp. 3–27.

McCoy, E.D., Britt, E.J. Catenazzi, A., et al. (2013). Fire and herpetofaunal diversity in the Florida scrub ecosystem. *Natural Areas Journal*, **33**, 500–10.

McCoy, E.D., Osman, N., Hauch, B., et al. (2014). Increasing the chance of successful translocation of a threatened lizard. *Animal Conservation*, **17**(Supplement S1), 56–64.

McCoy, E.D., Sutton, P.E., and Mushinsky, H.R. (1999). The role of guesswork in conserving the threatened sand skink. *Conservation Biology*, **13**, 190–4.

Moreno-Arias, R.A., Ardila-Robayo, M.C., Martinez-Barreto, W., et al. (2013). Ecologiapoblacional de la babilla (*Caiman crocodilus fuscus*) en el Valle del Rio Magdalena (Cundinamarca, Colombia). *Caldasia*, **35**, 25–32.

Morrison, M.L., and Hall, L.S. (2002). Standard terminology: toward a common language to advance ecological understanding and application. In J.M. Scott, P.J. Heglund, M.L. Morrison, et al. (eds) *Predicting Species Occurrences: Issues of Accuracy and Scale*. Washington, DC: Island Press, pp. 43–52.

Morrison, M.L., Marcot, B.J., and Mannan, R.W. (2006). *Wildlife–Habitat Relationships: Concepts and Applications*. Washington, DC: Island Press.

O'Connor, R.J. (2002). The conceptual basis of species distribution modeling: time for a paradigm shift. In J.M. Scott, J. M., P.J. Hugland, M.L. Morrison, et al. (eds) *Predicting Species Occurrences: Issues of Accuracy and Scale*. Washington, DC: Island Press, pp. 25–33.

Paulissen, M.A. (1988). Ontogenetic and seasonal shifts in microhabitat use by the lizard *Cnemidophorus sexlineatus*. *Copeia*, **1988**, 1021–9.

Pianka, E.R. (1973). The structure of lizard communities. *Annual Review of Ecology and Systematics*, **4**, 53–74.

Plummer, M.V. (1981). Habitat utilization, diet and movements of a temperate arboreal snake (*Opheodrys aestivus*). *Journal of Herpetology*, **15**, 425–32.
Reagan, D.O. (1974). Habitat selection in the three-toed box turtle, *Terrapene carolina triunguis*. *Copeia*, **1974**, 512–27.
Reese, D.A., and Welsh, H.H. (1998). Habitat use by the western pond turtle in the Trinity River, California. *Journal of Wildlife Management*, **62**, 842–53.
Reinert, H.K. (1984). Habitat variation within sympatric snake populations. *Ecology*, **65**, 673–82.
Roosenburg, W.M., Haley, K.L., and McGuire, S. (1999). Habitat selection and movements of diamondback terrapins, *Malaclemys terrapin*, in a Maryland estuary. *Chelonian Conservation Biology*, **3**, 425–9.
Scott, J.M., Heglund, P.J., Morrison, M.L., et al. (2002). *Predicting Species Occurrences: Issues of Accuracy and Scale*. Washington, DC: Island Press,.
Shine, R., Shine, T., and Shine, B. (2003).Intraspecific habitat partitioning by the sea snake *Emydocephalus annulatus* (Serpentes, Hydrophiidae): the effects of sex, body size, and colour pattern. *Biological Journal of the Linnean Society*, **80**, 1–10.
Singh, S., Smyth, A.K., and Blomberg, S.P. (2002). Thermal ecology and structural habitat use of two sympatric lizards (*Carlia vivax and Lygisaurus foliorum*) in subtropical Australia. *Austral Ecology*, **27**, 616–23.
Sutherland, W.J. (1996). *Ecological Census Techniques: A Handbook*. Cambridge: Cambridge University Press.
Thomas, D.L., and Taylor, E.J. (1990). Study designs and tests for comparing resource use and availability. *Journal of Wildlife Management*, **54**, 322–30.
Thomas, D.L., and Taylor, E.J. (2006). Study designs and tests for comparing resource use and availability II. *Journal of Wildlife Management*, **70**, 324–36.
Tiebout, H.M., III, and Gary, J.R. (1987). Dynamic spatial ecology of the water snake, *Nerodia sipedon*. *Copeia*, **1987**, 1–18.
Tremblay, Y., Bertrand, S., Henry, R.W., et al. (2000). Analytical approaches to investigating seabird–environment interactions: a review. *Marine Ecology Progress Series*, **391**, 153–63.
Underwood, A.J. (2009). Components of design in ecological field studies. *Annales Zoologici Fennici*, **46**, 93–111.
Whitmore, R.C. (1981). Applied aspects of choosing variables in studies of bird habitats. In D.E. Capen (ed) *The Use of Multivariate Statistics in Studies of Wildlife Habitat*. USDA Forest Service General Technical Report RM 87.Fort Collins, CO: USDA Forest Service, pp. 38–41.
Witherington, B. (2002). Ecology of neonate loggerhead turtles inhabiting lines of downwelling near a Gulf Stream front. *Marine Biology*, **140**, 843–53.
Wood, D.W., Bjorndal, K.A., and Ross, S.T. (2000). Relation of temperature, moisture, salinity, and slope to nest site selection in loggerhead sea turtles. *Copeia*, **2000**, 119–28.

20

Water quality and toxicology

Christine Bishop

20.1 Introduction

Globally, environmental contaminants are considered one of the six top threats to reptile populations (Gibbons et al., 2000). Wildlife, such as reptiles, are also bioindicators of the contamination and health of ecosystems. The presence or absence of species, their populations, or communities and the health of the animals within an ecosystem provide a more sensitive and reliable indicators of environmental conditions than do chemical and physical measurements alone (Hoffman et al., 2003). The health impact of pollutants on all organisms is, however, dependent on the dose and sensitivity of the species (Hoffman et al., 1995). Comparisons in sensitivity between birds and reptiles, to the extent that the same chemicals have been tested on both phyla, indicate reptiles are often as or more sensitive than birds (Weir et al., 2010). However, toxicological comparisons among reptile species are sparse (Gardner and Oberdorster, 2006). Since the number of reptile species and chemicals examined in wild populations or controlled dose–response studies is relatively low, the determination of effects in reptiles usually requires research to detect the level of exposure and the sensitivity of the species on a case by case basis.

Reptiles are well distributed throughout the globe in aquatic and terrestrial ecosystems and in wild, urban, and agricultural landscapes. Therefore, they can be exposed to many types of chemicals throughout their often much extended lifetime. Many reptile species live within relatively small home ranges and have high site fidelity to their territories, and therefore contaminants in their tissues will be indicative of local point sources. Others with extensive home ranges such as larger snakes and/or wide-ranging migratory marine turtles still tend to establish predictable movement patterns throughout their lifetime allowing contaminant exposure to be repeatable and representative of their selected habitats. There are extensive literature reviews and detail on biological endpoints and methods in ecotoxicology including those relating to reptiles, and they are invaluable when designing studies and are highly recommended (Hoffman et al., 1995, 2003; Hose and Guillette, 1995; Sparling et al., 2000, 2010; Campbell and Campbell, 2002; Gardner and Oberdorster, 2006; Mitchelmore et al., 2006; Weir et al., 2010). Here, a brief overview of ecotoxicological methods for measuring pollution in the environment is followed by methods used to date to detect contaminant levels and effects in reptiles.

20.2 Measurement of exposure and interpreting concentrations

Reptiles can be exposed to pollution via ingestion (food, water, soil/sediment), inhalation, and dermal absorption (direct contact or indirect substrate absorption) (Weir et al., 2015). As ectothermic animals, when body temperature rises, metabolism increases and rates of inhalation, dermal absorption, and consumption increase; therefore absorption of contamination may coincidentally rise. Many reptiles are carnivorous, which means they can be readily exposed to chemicals that bioaccumulate in food webs, especially lipophilic organic contaminants. However, reptiles that are herbivorous during all or part of their life may be more likely exposed to chemicals that do not particularly bioconcentrate up the food chain (e.g. trace metals, radionuclides, polyaromatic hydrocarbons (PAHs), most current-use pesticides, fluoride, and selenium) but are nonetheless toxic when exposure occurs through consumption of soil, sediment, invertebrates, and plant tissues. Because many reptiles are omnivorous, measurement of stable isotopes as an indicator of trophic position often provides insight on route of exposure that otherwise would not be available (Van Dyke et al., 2013). If contamination is to be linked to a source, ideally, quantitative data on exposure will compliment levels and effects measured in reptiles both in the wild and in controlled dose–response studies. Study site selection should include reference sites, preferably several since unexpected contamination in reference sites can occur. Limnological characteristics of water and sediment can influence bioavailability of contaminants and therefore measurement of these parameters is useful when also measuring contamination (Sprague, 1985). Factors such as site age, depth of the water body and proximity to point sources, and wind direction can all critically influence contaminant deposition within wildlife habitats.

20.3 Field collections to measure contamination in air, water, sediment, biota

The first question usually asked about toxicological sampling is how best to obtain a representative sample? In environmental media and among tissues within an organism, concentrations of contaminants vary based on many factors. Collection strategy should be based on study design and goals, which define the required data samples will provide. Sampling biota detects the levels accumulated in the animal and is a measure of relative exposure if the compound of interest bioconcentrates in food webs, but not all do so. Therefore, sampling air, water, food, soil, or sediment can reveal exposure where the compound does not necessarily bioconcentrate in the animal. Sampling those media can also confirm environmental contamination on site which can help to trace the source and this is, of course, relevant to regulatory action.

The second question often asked is how much sample mass or volume will be needed to measure contaminant concentrations or a biomarker? Ideally, this is determined initially by the concentration expected or predicted in the environment and hence the media to be analysed. This is dependent on the recovery efficiency and the method detection limit. This tells you how much sample is needed to even be recognized by the instrument being used. Consult with the chemists who will analyse the samples prior to collection.

If previous abiotic or biotic sampling has been done at the site, then those findings serve as a guide. If no previous results are available, literature values for a similar exposure scenario and media for analysis can be helpful. Before extensive sampling and analysis proceeds, it is valuable to conduct a range-finding analysis to determine the highest concentrations present. This means selection of the site or tissue predicted to be the most contaminated, chemical extraction, and then analysis of one or a few pooled samples. Proceeding cautiously can save a lot of time and money. For sampling of tissues or the whole body of an animal, most analytical laboratories prefer a minimum amount for analysis of at least 5 g and, for whole blood, 2 ml or more (or plasma = 1 ml, usually about 50% of the volume of centrifuged whole blood). If that size of sample is not possible due to small body size of the animal, blood samples (or other tissues) from the same site may need to be 'pooled' to create an adequate volume or gram weight for analysis. Detection limits to be expected by current analytical chemistry standards would be 1–10 parts per billion for most organic and inorganic environmental contaminants.

20.3.1 Water, sediment, and biota sample container preparation for trace analyses

Teflon™, glass, or polyethylene sample containers can be used for sampling soil, sediment, biota, or water. For organics, Teflon™ or glass is preferred. Sample bottles should be cleaned with non-phosphate soap and rinsed with hexane and acetone if the samples are for organic chemical analysis, whereas bottles are rinsed with acid if trace metals analysis will be performed (e.g. Al, Cd, Cr, Cu, Ni, Pb, Zn, As, Al, Se) (Laxen and Harrison, 1981). Foil liners (pre-rinsed with hexane and acetone) are applied to the mouth of sample jars to avoid contact with plastic lids. Once collected, samples are kept away from sunlight, on ice in the field, and then transferred to a refrigerator (water) or a −5°C freezer for short-term archiving; they should be kept at −20°C for longer term archive prior to analysis (soil, sediment, biota).

20.3.2 Sediments

Surficial or bottom sediments are usually collected using a grab sediment sampler or core sampler. Multiple grab samples can be collected across a transect in the area where the animals are most likely exposed. Sediment contamination is notoriously patchy in its distribution; therefore, multiple samples analysed individually or as a pooled sample are recommended, especially when range finding to determine the extent of contamination in a site. Recent and/or the most biologically available contamination in core samples are generally in the top 2–5 cm. For organics and trace metals, samples can be pooled and mixed thoroughly into large buckets that have been chemically cleaned or can be lined with polyethylene bags. After thorough mixing in the field, a subsample (e.g. ~500 g) is stored in a Teflon™ jar or an appropriately cleaned glass jar.

20.3.3 Water

Water sampling can be conducted with an outstretched arm or a hand-held sampling device to collect water samples approximately 20 cm below the surface, if the depth allows at a site. This avoids sampling surface water film. Water samples are stored on ice then preserved for chemical analyses. Whole water unfiltered samples used for total phosphorus analysis are preserved with 1 ml of 30% sulphuric acid per 100 ml of sample. For trace metal analysis, water is preserved to avoid adsorption to any metal contamination from the sampling vessel with 1 ml of acid (e.g. 1:1 nitric acid) per 250 ml of sample (Batley and Gardner, 1977). For mercury analysis, the addition of an oxidant or complexing agent is also necessary (Batley and Gardner, 1977). For organic contaminants (e.g. pesticides, brominated fire retardants, dioxin-like chemicals such as polychlorinated biphenyls), an organic solvent is added to the water sample if it is to be stored for days or longer in a refrigerator.

20.3.4 Water and sediment chemistry

Limnological parameters indicative of productivity within the environment are valuable to measure in a toxicological study because they may influence the toxicity of environmental contaminants and/or cause toxic effects (Guillette and Edwards, 2005). Water chemistry can change the bioavailability of many chemicals within the water column and sediments (Sprague, 1985). A typical list of limnological analytes includes: chloride, major ions (Mg, Ca, Na, and K), total phosphorus (TP), nitrate (NO_3-N), nitrite (NO_2-N), ammonia-N (NH_3-N), total Kjeldahl nitrogen (TKN), specific conductivity, turbidity, pH, dissolved inorganic carbon, and dissolved organic carbon. In sediments, total organic carbon (TOC), TKN, NO_3, NO_2, and total phosphorus are usually measured.

20.3.5 Passive integrative samplers for air and water

Passive sampling devices can be used for aqueous and air monitoring (Gangfeng and Pawliszyn, 2007). Passive samplers are left in/on the study site for extended periods of time (days, weeks, months) to facilitate absorption of chemicals from the environment. Rather than a snapshot of contamination derived from a daily sample, the passive sampler integrates exposure over time and reduces cost when fewer samples are necessary. Concentrations of hydrophilic organic contaminants, such as pesticides, pharmaceuticals, and personal care products, can be measured using a passive in-situ sampling device that integratively concentrates trace levels of complex mixtures and enables the determination of their time-weighted average water concentrations. Estimation of the ambient concentrations of chemicals of interest is achieved by using appropriate uptake models and determination of sampling rates for appropriate exposure conditions (Alvarez et al., 2004). Validation indicates that the passive samplers are comparable to concentrations detected in traditional grab water samples (Alvarez et al., 2004). Importantly, the design and quantification of passive sampling devices to be used requires appropriate calibration methods (Gangfeng and Pawliszyn, 2007).

20.4 Measurement of levels and effects of environmental contamination in reptiles

20.4.1 Measurement of contaminants in reptiles

Within species, factors such as size, age, and sex of reptiles influence the contaminant exposure and degree of contamination (Hebert et al., 1993). In carnivorous or omnivorous reptiles, contamination tends to be higher in older, larger reptiles (Hebert et al.,

1993). For lipophilic compounds, males of the same size and age as females will likely be more contaminated because they cannot depurate contamination within lipid-rich yolk of eggs or embryos. Among species that switch from carnivorous to herbivorous diets as adults, the juvenile animals, especially those approaching adulthood, would likely be the most contaminated within the population.

20.4.2 Eggs

Inorganic and organic contaminants will be transferred from the reptilian female to the embryo and/or egg. Because reptile eggs absorb water throughout development, it is also possible that highly hydrophilic compounds can pass from soil to embryo. From one to multiple eggs can be collected from a clutch for analysis. There is little variation in lipid concentrations among eggs in a reptile clutch; therefore, random sampling or collection of just one egg can be generally representative of levels in the clutch. For organics, egg contents are analysed and percentage moisture and lipid should be part of that analysis. Eggs can be stored in a refrigerator. For analysis, eggshells are removed and egg contents placed into a pre-cleaned glass jar. Eggshells can be analysed for trace metals and other elements. Egg contents are preferable for organic contaminant detection due to the high lipid content in eggs. However, when only the eggshells are available, the chorioallantoic membrane can be analysed for a wide variety of contaminants (Cobb et al., 2003).

20.4.3 Internal organs and bone

Where whole animals are available for contaminant analysis, for example, due to road-kill or harvesting, contamination can be measured in the whole body or in any organ of the animal (e.g. liver, brain, gonads, kidney), although contaminants sequester in different tissues depending on the structure of the chemical (Hopkins et al., 2013a; Faust et al., 2014).

Sampling specific organs will more readily detect the chemical of interest and can provide important information about the organ that will experience ill-effects of the contaminant. Organic chemicals, especially chlorinated and brominated compounds, are commonly measured in organs with high lipid contents, such as fat, and liver (Weir et al., 2013), but can be readily measured in whole blood and/or plasma. Hepatocytes have been utilized to examine the cytochrome P4501A system in reptiles in response to halogenated aromatic hydrocarbons through the measure of ethoxyresorufin-O-deethylase activity (Hecker et al., 2006). Because muscle is often very lean in reptiles, organic chemicals do not bioaccumulate at high concentrations in this tissue; however, it is often of interest to analyse muscle due to consumption of reptiles by humans. Bone, kidney, lung, and liver sequester trace metals and radionuclides. At the time of sampling for chemical analysis, collection of a subsample of tissue for many types of biomarkers of effects can also be performed.

20.4.4 Blood and plasma

Many biomarkers can be measured in whole blood or plasma. Samples of reptile blood are often collected with lithium heparin vacutainers from the caudal vein between vertebrae (needle injection approximately two-thirds of the way down the tail along the midline of the ventral surface) (Bishop and Rouse, 2006), but the maximum amount of blood that can be drawn from any individual is 7–10% of blood volume or approximately 1% of total body mass. Clinical parameters, including red blood cell count, haemoglobin, haematocrit, white blood cell count, albumin, alkaline phosphatase, calcium, cholesterol, creatinine, total globulin, iron, magnesium, total protein, and sodium, are a typical blood screen, which provides useful indicators of the health of an animal and can be sensitive to contaminant exposure. Hormones (sexual, vitellogenin, thyroid hormones), vitamins (e.g. vitamin A: retinol, retinyl palmitate), and corticosterone and related compounds can also be measured in blood and/or plasma. Specific indicators of exposure to environmental contaminants such as cholinesterase activity (an indicator of organophosphate or carbamate pesticide exposure; Sanchez-Hernandez, 2003) and δ-aminolevulinic acid dehydratase (ALAD) (an indicator of lead exposure) can be measured in blood.

20.4.5 Scales, claws, tails, scutes

Recently, keratin-based body parts, such as scales, claws, tails, shed skins, and scutes, have been collected and analysed for contaminant levels (e.g. trace metals in tail tips, claws, and shed skins; Hopkins et al., 2013a; Faust et al., 2014), and biomarkers such as corticosterone and stable isotopes have been measured in turtle claws (Baxter-Gilbert et al., 2014).

20.5 Measurement of effects of environmental contaminants on reptiles

There are two general situations that lead to ecotoxicological research and monitoring:

(a) a chemical exposure is suspected and the health of the individual or population is evaluated;
(b) a decline in a species occurrence or the population size or occurrence in a local area is in decline and/or individuals in a population are showing signs of ill health or survival.

Therefore, an investigation of the cause(s) is initiated. Environmental contamination, among many potential factors, is hypothesized to be the cause of those problems.

Ecotoxicological studies of reptiles to determine cause–effect linkages between contamination and biological responses benefit from complimentary studies on wild populations and quantification of the range of concentrations to which reptiles are affected through controlled dose–response studies. Toxicological studies on wild populations should include an evaluation of diseases, parasitic infections and other stress factors that may trigger or combine with toxic chemical exposure to induce health effects in reptiles. A useful online resource for wildlife diseases is the Field Manual of Wildlife Diseases (Franson et al., 2015).

Toxicological endpoints are ideally chosen based on the known mechanism of action of a particular chemical (e.g. neurological effects of mercury, lead, cholinesterase-inhibiting pesticides); however, a specific biomarker for a contaminant(s) or a complex mixture of chemicals is often unknown; therefore, key health indicators and exposure profiles are examined. Aspects of reproduction, growth and morphological development, behaviour, immune or endocrine function, or DNA damage are often used as indicators of the health of individuals. Population size, occurrence and trends, and/or annual survivorship of adults and juveniles are important indicators of population level effects of one or more stressors. A wide variety of biological responses to contaminant exposures have been examined in reptiles, although it remains a relatively young science with no standardized tests recognized by regulatory agencies and relatively few species studied. As research progresses, adapting methods from avian and/or aquatic toxicological investigations in fish or amphibians may also expand the approaches used in reptiles.

The early development of reptiles, with its myriad of cell division and organ, endocrine, sexual, and neurological development, is often sensitive to contaminant exposure (Hose and Guillette, 1995). A standard measure of reproduction is to evaluate productivity as an important and basic biological parameter in all species and is one of the most often examined and important endpoints in toxicology (Hoffman, 2003). Hatching success, infertility, and/or production of healthy juveniles, embryonic (e.g. use of ultrasound: Bishop and Rouse, 2006) and hatchling development, morphological deformities (Bishop et al., 1998), sex ratio, and secondary sexual characteristics (Milnes et al., 2005) are sensitive and important aspects of reproductive fitness that can be measured in artificially incubated eggs (e.g. dioxin-like compounds: Bishop et al., 1998; mercury: Hopkins et al., 2013b) or could be evaluated in wild hatchling reptiles.

As endpoints potentially associated with contaminant exposure, behaviour (Hopkins and Winne, 2006), growth (DuRant et al., 2007), metabolic rates, food consumption, body condition, gonadosomatic index (Hopkins et al., 2002), bone composition (Lind et al., 2004), histological evaluation of tissues such as liver, testes, and immune and endocrine tissues (e.g. testes; Cakici and Akat, 2012) have been measured in crocodilians, snakes, lizards, and hatchling turtles in wild populations. Many have also been measured in lizard and snake species that lend themselves to captivity and dose–response studies (Hopkins et al., 2002; Talent et al., 2002; Mann et al., 2007; Neuman-Lee et al., 2015).

Immune and endocrine responses can be measured in reptiles and are sensitive to a variety of pollutants. Sexual and thyroid hormones, vitellogenin (Rainwater et al., 2008), and vitamin A compounds (e.g. retinol and retinyl palmitate) can be sensitive to contaminants. Corticosterone, a measure of short-term and chronic stress in an organism can be measured in plasma or keratin-based tissues (e.g. Baxter-Gilbert et al., 2014) and can be a sensitive indicator of pollution effects on the hypothalamic–adrenal axis function (Wikelski et al., 2002). The integrity of DNA can also be measured in reptiles (Caliani et al., 2013). In all cases of biomarker measurements, expect high variation among individuals, and variable response among species, gender, season, and even time of day; therefore, maximize sample size and standardize sampling conditions where possible.

20.6 Population level effects of environmental contaminants in reptiles

When ill health of reptiles is detected and a correlation with environmental contaminant exposure occurs, the wildlife toxicologist will eventually be asked: 'is there a population level effect?' Another speculation may be whether the loss of some reptiles in a population, particularly juveniles who have a naturally low annual survivorship (Congdon et al., 1994), due to poisoning by a pollutant is truly cause for concern if the population or some fraction of it appears to persist. Although the unnecessary or unnatural loss of animals and their inherent value should be reason enough to question the cost–benefits of polluting, there are demographic modelling tools (Selcer, 2006) that enable quantification of the impacts of the loss of individuals to the long-term survival of a population. In particular, any factor that can kill or affect reproduction in mature adult reptiles will likely present significant effects on the population survival (Congdon et al., 1994). However, even the effects of chemical exposure on juveniles can have a population level effect (e.g. PCB toxicity to juvenile snapping turtles; Salice et al., 2014), which can only be quantified through the use of integrative demographic modelling. Nonetheless, the robust and credible data generated from a long-term study on population demography and health of a reptile population pre- and post-contaminant exposure (e.g. Wikelski et al., 2002) strengthens the credibility of any modelling approach and the quantification of a contaminant effect within that population.

20.7 Summary

Determination of exposure and detection of biological endpoints in reptiles in the wild combined with dose–response studies are core to answering questions about causation in ecotoxicology. The initial study design and the interpretation of the outcomes should be evaluated within a specific set of criteria. Cormier et al. (2010) summarized the six fundamental characteristics of causation: time order, co-occurrence, preceding causation, sufficiency, interaction, and alteration. The cause precedes the effect (time order). The cause co-occurs with the unaffected entity in space and time (co-occurrence). Causes and their effects are the result of a web of causation (preceding causation). The intensity, frequency, and duration of the cause are adequate and the susceptible entity can exhibit the type and magnitude of the effect (sufficiency). The cause effectively interacts with the entity in a way that induces the effect (interaction). And, the entity is changed by the interactions with the cause (alteration). As these characteristics of evidence-based research in ecotoxicology are fulfilled either within a research or a risk-assessment approach, the more credible a case will be to regulate anthropogenic stressors that threaten reptiles.

References

Alvarez, D.A., Petty J.D., Huckins, J.N., et al. (2004). Development of a passive, in situ, integrative sampler for hydrophilic organic contaminants in aquatic environments. *Environmental Toxicology and Chemistry*, **23**, 1640–8.

Batley, G.E., and Gardner, D. (1977). Sampling and storage of natural waters for trace metal analysis. *Water Research*, **11**, 745–56.

Baxter-Gilbert, J.H., Riley, J.L., Mastromonaco, G.F., et al. (2014). A novel technique to measure chronic levels of corticosterone in turtles living around a major roadway. *Conservation Physiology*, **2**, cou036.

Bishop, C.A., and Rouse J.D. (2006). Polychlorinated biphenyls and organochlorine pesticides in plasma and embryonic development in Lake Erie water snakes (*Nerodia sipedon insularum*) from Pelee Island, Ontario, Canada (1999). *Archives of Environmental Contamination and Toxicology*, **51**, 452–7.

Bishop, C.A., Ng, P., Pettit, K.E., et al. (1998). Environmental contamination and developmental abnormalities in eggs and hatchlings of the common snapping turtle (*Chelydra serpentina serpentina*) from the Great Lakes-St. Lawrence River basin (1989–91). *Environmental Pollution*, **99**, 1–14.

Cakici, O., and Akat, E. (2012). Histopathological effects of carbaryl on testes of snake-eyed lizards, *Ophisops elegans*. *Environmental Science and Pollution Research*, **19**, 64–71.

Caliani, I., Campani, T., Giannetti, M., et al. (2013). First application of comet assay in blood cells of Mediterranean loggerhead sea turtle (*Caretta caretta*). *Marine Environmental Research*, **96**, 68–72.

Campbell, K.R., and Campbell, T.S. (2002). A logical starting point for developing priorities for lizard and snake ecotoxicology: a review of available data. *Environmental Toxicology and Chemistry*, **21**, 894–8.

Cobb, G.P., Bargar, T.A., Pepper, C.B., et al. (2003). Using chorioallantoic membranes for nonlethal assessment of persistent organic pollutant exposure and effect in oviparious wildlife. *Ecotoxicology*, **12**, 31–45.

Congdon, J.D., Dunham, A.E., and Van Loben Sels, R.C. (1994). Demographics of common snapping turtles (*Chelydra serpentina*): implications for conservation and management of long-lived organisms. *American Zoologist*, **34**, 397–408.

Cormier S.M., Suter, G.W., and Norton, S.B. (2010). Causal characteristics for ecoepidemiology. *Human and Ecological Risk Assessment*, **16**, 53–73.

DuRant, S.E., Hopkins, W.A., and Talent, L.G. (2007). Energy acquisition and allocation in an ectothermic predator exposed to a common environmental stressor. *Comparative Biochemistry and Physiology Part C: Toxicology and Pharmacology*, **145**, 442–8.

Faust, D.R., Hooper, M.J., Cobb, G.P., et al. (2014). Inorganic elements in green sea turtles (*Chelonia mydas*): relationships among external and internal tissues. *Environmental Toxicology and Chemistry*, **33**, 2020–7.

Franson, J.C., Friend, M., Gibbs, S.E.J., et al. (eds). (2015). Field manual of wildlife diseases. *U.S. Geological Survey Techniques and Methods*, **15**. Available at: http://pubs.usgs.gov/tm/15/.

Gangfeng, O., and Pawliszyn, J. (2007). Configurations and calibration methods for passive sampling techniques. *Journal of Chromatography A*, **1168**, 226–35.

Gardner, S.C., and Oberdorster, E. (eds). (2006). *Toxicology of Reptiles*. Boca Raton, FL:CRC Press.

Gibbons, J.W., Scott, D.E., Ryan, T.R., et al. (2000). The global decline of reptiles, déjà vu amphibians. *Bioscience*, **50**, 653–66.

Guillette, L.J., and Edwards, T.M. (2005). Is nitrate an ecologically relevant endocrine disruptor in vertebrates? *Integrative and Comparative Biology*, **45**, 19–27.

Hebert, C.E., Glooschenko V., Haffner, G.D., et al. (1993). Organic contaminants in snapping turtle (*Chelydra serpentina*) populations from southern Ontario, Canada. *Archives of Environmental Contamination & Toxicology*, **2492**, 35–43.

Hecker, M., Murphy. M.B., Giesy, J.P., et al. (2006). Induction of cytochrome P4501A in African brown house snake (*Lamprophis fuliginosus*) primary hepatocytes. *Environmental Toxicology and Chemistry*, **25**, 496–502.

Hoffman, D.J. (2003). Wildlife toxicity testing. In D.J. Hoffman, B.A. Rattner, G.A. Burton, et al. (eds) *Handbook of Ecotoxicology*. 2nd ed. Boca Raton, FL: Lewis Publishers/CRC Press, pp. 75–100.

Hoffman, D.J., Rattner, B.A., Burton, G.A., et al. (eds). (1995). *Handbook of Ecotoxicology*. 1st ed. Boca Raton, FL:Lewis Publishers/CRC Press,.

Hoffman, D.J., Rattner, B.A., Burton, G.A., et al. (eds). (2003). *Handbook of Ecotoxicology*. 2nd ed. Boca Raton, FL:Lewis Publishers/CRC Press.

Hopkins, B.C., Hepner, M.J., and Hopkins, W.A. (2013a). Non-destructive techniques for biomonitoring of spatial, temporal, and demographic patterns of mercury bioaccumulation and maternal transfer in turtles. *Environmental Pollution*, **177**, 164–70.

Hopkins, B.C., Wilson, J.D., and Hopkins, W.A. (2013b). Mercury exposure is associated with negative effects on turtle reproduction. *Environmental Science and Technology*, **47**, 2416–22.

Hopkins, W.A., and Winne, C.T. (2006). Influence of body size on swimming performance of four species of neonatal natricine snakes acutely exposed to a cholinesterase-inhibiting pesticide. *Environmental Toxicology and Chemistry*, **25**, 1208–13.

Hopkins, W.A., Roe, J.H., Snodgrass, J.W., et al. (2002). Effects of chronic dietary exposure to trace elements on banded water snakes (*Nerodia fasciata*). *Environmental Toxicology and Chemistry*, **21**, 906–13.

Hose, J.E., and Guillette, L.J. (1995). Defining the role of pollutants in the disruption of reproduction in wildlife. *Environmental Health Perspectives*, **103(Suppl 4)**, 87–91.

Laxen, D.P.H., and Harrison, R.M. (1981). Cleaning methods for polythene containers prior to the determination of trace metals in freshwater samples. *Analytical Chemistry*, **53**, 346–50.

Lind, P.M., Milnes, M.R., Lundberg, R., et al. (2004). Abnormal bone composition in female juvenile American alligators from a pesticide-polluted lake (Lake Apopka, Florida). *Environmental Health Perspectives*, **112**, 359–62.

Mann, R.M., Sánchez-Hernández, J.C., Serra, E.A., et al. (2007). Bioaccumulation of Cd by a European lacertid lizard after chronic exposure to Cd-contaminated food. *Chemosphere*, **68**, 1525–34.

Milnes, M.R., Bryan, T.A., Medina, J.G., et al. (2005). Developmental alterations as a result of in ovo exposure to the pesticide metabolite *p,p'*-DDE in *Alligator mississippiensis. General and Comparative Endocrinology*, **144**, 257–63.

Mitchelmore, C.L., Rowe, C.L., and Place, A.R. (2006). Tools for assessing contaminant exposure and effects in reptiles. In S.C. Gardner and E. Oberdorster (eds) *Toxicology of Reptiles*. Boca Raton, FL:CRC Press, pp. 63–122.

Neuman-Lee, L.A., Carr, J., Vaughn, K., et al. (2015). Physiological effects of polybrominated diphenyl ether (PBDE-47) on pregnant gartersnakes and resulting offspring. *General and Comparative Endocrinology*, **219**, 143–151.

Rainwater, T.R., Selcer, K.W., Nespoli, L.M., et al. (2008). Plasma vitellogenin in Morelet's crocodiles from contaminated habitats in northern Belize. *Environmental Pollution*, **153**, 101–9.

Salice, C.J., Rowe, C.L., and Eisenreich, K.M. (2014). Integrative demographic modeling reveals population level impacts of PCB toxicity to juvenile snapping turtles. *Environmental Pollution*, **184**, 154–60.

Sanchez-Hernandez, J.C. (2003). Evaluating exposure to cholinesterase-inhibiting agrochemicals by serum butyrylcholinesterase activity. *Environmental Toxicology and Chemistry*, **22**, 296–301.

Selcer, K.W. (2006). Reptile ecotoxicology: studying the effects of contaminants on populations. In S.C. Gardner and E. Oberdorster (eds) *Toxicology of Reptiles*. Boca Raton, FL:CRC Press, pp. 267–90.

Sparling, D.W., Linder, G.L., and Bishop, C.A. (eds). (2000). *The Ecotoxicology of Amphibians and Reptiles*. 1st ed. Pensacola, FL:Society of Environmental Toxicology and Chemistry Press.

Sparling, D.W., Linder, G.L., Krest S., et al. (eds). (2010). *The Ecotoxicology of Amphibians and Reptiles*. 2nd ed. Pensacola, FL:Society of Environmental Toxicology and Chemistry Press.

Sprague, J.B. (1985). Factors that modify toxicity. In G.M. Rand and S.R. Petrocelli (eds) *Fundamentals of Aquatic Toxicology: Methods and Applications*. New York: Hemisphere Publishing Corporation, pp. 124–63.

Talent, L.G., Dumont, J.N., Bantle, J.A., et al. (2002). Evaluation of western fence lizards (*Sceloporus occidentals*) and eastern fence lizards (*Sceloporus undulatus*) as laboratory reptile models for toxicological investigations. *Environmental Toxicology and Chemistry*, **21**, 899–905.

Van Dyke, J.U., Hopkins, W.A., and Jackson, B.P. (2013). Influence of relative trophic position and carbon source on selenium bioaccumulation in turtles from a coal fly-ash spill site. *Environmental Pollution*, **182**, 45–52.

Weir, S.M., Dobrovolny, M., Torres, C., et al. (2013). Organochlorine pesticides in squamate reptiles from Southern Arizona, USA. *Bulletin of Environmental Contamination and Toxicology*, **90**, 654–9.

Weir, S.M., Suski, J.G., and Salice, C.J. (2010). Ecological risk of anthropogenic pollutants to reptiles: evaluating assumptions of sensitivity and exposure. *Environmental Pollution*, **158**, 3596–606.

Weir, S.M., Yu, S., Talent, L.G., et al. (2015). Improving reptile ecological risk assessment: oral and dermal toxicity of pesticides to a common lizard species (*Sceloporus occidentalis*). *Environmental Toxicology and Chemistry*, in press.

Wikelski, M., Wong, V., Chevalier, B., et al. (2002). Galapagos Islands: marine iguanas die from trace oil pollution. *Nature*, **417**, 607–8.

21

Richness, diversity, and similarity

C. Kenneth Dodd, Jr.

21.1 Introduction

Determining how many species are present, their abundance, their importance, and how communities are related are critical questions in ecology and conservation biology. Ecologists use measures of diversity and similarity in order to understand community structure and function, particularly in terms of habitat use, food webs, predator–prey relationships, estimating how many species can coexist within a community, energy flow, and nutrient cycling. Conservation biologists need to estimate diversity to identify areas for protection and management (Scott et al., 1987; Snodgrass et al., 2000), and to assess the effects of habitat change through time. Knowledge of these variables also is practical for consultants or others doing rapid assessments, such as in the preparation of environmental impact assessments. Species diversity (richness, heterogeneity, evenness) is fairly simple to estimate, yet the use of these indices in reptile studies is generally less than in many other areas of field research, and many reptile researchers have yet to explore the utility of measures of similarity.

Many indices are used to express various aspects of diversity and similarity; some of the most common are listed in Table 21.1. Krebs (1999), Clarke and Warnick (2001), Magurran (2003), and Magurran and McGill (2011) provide good discussions of some diversity and similarity indices, and these references can serve as a starting point for understanding what the indices do, how they are calculated, and where to find more information about them. Whatever index is chosen, assumptions surrounding the computation must be understood as well as the biases of the index and the way the index should be interpreted. Biologists commonly sacrifice rigor (by violating assumptions) for ease of computation or precedence of use by other biologists. Some of the easiest indices to use (e.g. Margalef) are also some of the most informative, whereas other indices (e.g. Shannon) are popular but not very informative.

21.2 Data transformation

Measures of diversity and similarity give differing weight to abundance, that is, some indices are more sensitive to the presence of rare species, whereas others are more sensitive to the abundance of the more common species. Thus, using the same data

Reptile Ecology and Conservation. Edited by C. Kenneth Dodd, Jr. © Oxford University Press 2016.
Published 2016 by Oxford University Press.

Table 21.1 *Measures of diversity and similarity give differing weight to abundance, that is, some indices are more sensitive to the presence of rare species, whereas others are more sensitive to the abundance of the more common species. Transformation offers a means to validate the statistical assumptions of parametric techniques prior to using them. Transformation normalizes data, and allows for variable weighting in studies of reptile diversity. A second reason for transforming data is that sampling usually involves large numbers of zero captures. Zeros present complications in data analysis, especially when using parametric measures which assume normal distributions. As you move to the right in the table, rare species assume greater importance in the analysis of community similarity/dissimilarity. Reprinted from Dodd (2010).*

No transformation	Square root	Double square root	Log(1 + x)*	Presence/absence
N	√	√√		
0	0	0	0	0
1	1	1	0.3	1
10	3.3	1.7	1	1
100	10	3.3	2	1
1000	33	5.7	3	1
10,000	100	10	4	1

**Because the log(0) = ∞, 1 is added to the value inasmuch as the log(1 + y) = 0 when y = 0.*

set but analysing it with different indices could result in very different profiles of community diversity (Bloom, 1981). Inclusion of rare species in diversity and similarity estimates might indicate a community is far richer than it normally is, especially if the species is migratory or an extremely unusual find. On the other hand, synchronized hatching could heavily weight a reptile community index towards one particular species.

One way to change weighting would be to transform the data. Table 21.2 shows how data transformation can weight the 'importance' of the less abundant or rare species, and thus allow them to have more influence in perceptions of community patterns. Data transformation also helps in clustering and ordination methods. Thus, the objectives of the researcher become very important in the selection of diversity and similarity indices. Knowing the limitations and assumptions of the indices helps avoid confusion and aids in the interpretive process.

A second reason for transforming data is that sampling usually involves large numbers of zero captures, that is, all species are not caught on each day of sampling. Zeros present complications in data analysis, especially when using parametric measures which assume normal distributions. Transformation offers a means to validate the statistical assumptions of parametric techniques prior to using them. One must always remember to verify that the data transformation actually corrects the original issue with the data set. If it does not, transformation is not warranted. An alternative approach would be to avoid parametric statistics, and many of the diversity and similarity indices commonly used are based on non-parametric techniques.

Table 21.2 *Indices and measures of species diversity commonly used in ecological studies. See Magurran (2003) for details of use and computation. Reprinted from Dodd (2010).*

Species richness
Rarefaction
Menhinick (D_{Mn})
Jacknife
Richness (S)
Boostrap
Margalef (D_{Mg})
Species–area curve estimates
McIntosh (U)
Diversity and heterogeneity
α (logarithmic)
Brillouin's index (HB)
λ (log normal)
Fisher's α
Simpson's index (1-D)
Shannon–Wiener function (H')
Equitability
Simpson's (D')
Berger–Parker (d)
Camargo (E')
Hill's ratio (N_1)
Smith and Wilson (E_Q)
Pielou's (E)
Modified Nee
McIntosh (D)

21.3 Species diversity

21.3.1 Sampling considerations

A number of considerations are important to remember if one of the goals of sampling is to develop an estimate of species diversity within an area. First, estimates of diversity are only as accurate as the detection probabilities of finding individual species during the sampling period. This is especially important inasmuch as most species have heterogeneous detection probabilities due to a variety of both natural and observer-related causes (Boulinier et al., 1998). Second, all sampling techniques have biases, as numerous chapters in this volume discuss. Estimates of species diversity will be biased to the extent that sampling techniques are biased. Examples of such biases include:

- Methods. If the techniques used will not detect the species, estimates of diversity within an area will be inaccurate. If the goal is to estimate reptile community diversity, then multiple and appropriate techniques must be employed, especially if rare species are suspected to be in the study area.
- Timing. Sampling must extend throughout the activity period of all the species in question, or at least it must include representative subsamples throughout the

activity period. For example, the timing of sampling has been shown to alter estimates of species richness by changing the shape of the species accumulation curve (e.g. Ouboter et al., 2007). If continuous sampling is not possible, biologists might consider randomization (e.g. by randomizing the start of weekly sampling periods throughout an activity season).

- Location. Estimates of community diversity will only be accurate if all habitats (arboreal, aquatic, terrestrial, fossorial) are sampled. Species diversity estimates should not be extrapolated beyond the area sampled (i.e. the statistical area of inference).
- Stochastic events. Weather conditions (heat, cold, drought, precipitation) and unexpected disturbances (catastrophic storms, earthquakes) all have the potential to influence sampling and thus estimates of diversity. Biologists need to apply caveats, where necessary, to their estimates to acknowledge the limitations of their data.

21.3.2 Species richness

The number of species within a community, species richness, is the simplest way to describe local and regional diversity (Magurran, 2003; Magurran and McGill, 2011). Although species richness is a natural measure of diversity, it is also an elusive quantity to measure accurately (May, 1988; Gotelli and Colwell, 2001), especially if species have different detection probabilities during the sampling period (Durso et al., 2011). The number of species observed is frequently an underestimate of actual species richness if based on counts conducted over limited periods of time. Species richness estimates are very sensitive to the sample size and methods of sampling on which they are based (Smith and van Belle, 1984).

Species richness does not imply anything about abundance, although the term frequently is used inaccurately as a synonym of diversity (see Section 21.3.4). In some situations, species richness can be estimated in advance using field guides or museum specimens, coupled with knowledge of habitat requirements; this method is termed *interpolation*. For example, the number of species found within a forested area of a national park might be predicted from published literature and museum records. Interpolation can be very inaccurate, depending on scale. Knowledge of species richness in one watershed might be useful for predicting species occupancy in an adjacent watershed, but not on a regional scale. Thus, interpolation does not supplant the necessity of field sampling for an accurate understanding of richness since myriad factors may influence occupancy at a particular location.

The number of species within an area results from a complex interaction of resource availability, habitat structure complexity, biogeography, land-use history, and phylogenetic history. Habitats with large numbers of reptile species include arid deserts, island archipelagos, and areas with diverse topography with favourable thermal conditions (Pianka, 1966, 1967; Powney et al., 2010; Meiri, 2014). As might be expected, reptile species richness is inversely correlated with latitude and, usually, elevation (McCain, 2010). Even within worldwide arid desert zones, for example, reptile species richness changes spatially (Pianka, 1986), so sampling considerations become paramount when estimating how many species are present. An estimate of species richness is only as accurate as the reliability of the sampling methods on which the estimate is based.

21.3.3 Species accumulation curves

One measure of species richness in a region is the rate at which new species are added to an inventory (Soberón and Llorente, 1993). If repetitive samples are taken using standardized techniques, the rate at which species are detected and the point at which detection of new species levels off should give an indication of the number of species within an area. Such information is useful when little or nothing is known a priori concerning species richness.

Of course, sampling every species within a habitat may not be temporally or logistically possible, and rarely would all species be captured with 100% confidence within a limited sampling period. Thus, it usually becomes necessary to estimate statistically the number of species within the area of interest. This is done by plotting the increasing number of species captured (on the Y-axis, termed the *ordinate*) by either effort (assuming sampling is regularly spaced through time) or by the cumulative number of individuals observed or captured (on the X-axis, termed the *abscissa*). Effort is defined quantitatively, such as by trap-days, hours searched, or number of pitfalls checked. Thus, an estimate of species richness is derived through cumulative sampling, preferably over a substantial period of time.

Species accumulation curves can be used in two primary ways. First, estimates of the number of reptile species within the sampling area during the time of the survey can be derived (Thompson et al., 2003; Dodd et al., 2007), especially when that number is imprecisely known, such as during an inventory of tropical forest or desert habitats. Second, estimates can be derived of the amount of effort needed to sample a particular area with a preset degree of confidence (usually within 80–90% of the asymptote) (Thompson and Thompson, 2007; Thompson et al., 2007). During the initial phases of sampling, a species accumulation curve will rise steeply, especially in species-rich communities (Cam et al., 2002). When the curve levels off, it is likely that additional sampling will not detect many more species, and thus sampling could be curtailed. This technique helps avoid unnecessary sampling, especially when it is logistically difficult or expensive.

Species richness estimates are of two types, *extrapolative* (inferring species richness based on sub-samples of a data set) and *interpolative* (inferring species richness based on comparisons with other areas or data sets). The former method is widely used, with three distinct classes of statistical approaches (Chazdon et al., 1998): (1) extrapolation of either species accumulation curves or species–area curves to an asymptotic value; (2) fitting data on the relative abundance of species in a single sample to a parametric distribution (e.g. log-series, log-normal, Poisson log-normal); (3) non-parametric estimators.

The failure to detect rare species can dramatically underestimate the true local species richness. If, however, a limited fraction of a specific taxonomic group is sampled quantitatively, sampling bias can be reduced by using statistical extrapolation to estimate species richness (Colwell and Coddington, 1994). Several non-parametric estimators have either been developed specifically for estimating species richness from samples, adapted to do so from mark–recapture applications, or were developed for the general class-estimation problem (Smith and van Belle, 1984; Colwell and Coddington, 1994). These non-parametric estimators only require the number of samples in which each species is found, rather than any parametric information about their abundance (Brose et al., 2003). Some of them can be reduced to a very simple form: $S_{estimated} = S_{observed} + R$, where R is an estimate based on whether rare species are present or undetected in the samples. Overall, non-parametric estimators appear to be less biased and more precise than the other two approaches.

The shapes of species accumulation curves are influenced by both abundance and diversity (Thompson and Withers, 2003). If rare species are present, or if there are few species with high abundance, accumulation curves have low shoulders and long trajectories to the asymptote. Conversely, areas with large numbers of abundant species have steep trajectories and reach asymptotes quickly. Species diversity is positively correlated with the initial slope of the trajectory of the accumulation curve.

Both incidence-based (ICE; i.e. occupancy) and abundance-based (ACE; i.e. incorporating diversity) species accumulation curves can be generated. A computer program (such as Mao Tau or EstimateS; see Section 21.5) might use 1000 iterations of a data set to predict the expected range of shapes of the accumulation curve. This allows biologists to incorporate confidence intervals (95%) around the accumulation curve, which provides a better estimate of the actual numbers of species likely within the area than a simple curve. Examples of species accumulation curves based on intensive pitfall and other sampling techniques for herpetofauna are shown in Figure 21.1.

Several examples of the use of species accumulation curves in reptile field research illustrate their usefulness. G.G. Thompson and colleagues used species accumulation curves to assess species richness of reptiles in the Australian deserts, and to examine the amount of trapping effort necessary to adequately sample the fauna (Thompson and Withers, 2003; Thompson et al., 2003; Thompson and Thompson, 2007; Thompson et al., 2007). Dodd et al. (2007) used species accumulation curves to compare changes in species presence through time as a consequence of habitat changes. They showed that long-term changes in habitat management resulted in decreases in species richness within the amphibian community, but not within the reptile community. Species accumulation curves have even been used in understanding the nature of reptile taxonomic diversity (Pincheira-Donoso et al., 2013).

21.3.4 Heterogeneity

Heterogeneity (also sometimes termed 'species diversity') provides a way of expressing both the number of species and a measure of counts or abundance into a single index. What is more diverse: a community with many rare species or one with fewer but much more abundant species? What if two communities have the same number of species, but at different abundances? In computations, actual (rarely measured) or relative (based on counts) abundances are combined with species richness to measure the heterogeneity of the community. The result is an index which allows the observer to compare the heterogeneity of one or more communities (e.g. Dodd, 1992). An index by itself tells little; it gains value when it is used comparatively.

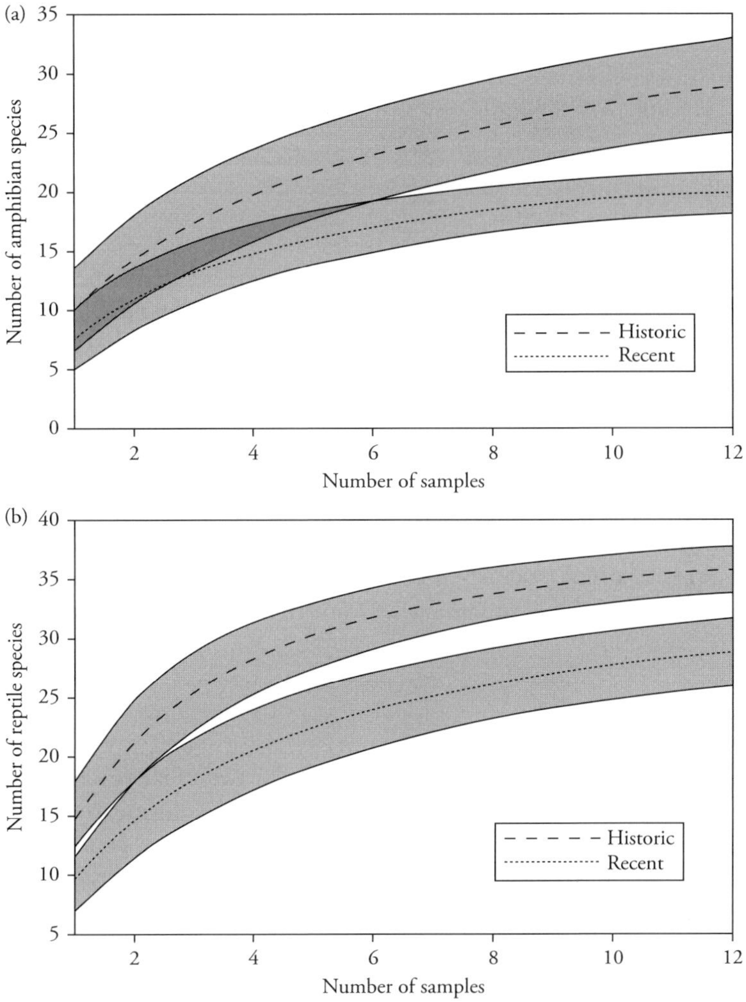

Figure 21.1 Species accumulation curves, with 95% confidence intervals, for amphibians (a) and reptiles (b) sampled at 12 study sites at St. Marks National Wildlife Refuge, Florida, USA. The historic curve shows relationships in 1977–1979, whereas the recent curve shows the relationships in 2002–2005. These species accumulation curves suggested that fewer species of amphibians were expected in the 2000s than in the 1970s. In the 1970s, the curves predicted that 29 species of amphibians might be present throughout the sampling sites, but only 19 in the 2000s. The actual values were 29 (1970s) and 24 (2000s). Reprinted from Dodd et al. (2007).

Reptiles are sampled as one might for other measures of diversity, that is, via a variety of techniques that provide capture histories through time. Abundance is usually and somewhat erroneously expressed in terms of counts taken during the course of a sampling period. Counts and abundance are not the same thing. A count is simply that: the number of individuals captured or observed during the course of a sampling period.

On the other hand, true abundance is estimated by the equation: $\hat{N} = C/\hat{p}$ where $\hat{N}$ is abundance, C is the number of animals counted, and $\hat{p}$ is the probability of detecting an individual. Thus, counts may or may not be a relative index of abundance and, hence, statistical indices based on counts may or may not be accurate. Given the unlikelihood of obtaining true estimates of abundance for all members of a community, counts will have to suffice if diversity indices are used.

21.3.5 Evenness and dominance

Evenness and dominance refer to the influence a species has in numbers on the community, particularly the most abundant species. Evenness is an important variable to measure, since high evenness is generally considered synonymous with high diversity (Magurran, 2003; Magurran and McGill, 2011); the most diverse community is thought of as one that has a large number of species and a large number of individuals uniformly abundant among samples. Like other measures of the proportional abundance of species, dominance and evenness indices incorporate estimates of richness and abundance. The result is a number that can be used to interpret whether a community is dominated by one or more species, or if species are more equitably distributed within the community. The biologist then uses this information to examine the data and identify the species responsible for dominance. Representative evenness/dominance indices are listed in Table 21.1.

21.4 Similarity

Measures of similarity are used to examine data from a number of sampling areas in order to compare community similarity, or more correctly, dissimilarity among those sampling areas. The result is a matrix of numbers representing paired comparisons which can be difficult to interpret without a visual context. However, these matrix-based comparisons can be fitted into a graphic depiction of similarity, aligning those communities or sampling areas most similar to one another into a cluster dendrogram. Cluster dendrograms then can be compared to see how communities change through time and how variables change from one community to another. One of the advantages of using similarity measures is that, depending on the measure, various types of data can be compared, such as species richness, abundance, biomass, or ecological parameters. Thus, these measures can be used to compare species diversity as well as some of the factors that might influence community structure or function.

Distance coefficients are often used to calculate the indices of similarity for some of the most commonly used approaches (e.g. Bray–Curtis, Canberra), and computationally these are actually measures of dissimilarity. Thus, the measure of similarity is the reciprocal of the calculated value. When communities are exactly equal in the variables being compared, $d = 0$; d approaches 1 as the communities become more dissimilar. Distance coefficients are computed in two ways, termed the *Euclidean* and *Manhattan* metrics. Think of walking between two points not in a straight line in a city. Because of buildings and road structure, walking a direct line (*Euclidean*) is not possible, so it becomes necessary to go up and down streets in a somewhat non-direct pathway (*Manhattan*). Using a similar logic, popular metrics such as the Bray–Curtis similarity index use the Manhattan metric to estimate dissimilarity among all the possible comparisons within the data matrix, although other indices employ a somewhat less informative Euclidean metric.

Of course, it helps to know which species are responsible for perceptions of similarity or dissimilarity. The routine (SIMPER) in the statistical program Primer-E can be used to compute the overall percentage contribution that each species (or variable) makes to the dissimilarity between groups in a dendrogram cluster analysis (see Section 21.5). The outcome is a list of species (or variables) in a decreasing order of importance in discriminating between all possible pairs of dissimilarity coefficients within the cluster (Clarke and Gorley, 2006).

Similarity indices are just like any other indices, that is, they have no real significance except comparatively. The indices will not tell the observer why certain communities are more similar or dissimilar to one another, or tell him or her why communities have changed in diversity through time. Instead, they help direct the observer towards the reasons for change, and allow for the development of hypotheses which can be tested using other data, future experimentation, or habitat manipulation. Some of the common similarity indices (see Krebs, 1999; Clarke and Warnick, 2001; Magurran, 2003; Magurran and McGill, 2011) include:

- Jaccard (C)—uses binary data in a simplistic formula ($C_{jk} = p/(p + m)$, where p = the number of species in both communities j and k, and m = the number of species in one community but not the other). This is not a distance measure.
- Morisita (C_λ)—a true similarity index formulated for counts of individuals only.
- Horn—a modification of the Morisita C_λ that uses proportions instead of counts (Horn, 1966). It thus is not affected by sample size, and can be used for variables other than count or abundance data. However, it is not as robust as Morisita.
- Bray–Curtis (B)—a distance measure of dissimilarity, whereby coefficients are weighted towards abundant species, with rare species adding very little to the value. In addition to counts, matrices of catch-per-unit-effort or environmental variables can be analysed for dissimilarity. See Beals (1984) and Clarke et al. (2006).
- Canberra (C)—another distance measure of dissimilarity, but one that is not as influenced by abundant species. This index gives more weight to rare species than the Bray–Curtis Index.
- Other similarity/dissimilarity indices: squared Euclidean distance, percentage similarity, Mountford, Dice, Kendall, simplified Morisita, Sørensen, product moment correlation coefficient, Raup–Crick, and Whittaker.

In an example, Dodd and colleagues conducted field sampling for reptiles at St. Marks National Wildlife Refuge, Florida (USA), 30 years after similar sampling had occurred there; we used the same techniques at exactly the same sites. Based on capture results, we computed Bray–Curtis similarity indices for each community and sampling period, square root transformed the Bray–Curtis indices, and examined cluster dendrograms to visually compare community similarity. Changes in the dendrograms indicated significant changes in community similarity through time (Figure 21.2). We re-examined weather conditions, habitat types, succession stages, and management practices during the two sampling periods. This led us to hypothesize four possible causes, acting interactively, that led to minor changes in species diversity within the reptile community (Dodd et al., 2007).

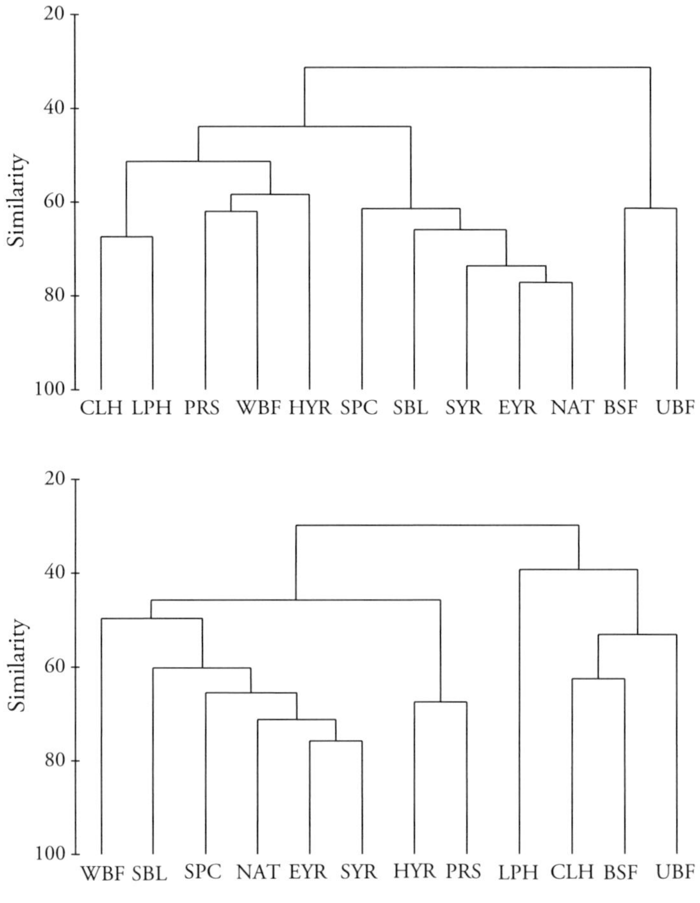

Figure 21.2 Cluster dendrograms based on square-root transformed Bray–Curtis similarity index values showing the relationship among habitat types and the reptile community at 12 study sites (coded on the x-axis) at St. Marks National Wildlife Refuge, Florida, USA. The cluster at top shows relationships in 1977–1979, whereas the cluster at the bottom shows the relationships in 2002–2005. For reptiles, three sites became more similar to one another (BSF, SBL, SPC), three sites (UBF, LPH, WBF) became less similar, and the remainder were essentially unchanged. Unlike the amphibian fauna, there were no discernable patterns among the reptile communities related to geography, major habitat type, or management practice either in the 1970s or the 2000s. Reprinted from Dodd et al. (2007).

21.5 Software

The program EstimateS does most computations required for species accumulation curves and non-parametric analyses of species richness, diversity, and dominance. A detailed description of the estimators computed can be found in Colwell and Coddington (1994), Colwell et al. (2012), and Colwell (2013). EstimateS (version 9.1.0) computes randomized species accumulation curves, statistical estimators of true species richness (*S*) and a statistical estimator of the true number of species shared between pairs of samples, based on species-by-sample (or sample-by-species) incidence or abundance matrices. It can be used to compute an expected species accumulation curve, the Mao Tau (with 95% confidence limits). The Mao Tau is a sample-based rarefaction curve which provides a graphic estimate of expected species accumulation (Colwell et al., 2004). The program also can be used to compute both ICE and ACE coverage estimates of species richness among sampling sites. The derivation and use of these estimators is discussed by Colwell (2013).

EstimateS further allows computation of Fisher's alpha, Shannon, and Simpson diversity indices; the Chao, Jacknife, ICE, ACE and other species richness estimators for abundance and incidence data; modified versions of the Sørensen and Jaccard similarity indices based on abundances, including the effects of unseen shared species (Chao et al., 2005), and classic Jaccard, Bray–Curtis, and Morisita–Horn (both incidence-based and abundance-based) similarity estimators. EstimateS can be downloaded free of charge at: http://viceroy.eeb.uconn.edu/estimates/.

Ecological software to accompany Kreb's *Ecological Methodology* is available from Exeter Software (Setauket, New York, USA) (http://www.exetersoftware.com/cat/ecometh/ecomethodology.html). Version 7.2 sells for US$150 (March 2015). The software covers the topics and analyses discussed in the book, which include various measures of richness, heterogeneity, and equitability, in addition to a wide range of other ecological analyses. These include: binary coefficients, Euclidean distance coefficients, Bray–Curtis metric, Canberra metric, percentage similarity, Morisita's index of similarity, Horn's index, species richness (rarefaction method, jackknife method for counts), logarithmic series, lognormal distribution, Simpson's index of diversity, Shannon–Wiener measure, Brillouin's index of diversity, and evenness measures. The program comes with a manual describing the analyses, and includes some information on the use and theory behind the various indices.

Primer 7 for Windows is another powerful tool for analysis of ecological data, including both diversity and dominance indices. Information on the program can be obtained from the website (http://www.primer-e.com/). Primer 7 allows univariate, graphical, and multivariate analyses of species, abundance, biomass, and physio-chemical data. The program facilitates grouping data into clusters, allows identification of species that are responsible for discrimination among two sample clusters, and graphs species abundance distributions through ordination and multidimensional scaling plots. As of March 2015, costs vary (US$250 for a student to US$1000 for a single-user license for academic research; available from Primer-E, Ivybridge, UK). Primer-E holds advanced workshops; information on forthcoming workshops may be obtained from the website.

EcoSim Professional Version 1.2d allows you to test for community patterns with non-experimental data. EcoSim performs Monte Carlo randomizations to create 'pseudo-communities', then statistically compares the patterns in these randomized communities with those in the real data matrix. These null model tests have applications in both applied and basic ecology. For example, the program can be used to compare lizard species richness and equitability in unaltered versus altered habitats. The cost is US$50 (March 2015). Information on the program is available at http:// garyentsminger.com/ ecosim/index.htm.

21.6 Summary

Indices of diversity and similarity offer a means of critically examining current ecological patterns and changes in community composition, especially when estimates of site occupation across a sufficient number of habitats are not available. In particular, the Bray–Curtis similarity index has proven useful in assessing the effects of habitat changes on herpetofauna and other taxa during monitoring programmes (Ryan et al., 2002; Pawar et al., 2004; Dodd et al., 2007), comparing stream and forest faunas (Parris and McCarthy, 1999; Huang and Hou, 2004), assessing the effects of disturbances (Brown, 2001; Driscoll and Henderson, 2008), measuring the success of restoration efforts (Ruiz-Jean and Aide, 2005), prioritizing areas for conservation (Seymour et al., 2001), assessing dietary differences (Whitfield and Donnelly, 2006), and in analyses of geographic differences in diversity patterns (Urbina-Cardona and Londroño-Murcia, 2003; Menegon and Salvidio, 2005).

An index-based approach offers insights into potential, if not definitive causes of community change. Once potential causes are identified, research can be designed to test hypotheses related to changes in species composition and relative abundance. Community ecology, conservation, and monitoring programmes should incorporate an evaluation of habitat variables into data-collection protocols, and researchers must be aware of the potential importance of stochastic or periodic environmental disturbances, such as storms and flooding, when interpreting species presence/not detected and abundance data. Counting individual animals or determining the percentage of site occupancy neglects much information needed to understand changes in community composition through time. The use of diversity and similarity indices offers further insight into the comparative structure of reptile communities. When combined with landscape and GIS analyses (Chapter 22), they provide important tools in understanding reptile species richness and conservation at multiple spatial scales.

References

Beals, E.W. (1984). Bray–Curtis ordination: an effective strategy for analysis of multivariate ecological data. *Advances in Ecological Research*, **14**, 1–56.

Bloom, S.A. (1981). Similarity indices in community studies: potential pitfalls. *Marine Ecology Progress Series*, **5**, 125–8.

Boulinier, T., Nichols, J.D., Sauer, J.R., et al. (1998). Estimating species richness: the importance of heterogeneity in species detectability. *Ecology*, **79**, 1018–28.

Brose, U., Martinez, N.D., and Williams, R.J. (2003). Estimating species richness: sensitivity to sample coverage and insensitivity to spatial patterns. *Ecology*, **84**, 2364–77.

Brown, G.W. (2001). The influence of habitat disturbance on reptiles in a Box-Ironbark eucalypt forest of south-eastern Australia. *Biodiversity & Conservation*, **10**, 161–76.

Cam, E., Nichols, J.D., Sauer, J.R., et al. (2002). On the estimation of species richness based on the accumulation of previously unrecorded species. *Ecography*, **25**, 102–8.

Chao, A., Chazdon, R.L., Colwell, R.K., et al. (2005). A new statistical approach for assessing similarity of species composition with incidence and abundance data. *Ecology Letters*, **8**, 148–59.

Chazdon, R.L., Colwell, R.K., Denslow, J.S., et al. (1998). Statistical methods for estimating species richness of woody regeneration in primary and secondary rainforests of northeastern Costa Rica. In F. Dallmeier and J.A. Comisky (eds) *Forest Biodiversity Research, Monitoring and Modelling: Conceptual Background and Old World Case Studies*. Smithsonian Institution/Man and Biosphere No. 20. Washington, DC: Smithsonian Institution, pp. 285–309.

Clarke, K.R., and Gorley, R.N. (2006). *PRIMER v6: User Manual/Tutorial*. Plymouth, UK: Primer-E.

Clarke, K.R., and Warnick, R.M. (2001). *Change in Marine Communities: An Approach to Statistical Analysis and Interpretation*. 2nd ed. Plymouth, UK: Primer-E Ltd.

Clarke, K.R., Somerfield, P.J., and Chapman, M.G. (2006). On resemblance measures for ecological studies, including taxonomic dissimilarities and a zero-adjusted Bray–Curtis coefficient for denuded assemblages. *Journal of Experimental Marine Biology and Ecology*, **330**, 55–80.

Colwell, R.K. (2013). EstimateS: statistical estimation of species richness and shared species from samples. Version 9. User's guide and application. Available at: http://purl.oclc.org/estimates.

Colwell, R.K., and Coddington, J.A. (1994). Estimating terrestrial biodiversity through extrapolation. *Philosophical Transactions of the Royal Society of London B: Biological Sciences*, **345**, 101–18.

Colwell, R.K., Chao, A., Gotelli, N.J., et al. (2012). Models and estimators linking individual-based and sample-based rarefaction, extrapolation and comparison of assemblages. *Journal of Plant Ecology*, **5**, 3–21.

Colwell, R.K., Mao, C.X., and Chang, J. (2004). Interpolating, extrapolating, and comparing incidence-based species accumulation curves. *Ecology*, **85**, 2717–27.

Dodd, C.K., Jr. (1992). Biological diversity of a temporary pond herpetofauna in north Florida sandhills. *Biodiversity and Conservation*, **1**, 125–42.

Dodd, C.K., Jr. (ed). (2010). *Amphibian Ecology and Conservation: A Handbook of Techniques*. Oxford: Oxford University Press.

Dodd, C.K., Jr., Barichivich, W.J., Johnson, S.A., et al. (2007). Changes in a northwestern Florida Gulf Coast herpetofaunal community over a 28-y period. *American Midland Naturalist*, **158**, 29–48.

Driscoll, D.A., and Henderson, M.K. (2008). How many common reptile species are fire specialists? A replicated natural experiment highlights the predictive weakness of a fire succession model. *Biological Conservation*, **141**, 460–71.

Durso, A.M., Willson, J.D., and Winne, C.T. (2011). Needles in haystacks: estimating detection probability and occupancy of rare and cryptic snakes. *Biological Conservation*, **144**, 1508–15.

Gotelli, N.J., and Colwell, R.K. (2001). Quantifying biodiversity: procedures and pitfalls in the measurement and comparison of species richness. *Ecology Letters*, **4**, 379–91.

Horn, H.S. (1966). Measurement of overlap in comparative ecological studies. *American Naturalist*, **100**, 419–24.

Huang, C.-Y., and Hou, P.-C.L. (2004). Density and diversity of litter amphibians in a monsoon forest of southern Taiwan. *Zoological Studies*, **43**, 795–802.

Krebs, C.J. (1999). *Ecological Methodology*. 2nd ed. Menlo Park, CA: Benjamin-Cummings.

Magurran, A.E. (2003). *Measuring Biological Diversity*. Oxford: Blackwell Publishers.

Magurran, A.E., and McGill, B.J. (2011). *Biological Diversity: Frontiers in Measurement and Assessment*. Oxford: Oxford University Press.

May, R.M. (1988). How many species are there on Earth? *Science*, **241**, 1441–9.

McCain, C.M. (2010). Global analysis of reptile elevational diversity. *Global Ecology and Biogeography*, **19**, 541–53.

Meiri, S. (2014). Lizard and amphisbaenians species richness. Available at: http://shaimeirilab.weebly.com/research-interests.html (accessed 9 March 2015).

Menegon, M., and Salvidio, S. (2005). Amphibian and reptile diversity in the southern Udzungwa Scarp Forest Reserve, south-eastern Tanzania. In B.A. Huber, B.J. Sinclair, and K.-H. Lampe (eds). *African Biodiversity. Molecules, Organisms, Ecosystems*. New York: Springer, pp. 205–12.

Ouboter, P.E., Jairam, R., and Wan Tong You, K. (2007). Additional records of amphibians and reptiles from Nassau Mountain, Suriname. In L.E. Alonso and J.H. Mol (eds) *A Rapid Biological Assessment of the Lely and Nassau Plateaus, Suriname (with additional information on the Brownsberg Plateau)*. RAP Bulletin of Biological Assessment 43, pp. 126–9.

Parris, K.M., and McCarthy, M.A. (1999). What influences the structure of frog assemblages at forest streams? *Australian Journal of Ecology*, **24**, 495–502.

Pawar, S.S., Rawat, G.S., and Choudhury, B.C. (2004). Recovery of frog and lizard communities following primary habitat alteration in Mizoram, northeast India. *BMC Ecology*, **4**(1), 10.

Pianka, E.R. (1966). Convexity, desert lizards and spatial heterogeneity. *Ecology*, **47**, 1055–9.

Pianka, E.R. (1967). On lizard species diversity—North American flatland deserts. *Ecology*, **48**, 333–51.

Pianka, E.R. (1986). *Ecology and Natural History of Desert Lizards: Analyses of the Ecological Niche and Community Structure*. Princeton, NJ:Princeton University Press.

Pincheira-Donoso, D., Bauer, A.M., Meiri, S., et al. (2013). Global taxonomic diversity of living reptiles. *PLoS One*, **8**, e59741.

Powney, G.D., Grenyer, R., Orme, C.D.L., et al. (2010). Hot, dry and different: Australian lizard richness is unlike that of mammals, amphibians and birds. *Global Ecology and Biogeography*, **19**, 386–96.

Ruiz-Jean, M.C., and Aide, T.M. (2005). Restoration success: how is it measured? *Restoration Ecology*, **13**, 569–77.

Ryan, T.J., Philippi, T., Leiden, Y.A., et al. (2002). Monitoring herpetofauna in a managed forest landscape: effects of habitat types and census techniques. *Forest Ecology and Management*, **167**, 83–90.

Scott, J.M., Csuti, B., Jacobi, J.D., et al. (1987). Species richness. A geographic approach to protecting future biological diversity. *Bioscience*, **37**, 782–8.

Seymour, C.L., De Klerk, H.M., Channing, A., et al. (2001). The biogeography of the Anura of sub-equatorial Africa and the prioritisation of areas for their conservation. *Biodiversity and Conservation*, **10**, 2045–76.

Smith, E.P., and van Belle, G. (1984). Nonparametric estimation of species richness. *Biometrics*, **40**, 119–29.

Snodgrass, J.W., Komorosski, M.J., Bryan, L., Jr., et al. (2000). Relationships among isolated wetland size, hydroperiod, and amphibian species richness: implications for wetland regulation. *Conservation Biology*, **14**, 414–19.

Soberón, M.J., and Llorente, B.J. (1993). The use of species accumulation functions for the prediction of species richness. *Conservation Biology*, 7, 480–8.

Thompson, G.G., and Thompson, S.A. (2007). Using species accumulation curves to estimate trapping effort in fauna surveys and species richness. *Austral Ecology*, **32**, 564–9.

Thompson, G.G., and Withers, P.C. (2003). Effect of species richness and relative abundance on the shape of the species accumulation curve. *Austral Ecology*, **28**, 355–60.

Thompson, G.G., Thompson, S.A., Withers, P.C., et al. (2007). Determining adequate trapping effort and species richness using species accumulation curves for environmental impact assessments. *Austral Ecology*, **32**, 570–80.

Thompson, G.G., Withers, P.C., Pianka, E.R., et al. (2003). Assessing biodiversity with species accumulation curves; inventories of small reptiles by pit- trapping in Western Australia. *Austral Ecology*, **28**, 361–83.

Urbina-Cardona, J.N. and Londoño-Murcia, M.C. (2003). Distribución de la comunidad de herpetofauna asociada a cuatro áreas con diferente grado de perturbación en la Isla Gorgona, Pacífico Colombiano. *Revista de la Academia Colombiana de Cienias Exactas y Fisicas Naturales*, 27, 105–13.

Whitfield, S.M. and Donnelly, M.A. (2006). Ontogenetic and seasonal variation in the diets of a Costa Rican leaf-litter herpetofauna. *Journal of Tropical Ecology*, **22**, 409–17.

22

Landscape ecology, biogeography, and GIS methods

Monika Böhm and Viorel D. Popescu

22.1 Introduction

22.1.1 Landscape ecology, biogeography, and macroecology

Landscape ecology examines 'the effects of the spatial configuration of mosaics on a wide variety of ecological phenomena' (Wiens et al., 1993). Landscape composition and configuration across space have wide-ranging effects on species. It determines where the right climatic, elevation, or soil conditions occur to suit the physiological requirements of a species (Kearney and Porter, 2004). It also affects where a species can feed, breed, and how they can avoid mortality from predators or inter-species competition. In its simplest form, landscape ecology aims to examine the distribution of habitat and its effects on ecological processes (Lindenmayer et al., 2008).

Because habitat loss is the overriding cause of biodiversity loss, including in reptiles (Böhm et al., 2013), knowledge of habitat distribution across space, as well as changes through time, are essential to management and conservation initiatives. While landscape ecology research is often species- or landscape-specific, generalizing patterns across landscapes and species is another important field gaining momentum in ecology and conservation. Biogeography and macroecology analyse patterns between species (e.g. species richness, range size, threat status) and the environment over broad spatial (e.g. regional, continental, global) or temporal scales (e.g. evolutionary timescales).

This broad-scale view—as is also the case with landscape ecology—results from the realization that looking at small-scale processes alone often fails to fully explain observed patterns in the abundance or distribution of species. The aim of broad-scale analyses is to find generalizations across larger spatial or temporal scales, a critical perspective in conservation, since it is often impossible to study all landscapes and species to the detail required for their effective conservation. Other threats, especially climate change, are likely to exacerbate landscape and ecosystem changes (Thomas et al., 2004). Thus, general conclusions from broadly observed patterns are often the primary focus of global conservation policy and decision-making, and can help steer conservation planning towards the most vulnerable species, landscapes, or ecosystems in the face of environmental change. In contrast, insights from landscape ecology studies focused on specific regions, species, or communities are critical for informing management or

Published 2016 by Oxford University Press.

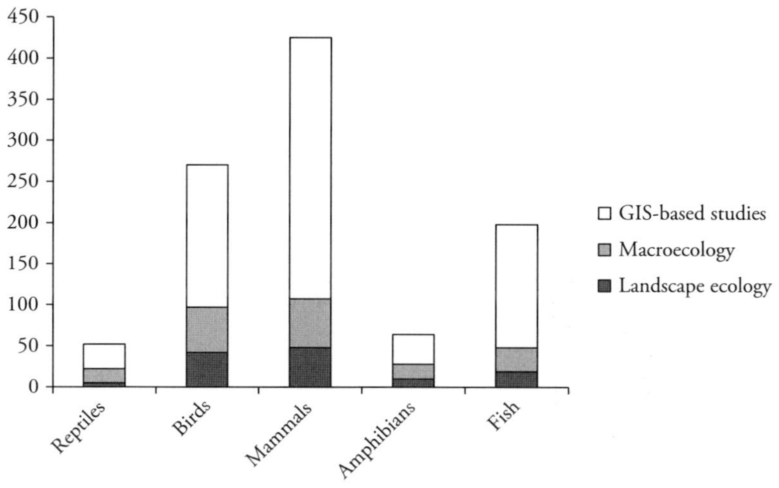

Figure 22.1 Number of journal articles published on the topics of landscape ecology, macroecology and GIS-based studies, by vertebrate taxon group. (Source: Thomson Reuters Web of Science; accessed 25 April 2015).

conservation decisions at local and regional scales (e.g. habitat restoration or population augmentation).

Reptiles are still scarcely represented in landscape ecology, biogeography, and macroecology compared to other vertebrate taxa (Figure 22.1). Yet technological advances have brought about a wealth of spatial data, from locality data taken by global positioning systems (GPS) to high-resolution satellite imagery and aerial photography. Faster and more powerful computers are able to handle complex spatial analyses and store large datasets. Software developments for spatial analyses, that is, geographic information systems (GIS), have produced a large suite of tools to manipulate and analyse data. Given these developments, we can become more spatially explicit in our problem-solving: why does a species occur in one place, but not another? Which environmental conditions are important to a species? What are the hotspots of species richness? Where should we focus protected areas and conservation funding?

In this chapter, we introduce recent developments in GIS, landscape ecology, macroecology, and biogeography, and list important sources of data and applications that help to tackle complex biological and ecological questions spanning many spatial and temporal scales.

22.1.2 Geographic information systems (GIS)

A GIS is a family of software that allows us to visualize, store, manipulate, analyse, and model spatial data (i.e. georeferenced data). Spatial data come in vector or raster format. Vector data include point data, lines, and polygons (e.g. coordinates, transect lines, or habitat ranges, respectively; Figure 22.2). Vector data are associated with additional data attributes, which provide information such as the number of individuals

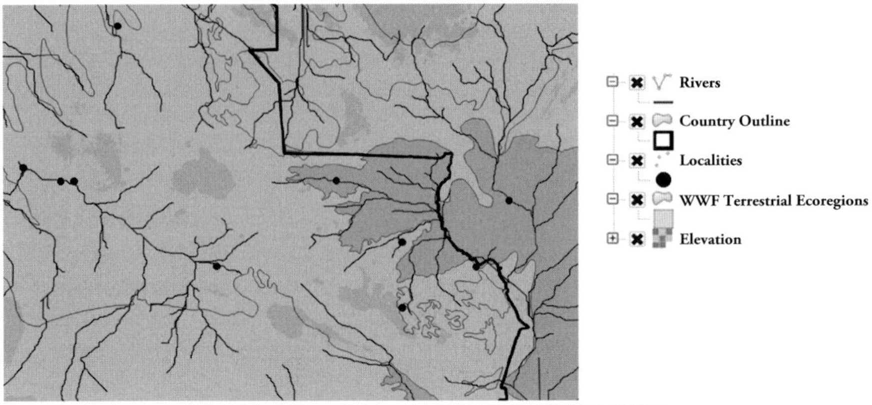

	id	Date	Recorder ID	No. ind
0	10	2010-11-24	2	1
1	9	2010-11-22	2	2
2	8	2008-11-10	3	2
3	7	2008-10-04	3	9
4	6	2011-11-23	2	7
5	5	2011-12-01	2	8
6	4	2011-11-05	2	7
7	3	2010-11-11	2	1
8	2	2011-11-01	2	7
9	1	2012-12-10	1	5
10	15	2014-02-04	1	2
11	14	2014-02-01	1	5
12	13	2013-01-05	1	8
13	12	2013-01-04	1	9
14	11	2013-01-04	1	3

Figure 22.2 Different types of vector data (points, lines, polygons) and raster data overlaid onto each other, showing locality point data (black dots) for an imaginary species collected along rivers (line vector data) in the border region of Brazil and Bolivia (thick black line is the polygon outline, with each country being represented by a polygon). The background shading depicts elevation (raster data), while the grey lines are ecoregion boundaries (again a polygon layer). For example, the ecoregion shaded in darker grey on the right is the Pantanal. See Table 22.1 for commonly used data sources available online. An example of an attribute table shows data collected for each species locality data point. In this example, each row represents a single point.

sampled at a point locality, the name of a river or a road displayed as a line, or the type of habitat represented by a polygon.

Rasters are continuous matrices of grid cells, with each cell containing a single value summarizing the landscape feature it represents (e.g. mean elevation, or a code defining the prevalent habitat type in the grid cell, such as 1 for tropical rain forest, 2 for agricultural lands). The spatial resolution of a raster is reflected in its grid cell size: finer grids with smaller grid cells (e.g. 1–100 m^2) capture a high degree of spatial complexity and detail, while coarser grids, with larger cells (e.g. 1–100 km^2) provide a more generalized view of the landscape, at the cost of losing detail. Unlike vectors, rasters do not represent the exact boundaries of a spatial object, but their continuous nature allows us to carry out mathematical operations on cell values and model surfaces across space.

Table 22.1 *Global and regional open-source data layers and providers to get started with spatial analysis at the landscape and macroecological level. This is by no means an exhaustive list and web searches are likely to uncover many more data sources and GIS resources.*

Data type	Variables/layers	Resolution	Timeframe	URL
Geographical	World maps and country outlines	NA	NA	https://www.freevectormaps.com/world-maps
	Hydrological variables	3 arcseconds 15 arcseconds 30 arcseconds 5 arcminutes		http://hydrosheds.cr.usgs.gov/hydro.php
	Protected areas	NA	Continuously updated	http://www.protectedplanet.net/
Climatic	Temperature (monthly min, max, mean) Precipitation Bioclimatic variables	10 arcminutes 5 arcminutes 2.5 arcminutes 30 arcseconds	Past, current, future	http://www.worldclim.org/
	Climatic observations and scenarios	Various	Various	http://www.ipcc-data.org/
	Atmospheric and oceanic variables	Various	Past, current, future	http://www.ncdc.noaa.gov/
Topological	Altitude	30 arcseconds 10 arcminutes 5 arcminutes 2.5 arcminutes 30 arcseconds	NA	http://earthexplorer.usgs.gov/ http://www.worldclim.org/
Biological	Net primary productivity (NPP)	0.25 decimal degrees	1995	http://sedac.ciesin.columbia.edu
	Global Mangrove Forest Distributions	30m	2000	http://sedac.ciesin.columbia.edu
	Ecoregions: terrestrial, freshwater, marine	NA	2001	https://www.worldwildlife.org/publications/terrestrial-ecoregions-of-the-world
			2013	http://www.feow.org/downloads
			2007	http://www.worldwildlife.org/publications/marine-ecoregions-of-the-world-a-bioregionalization-of-coastal-and-shelf-areas

continued

Table 22.1 *Continued*

Data type	Variables/layers	Resolution	Timeframe	URL
	Land cover	30 arcseconds (~1 km)	1992—2000 consensus	http://www.earthenv.org/landcover.html
		1 km	1992/1993	http://edc2.usgs.gov/glcc/glcc.php
		1 km	2000	http://www.eea.europa.eu/data-and-maps/data/global-land-cover-2000-europe
		500m	2001–2012	https://lpdaac.usgs.gov/dataset_discovery/modis/modis_products_table/mcd12q1
		300m	2004–2006	http://due.esrin.esa.int/page_globcover.php
Pressures and threats	Human population density		1990, 1995, 2000	http://sedac.ciesin.columbia.edu
	Accessibility	30 arcseconds	2000	http://forobs.jrc.ec.europa.eu/products/gam/
	Human appropriation of NPP (HANPP)	0.25 decimal degrees	1995	http://sedac.ciesin.columbia.edu
	Tree cover and change	30 metres	2000-present	http://data.globalforestwatch.org/
	Millennium Ecosystem scenarios	Various	1995–2100	http://sedac.ciesin.columbia.edu/data/collection/ma
Species data	Species distributions	NA	updated periodically	http://www.iucnredlist.org/technical-documents/spatial-data
			2015	http://www.gardinitiative.org/

Both raster and vector data relevant to ecology and conservation have become widely available and are, in many cases, open-source (see Sillero and Tarroso, 2010). Similarly, there is a wide choice of GIS packages that allow these data to be stored, visualized, manipulated, and analysed, often featuring graphical user interfaces to facilitate software use. While prices for commercial packages vary depending on the licenses acquired and functionalities included, there is an ever-increasing number of open-source GIS software available (Table 22.2). Many of these allow users to develop their own functionalities that, in turn, may become available open-source (e.g. Quantum GIS and its plugin repository at http://plugins.qgis.org/plugins/). Additionally, tools to aid spatial data visualization and analysis have also been developed for other software environments, most prominently R, a freely available environment for statistical computing (http://www.r-project.org/index.html). However, R may require the writing of scripts, and some understanding of programming languages is required.

22.2 Landscape ecology concepts applied to reptile ecology and conservation

22.2.1 Landscape composition and configuration

Landscapes can be perceived as mosaics of habitats with varying degrees of heterogeneity in their composition or configuration (e.g. continuous boreal forest with little variation in tree species composition vs. rural landscapes with many native and disturbed habitats). Landscapes can also be defined more simply as patches of suitable habitat within a matrix of less suitable or unsuitable habitat. Habitat suitability varies across species, but it may also vary within species, for example, with developmental stage, such as between juveniles and adults (e.g. sand lizards using microhabitats differently depending on age group; Stellatelli et al., 2013). The size and quality of available habitat patches in the landscape are intrinsically linked to species conservation as they affect population densities and persistence, and extinction risk (Hanski, 1999; Lindenmayer et al., 2008). GIS can help delineate habitat patches, evaluate their size, shape, and connectivity, and, in doing so, aid conservation efforts. For example, rocky outcrops are vital habitats for species such as the New Zealand Grand Skink (*Oligosoma grande*), a species of conservation concern (Gebauer et al., 2013; Harris et al., 2014). Recent studies defined the occurrence of such outcrops from aerial photography and GPS-captured occurrence records from field studies.

22.2.2 Structural and functional connectivity

Connectivity is the degree to which the landscape facilitates or impedes movement among habitat patches. Landscape connectivity can be structural/physical, defined by the spatial arrangement of patches, as well as functional, defined by the likelihood of movement of individuals among patches. Assessing the connectivity between habitat patches and the type and quality of the matrix of non-habitat is an important consideration in ecological and conservation studies of reptiles. GIS can be used to map and identify corridors connecting high suitability habitat patches or non-habitat matrices of

Table 22.2 *Popular commercial and open-source GIS software packages. Also consult http://freegis.org and http://opensourcegis.org.*

	Package	Developer	URL	Capabilities
Commercial	ArcGIS	ESRI	http://www.esri.com	Complete GIS: spatial analysis and modelling
	IDRISI	Clark Labs, Clark University	http://www.clarklabs.org	Raster-based GIS and image processing
	MapInfo	PB MapInfo Corporation	http://www.mapinfo.com	Complete GIS: spatial analysis and modelling
	Manifold	CDA International	http://www.manifold.net	Database and mapping functionality
Free and open-source software/environments	Quantum GIS	GNU Project	http://www.qgis.org	Complete GIS with increasingly complex plugins becoming available; allows access to GRASS tools
	GRASS GIS	Open Source Geospatial Foundation	http://grass.osgeo.org	Complete GIS comparable to commercial packages
	DIVA-GIS	R.J. Hijmans et al.	http://www.diva-gis.org	Raster-based climate modelling
	OpenJUMP GIS	The JUMP Project	http://www.openjump.org/	Java-based vector GIS
	MapServer	University of Minnesota	http://mapserver.org/	Spatially enabled Internet applications
	ArcExplorer	ESRI	http://www.esri.com	Viewer with display and query capabilities
	R	The R Project	http://www.r-project.org	Statistical environment which allows increasingly GIS functionality through contributed packages

varying quality. In its simplest form, we can estimate physical and structural connectivity using Euclidean (or straight-line) distance between patches. A simple measure of connectivity was proposed by Hanski (1999) and forms the basis of metapopulation theory—the dynamics of populations arranged in distinct habitat patches within a non-habitat matrix. Here,

$$connectivity = \sum_{i \neq j} e^{-d_{ij}} A$$

where d is the distance between patch i and patch j and A_j is the carrying capacity of patch j. Thus, this index takes into account distance between patches as well as patch size. Such an index may work well for measuring connectivity between populations or subpopulations confined to distinct landscape features (e.g. pools of water, discrete rocky outcrops).

Simple connectivity measures assume that the non-habitat matrix has no effect on the movement of individuals between patches. In reality, permeability of non-habitat is likely to vary across space, based on prevailing habitat features affecting the ability of animals to migrate and disperse; therefore, it is necessary to define characteristics of the landscape that facilitate or oppose dispersal across space (e.g. turtle population structure in relation to roads; Patrick and Gibbs, 2010). One approach is to develop connectivity measures specific to the species of interest, because the way in which species perceive the environment may differ dramatically based on features such as body size, crypsis, or thermal suitability. Consequently, there is no single connectivity index to choose from, but a multitude reflecting the environment and species in question.

The permeability of the landscape to species movements can be assessed using least-cost path analysis. Least-cost path analysis calculates a cost surface based on habitat qualities that impede or facilitate movement of a species (e.g. elevation, high UV, or rivers): the lower the cost, the more likely it is for a species to disperse along this path. Cost surfaces do not take into account other landscape features important to a species, such as habitat patch size. In a study on Florida Scrub Lizards (*Sceloporus woodi*), least-cost surfaces were generated by classifying habitat types relative to the movement abilities of the lizards, an approach that was a better predictor of genetic variation in the lizards than simple Euclidean distance (Hokit et al., 2010). Similarly, least-cost surfaces have been used to identify road mortality hotspots for four species of turtles in New York State, and inform mitigation strategies (Patrick et al., 2012).

Latest developments in evaluating functional connectivity of landscapes draw on network analysis, a branch of graph theory which analyses flow and connectivity. In the case of landscapes, a network consists of discrete habitat patches ('nodes') connected via links along which dispersal or gene flow occurs. This approach has been applied to New Zealand Grand Skinks (O. grande), for assessing the effects of reductions in vegetation matrix quality on connectivity and thus metapopulation dynamics (Harris et al., 2014).

22.2.3 Landscape thresholds and conservation management decision-making

Landscape thresholds, which combine aspects of landscape configuration, composition, and connectivity, have become an important tool for defining critical thresholds in resource distribution that would entail significant ecological responses of species. Specifically, habitat loss may reach certain levels, or thresholds, beyond which species occupancy may be compromised due to changes in structural and functional connectivity. Thresholds may be examined at the level of individual species (Betts and Villard, 2009) or communities, that is, through species richness (Radford et al., 2005). Landscape thresholds have been identified for some amphibian species in response to anthropogenic habitat alteration (reviewed in Popescu and Gibbs, 2010), but have not been widely adopted in reptile studies. Because they have the potential to offer specific management recommendations (i.e. retain a specific percentage of a forest habitat within a certain radius from rocky outcrops to ensure population persistence), research linking habitat change to population response deserves further exploration.

22.2.4 Edge effects

An important consideration in landscape ecology, in addition to patch size, patch quality and connectivity, is the 'edge effect'. Edges encompass biotic and abiotic differences, in comparison to core habitat, due to the interaction of two habitat types (Murcia, 1995), and often have different environmental conditions, such as temperature or moisture, to which reptiles are particularly sensitive (Lehtinen et al., 2003). The effects of edges on reptiles have proven to be variable, being found to affect community structure in some studies (Lehtinen et al., 2003), but not in others (Dixo and Martins, 2008). Edges can be identified in GIS based on habitat boundaries, and edge effects can be investigated, for example, by comparing habitat patch size with the length of habitat edges (e.g. the larger the ratio of edge to patch area, the larger the edge effect); software such as Fragstats (see Table 22.3) can compute a range of such landscape metrics.

22.3 GIS for species conservation

22.3.1 Modelling and mapping species distributions

Knowledge of factors which correlate with species presence or abundance is important for defining distribution patterns of species, which consequently influence conservation and management decision. Reptile occurrence and abundance are strongly influenced by vegetation type and structure, soil type, climate, and other environmental factors, the effects of which have often been found to override the influence of habitat patch size and shape (Jellinek et al., 2004; Schutz and Driscoll, 2008).

The emergence of GIS has greatly enhanced our ability to quantitatively describe environmental factors with which a species associates and predict species occurrence and abundance. For example, for known occurrence points of a species, we can easily extract climatic variables, elevation, habitat type, or soil type (Kearney and Porter, 2004). Given the many data gaps that persist in our knowledge of reptile distributions, we can use these correlative variables to predict where species may occur in space or, given scenarios of climate change or land use change, where they may occur in the future; these are ecological niche modelling exercises, which can be addressed using methods such as ensemble species distribution modelling (Araújo and New, 2007). For example, Raxworthy et al. (2003) related known occurrence records for Madagascan chameleon species to a suite of spatial data layers describing the ecological landscape, including land cover, a range of variables on precipitation, temperature and cloud cover, and topographical data (e.g. elevation, slope, aspect, flow accumulation and direction). This approach provided informative distribution data for the species under study and offers an innovative way for discovering unknown distributional areas of species.

Species distribution and locality data also aid conservation assessments (e.g. IUCN Red List of Threatened Species), and conservation decision-making. For many smaller-scale landscape studies, these data are often collected during field observations. However, for larger-scale studies, species locality data have traditionally come from georeferenced literature records and museum specimens via online repositories. The availability of large data repositories of species occurrence records, such as the Global Biodiversity Information Facility (GBIF; http://www.gbif.org/), iNaturalist (http://www.inaturalist.org/), or georeferenced photo records (e.g. Flickr (https://www.flickr.com) or Picasa (https://picasaweb.google.com)), has increased our ability to access and share locality information of species. For example, GBIF records were recently combined with species occurrences published in the literature and expert data to produce an updated atlas of European reptiles and amphibians (Sillero et al., 2014). However, care should be taken when using these data due to quality issues that may affect the accuracy of resulting distribution maps (see Section 22.5).

Additionally, spatial tools are increasingly being developed to aid species distribution mapping for conservation. For example, GeoCAT, developed by the Royal Botanic Gardens at Kew, allows users to upload locality data from their own records as well as online databases, calculate range-based measurements for IUCN Red List assessments, and allow output of locality records for further analysis or sharing with collaborators (http://geocat.kew.org; Bachman et al., 2011).

22.3.2 Landscape ecology for reptile conservation

Knowledge of species distributions and habitat associations is vital to determine the most appropriate conservation and management actions. Many studies rely heavily on remote sensing and GIS to determine habitat suitability. For example, GIS has been used to determine the most suitable reintroduction sites for species of conservation concern, or to determine sites for assisted translocations (Dade et al., 2014). Although there is much controversy about assisted translocations due to the dangers of introducing species to new environments, some species with poor dispersal ability may rely on this approach for survival, specifically under scenarios of climate change. For example, a composite index of habitat suitability was created to map habitat for the critically endangered Western Swamp Tortoise (*Pseudemydura umbrina*), thus facilitating the conservation decision-making process with the use of spatially explicit data and GIS (Dade et al., 2014).

Landscape genetics combines landscape ecology with population genetics, investigating the effects of global change on evolutionary processes, patterns of genetic diversity and gene flow (Manel and Holderegger, 2013). In essence, landscape genetics correlates spatial heterogeneity of landscapes with gene flow, using a number of methodological approaches, such as Mantel tests, resistance surfaces, and network theory. Mantel tests, for example, relate matrices of genetic distance to matrices of Euclidean distances (e.g. distances between discrete habitat patches). Resistance surfaces assign values of permeability to landscape features, that is, reflecting the degree to which landscape features impede or enhance gene flow (Spear et al., 2010). For example, genetic variability across a landscape of rocky outcrops was studied in the Ornate Dragon (*Ctenophorus ornatus*) using Mantel tests. This research determined that there was significant genetic differentiation between discrete rocky outcrop populations and significant effects of isolation across geographic regions (Levy et al., 2013).

Spatial analyses relying on empirical information on animal movements and habitat associations have been used to inform conservation strategies for mitigating one of the most prevalent threats to reptile population persistence—road mortality (Steen and Gibbs, 2004). For example, using analysis of movement at three spatial scales, Beaudry et al. (2008) identified road mortality hotspots for two threatened turtles in North America (Spotted Turtle, *Clemmys guttata*, and Blanding's Turtle, *Emydoidea blandingii*), and highlighted the best locations and timing for implementing mitigation strategies. Other studies combined spatial analyses with empirical movement data (i.e. road crossing speed) to identify road mortality risk for Hermann's Tortoises (*Testudo hermanni boettgeri*; Iosif et al., 2013) and mortality hotspots for turtles (Patrick et al., 2012) across large geographic extents (1000s of km^2).

22.3.3 Macroecology and biogeography for reptile conservation

When steering global conservation action, broad-scale analyses can help us find answers to some key questions: Where do we best target conservation action (e.g. where are most of our threatened species found)? Where do we best target research to address data gaps (e.g. where are areas of high data deficiency)? Can we maximize conservation outcomes for a large number of species (e.g. are patterns we see congruent between species groups)? The recent global assessment of extinction risk of a random sample of 1500 reptiles (Böhm et al., 2013) has begun to address some of these questions for reptiles. By overlaying a grid (here, hexagonal grid cells of 7770 km^2) onto the aggregated species distribution and calculating the proportion of species in each grid cell (Figure 22.3), Böhm et al. (2013) identified (a) species richness in the sample to be highest in tropical regions; (b) localized centres of threatened species richness across the globe; and (c) centres of data deficiency.

Patterns of species richness are often used to define hotspots of biodiversity, although these hotspots are generally defined using a restricted number of species groups. Whether or not richness patterns between species groups are congruent greatly affects the delineation and effectiveness of any such hotspots. Reptiles have been scarcely addressed in such large-scale analyses. For example, lizard species richness in Australia was found to be generally uncorrelated with that of other vertebrate taxa because different environmental factors were predictors of lizard richness: richness was highest in dry and hot regions (Powney et al., 2010).

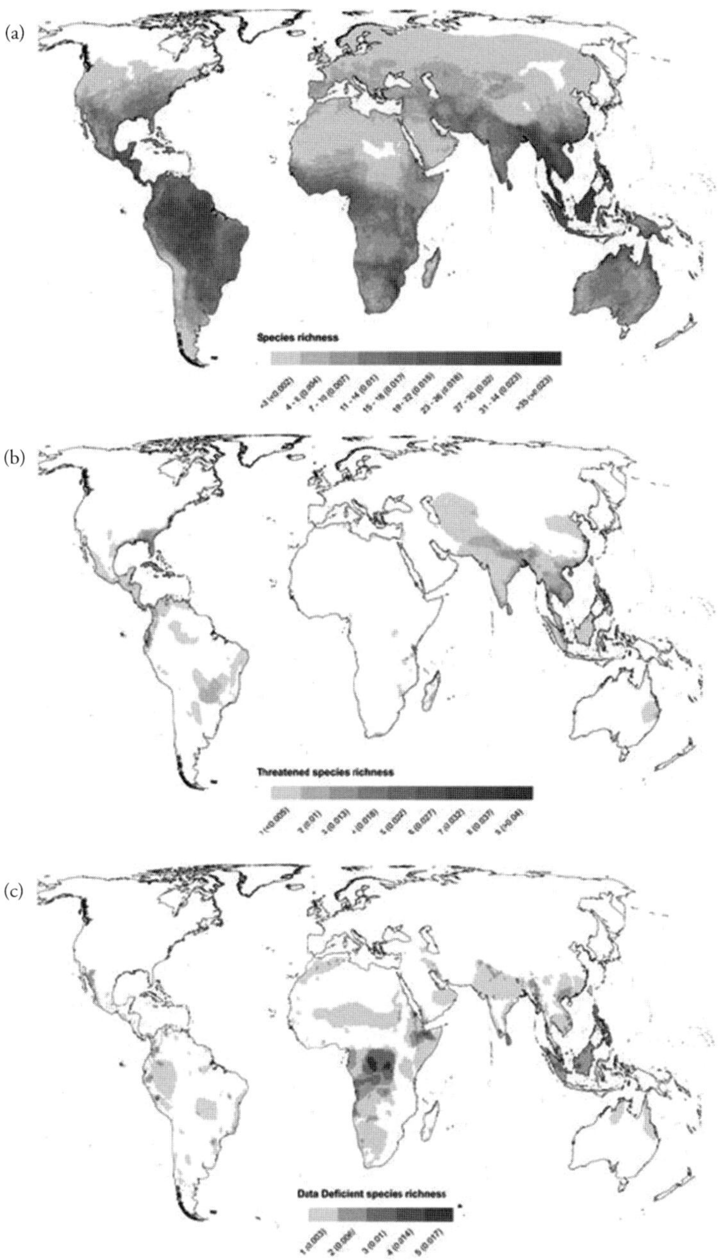

Figure 22.3 Global terrestrial and freshwater species richness distributions of a sampled reptile assessment (Böhm et al., 2013): (a) species richness of all species in sample (N = 1484); (b) distribution of threatened (CR, EN, VU) species in the sample (N = 219); (c) distribution of data deficient species in the sample (N = 288). Updated from maps shown in Böhm et al. (2013) to reflect latest Red List category changes in the data. All maps show number of species and proportion of species in sample per grid cell.

22.4 Spatial statistics: the analysis of spatially correlated data

Spatial data are likely to violate data independence assumptions because measurements taken at geographically close locations are generally more alike than measurements taken at geographically distant locations (spatial autocorrelation). The risk of ignoring spatial autocorrelation in the analysis of spatial data is that we may obtain significant results when these are only a reflection of underlying spatial effects (Type I error). There are a number of tools and packages available providing user-friendly options for analysing spatial data, including analysis options in the freely available statistical environment R or other specialist open-access software (see Table 22.3). However, it is often difficult

Table 22.3 *Spatial statistics functionality in commonly used GIS software (including QGIS plugins, see https://plugins.qgis.org) and other useful spatial statistics software applications, including relevant libraries in R.*

Functionality	Description	Package(s)
Data extraction/ manipulation	Manipulation of spatial features (e.g. clip, intersect, union, buffer); data extraction from raster data (e.g. zonal statistics)	ArcGIS, Quantum GIS (QGIS), Geospatial Modelling Environment[1], rgeos (R), raster (R)
	Counting points in polygons; generating random points/shapes	Geospatial Modelling Environment[1]
	Sampling of polygon attributes and raster values	Point Sampling Tool (QGIS)
Spatial models	Spatial and non-spatial relationships (regressions), spatial models (e.g. kriging)	ArcGIS, Spatial Analysis in Macroecology (SAM)[2], raster (R)
Analysing spatial patterns	Nearest neighbour analysis and statistics, point distances	ArcGIS, QGIS, Fragstats[3], rgeos (R)
	Clustering, spatial autocorrelation, Ripley's K-function	ArcGIS, Spatial Analysis in Macroecology (SAM)[2], spdep (R), splancs (R)
	Patch-based metrics (e.g. edge length, patch size, isolation, edge contrast, etc.)	Fragstats[3]
Mapping clusters	Identification of significant hot and cold spots and spatial outliers	ArcGIS
Measuring geographic distributions	Distributional characteristics, such as centre, compactness, orientation	ArcGIS
Animal movement analysis and networks	Minimum convex polygons	QGIS, AniMove for QGIS, rgeos (R), adehabitat (R)
	Movement path analysis, least cost path analysis, network analysis	ArcGIS, AniMove for QGIS, igraph (R)
Land cover analysis	Calculation of metrics from raster and vector layers; overlays	LecoS: Landscape Ecology Statistics (QGIS)
Habitat analyses	Habitat preference, etc.	adehabitat (R)
Interface to the Global Biodiversity Information Facility (GBIF)	Searching for and retrieving species occurrence records directly from GBIF	rgbif (R)

[1] *http://www.spatialecology.com/gme/index.htm*
[2] *http://www.ecoevol.ufg.br/sam*
[3] *http://www.umass.edu/landeco/research/fragstats/fragstats.html*

to decide *a priori* which analysis method is best because not all spatial methods have been shown to improve inference over non-spatial methods (Bini et al., 2009).

Depending on the question under investigation, spatial autocorrelation can be analysed in a multitude of ways. Most prominent are indices for global spatial autocorrelation (e.g. Moran's *I*) and local spatial clustering (K functions, Getis–Ord local *G*), tests of spatial autocorrelation (Mantel and paired Mantel tests), and correlations estimating effective degrees of freedom based on spatial autocorrelation in the data.

Apart from reflecting the degree of spatial autocorrelation in a dataset, analysis of spatial clusters can help us to investigate how a species uses its environment (e.g. analysing the placement of burrows). In its simplest form, a univariate K-function, $K(r)$, of a point pattern is defined as the expected number of points within a distance r of an arbitrary point; these K-functions are considered robust in cases where a point pattern is incomplete (i.e. where there are missing data). Using this method established, for example, that Desert Tortoise (*Gopherus agassizii*) burrows are aggregated across the landscape at multiple spatial scales and that tortoises are spatially associated with burrows (Duda et al., 2002), suggesting best surveying techniques for this species.

Spatial autocorrelation can be accounted for in advanced modelling techniques through autologistic regression and geographically weighted regression (GWR), or as spatial autocorrelation structures in generalized linear mixed models or generalized least squares models. Autologistic regression models provide an extension to logistic (presence/absence) models by including an auto-covariate to account for spatial autocorrelation within the data. For example, autologistic regression was used to investigate patterns of turtle nest predation (*Kinosternon subrubrum, Pseudemys concinna floridana*, and *Trachemys scripta*), where it was assumed that a predator preying on one nest was more likely to search for and find neighbouring nests (Burke et al., 1998). GWR considers local spatial relationships by creating a local regression equation for each data point, thus allowing the relationship between predictor and response variables to vary across space. For example, Powney et al. (2010) used GWR to explore geographical patterns of lizard species richness in Australia, showing that richness is predicted by different environmental factors than in other vertebrates.

22.5 Shortcomings and future directions

Despite the many research opportunities they provide, GIS and spatial data come with a set of limitations. It is important to be aware of these in order to produce robust analyses and the best possible outcomes for conservation:

1. Although technology is rapidly advancing, data availability is still somewhat lagging behind. This is especially true for data that capture rapidly occurring landscape change. Updating large-scale global data sets at high resolutions is time-consuming, and there is a considerable temporal data gap in many spatial data layers (e.g. updated every 10 years; or data for many years are aggregated into a single data layer).
2. Because researchers are looking for the most up-to-date information, many analyse remotely sensed data (e.g. Landsat) by implementing their own classification

system. Consequently, there is a multitude of differently classified data available, often designed to best represent certain study species, which limits comparability between studies.

3. For many herpetologists, remotely sensed data are often still at too coarse a scale to allow the accuracy needed to depict habitat types or habitat change over time, and relate this to specific reptile populations. Similarly, global databases such as GBIF (see Section 22.3.1) often include spatially and taxonomically inaccurate data; therefore, great care needs to be taken when using these data.
4. Spatial data can only provide part of the puzzle of what determines reptile distributions and abundance. There are other important factors for which it is more difficult to obtain spatial data or for which spatial data do not exist (e.g. interspecific interactions, certain threat processes such as overharvesting). In addition, it is important that field data underpin or validate any model approaches, both for species occurrences and environmental data related to these occurrences.

Many of these limitations are likely to be overcome or at least minimized with technological advances in the gathering and processing of spatial data. Technological advances have recently led to the first use of remotely sensed data from airborne LiDAR (light detection and ranging) sensors in ecological studies of reptiles, with the development of digital vegetation surfaces based on satellite data with a pixel size <1 m resolution (Sillero and Goncalves-Seco, 2014).

Reptiles are still widely overlooked in conservation decision-making unless they are directly targeted by endangered species legislation (e.g. Endangered Species Act (ESA) in the United States, Species at Risk Act (SARA) in Canada, Habitats Directive in Europe). Since population data for status assessments of many species are often lacking, many conservation assessments derive from knowledge of reptile species distributions. More and more data are becoming available on reptile distribution, not the least through the work of initiatives such as the IUCN Global Reptile Assessment and collaborative efforts to map the distribution of all reptiles (http://www.gardinitiative.org/). Both initiatives are set to produce large spatial datasets of reptile distributions, which together with the ever-increasing availability of large-scale environmental and threat data will further aid future conservation assessments. For example, reptile distribution maps in conjunction with correlates of extinction risk may allow us to be more predictive about extinction risk and to provide more timely assessments for species. With increased research attention on species-independent threat mapping (e.g. Murray et al., 2014), future assessments of extinction risk may be increasingly founded on objective spatial data on threat processes (e.g. forest cover change (Hansen et al., 2013); climate change (IPCC, 2013)).

GIS and spatial analyses for studying reptile ecology and conservation are increasing, but it is paramount that GIS literacy and proficiency is increased through collaborative efforts and capacity building. Since conservation decisions are often based on spatial data (i.e. species and threat distributions), there is a dire need to better understand how reptiles interact with their environment, and how landscape or climatic changes are likely to shape reptile distributions locally, regionally, and globally.

References

Araújo, M.B., and New, M. (2007). Ensemble forecasting of species distributions. *Trends in Ecology & Evolution*, **22**, 42–7.

Bachman, S., Moat, J., Hill, A.W., et al. (2011). Supporting Red List threat assessments with GeoCAT: geospatial conservation assessment tool. *ZooKeys*, **150**, 117–26.

Beaudry, F., de Maynadier, P.G., and Hunter, M.L. (2008). Identifying road mortality threat at multiple spatial scales for semi-aquatic turtles. *Biological Conservation*, **141**, 2550–63.

Betts, M.G., and Villard, M.-A. (2009). Landscape thresholds in species occurrence as quantitative targets in forest management: generality in space and time? In M.-A. Villard and B.G. Jonsson (eds) *Setting Conservation Targets for Managed Forest Landscapes*. Cambridge: Cambridge University Press, pp. 185–206.

Bini, L.M., Diniz-Filho, J.A., Rangel, T.F., et al. (2009). Coefficient shifts in geographical ecology: an empirical evaluation of spatial and non-spatial regression. *Ecography*, **32**, 193–204.

Böhm, M., Collen, B., Baillie, J.E.M., et al. (2013). The conservation status of the world's reptiles. *Biological Conservation*, **157**, 372–85.

Burke, V.J., Rathbun, S.L., Bodie, J.R., et al. (1998). Effect of density on predation rate for turtle nests in a complex landscape. *Oikos*, **83**, 3–11.

Dade, M.C., Pauli, N., and Mitchell, N.J. (2014). Mapping a new future: using spatial multiple criteria analysis to identify novel habitats for assisted colonization of endangered species. *Animal Conservation*, **17**, 4–17.

Dixo, M., and Martins, M. (2008). Are leaf-litter frogs and lizards affected by edge effects due to forest fragmentation in Brazilian Atlantic forest? *Journal of Tropical Ecology*, **24**, 551–4.

Duda, J.J., Krzysik, A.J., and Meloche, J.M. (2002). Spatial organization of desert tortoises and their burrows at a landscape scale. *Chelonian Conservation and Biology*, **4**, 387–97.

Gebauer, K., Dickinson, K.J.M., Whigham, P.A., et al. (2013). Matrix matters: differences of grand skink metapopulation parameters in native tussock grasslands and exotic pasture grasslands. *PLoS One*, **8**, e76076.

Hansen, M.C., Potapov, P.V., Moore, R., et al. (2013). High-resolution global maps of 21st-century forest cover change. *Science*, **342**, 850–3.

Hanski, I. (1999). *Metapopulation Ecology*. Oxford: Oxford University Press.

Harris, K.M., Dickinson, K.J.M., and Whigham, P.A. (2014). Functional connectivity and matrix quality: network analysis for a critically endangered New Zealand lizard. *Landscape Ecology*, **29**, 41–53.

Hokit, D.G., Ascunce, M., Ernst, J., et al. (2010). Ecological metrics predict connectivity better than geographic distance. *Conservation Genetics*, **11**, 149–59.

Iosif, R., Rozylowicz, L., and Popescu, V.D. (2013). Modeling road mortality hotspots for Eastern Hermann's tortoise in Romania. *Amphibia-Reptilia*, **34**, 163–72.

IPCC (2013). Climate change 2013: the physical science basis. In T.F. Stocker, D. Qin, G.-K. Plattner, et al. (eds). *Working Group I Contribution to the Fifth Assessment Report of the Intergovernmental Panel on Climate Change*. Cambridge: Cambridge University Press, p. 203.

Jellinek, S., Driscoll, D.A., and Kirkpatrick, J.B. (2004). Environmental and vegetation variables have a greater influence than habitat fragmentation in structuring lizard communities in remnant urban bushland. *Austral Ecology*, **29**, 294–304.

Kearney, M., and Porter, W.P. (2004). Mapping the fundamental niche: physiology, climate, and the distribution of a nocturnal lizard. *Ecology*, **85**, 3119–31.

Lehtinen, R.M., Ramanamanjato, J., and Raveloarison, J.G. (2003). Edge effects and extinction proneness in a herpetofauna from Madagascar. *Biodiversity and Conservation*, **12**, 1357–70.

Levy, E., Tomkins, J.L., Lebas, N.R., et al. (2013). Contrasting effects of landscape features on genetic structure in different geographic regions in the ornate dragon lizard, *Ctenophorus ornatus*. *Molecular Ecology*, **22**, 3904–15.

Lindenmayer, D., Hobbs, R.J., Montague-Drake, R., et al. (2008). A checklist for ecological management of landscapes for conservation. *Ecology Letters*, **11**, 78–91.

Manel, S., and Holderegger, R. (2013). Ten years of landscape genetics. *Trends in Ecology & Evolution*, **28**, 614–21.

Murcia, C. (1995). Edge effects in fragmented forest: implication for conservation. *Trends in Ecology & Evolution*, **10**, 58–62.

Murray, K.A., Verde Arregoitia, L.D., Davidson, A., et al. (2014). Threat to the point: improving the value of comparative extinction risk analysis for conservation action. *Global Change Biology*, **20**, 483–94.

Patrick, D., and Gibbs, J. (2010). Population structure and movements of freshwater turtles across a road-density gradient. *Landscape Ecology*, **25**, 791–801.

Patrick, D.A., Gibbs, J.P., Popescu, V.D., et al. (2012). Multi-scale habitat-resistance models for predicting road mortality 'hotspots' for turtles and amphibians. *Herpetological Conservation and Biology*, **7**, 407–26.

Popescu, V.D., and Gibbs, J.P. (2010). Landscape ecology and GIS methods. In C.K. Dodd, Jr. (ed) *Amphibian Ecology and Conservation: A Handbook of Techniques*. Oxford: Oxford University Press, pp. 339–60.

Powney, G.D., Grenyer, R., Orme, C.D.L., et al. (2010). Hot, dry and different: Australian lizard richness is unlike that of mammals, amphibians and birds. *Global Ecology and Biogeography*, **19**, 386–96.

Radford, J.Q., Bennett, A.F., and Cheers, G.J. (2005) Landscape-level thresholds of habitat cover for woodland-dependent birds. *Biological Conservation*, **125**, 317–37.

Raxworthy, C.J., Martinez-Meyer, E., Horning, N., et al. (2003). Predicting distributions of known and unknown reptile species in Madagascar. *Nature*, **426**, 837–41.

Schutz, A.J., and Driscoll, D.A. (2008). Common reptiles unaffected by connectivity or condition in a fragmented farming landscape. *Austral Ecology*, **33**, 641–52.

Sillero, N., and Goncalves-Seco, L. (2014). Spatial structure analysis of a reptile community with airborne LiDAR data. *International Journal of Geographical Information Science*, **28**, 1709–22.

Sillero, N., and Tarroso, P. (2010). Free GIS for herpetologists: free data sources on Internet and comparison analysis of proprietary and free/open source software. *Acta Herpetologica*, **5**, 63–85.

Sillero, N., Campos, J., Bonardi, A., et al. (2014). Updated distribution and biogeography of amphibians and reptiles of Europe. *Amphibia-Reptilia*, **35**, 1–31.

Spear, S.F., Balkenhol, N., Fortin, M.J., et al. (2010). Use of resistance surfaces for landscape genetic studies: considerations for parameterization and analysis. *Molecular Ecology*, **19**, 3576–91.

Steen, D.A., and Gibbs, J.P. (2004). Effects of roads on the structure of freshwater turtle populations. *Conservation Biology*, **18**, 1143–48.

Stellatelli, O.A., Vega, L.E., Block, C., et al. (2013). Effects of tree invasion on the habitat use of sand lizards. *Herpetologica*, **69**, 455–65.

Thomas, C.D., Cameron, A., Green, R.E., et al. (2004). Extinction risk from climate change. *Nature*, **427**, 145–8.

Wiens, J.A., Stenseth, N.C., Horne, B.V., et al. (1993). Ecological mechanisms and landscape ecology. *Oikos*, **66**, 369–80.

Part 5

Experimental Applications, Physiological Ecology, and Genetics

23

Experimental applications

Stephen J. Mullin

23.1 Introduction

Over the past quarter of a century, advances in the understanding of reptile life-history traits have made it possible to test important predictions of ecological theory. These knowledge gains have facilitated additional studies involving sample sizes (e.g. Steen et al., 2013) that rival those reported for amphibians. Coupled with long-term monitoring, the application of experimental techniques allows researchers to distinguish among specific biotic and abiotic parameters that influence a wide range of behavioural, physiological, and ecological traits for a given reptile species.

In areas where individual species or communities of reptiles have experienced population declines, implementing effective strategies for their conservation requires an understanding of the ecological processes underlying the declines (Pike et al., 2010). Experimental applications have the potential to identify some of these processes, especially if reptiles can be studied *in situ* or in simulated habitats where those environmental parameters predicted to influence population ecology can be precisely monitored. Well-designed experiments having appropriate controls are useful for identifying the environmental conditions under which the focal species experiences greater longevity and/or reproductive output (Downes and Hoefer, 2007).

The design of any study dealing with the ecology or conservation of reptiles should allow the researcher to easily quantify life-history traits that are influenced by manipulation of features within a habitat. Manipulations of this nature often cannot occur without confining the focal species within a parcel of habitat (or a simulation of it). In lieu of altering environmental parameters within an enclosure, individuals can be experimentally manipulated through tactics like diet or refuge supplementation, translocation, or individual restraint. Among various types of experimental manipulation, an advantage to using either enclosures or tethered subjects is that ambient parameters can be closely monitored and, along with the response variable(s), measured at appropriate sampling frequencies.

After presenting guidelines for selecting reptile species that are ideally suited for use in enclosures, I then summarize conceptual and practical guidelines for using enclosures to answer questions about reptile ecology and conservation. If the effects of inter- or intraspecific interactions are of interest, alternatives to using enclosures involve the restraint of a stimulus animal ('tethering'), or the presentation of stimuli

Reptile Ecology and Conservation. Edited by C. Kenneth Dodd, Jr.
Published 2016 by Oxford University Press.

to free-ranging subjects, within a natural setting and documenting interactions that ensue. Although enclosures are absent from these experimental designs, focal subjects might have their movements restricted and responses can be easily quantified. I also provide a summary of a few techniques in which habitat features or individual subjects are manipulated.

Although valuable in their own right, I exclude studies where habitat manipulations occur in an experimental context but individuals representing the target species are not forced to interact with them (i.e. situations where subject movements are not limited to a specific habitat; Grillet et al., 2010; Sutton et al., 2013). I also exclude studies where the enclosure size is less than 100 cm in two of its dimensions (or one dimension for a circular enclosure), as such space could easily double as an environment for husbandry. For simplicity, I refer to structures deployed *in situ* at field sites (whether intended to contain study subjects or exclude other biota) as 'habitat enclosures', whereas 'mesocosms' (Odum, 1984) denote naturalistic containers maintained in a zoo, laboratory, or greenhouse environment. I use 'cages' as a more general term for both habitat enclosures and mesocosms.

23.2 Selecting species

By their very nature, experiments involving cages or tethered subjects constrain the possible movements of the studied species. Even among reptiles that exhibit short daily movements, a researcher must acknowledge that the study subjects will react in some way to having their movements restricted (Thaker et al., 2009a). An understanding of life-history parameters such as territory or home range size, foraging mode, or individual density should be used to guide the choice of taxa and the experimental context in which they are studied.

23.2.1 Terrestrial species

For turtles, cages involving terrestrial habitats are restricted to tortoises (Testudinidae) or species within the Emydidae and Geoemydidae. In contrast, there are many more opportunities available among squamates. A concern is that lizards and snakes generally have higher movement rates than turtles (measured in either frequency or speed) and, therefore, require larger cages for experimental studies. To avoid introducing bias from extended periods of restraint (and exposure-related stress), tethering experiments are necessarily short in duration. Larger species are likely to range over greater distances (Reed, 2003), so the logistical feasibility of studying the largest members among these clades (e.g. *Chelonoidis, Eunectes, Varanus*) is limited. Among squamates, the frequency and range of movement of some small species are great enough as to limit the opportunity of using these taxa in studies involving cages.

23.2.2 Aquatic and semi-aquatic species

Similar to the features unique to terrestrial systems, a number of hydrophysical characteristics need maintenance when working with aquatic environments. Although the density of water ensures that its temperature will not fluctuate rapidly, parameters like dissolved oxygen, turbidity, and accretion of nitrogenous wastes can influence aquatic reptiles and should be monitored (Mullin and Mushinsky, 1997; Schneider et al., 2014). If simulating a lotic environment, the researcher should decide whether or not variables like flow rates, depth profile, or substrate variability are incorporated into cage design. If habitat enclosures are established in the field, then one should be prepared to address flooding and tidal fluctuations (and any corresponding changes in salinity if working in an estuarine system), or other physical insults to the boundaries of the cage. Crossland et al. (1992) defined an aquatic mesocosm as any structure exceeding 15 m^3 or 15 m in a single dimension. Not all experimental studies of reptiles would necessarily require such a size but, as is the case for terrestrial cages, the constraints on subject movement when enclosing aquatic habitat must be acknowledged. Taxa that range widely (e.g. *Podocnemis* (Vogt, 2008), *Pelamis* (Brischoux et al., 2012)), or attain large adult size (e.g. *Dermochelys, Crocodylus*), would make poor choices for study in this context. Tethering subjects in aquatic habitats should be avoided to prevent physiological stress or drowning.

23.2.3 Additional considerations

Individual reptiles might shift among different habitat types throughout their lives as a function of various intrinsic (e.g. ontogeny, sex) or extrinsic (e.g. season, density of heterospecifics) factors (Subalusky et al., 2009; Newbold and MacMahon, 2014). Therefore, use of cages or tethering in an experimental context should occur only after the habitat parameters relevant to that particular life-history stage have been quantified. Another consideration is the source of the animals used in such studies. Whereas wild-caught individuals are likely best for examining questions related to the ecology of a species, obtaining sufficient numbers to satisfy sample size concerns might be challenging, especially if the species is of conservation interest. Coordinating the timing of the collection effort is an important consideration: on one hand, physiological differences (e.g. body condition, reproductive state) can be minimized if subjects are collected and tested in the experimental cages within a relatively brief window of time. On the other hand, constraining the collection effort to a discrete time period might force the researcher to work with a sample population that includes subjects that differ in other ways (e.g. body size, sex; see Dorcas and Willson, 2009).

Conservation strategies for reptiles might involve head-starting and such programmes can supply a relatively large sample of similar-aged individuals. After exceeding a threshold age or body size, subjects can be studied in a cage prior to being released (Godwin et al., 2011). If the success of the head-starting effort is improved by the use of naturalistic rearing environments (McCann, 2011), these cages can be co-opted for use in experimental examinations of individual or group responses to particular environmental traits. Another advantage to coupling head-starting programmes with experimental studies is that the researcher can randomly assign individuals having a known history to different treatment groups, thereby controlling for clutch/litter effects while simultaneously avoiding pseudoreplication.

23.3 Studies using cages

Experimental designs should maximize the inferential power of their findings, but this is necessarily constrained by using cages that confine subjects to a particular area. Funding limitations will typically restrict the number of cages that can be used in a given study and the labour needed to maintain them. These factors might be further confounded by the environment in which the cages are constructed and the duration of subject confinement (e.g. the need to provide food and water). Fortunately, limits to the extent of inference attributable to cage size or number can be partially offset by greater sampling frequencies than would be possible for free-ranging animals. Providing that response variables can be measured (or video-recorded) with minimal disturbance to the focal subjects, experiments using cages can provide the researcher with a wide range of census intervals.

Whether based in a field or lab setting, experimental cages should mimic as many environmental features as possible while allowing researchers to manipulate one or more defined parameters. This is a potential paradox for the design of any such experiment—resources need to be available for the target species at levels that allow subjects to act naturally in the face of confinement (e.g. refuges, basking sites). Too much realism within the cage, however, and the influence of the manipulated variable might be difficult to quantify or biased by an unrelated feature of the cage. Thus, it is incumbent upon researchers to understand the life-history traits of their species and key features of its habitat before proceeding with experimental applications (Ford, 1995; Chapter 19). This information will identify resources that should be included with cage design and provide insight to the range of values needed to address the ecological variable of interest (Mullin et al., 1998). If using discrete categories of an independent variable to test subjects, this step can help researchers determine the intervals between each category that are predicted to produce differences in subject response (Downes and Hoefer, 2007). If logistical constraints prevent certain features of the natural environment from being included in cage design (e.g. food availability), researchers should alter the experimental protocol in a manner that negates the need for those features (e.g. subjects spending a shorter trial time in the cage).

23.3.1 Lab-based mesocosms

Although mesocosms might be physically smaller than habitat enclosures, they offer the benefits of increased control of conditions within them. For example, if mesocosms are erected indoors, changes in weather conditions are unlikely to influence subject responses, even for experiments of extended duration. Working indoors allows the study subjects to be habituated to altered conditions (e.g. reverse photoperiod) that might make the logistics of completing the experiment more convenient for researchers. If space for multiple mesocosms is available, there should be less variability among them (compared to habitat enclosures) in the environmental conditions that are not the focus of study. Given that cues from conspecifics might influence the performance of test subjects used in subsequent trials, it is also easier to remove those cues between repeated uses of mesocosms (e.g. cleaning, replacing substrate; Greene and Mason, 2003). These advantages reduce the potential sources of bias in the response variables of interest.

The nature of the variable examined might require an amount of precision that, while reflecting levels that the study species encounters in nature, simply cannot be maintained in experimental cages constructed in field settings (Fornell, 2008). Similarly, if differences in the expression of the dependent variable(s) are so subtle that they could not be detected in a field-based cage, then using mesocosms with monitoring equipment that can continuously record those subtleties might be the preferred strategy (Todd, 2005). Mesocosms located indoors are also protected from violent weather events, unintended trespass, or attempts to purloin monitoring equipment or the study organisms themselves. The likelihood of subjects escaping a mesocosm might be similar to studies involving habitat enclosures, but the building might hold the animal long enough to be recaptured.

23.3.2 *In situ* habitat enclosures

Using habitat enclosures in field settings provides the advantage of increased realism because the focal animals experience a wider range of ambient conditions, especially when compared to mesocosms kept indoors. Assuming they are not important to the questions being asked, variables like photoperiod, light intensity and quality, temperature, precipitation, and humidity can fluctuate naturally, which might prompt the most realistic responses from the study subjects to other variables being manipulated within the habitat enclosures. Not only can habitat enclosures admit abiotic elements, but they can also be constructed to contain the study species while allowing the entry of other biota (Figure 23.1(a); Sabo and Power, 2002). If other organisms are potential food for the focal species, the researcher can minimize the amount of time spent providing husbandry for the study subjects. This is especially beneficial for studies involving long-term confinement within the habitat enclosure (Amaral et al., 2012).

Whereas the size of a lab-based mesocosm is limited by the dimensions of the room or building that surrounds it, a habitat enclosure is typically larger. In turn, this lowers the stocking density of the cage (see Table 23.1), which reduces the likelihood that response variables are biased by physical confinement or an artificially high frequency of interaction with other individuals (if more than one subject is simultaneously tested within the same cage). Being situated outside allows researchers to erect multiple cages (Figure 23.1(b)); this can be done while holding other variables relatively constant, or positioning each habitat enclosure along an environmental gradient of interest (Niewiarowski and Roosenburg, 1993). Using multiple habitat enclosures permits simultaneous replication, which might be important if subject age (Downes and Hoefer, 2007) or similarity of abiotic elements (Langkilde et al., 2005) is critical to the experimental design.

Concerns associated with the outdoor siting of habitat enclosures include damage to the structure (from either natural or human elements), escape of the study subjects, or, conversely, intrusion by non-target organisms that then cannot escape and possibly bias the responses of, or depredate, the focal species. Even when ambient conditions are integral components of experimental design, conditions sometimes change in unexpected ways that influence subject responses to the variables of interest. Weathering of the materials used in constructing a habitat enclosure might force the researcher to commit

Figure 23.1 Examples of habitat enclosures constructed in the field to examine the responses of reptiles to manipulated environmental variables. (a) An enclosure (1300 × 700 × 200 cm) along a river in Mendocino County (California) that serves as both the source of invertebrate prey for Western Fence Lizards (*Sceloporus occidentalis*), and as a barricade preventing escape of test subjects. The enclosure walls are constructed from bird netting, Visqueen plastic, and polypropylene netting. Photograph by M. Power. (b) An enclosure (2700 × 2700 × 250 cm) containing multiple partitions (45 cm tall), thereby allowing replicate examinations of multiple individuals from the same cohort of head-started Desert Tortoises (*Gopherus agassizii*). The enclosure boundary includes corrugated steel and chain-link fencing and is topped with heavy-gauge bird netting. Photograph by T.D. Tuberville.

Table 23.1 *A sampling of ecological studies involving reptiles maintained within cages. Studies are listed chronologically within each setting.*

Reference	Taxon	Enclosure size (L × H × W, cm)	Manipulated variable(s)	Duration of study	Sampling interval	Stocking density (subjects/m^2)	Response metric(s)
		Indoor mesocosms					
Schuett and Gillingham (1988)	*Agkistrodon*	240 × 240 × 240	Sex ratio	15 min	Constant	0.35–1.04	Courtship, mating
Kaufman et al. (1994)[a]	*Varanus*	300 × 400 × 60	Prey stimulus	Variable (<1 day)	Constant	0.000008[b]	Prey choice
Mullin et al. (1998)	*Pantherophis*	225 × 225 × 200	Habitat complexity, prey type	2 hr	Constant	0.2[b]	Time to capture prey
Greene and Mason (2003)	*Boiga*	150 × 150 × 150	Sex ratio	2 days	10 min (1 ×)	0.89	Combat/courtship frequency and intensity
Mann and Meek (2004)[a]	*Corucia*	1000 × 750 × 320	None	1 yr	1 ×/week	0.12	Habitat selection, activity, body temperature
Fornell (2008)	*Lampropeltis*	200 × 200 × 75	Stimulus type, habitat complexity	90 min	Constant	0.25[b]	Time to ingest prey
Schneider et al. (2014)[a]	*Crocodylus*	620,000 cm^2 × 200 (irregular border)	None	1 year	Constant during 3 periods	0.000003	Reproductive frequency
		Field-based enclosures					
Hammerson (1987)	*Coluber*	350 × 350 × ?	None	24 hr	1 ×/15 min	0.08[b]	Body temperature
Kingsbury (1989)	*Coleonyx*	500 × 100 (circular)	None	4 hr	3 ×	0.2	Activity, location
Charland and Gregory (1990)	*Crotalus*	600 × 600 × 100	Reproductive status	8 weeks	1 ×/15 min	1.55	Body temperature
Niewiarowski and Roosenburg (1993)	*Sceloporus*	20,000 × 10,000 × 75	Site, translocated subjects	43 days	1 ×/10 day	0.0023	Growth rate

continued

Table 23.1 *Continued*

Reference	Taxon	Enclosure size (L × W × H, cm)	Manipulated variable(s)	Duration of study	Sampling interval	Stocking density (subjects/m^2)	Response metric(s)
Rostal et al. (1998)	*Lepidochelys*	2100 × 900 × 280	None	14 months	60 hr/month	0.15	Hormone levels, mating frequency, vitellogenesis
Lee and Mills (2000)	*Agkistrodon*	3000 × 2000 × 185	None	24 hr	Constant	not reported	Body temperature
Buhlmann and Coffman (2001)	*Trachemys*	2500 × 2500 × 150	None	18 weeks	1×/week	0.000007	Hatching success
Downes (2001)	*Lampropholis*	180 × 180 × 160	Presence of predator scent	455 days	1×/12 weeks	6.17	Activity, tongue-flick rate, feeding rate, sprint speed, clutch mass
Sabo and Power (2002)	*Sceloporus*	1300 × 700 × 200	Prey immigration, predator presence	5 months	3×	0.03	Body size, prey availability
Ferguson et al. (2003)[a]	*Furcifer*	200 × 200 (circular)	Diet	12 hr	1×/20 min	0.32[b]	Habitat selection, UV exposure
Himes (2003)	*Nerodia*	122 × 122 × 91	Subject density, prey density	14 weeks	1×	0.67–1.34	Body size
Kissner and Weatherhead (2005)	*Nerodia*	600 × 500 × 75	None	6 months	Daily (after hibernation)	1.33–2.67	Survivorship, growth, sprint speed
Langkilde et al. (2005)	*Eulamprus*	200 × 200 × 100	Species ratio	10 weeks	3×	0.5	Shelter selection, activity, food eaten, hormone levels
Meek (2005)	*Anguis*	620 × 620 × ?	None	Not reported	n/a	0.16	Body temperature, behaviour
Todd (2005)	*Hoplodactylus*	165 × 155 × 96	None	1 yr	Not reported	3.91	Mating behaviour and frequency
Tuberville et al. (2005)	*Gopherus*	5642 × 92 (circular)	None	2 yr	2–3×/week	0.00000001	Site fidelity, activity area
Isaac and Gregory (2004)	*Natrix*	800 × 800 × 130	Reproductive status		1×/15 min		Body temperature
Willson and Brooks (2006)	*Pantherophis*	3600 × 100 (circular)	Reproductive status	8 days	1×/hr	0.014	Body temperature

Downes and Hoefer (2007)	*Lampropholis*	400 × 200 × 160	Density of invasive vegetation	1 yr	1×/3 mo	1.25	Activity, thermal preference, sprint speed, body size
Lin et al. (2007)[a]	*Trimeresurus*	1200 × 240 × 300	Habitat complexity, prey availability	24 hr	1×	0.17–0.35	Habitat selection, activity
Rodda et al. (2007)	*Boiga*	22,400 × 22,400 × 160	Translocated subjects	19 weeks	1×/day	0.0009[c]	Search and barrier effectiveness, trap success, size distribution
Daly et al. (2008)	*Ctenophorus*	400 × 200 × 50	Species ratio	48 hr	3×/day	0.25	Habitat selection
Dorcas et al. (2011)	*Python*	3100 × 2500 × 250	None	7 months	3×/week	0.013	Body temperature, behaviour, habitat selection
Godwin et al. (2011)[a]	*Drymarchon*	10,000 × 10,000 × ?	None	2 years	3–5×/week	0.0001[b]–0.001[c]	Activity, habitat selection, home range
McCann (2011)[a]	*Gopherus*	1300 × 750 × 30	Burrow orientation	20 weeks	1×/week	0.24	Burrow use, social interactions
Amaral et al. (2012)	*Podarcis*	150 × 100 × 100	Pesticide application	413 days	1×/3 mo	2	Growth, locomotor performance, oxidative stress
Newbold and MacMahon (2014)	*Phrynosoma*	2800 × 2800 × 80	Habitat complexity	3 day	4×/day	0.006	Habitat selection, activity
Nagy et al. (2015)	*Gopherus*	520 × 183 (hexagonal)	None	4 months	2×/week	0.0002	Survivorship

[a]*Studies that included collaborative effort with zoological park or conservation organization.*

[b]*Density represents one subject of focal species per cage.*

[c]*Density reported only for translocated or head-started subjects, and excludes resident population in cage.*

more time and funds to maintaining the integrity of the cage. Furthermore, elements forming the periphery of the habitat enclosure will change the abiotic characteristics within them. For instance, the thermal profiles for the areas immediately adjacent to metal walls having any sun exposure will likely be higher on account of the conduction of heat to the substrate. Studies examining this particular abiotic variable can incorporate different layers of shade cloth as part of the habitat enclosure walls or ceiling, or suspended above the entire structure, so as to establish discrete categories of access to solar radiation (Langkilde et al., 2005; Andersson et al., 2010).

23.3.3 Cage construction and siting

Given that cages are intended to expose subjects to experimental levels of certain independent variables, priority should be assigned to ensuring that the subjects remain within them. One solution when working with reptiles that burrow is to use a structure having an integrated floor. For example, cattle tanks are often used for amphibian studies (reviewed in Harper et al., 2010), and their size might be appropriate for application in some reptile systems (Andersson et al., 2010). To provide aquatic habitat, wading or above-ground pools can be used (Mullin and Mushinsky, 1997). Still larger mesocosms maintained in the lab should have their own flooring securely anchored to the bottom of the walls. This added substrate can be easily cleaned and checked for potential points of escape. It can minimize exposure of the study subjects to compounds found in the floor of the lab or greenhouse (e.g. wax, calcium hydroxide, petroleum-based products). Preventing the escape of burrowing reptiles from field-based habitat enclosures typically involves installing the lower portion of the cage wall below grade (Amaral et al., 2012; Newbold and MacMahon, 2014), but occasionally more substantial measures are taken (e.g. Lee and Mills, 2000; Rodda et al., 2007).

Reptiles that climb present multiple challenges in confined cages. Although some materials allow researchers to construct a cage in a number of sizes or shapes, walls made with silt or hardware cloth, screening, or fencing are far from escape-proof. If these materials are used, researchers nearly always add a smooth lining to the interior walls, such as plastic sheeting (Daly et al., 2008), tin flashing (Figure 23.2; Ferguson et al., 2003), or glass (Meek, 2005). Even then, additional steps are often incorporated into cage design to minimize the likelihood of escape, including a negative slope to the upper portion of the wall (Rodda et al., 2007; Dorcas et al., 2011) or some type of fastened ceiling (Greene and Mason, 2003). The advantage to having a complete ceiling on any habitat enclosure is that it minimizes the risk of the study subjects being taken by predators (McCann, 2011). Even if the area available within a habitat enclosure is appreciable, walls without an affixed ceiling represent mere suggestions of confinement to reptiles having sufficient motivation to leave (Godwin et al., 2011). Specifically for lizards, any elevated substrate (e.g. rocks, logs) within an open-topped cage should be placed centrally so that subjects cannot jump from the substrate onto or over the walls.

Habitat enclosures associated with permanent water features are of interest for two reasons. On one hand, water might be an important ecological resource for the species and the habitat enclosure could be intentionally constructed to surround a water body (Lee and Mills, 2000; Kissner and Weatherhead, 2005) or include a portion of it (Rostal

Figure 23.2 A mesocosm (200 × 200 × 75 cm) maintained in the lab for examining foraging behaviours in Prairie Kingsnakes (*Lampropeltis calligaster*) as a function of cues available from their prey and habitat structure. The video camera suspended above the enclosure was mounted on a motorized tripod head that could be controlled remotely, thereby allowing the researcher to continually follow a focal subject without disturbance, in order to quantify head orientation and exact timing of specific behaviours (e.g. tongue-flick rates, predatory strikes). The ropes attached to the enclosure walls allowed the researcher to lift them off a heated subfloor such that, between successive trials, chemosensory information could be removed and natural substrate replaced. Photograph by S.J. Mullin.

et al., 1998; Himes, 2003). Conversely, if the focal species has an aversion to water, the boundary of the habitat enclosure can include a permanent water feature which itself prevents escape (Figure 23.1(a); Sabo and Power, 2002). This type of experimental design has been extended in a manner similar to examinations of biogeography and community ecology (Schoener and Schoener, 1983). Providing that the probability of dispersal is near zero, small islands can be treated as closed systems for reptiles that avoid open water. In this context, the density of the study species or the resources on which it depends can be experimentally manipulated to assess their effects on the response metrics of interest. Because constructing cages is not needed, researchers will necessarily accept all ambient conditions that the islands experience during the course of study (e.g. Schoener et al., 2004) and the possibility that predators might remove experimental subjects (although predator introduction itself might be the manipulated variable; Schoener et al., 2002).

The sensitivity of the vomeronasal system in reptiles (Mason and Parker, 2010) is greater than that of most amphibians and many other vertebrates. Therefore, the materials used to construct a cage should be inert whenever possible. Volatile compounds such as those found in adhesives or oil-based paints should be avoided or sealed such that animals cannot interact with them. Given that many reptiles can detect scent trails left by conspecifics (Scott et al., 2013), those odours and the cleaning solutions used to remove them should be rinsed thoroughly from the substrate so as to not bias subject responses in successive trials (Greene and Mason, 2003). Alternatively, residual chemosensory stimuli can be removed via replacement of the entire substrate within the cage (Figure 23.2; Mullin et al., 1998).

Edge effects occur when a study subject interacts with the walls of a cage, which could bias values of the response variables (Krebs, 1998) especially in studies concerning predator–prey interactions where the wall limits escape ability. Increasing cage size reduces the amount of wall relative to the rest of the interior area, thus minimizing edge effects, but this necessarily increases costs of construction and maintenance. A circular cage design minimizes edge effects because there are no points within the confined area where two walls converge; this design can be used in tandem with larger cages.

23.3.4 The utility of zoological parks

A solution to concerns associated with cage-based experiments (e.g. degree of habitat realism, costs of maintenance) is available in the major zoological parks having naturalistic exhibit spaces that allow the public to learn about animals in the context of their native habitat. Several zoos enhance that degree of realism by using larger cages that display more than one (sympatric or ecologically similar) species. Whether used to exhibit a single individual or a small reptile community, naturalistic mesocosms offer ample opportunities for collaborations between zoo staff and researchers wishing to examine responses to specific environmental parameters (e.g. Rose et al., 2014). Because particular features of each mesocosm can be controlled in a meta-analysis (Arnqvist and Wooster, 1995), experiments involving displays can be replicated at multiple zoos. Depending on the source of the animals at each zoo, such an experimental design might reveal inter-population differences in the species' biology across its range.

Zoos often participate in captive breeding programmes intended to increase the population size for species of conservation concern. Such programmes can be the source of individuals used in experiments involving mesocosms, whether or not those experiments involve the zoo's own exhibit or reserve spaces. Regardless of the exact setting, a researcher's experiment can help further the conservation mission of the zoo by manipulating those variables predicted to provide the greatest benefit to individual fitness. Collaborative efforts involving zoo personnel have produced both conservation successes (e.g. Schneider et al., 2014) and novel insights about reptile biology (e.g. Kaufman et al., 1996).

23.4 Manipulative applications

In situations where using cages is not feasible, opportunities exist to experimentally examine the ecology of reptiles by forcing subjects to interact with manipulated features of that habitat, often on larger spatial scales than are available in cages. Such studies often involve a habitat specialist that is reluctant to disperse across an ecotone into adjacent habitats, individuals that are unlikely to abandon their territories, or require the researcher to retrieve the study subjects after a defined period of interaction with features that occur within a specific microhabitat.

23.4.1 Manipulating habitat

Researchers can choose to intentionally alter features of the habitat in which a species occurs so as to better understand the conditions that maximize individual survivorship and reproductive output. Some of this work has occurred at the landscape scale within forested areas, with tactics ranging from removal of non-native vegetation (Virkki et al., 2012) to application of wildfire regimes or specific harvest strategies (Currylow et al., 2013; Steen et al., 2013). While these studies are insightful, controlled manipulation of habitat variables at a local level might improve understanding of effects on reptile ecology, providing that it is easy to monitor the responses of individual subjects.

Research that addresses the importance of particular habitat features typically considers a specific conservation objective for a single reptile taxon (e.g. Grillet et al., 2010). Although a manipulative study might focus on a single species, researchers should quantify the responses of syntopic taxa to the altered feature in the habitat because there might be additional benefits arising from applying the modification in a management context (Shoemaker and Gibbs, 2010; Nordberg and Schwartkopf, 2015). In other situations, the species might be sufficiently rare within the manipulated habitat that researchers are left to quantify the occurrence or responses of surrogate species that share some dimensions of the ecological niche (Pike et al., 2010).

Habitat manipulation might not focus on species of conservation concern and, instead, involve multiple reptile taxa or have the potential to inform management strategies should a future need arise. Previous studies have examined the impact of modifying vegetation on locomotory and foraging performance (Mullin and Cooper, 2002; Shine et al., 2002), and evaluated changes in survivorship following the installation of artificial hibernacula (Chapter 29) or road-crossing structures (Dodd et al., 2004). Given the need for comparisons to reference or control sites (which might necessarily be located some distance from the manipulated habitat), the investment in commodities, equipment, and monitoring time in these types of studies can be substantial (e.g. Shine and Bonnet, 2009), which might limit the number of replicates included in the experimental design.

23.4.2 Manipulating individuals

Reptiles that maintain small territories or those that adopt an ambush foraging strategy are ideal for individual manipulation. Study subjects that tolerate approach by researchers (close enough to confirm individual identity) can often be encountered in the same location during repeated visits (Shine and Sun, 2003). In these situations, individuals can be subject to experimental manipulations that elucidate behavioural, physiological, or ecological responses. For instance, the presentation of specific stimuli can identify those cues that elicit territorial behaviour or contribute to foraging success (Figure 23.3(a)). Alternatively, resources used by the study species can be supplemented

Figure 23.3 Using tethered objects to experimentally determine responses of focal reptiles. (a) Presentation of a visual stimulus to a Shedao Pit Viper (*Gloydius shedaoensis*) in ambush posture. The stimulus was suspended at the end of a 190 cm pole, and could remain stationary or moved in a manner simulating prey behaviour so as to determine specific cues that elicit predatory behaviour. Photograph by M. Kearney. (b) Presentation of a tethered Collared Lizard (*Crotaphytus collaris*) male to a free-ranging conspecific. The stimulus individual was suspended on a 300 cm pole and placed with line of sight of focal subjects. Photograph by Teresa Baird.

to assess the effects on energy budget, movement rates, and long-term fitness. The application of this technique can be limited to water (Davis and DeNardo, 2009), or expanded to include whole prey items (Wasko and Sasa, 2012). In these latter examples, radio-telemetry facilitated frequent subject location so that response metrics could be quantified at specific time intervals.

For reptiles that do not exhibit extended site fidelity, there are opportunities to manipulate features of the environment in an experimental context. Taxa offering the greatest potential have temporally or spatially constrained periods of activity where they can be easily observed. Such species include a variety of turtles that nest simultaneously (Witherington, 1992) and snakes that aggregate at foraging sites or communal dens (Shine et al., 2004, 2005).

Environmental variables available for manipulation include con- and heterospecifics, and researchers working with smaller reptiles can experimentally present a tethered stimulus animal to one or more respondents. The need to physically introduce the tethered subject (Figure 23.3(b)) limits such studies to smaller lizards, and excludes snakes (impractical morphology), most turtles and crocodilians (impractical body mass). These types of experiments are best suited for examining lizards that respond with such intensity to the stimulus as to allow relatively close approach by the researcher (<3 m; e.g. Baird, 2013). If the introduction of tethered individuals is too disruptive to the focal animals, presentation of chemical stimuli from them can be used to test concepts in behavioural ecology (Hews et al., 2011). Studies addressing similar questions have involved supplementing subject hormone levels via injections or adhesive patches (Knapp and Moore, 1997). Variables measured in response to a tethered individual generally include frequency and intensity of expressed behaviours (e.g. tongue-flicks, courtship, aggression) or post-encounter hormone levels (Thaker et al., 2009b; Whiting et al., 2009). If presented in an experimental fashion that controls for other biotic and abiotic variables, proximate responses can be used to address ultimate questions (e.g. signal evolution, fitness) with application to the conservation of the focal species.

Behavioural responses to the presence or approach of researchers themselves have both empirical and applied utility for understanding reptile ecology. Holding other features constant (e.g. clothing, hair), researchers can vary the distance, direction, and speed of their approach to simulate a predatory attack. In addition to elucidating sensitivity to changes in threat stimulus, combining multiple variables in a study design can allow researchers to examine behavioural prioritization (e.g. wariness as influenced by prey availability or presence of conspecifics). Addressing behavioural sensitivity in this manner can inform management initiatives intended to minimize impacts on reptiles when human interaction is unavoidable (reviewed in Cooper and Blumstein, 2015). The need to regularly service animals under their care means that personnel at zoological parks can play roles in experiments that address behavioural topics ranging from target conditioning to environmental enrichment (reviewed in Burghardt, 2013).

23.5 Conclusions

Successful conservation efforts have evolved from those that focus only on a single species of concern to those that target entire ecosystems; the prevailing logic in this transition being that providing stewardship for all abiotic and biotic elements within a sufficiently large habitat will necessarily ensure the survivorship of even the most sensitive species. I equate this transition in conservation tactics to Odum's (1984) recognition that mesocosms permit the simultaneous investigation of individual or population-level responses alongside those of an entire ecosystem. Among the greatest strengths of experimental manipulations is their ability to combine ecological realism and relevance to conservation practice with quantified assessment of responses to specific and oftentimes multiple features within the environment. Nobody can dispute the importance of understanding an organism's natural history. Where identifying the most successful conservation strategy is concerned, however, the benefits that come from experimental examination of its ecology (precision, accuracy, control, repeatability) are obvious.

By virtue of having greater species diversity and local abundance and smaller body sizes, it is not surprising that squamates are featured most often in experimental studies of reptile ecology. Given an appropriate choice of species, sample sizes (within one experimental unit or available for an entire study) can rival those reported for similar studies involving amphibians (Harper et al., 2010). Improvements in techniques, ranging from field sampling to trend analysis, continue to expand opportunities available to test responses of reptiles in an experimental context. Where conservation strategies are concerned, the longevity inherent to many reptile species will necessarily delay any determination of the efficacy of an experimental manipulation.

Future opportunities where experimental applications can target reptile conservation include relocation of populations threatened with local extirpation (Tuberville et al., 2005). Not only can cages or specific stimuli be installed on recipient sites, but any experimental modifications to the habitat can mature prior to release of the focal species such that subjects are presented with as realistic a setting as possible. Under these circumstances, even if subjects venture outside the prepared areas of the recipient site (Godwin et al., 2011), their chances of remaining within suitable habitat are likely to improve. Reptile ecologists should collaborate with non-profit conservation groups because many of these organizations acquire large tracts of land with specific conservation objectives in mind. An equally important source of collaborative potential exists in combining experimental studies with existing reserve and exhibit spaces in zoological parks.

Acknowledgements

My interest in the topics covered in this chapter has benefited from discussions with R.J. Cooper, J.W. Gibbons, W.H.N. Gutzke, M.S. Mills, H.R. Mushinsky, S.B. Reichling, G.H. Rodda, and current and former students in my lab. I am grateful to W.E. Cooper, Jr., A.M. Durso, D.K. Hews, and J.H.K. Pechmann for providing feedback that improved earlier drafts of this chapter.

References

Amaral, M.J., Bicho, R.C., Carretero, M.A., et al. (2012). The usefulness of mesocosms for ecotoxicity testing with lacertid lizards. *Acta Herpetologica*, 7, 263–80.

Andersson, M., Krockenberger, A., and Schwarzkopf, L. (2010). Experimental manipulation reveals the importance of refuge habitat temperature selected by lizards. *Austral Ecology*, **35**, 294–9.

Arnqvist, G., and Wooster, D. (1995). Meta-analysis: synthesizing research findings in ecology and evolution. *Trends in Ecology & Evolution*, **10**, 236–40.

Baird, T.A. (2013). Male collared lizards, *Crotaphytus collaris* (Sauria: Crotaphytidae), signal females by broadcasting visual displays. *Biological Journal of the Linnean Society*, **108**, 636–46.

Brischoux, F., Tingley, R., Shine, R., et al. (2012). Salinity influences the distribution of marine snakes: implications for evolutionary transitions to marine life. *Ecography*, **35**, 994–1003.

Buhlmann, K.A., and Coffman, G. (2001). Fire ant predation of turtle nests and implications for the strategy of delayed emergence. *Journal of the Elisha Mitchell Scientific Society*, **117**, 94–100.

Burghardt, G.M. (2013). Environmental enrichment and cognitive complexity in reptiles and amphibians: concepts, review, and implications for captive populations. *Applied Animal Behaviour Science*, **147**, 286–98.

Charland, M.B., and Gregory, P.T. (1990). The influence of female reproductive status on thermoregulation in a viviparous snake, *Crotalus viridis*. *Copeia*, **1990**, 1089–98.

Cooper, W.E., Jr., and Blumstein, D.T. (2015). *Escaping from Predators: An Integrative View of Escape Decisions*. Cambridge: Cambridge University Press.

Crossland, N.O., Heimbach, F., Hill, I.R., et al. (1992). *Summary and Recommendations of the European Workshop on Freshwater Field Tests*, Potsdam, Germany. Brussels: SETAC Europe.

Currylow, A.F., Macgowan, B.J., and Williams, R.N. (2013). Hibernal thermal ecology of eastern box turtles within a managed forest landscape. *Journal of Wildlife Management*, 77, 326–35.

Daly, B.G., Dickman, C.R., and Crowther, M.S. (2008). Causes of habitat divergence in two species of agamid lizards in arid central Australia. *Ecology*, **89**, 65–76.

Davis, J.R., and DeNardo, D.F. (2009). Water supplementation affects the behavioral and physiological ecology of Gila monsters (*Heloderma suspectum*) in the Sonoran Desert. *Physiological and Biochemical Zoology*, **82**, 739–48.

Dodd, C.K., Jr., Barichivich, W.J., and Smith, L.L. (2004). Effectiveness of a barrier wall and culverts in reducing wildlife mortality on a heavily traveled highway in Florida. *Biological Conservation*, **118**, 619–31.

Dorcas, M.E., and Willson, J.D. (2009). Innovative methods for studies of snake ecology and conservation. In S.J. Mullin and R.A. Seigel (eds) *Snakes: Ecology & Conservation*. Ithaca, NY:Cornell University Press, pp. 5–37.

Dorcas, M.E., Willson, J.D., and Gibbons, J.W. (2011). Can invasive Burmese pythons inhabit temperate regions of the southeastern United States? *Biological Invasions*, **13**, 793–802.

Downes, S. (2001). Trading heat and food for safety: costs of predator avoidance in a lizard. *Ecology*, **82**, 2870–81.

Downes, S., and Hoefer, A.M. (2007). An experimental study of the effects of weed invasion on lizard phenotypes. *Oecologia*, **153**, 775–85.

Ferguson, G.W., Gehrmann, W.H., Karsten, K.B., et al. (2003). Do panther chameleons bask to regulate endogenous vitamin D3 production? *Physiological and Biochemical Zoology*, **76**, 52–9.

Ford, N.B. (1995). Experimental design in studies of snake behavior. *Herpetological Monographs*, **9**, 130–9.

Fornell, A.M. (2008). *Foraging site selection in prairie kingsnakes (*Lampropeltis c. calligaster*): an experimental approach*. M.S. Thesis, Eastern Illinois University, Charleston, IL, USA.

Godwin, J., Wines, M., Stiles, J., et al. (2011). *Reintroduction of the Eastern Indigo Snake (*Drymarchon couperi*) into Conecuh National Forest*. Final report submitted to the Alabama Department of Conservation and Natural Resources, Montgomery, AL, USA.

Greene, M.J., and Mason, R.T. (2003). Pheromonal inhibition of male courtship behaviour in the brown tree snake, *Boiga irregularis*: a mechanism for the rejection of potential mates. *Animal Behaviour*, **65**, 905–10.

Grillet, P., Cheylan, M., Thirion, J.M., et al. (2010). Rabbit burrows or artificial refuges are a critical habitat component for the threatened lizard, *Timon lepidus* (Sauria, Lacertidae). *Biodiversity and Conservation*, **19**, 2039–51.

Hammerson, G.A. (1987). Thermal behaviour of the snake *Coluber constrictor* in west-central California. *Journal of Thermal Biology*, **12**, 195–7.

Harper, E.B., Pechmann, J.H.K., and Petranka, J.W. (2010). Field enclosures and terrestrial cages. In C.K., Dodd, Jr. (ed) *Amphibian Ecology and Conservation. A Handbook of Techniques*. Oxford: Oxford University Press, pp. 203–26.

Hews, D.K., Date, P., Hara, E., et al. (2011). Field presentation of male secretions alters social display in *Sceloporus virgatus* but not *S. undulatus* lizards. *Behavioral Ecology and Sociobiology*, **65**, 1403–10.

Himes, J.G. (2003). Intra- and interspecific competition among the water snakes *Nerodia sipedon* and *Nerodia rhombifer*. *Journal of Herpetology*, **37**, 126–31.

Isaac, L.A., and Gregory, P.T. (2004). Thermoregulatory behaviour of gravid and non-gravid female grass snakes (*Natrix natrix*) in a thermally limiting high-latitude environment. *Journal of Zoology*, **264**, 403–9.

Kaufman, J.D., Burghardt, G.M., and Phillips, J.A. (1994). Density-dependent foraging strategy of a large carnivorous lizard, the savanna monitor (*Varanus albigularis*). *Journal of Comparative Psychology*, **108**, 381–4.

Kaufman, J.D., Burghardt, G.M., and Phillips, J.A. (1996). Sensory cues and foraging decisions in a large carnivorous lizard, *Varanus albigularis*. *Animal Behaviour*, **52**, 727–36.

Kingsbury, B.A. (1989). Factors influencing activity in *Coleonyx variegatus*. *Journal of Herpetology*, **23**, 399–404.

Kissner, K.J., and Weatherhead, P.J. (2005). Phenotypic effects on survival of neonatal northern watersnakes *Nerodia sipedon*. *Journal of Animal Ecology*, **74**, 259–65.

Knapp, R., and Moore, M.C. (1997). Male morphs in tree lizards have different testosterone responses to elevated levels of corticosterone. *General and Comparative Endocrinology*, **107**, 273–79.

Krebs, C.J. (1998). *Ecological Methodology*. 2nd ed. Menlo Park, CA:Benjamin Cummings.

Langkilde, T., Lance, V.A., and Shine, R. (2005). Ecological consequences of agonistic interactions in lizards. *Ecology*, **86**, 1650–9.

Lee, J.R., and Mills, M.S. (2000). Design and construction of an outdoor enclosure for the study of snake thermal ecology. *Herpetological Review*, **31**, 24–6.

Lin, H.-C., Hung, H.-Y., Lue, K.-Y., et al. (2007). Diurnal retreat site selection by the arboreal Chinese Green Tree Viper (*Trimeresurus s. stejnegeri*) as influenced by temperature. *Zoological Studies*, **46**, 216–26.

Mann, S.L., and Meek, R. (2004). Understanding the relationship between body temperature and activity patterns in the giant Solomon Island Skink, *Corucia zebrata*, as a contribution to the effectiveness of captive breeding programmes. *Applied Herpetology*, **1**, 287–98.

Mason, R.T., and Parker, M.R. (2010). Social behavior and pheromonal communication in reptiles. *Journal of Comparative Physiology A*, **196**, 729–49.

McCann, M.J. (2011). *Burrow characteristics and social elements of captive juvenile bolson tortoises (*Gopherus flavomarginatus*) within a headstart enclosure in New Mexico*. M.S. Thesis, New Mexico State University, Las Cruces, NM, USA.

Meek, R. (2005). Null models and the thermal biology of the anguid lizard *Anguis fragilis*: evidence for thermoregulation? *Amphibia-Reptilia*, **26**, 445–50.

Mullin, S.J., and Cooper, R.J. (2002). Barking up the wrong tree: climbing performance of rat snakes and its implications for depredation of avian nests. *Canadian Journal of Zoology*, **80**, 591–5.

Mullin, S.J., and Mushinsky, H.R. (1997). Use of experimental enclosures to examine foraging success in water snakes: a case study. *Journal of Herpetology*, **31**, 565–9.

Mullin, S.J., Cooper, R.J., and Gutzke, W.H.N. (1998). The foraging ecology of the gray rat snake (*Elaphe obsoleta spiloides*). III. Searching for different prey types in structurally varied habitats. *Canadian Journal of Zoology*, **76**, 548–55.

Nagy, K.A., Scott Hillard, L., Tuma, M.W., and Morafka, D.J. (2015). Head-started desert tortoises (*Gopherus agassizii*): movements, survivorship and mortality causes following their release. *Herpetological Conservation and Biology*, **10**, 203–15.

Newbold, T.A.S., and MacMahon, J.A. (2014). Determinants of habitat selection by Desert Horned Lizards (*Phrynosoma platyrhinos*): the importance of abiotic factors associated with vegetation. *Journal of Herpetology*, **48**, 306–16.

Niewiarowski, P.H., and Roosenburg, W. (1993). Reciprocal transplant reveals sources of variation in growth rates of the lizard *Sceloporus undulatus*. *Ecology*, **74**, 1992–2002.

Nordberg, E.J., and Schwartzkopf, L. (2015). Arboreal cover boards: using artificial bark to sample cryptic arboreal lizards. *Herpetologica*, **71**, 268–73.

Odum, E.P. (1984). The mesocosm. *Bioscience*, **34**, 558–62.

Pike, D.A., Croak, B.M., Webb, J.K., et al. (2010). Subtle—but easily reversible—anthropogenic disturbance seriously degrades habitat quality for rock-dwelling reptiles. *Animal Conservation*, **13**, 411–18.

Reed, R.N. (2003). Interspecific patterns of species richness, geographic range size, and body size among New World venomous snakes. *Ecography*, **26**, 107–17.

Rodda, G.H., Savidge, J.A., Tyrrell, C.L., et al. (2007). Size bias in visual searches and trapping of brown treesnakes on Guam. *Journal of Wildlife Management*, **71**, 656–61.

Rose, P., Evans, C., Coffin, R., et al. (2014). Using student-centred research to evidence-base exhibition of reptiles and amphibians: three species-specific case studies. *Journal of Zoo and Aquarium Research*, **2**, 25–32.

Rostal, D.C., Owens, D.W., Grumbles, J.S., et al. (1998). Seasonal reproductive cycle of the Kemp's Ridley Sea Turtle (*Lepidochelys kempi*). *General and Comparative Endocrinology*, **109**, 232–43.

Sabo, J.L., and Power, M.E. (2002). River–watershed exchange: effects of riverine subsidies on riparian lizards and their terrestrial prey. *Ecology*, **83**, 1860–9.

Schneider, M., Klein, B., Krämer, D., et al. (2014). First observations on the courtship, mating, and nest visit behaviour of the Philippine crocodile (*Crocodylus mindorensis*) at the Cologne Zoo. *Journal of Zoo and Aquarium Research*, **2**, 123–9.

Schoener, T.W., and Schoener, A. (1983). Distribution of vertebrates on some very small islands. II. Patterns in species number. *Journal of Animal Ecology*, **52**, 237–62.

Schoener, T.W., Spiller, D.A., and Losos, J.B. (2002). Predation on a common *Anolis* lizard: can the food-web effects of a devastating predator be reversed? *Ecological Monographs*, **72**, 383–407.

Schoener, T.W., Spiller, D.A., and Losos, J.B. (2004). Variable ecological effects of hurricanes: the importance of seasonal timing for survival of lizards on Bahamian islands. *Proceedings of the National Academy of Sciences of the USA*, **101**, 177–81.

Schuett, G.W., and Gillingham, J.C. (1988). Courtship and mating of the copperhead, *Agkistrodon contortrix*. *Copeia*, **1988**, 374–81.

Scott, M.L., Whiting, M.J., Webb, J.K., et al. (2013). Chemosensory discrimination of social cues mediates space use in snakes, *Cryptophis nigrescens* (Elapidae). *Animal Behaviour*, **85**, 1493–500.

Shine, R., and Bonnet, X. (2009). Reproductive biology, population viability, and options for field management. In S.J. Mullin and R.A. Seigel (eds) *Snakes: Ecology & Conservation*. Ithaca, NY:Cornell University Press, pp. 172–200.

Shine, R., and Sun, L.X. (2003). Attack strategy of an ambush predator: which attributes of the prey trigger a pit-viper's strike? *Functional Ecology*, **17**, 340–8.

Shine, R., Brown, G.P., and Elphick, M.J. (2004). Field experiments on foraging in free-ranging water snakes *Enhydris polylepis* (Homalopsinae). *Animal Behaviour*, **68**, 1313–24.

Shine, R., Sun, L.X., Zhao, E., et al. (2002). A review of 30 years of ecological research on the Shedao Pitviper, *Gloydius shedaoensis*. *Herpetological Natural History*, **9**, 1–14.

Shine, R., Webb, J.K., Lane, A., et al. (2005). Mate location tactics in garter snakes: effects of rival males, interrupted trails and non-pheromonal cues. *Functional Ecology*, **19**, 1017–24.

Shoemaker, K.T., and Gibbs, J.P. (2010). Evaluating basking-habitat deficiency in the threatened Eastern Massasauga Rattlesnake. *Journal of Wildlife Management*, **74**, 504–13.

Steen, D.A., Smith, L.L., Conner, L.M., et al. (2013). Reptile assemblage response to restoration of fire-suppressed longleaf pine sandhills. *Ecological Applications*, **23**, 148–58.

Subalusky, A.L., Fitzgerald, L.A., and Smith, L.L. (2009). Ontogenetic niche shifts in the American Alligator establish functional connectivity between aquatic systems. *Biological Conservation*, **142**, 1507–14.

Sutton, W.B., Wang, Y., Schweitzer, C.J., et al. (2013). Lizard microhabitat and microclimate relationships in southeastern pine-hardwood forests managed with prescribed burning and thinning. *Forest Science*, **60**, 180–90.

Thaker, M., Lima, S.L., and Hews, D.K. (2009a). Acute corticosterone elevation enhances anti-predator behaviors in male tree lizard morphs. *Hormones and Behavior*, **56**, 51–7.

Thaker, M., Lima, S.L., and Hews, D.K. (2009b). Alternative antipredator tactics in tree lizard morphs: hormonal and behavioural responses to a predator encounter. *Animal Behaviour*, 77, 395–401.

Todd, A.C. (2005). The social mating system of *Hoplodactylus maculatus*. *New Zealand Journal of Zoology*, **32**, 251–62.

Tuberville, T.D., Clark, E.E., Buhlmann, K.A., et al. (2005). Translocation as a conservation tool: site fidelity and movement of repatriated gopher tortoises (*Gopherus polyphemus*). *Animal Conservation*, **8**, 349–58.

Virkki, D.A., Tran, C., and Castley, J.G. (2012). Reptile responses to Lantana management in a wet sclerophyll forest, Australia. *Journal of Herpetology*, **46**, 177–85.

Vogt, R.C. (2008). *Amazon Turtles*. Lima, Peru: Bíblios.

Wasko, D.K., and Sasa, M. (2012). Food resources influence spatial ecology, habitat selection, and foraging behavior in an ambush-hunting snake (Viperidae: *Bothrops asper*): an experimental study. *Zoology*, **115**, 179–87.

Whiting, M.J., Webb, J.K., and Keogh, J.S. (2009). Flat lizard female mimics use sexual deception in visual but not chemical signals. *Proceedings of the Royal Society of London B: Biological Sciences*, **276**, 1585–91.

Willson, R.J., and Brooks, R.J. (2006). Thermal biology of reproduction in female Eastern Foxsnakes (*Elaphe gloydi*). *Journal of Herpetology*, **40**, 285–9.

Witherington, B.E. (1992). Behavioral responses of nesting sea turtles to artificial lighting. *Herpetologica*, **38**, 31–9.

24

Body temperatures and the thermal environment

Keith A. Christian, Christopher R. Tracy, and C. Richard Tracy

24.1 Introduction

24.1.1 The importance thermal biology

Body temperature (T_b) influences most physiological processes, including metabolism, digestion, reproductive physiology, and locomotion. Consequently, an animal's T_b has ecological consequences, including its ability to escape predators, capture prey, grow, reproduce, and be susceptible to disease (Christian and Tracy, 1981; Huey, 1982). Reptiles and other ectothermic animals rely on their environments to achieve their T_b using (a) behaviour to select among thermal patches in the environment, and (b) behaviour (along with their morphology and physiological adjustments) to control heat exchanges within a thermal patch. However, thermoregulation is not an end in itself, but rather a means to influence behavioural and physiological performances that affect fitness (Huey, 1982).

Because thermoregulation is an interplay between the environment and the animal, thermoregulation is a multifactorial process leading to many possible outcomes and, thus, many potential strategies to achieve a desired T_b. Two animals in the same thermal patch can have very different T_bs, depending on how the thermal environment within the patch is exploited using behaviour, physiology, and morphology. Apart from this complex interplay giving reptiles a range of T_b options, this complexity introduces challenges for biologists trying to understand and quantify thermal biology.

24.1.2 The importance of using appropriate techniques to study thermal biology

To be able to quantify and compare reptiles of different species, during different seasons, or from different habitats, we need techniques that measure what we need to know. It is not always obvious what we need to measure, and indeed, as discussed in the next section, it took many years for reptile biologists to develop techniques that adequately measure the variables needed to answer questions about thermoregulation. The development of new technologies, such as small, easy to use data loggers, both provides new opportunities to collect a large amount of data very quickly and new ways to make mistakes if the underlying principals are not understood.

Reptile Ecology and Conservation. Edited by C. Kenneth Dodd, Jr. © Oxford University Press 2016.
Published 2016 by Oxford University Press.

Another reason we need appropriate techniques to study thermal biology is that, as upright endotherms, we have a biased perception of the thermal environment. A reptile on the ground experiences a very different microclimate from the one we experience standing well above the ground on our thick-soled shoes. On a cool, windy, sunny morning, a biologist may need to wear a jacket to observe lizards that have already retreated to shade after basking on a rock where wind speeds are low and environmental temperatures are high. Conversely, a biologist walking under the canopy of a humid rainforest would probably be sweating profusely in an air temperature (T_a) of 30°C, while the lizard she is studying is searching for a sun fleck so it can raise its T_b. The high humidity is of little consequence to the T_b of the lizard, but it is all sweaty biologists ever talk about among themselves. Thus, as upright endotherms, we need robust techniques to overcome our anthropocentric views of the thermal environment.

24.1.3 Historical perspective

The importance of physiological homeostasis has long been a concept in biology, but only 70 years ago, Cowles and Bogert (1944) published a study of desert lizards introducing the concept of behavioural homeostasis. 'This discovery revolutionized the philosophy and methodology of physiology and ecology' (Huey, 1982). Following this revelation, reptile biologists went into the field with quick-reading thermometers to further document thermoregulation. It was also easy to measure T_a at the same time, and any deviations between T_b and T_a were taken as data supporting the concept that the reptile was thermoregulating.

A wake-up call came 20 years later (Heath, 1964) with the report that beer cans filled with water and placed in the sun had temperatures throughout the day that were similar to those of lizards in the same environment. Heath's point was neither that beer cans thermoregulate nor that lizards do not. The objective of this elegant experiment was to demonstrate that thermal biologists badly needed better techniques to quantify thermoregulation, and by this time it was clear that reptiles were diverse in their thermoregulatory patterns and behaviour (Brattstrom, 1965).

A better way to quantify the thermal environment and to combine that information with characteristics of the animals came a few years later in the form of biophysical modelling (Bartlett and Gates, 1967; Norris, 1967). This technique involved measurements of the microclimate around the animal, measurements of the relevant characteristics of the animal, and combining these variables in an energy balance equation. This concept and the associated techniques were subsequently expanded and refined (Porter and Gates, 1969; Bakken and Gates, 1975; Tracy, 1975, 1976; Bakken 1976, 1981a). This approach highlighted the fact that the thermal environment is much more complex than T_a alone. In addition to other environmental temperatures (the ground and other nearby objects—even the sky), solar radiation (short-wave), thermal radiation (long-wave), and wind speed were all recognized as being important. Furthermore, the size and shape of the animal, its orientation with respect to the sun and wind, and the extent to which solar radiation is absorbed (absorptivity) were included in the energy balance equation. Thus, two reptiles that differ only in body size may come to different T_bs even if they are side-to-side in the environment (Spotila et al., 1973; Stevenson, 1985a). These revelations were both enlightening and confusing.

One of the most confusing aspects of biophysical models was what to call the temperature that resulted from solving the energy balance equation. It was not strictly an environmental temperature, and it was not strictly an animal's T_b, but rather it combined environmental variables with animal characteristics. Several different names were suggested, but these indices were mathematically equivalent (Bakken, 1981b), and the name 'operative temperature' (symbolized T_e) has become accepted. T_e is an index that represents the body temperature to which an animal would equilibrate in a given environment. As defined, it is a 'steady-state' (unchanging) temperature because the animal is assumed to have no mass, which allows the animal to achieve T_e instantaneously for a given set of environmental conditions and for a defined set of animal characteristics. The T_e of the animal changes instantaneously as environmental conditions change, for example, if the animal moves from a sunny to a shady position. Of course, instantaneous change is not realistic, particularly for larger animals, but this is not a problem if one keeps in mind that the index represents the consequences of the animal–environment interactions for each set of environmental conditions. T_e can also be considered an environmental 'driving force'.

Increasingly complex energy balance models that include the mass of the animal were subsequently developed to describe the animal–environment interactions more realistically (Porter et al., 1973; Spotila et al., 1973; Porter and James, 1979; Christian et al., 1983). These 'transient-state' models seek to reflect the changes in T_b over time. These models can also incorporate physiological processes, such as the ability to change blood flow to appendages, thereby controlling the rates of heating and cooling. When the more complex models are used, their output should be considered a 'predicted T_b' ($T_{b\ pred}$) rather than an operative temperature. Neither T_e nor $T_{b\ pred}$ is better than the other, but the choice depends on the specific questions being asked and the characteristics (particularly body size) of the animals being studied (see Section 24.5).

The next major advance in quantifying reptile thermoregulation came when researchers recognized that the question, 'How carefully does a reptile thermoregulate?' was not the right question because there are several aspects of thermoregulation that cannot be addressed when you start with that (seemingly) simple question (Hertz et al., 1993). We need to consider how variable the T_bs are (the 'precision' of thermoregulation), how closely an animal's T_bs are to its target, or set-point range (the 'accuracy' of thermoregulation), and whether the thermoregulatory behaviours of the animal result in more-accurate T_bs compared to a null distribution of T_bs that would be experienced by a lizard using its habitat randomly (the 'effectiveness' of thermoregulation) (Hertz et al., 1993). These different components are not necessarily correlated, so they must be addressed individually.

Hertz et al. (1993) proposed several indices of thermoregulation to address these different questions. Their indices incorporate the available thermal environment (hence, building on the T_e concept) and the concept of a set-point temperature range (T_{set}). By definition, the set-point range is determined in a laboratory thermal gradient rather than in the field. In the simple laboratory environment, the animal does not have to compromise its thermoregulatory behaviour to find food, avoid predators, or engage in activities other than thermoregulation. By convention, the T_{set} is not defined as a single T_b, but taken as the range of the central 50% of T_bs selected in the laboratory (but see Section 24.6).

The details of these indices are given in Hertz et al. (1993), but we give a brief description here. In addition to determining the T_{set} in the laboratory, the null distribution of T_es available in the animal's environment needs to be determined. This can be done by mapping the thermal environment, using computational energy-balance models (Porter et al., 1973; Waldschmidt and Tracy, 1983—see Section 24.2.2) or physical models (Bakken and Gates, 1975—see Section 24.2.3). The accuracy of thermoregulation can be quantified using the index d_b, which compares the T_bs of the animal to its T_{set}. If most T_bs are within the T_{set}, the index will indicate a high degree of thermoregulatory accuracy (as represented by a small mean d_b). However, a small d_b does not necessarily imply active thermoregulation because the thermal environment has not yet been taken into account; the T_es may simply all fall within the T_{set}. The thermal quality of the habitat can be quantified by the index d_e, which compares the T_es in the environment with the T_{set} of the animal. If there are many places in the animal's habitat in which the T_es are within the T_{set}, this index will indicate high thermal quality (as represented by a small mean d_e). If mean d_b < mean d_e, the T_bs of the animal are within the T_{set} more often than would be the case if it were randomly using its thermal environment (null model), therefore indicating thermoregulation. Thus, the effectiveness of thermoregulation, or the extent of departure from thermoconformity, is calculated as $d_e - d_b$ (Blouin-Demers and Weatherhead, 2001; Blouin-Demers and Nadeau, 2005).

These indices do not consider T_bs alone, but rather their function is to evaluate T_b in the context of the available thermal environment. An advantage of the indices is that they are scalable; they can be used in any comparison (e.g. different days, individuals, and seasons). Furthermore, when the indices are calculated for multiple individuals, it is possible to use the indices themselves as response variables to make comparisons using standard statistical techniques (Christian and Weavers, 1996).

The biggest disadvantage of these indices is that they are not intuitive or easily interpreted. It requires a lot of effort to interpret a d_b of 2.5 or a d_e of 5. With this in mind, Christian and Weavers (1996) proposed a more intuitive graphical technique and an associated index of exploitation (Ex) (Figure 24.1). The Ex index is simply calculated by expressing the amount of time that the animal spends within its T_{set} as a percentage of the amount of time that the animal could possibly be within its T_{set}. Thus, in the example of Figure 24.1(a), the lizard exploited all of the time that its T_{set} was available, resulting in an Ex = 100%. This index is undefined when environmental conditions do not allow the animal to attain T_{set} even though the animal may be exhibiting thermoregulatory behaviour (Figure 24.2(d)). Thus, this index is restricted to conditions in which T_{set} is attainable (Christian and Weavers, 1996; Blouin-Demers and Nadeau, 2005).

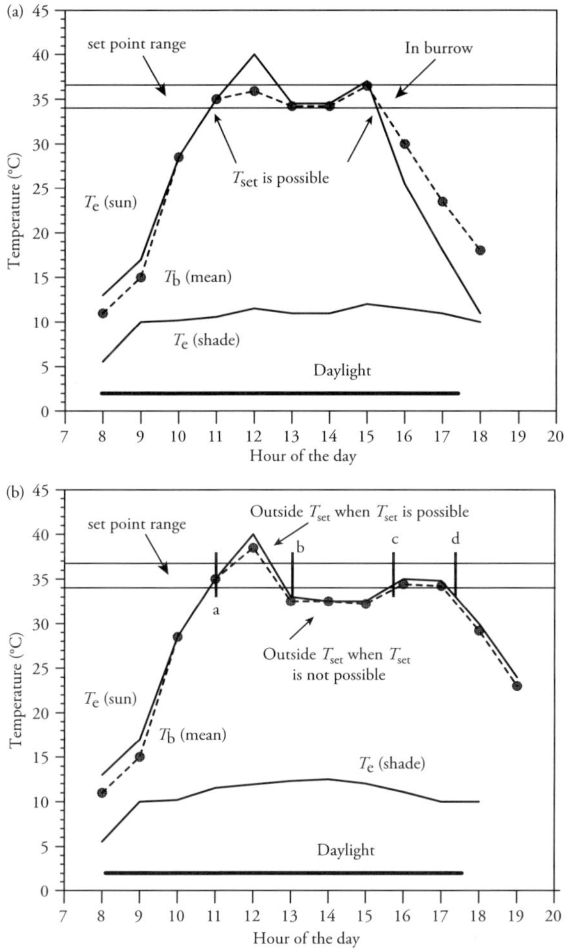

Figure 24.1 The index of exploitation (Ex) involves plotting the T_bs of the animal across the hours of the day and superimposing the set-point range (as a pair of parallel lines), the maximum T_e assuming that the animal remains in the sun all day (T_e (sun)), and the minimum T_e assuming that the animal remains in the coolest environment all day (T_e (shade)). Other T_es could be added as appropriate. It is fair to assume that all possible T_es in between these extremes could be attained with varying levels of shade. (a) *Varanus rosembergi* on a sunny spring day. A researcher first has to decide the appropriate time-scale to use to calculate Ex (and other indices of thermoregulation). It would be possible to calculate it over a 24 hour day, or from sunrise to sunset (illustrated by the "Daylight" bar). Christian and Weavers (1996) argued that the time-scale depends on the question being asked, but that the greatest insight into thermoregulation is gained from limiting the calculations to the period during which it is possible for the animal to attain its T_{set} range (the period designated as "Tset is possible"). This graph represents real data from a lizard that thermoregulated perfectly—that is, Ex = 100%. (b) Hypothetical data during a partly cloudy spring day. The periods between the labels a and b and between c and d represent times when the T_{set} would be possible, so the indices of thermoregulation would be calculated for these periods. During the period between b and c, however, it would not be possible for the animal to attain T_{set} (presumably because of cloud cover), so this period would be excluded. The small peak during which the T_b exceeds the T_{set} represents a period in which it would be possible to achieve the T_{set}, but the animal failed to do so, thus lowering the Ex index.

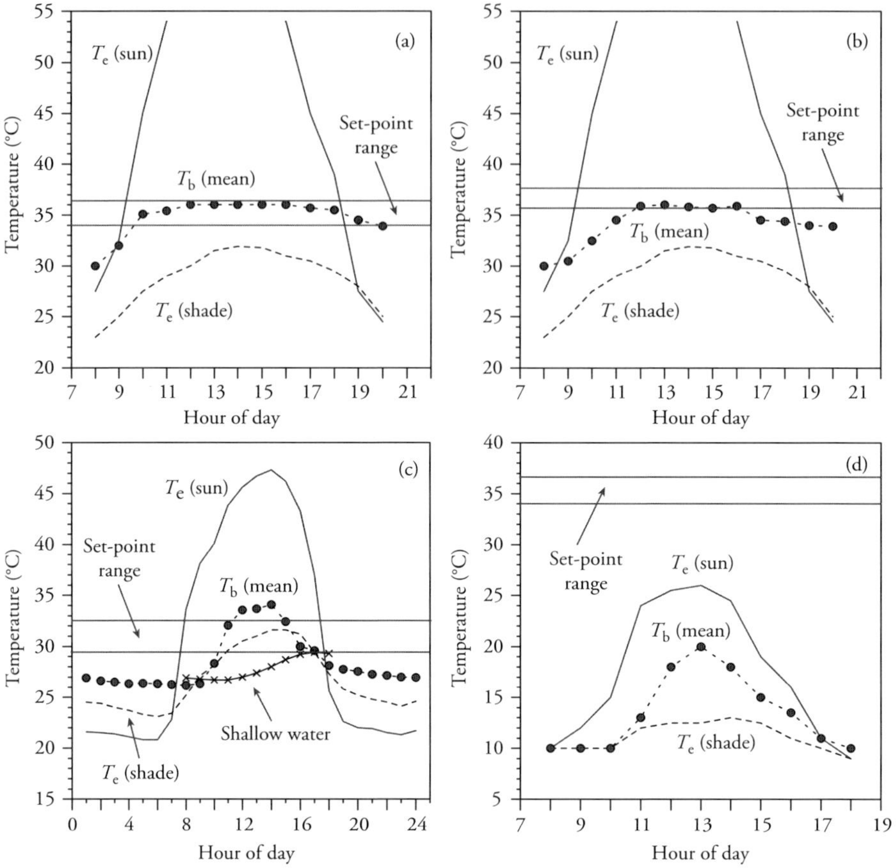

Figure 24.2 (a) The lizard *Varanus gouldii* thermoregulates carefully with Ex = 96%. (b) *V. panoptes* do not exploit the early morning and late afternoon thermal environments to the full extent possible, but during the middle of the day their T_bs are within the T_{set}. An Ex = 44% indicates imprecise thermoregulation, but these lizards nevertheless do not overheat because they avoid the extremely hot microenvironment in full sun. (c) Juvenile Saltwater Crocodiles (*Crocodylus porosus*) allow T_b to go outside both sides of the T_{set} range (Brien, 2015), resulting in an Ex = 33%. (d) *V. rosembergi* cannot achieve their T_{set} at any point during the day in winter (thus, Ex is undefined). Inexplicably, these lizards nevertheless came out of their burrows and basked (Christian and Weavers, 1996). This attempt to thermoregulate is not captured by the Ex index (Blouin-Demers and Nadeau, 2005).

24.2 Techniques for quantifying thermal biology

There are two ways to determine T_e for use in evaluating thermoregulation in reptiles: computational models and physical models.

24.2.1 Computational models

Computational models require information about the animal being studied and microclimates throughout the day. The critical animal characteristics include the size (mass, surface area, posture) and solar absorptivity. Metabolic rate and evaporative water loss are also components of an energy balance equation, but these are small relative to the other components for reptiles and are typically omitted (Tracy, 1982). Microclimatic variables include T_a, wind speed, short-wave radiation (direct solar, scattered solar, and reflected), ground temperature, thermal radiation from the ground (as determined by ground temperature), and thermal radiation from the sky (which can be calculated from air temperature).

The energy balance equation in its simplest form is: energy in = energy out. A slightly more expanded form is: absorbed radiation = energy exchange due to convection + radiant energy lost + energy conducted to or away from the animal. Each of these terms needs to be further expanded to incorporate the relevant animal characteristics and details of the energy exchange mechanisms. Detailed equations can be found in Tracy (1982), Porter and Tracy (1983), Bakken (1992), and other references cited in Section 24.1.

Direct and solar radiation can be measured with a pyranometer or calculated (Porter and Tracy, 1983). Solar radiation reflected off the ground can be measured with a pyranometer pointed downward. The T_a can be accurately measured using small thermocouples that are either shaded or painted bright white (Christian and Tracy, 1985) to avoid errors due to radiation. If there is no solar radiation in the environment (at night, in a burrow, or in a tree hollow), the small single channel data loggers such as iButtons (Maxim Integrated, San Jose, CA) or HOBO TidBits (Onset Computer Corporation, Bourne, MA) can be used to measure T_a. Soil surface temperature is best measured with an infrared thermometer. Wind speed is measured with an anemometer (cup, hot-wire, or sonic), preferably with the ability to integrate measurements over time. Care must be taken in any of these measurements to ensure the accuracy of the thermocouple or data logger. Calibration is essential, and there is no point in reporting temperatures to several decimal places if the temperature sensor has an accuracy of ±0.5°C.

24.2.2 Physical models

A physical representation of the animal with the appropriate characteristics can also be used to determine T_e (Bakken and Gates, 1975), and the result should be the same as determined by a computational model (Bakken, 1981b). Ideally, models can be cast in Wood's metal or made of hollow copper, then painted to match the absorptivity of the reptile (Bakken and Gates, 1975; Dzialowski, 2005; Bakken and Angilletta, 2014). Note that simply matching colours is not sufficient because the critical characteristic is absorptivity, not colour. The aim is for the model to achieve an equilibrium temperature quickly, so the materials used should have high conductivity.

The temperatures of physical models can be calibrated against the real animal (sometimes a dead specimen). Anatomically accurate models generally produce smaller errors (Walsberg and Wolf, 1996; Dzialowski, 2005; Clusella-Trullas et al., 2009; Bakken and Angilletta, 2014), but it is possible that some of the precise detail (scale patterns, exact shape) may not necessarily be required (Shine and Kearney, 2001). Many studies have used hollow copper tubes painted to match absorptivity (see Table 1 in Dzialowski, 2005) for their ease of production. These can result in errors larger than the 2°C threshold for acceptability suggested by Dzialowski (2005), and it is likely that legless copper tubes are more anatomically appropriate as snake models (Peterson, 1987; Dorcas and Peterson, 1998; Row and Blouin-Demers, 2006) than as lizard models (Walsberg and Wolf, 1996). Appendages increase the options for physiological control of heat exchange (which is not an issue for physical T_e models) and the surface area exposed to radiant and convective exchange (which is an issue for T_e models) (Tracy, 1982). We advocate a balance between detail and ease of use, but certainly the critical features of high thermal conductivity of the model material and absorptivity and surface area that are matched to the animal must be taken into account, and calibration against the real animal is crucial (Bakken and Gates, 1975; Shine and Kearney, 2001; Dzialowski, 2005; Bakken and Angilletta, 2014).

Once the design and construction materials have been decided, it may be worthwhile to produce a number of models (Dzialowski, 2005), and it may be worth considering making more than one type representing variable characteristics such as posture, absorptivity (for species that can change absorptivity), or size. They can be dispersed in the environment randomly, or placed in specific microhabitats (full sun, full shade) to monitor the extremes. The random dispersal of physical models can be a way of determining the availability of T_es throughout the day at a site (Hertz, 1992). Temperatures of the physical models can be taken manually (using an imbedded probe or an infrared thermometer) or they can be logged using inserted probes connected to a data logger or with single channel data loggers embedded inside the model and retrieved later.

24.3 Advantages and disadvantages of computational and physical models

Each approach has its advantages and disadvantages, so the choice will depend on factors such as resources available and the specifics of the kinds of questions being asked.

Computational models require access to micrometeorological equipment (or to relevant micrometeorological data) and an appropriate energy balance model. Although some of these instruments can be expensive, there are many options available, and a basic set of instruments is not extremely expensive. To develop a working energy-balance model from scratch requires a good understanding of the first principals involved (mechanisms of energy exchange) and how they fit together in an energy balance equation, but there are many resources that can help (Tracy, 1976, 1982; Bakken, 1981a; Porter and Tracy 1983).

Once the model has been constructed, it can be used not only to address the conditions you have measured in the field (to allow you to calculate indices of thermoregulation and interpret the behaviours you may have observed), but it can be used to explore hypothetical scenarios (climate change, land clearing, or other anthropogenic or natural changes to a habitat) or changes in the animal (different body sizes, different solar absorptivities). Thus, a robust computational model is an excellent tool for exploring animal–environment interactions that are occurring now, may have occurred in the past, or may occur in the future. It is also possible to add complexity to computational models (body mass, physiological control of heating and cooling rates) so that they can be used to explore thermal transients, which is particularly important for larger animals (Section 24.5).

Once constructed, physical models are easy to use. Physical models, however, are less useful than computational models for exploring hypothetical situations or thermal transients. The materials used to make physical models do not have the thermal properties of animal tissues. Thus, the equilibrium temperature of a T_e model should be the same as the equilibrium temperature that an animal would (eventually) assume in the conditions, but the rate at which the two would reach equilibrium temperature would be very different. Water-filled models certainly produce a lag in thermal responses, but it is doubtful that they do so in a way representative of real animals (Bakken and Angilletta, 2014). Thus, physical models are a good way to measure T_e under a set of environmental conditions, but they are not useful for determining $T_{b\,pred}$ or thermal transients.

24.4 Use of data loggers as surrogate physical T_e models

Although small data loggers are useful for some microclimate measurements, they do not have the critical characteristics to serve as substitutes for physical T_e models. Without the correct surface area, shape and absorptivity, the resulting measurements would be suspect and the use of data loggers as surrogate T_e models is not advised.

Vitt and Sartorius (1999) published empirical data suggesting that small data loggers like HOBOs or iButtons can be adequate surrogate T_e models. These results are difficult to evaluate completely because of the lack of information about the microclimatic conditions. However, they reported errors up to 6°C, which are greater than Dzialowski's (2005) criterion for acceptability and are not trivial, especially if one is documenting something as important as the consequences of climate change (Bakken and Angilletta, 2014). We also note that other studies have demonstrated effects of small differences in shape, size, and absorptivity (Walsberg and Wolf, 1996; Bakken and Angilletta, 2014), and the relative importance of wind, absorptivity, and shape can change in the broader context of environmental conditions. Thus, the reasonably close results of Vitt and Sartorius (1999) may have been due to chance and cannot be generalized to other conditions. These authors emphasized this point, but others have not heeded their warning (Sartorius et al., 1999; Price-Rees et al., 2013). Similarly, the use of physical models with an inappropriate surface area (Peterson et al., 1993) may result in acceptable errors in some, but not all, circumstances. However, under conditions of no solar radiation and little wind (underground, in a burrow, in a tree hollow, or under water), small data loggers can be appropriate T_e surrogates.

24.5 Thermal transients

The T_e, by definition, is an index that does not incorporate mass, so it does not include any information related to the heat capacity or 'thermal inertia' of the animal for which it is being used. This is not a shortcoming of the index, considering that it was designed to represent the driving force of the thermal environment on the T_b of an animal. However, large reptiles do exhibit thermal inertia, thus making it less straightforward to use the indices of thermoregulation described in Section 24.1.3. For example, consider two lizards, one large and one small, in the same environment in which T_e in the sun is 45°C and T_e in the shade is only 25°C. Assume that both stay in the sun until their respective T_bs = 38°C (which will happen more quickly for the small lizard), but then they move into the shade. The smaller lizard will start to cool very quickly, but the large lizard is more complex. Because of its thermal inertia, and because its skin temperature is likely to be several degrees warmer than its core T_b, the core T_b of the large lizard would continue to rise in the new cooler environment. Thus, the mass of the large lizard results in a lag related to its immediate past environment. This means that the information about the current T_e is insufficient to predict how the large lizard's T_b will change because its T_b is influenced both by the T_e of its new environment and the lagged effects of its previous environment (Christian et al., 2006).

24.5.1 How to account for thermal transients in large animals

To take account of the effects of thermal inertia, random movements, and different rates of heating and cooling, Christian et al. (2006) created a technique that calculates a null distribution of body temperatures that can be used in place of a null distribution based on T_e. Thus, the mean achievable temperature ($T_{b\ pred}$) is substituted for mean T_e in the calculations of the index d_e, which is a measure of the thermal quality of an animal's environment. This is achieved by combining the results of a computational energy balance model with a model of random interactions with the thermal environment (random movements to different places on a random schedule) over the course of a day. The extreme temperatures available in the environment ($T_{e\ min}$ and $T_{e\ max}$) are calculated across the day using a steady-state computational energy balance model with the appropriate microclimate data and animal characteristics. These temperatures set the bounds for $T_{b\ pred}$ and act as driving variables. After an animal moves to a new place (after a randomly determined period), its new $T_{b\ pred}$ is calculated every minute using T_e as the driving force and the thermal time constant to account for mass (Dzialowski and O'Connor, 2001; Christian et al., 2006). Here, the T_e is randomly selected as $T_{e\ min}$, $T_{e\ max}$, or any intermediate temperature (representing varying degrees of partial shade). After the randomly determined period has elapsed, the model randomly allocates a new place and time, and this process continues for a 24 hour day. This is repeated for 1000 iterations (days) in which the microclimate and animal characteristics remain the same, but the random movements result in a null distribution of $T_{b\ pred}$ over the day. The mean $T_{b\ pred}$ for each time period is used to calculate d_e (rather than using a mean T_e as originally described by Hertz et al., 1993). The d_b index and the other indices of thermoregulation are calculated as originally described (Hertz et al., 1993).

This technique modifies T_e to reflect the fact that a large moving animal in a complex thermal environment may not be able to achieve a steady-state T_b. That is, the range of achievable T_bs of a large reptile is narrower than the range of T_es in its environment, whereas a smaller animal should, by definition, be able to achieve the full range of T_es in the environment (notwithstanding that some of these T_es represent lethal temperatures).

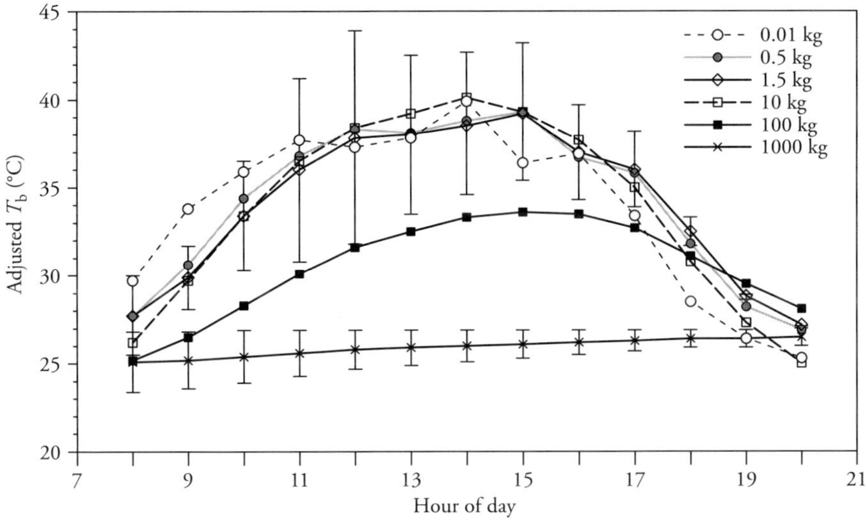

Figure 24.3 The $T_{b\ pred}$s of reptiles ranging in size from 0.01 kg to 1000 kg were calculated using the techniques described in Section 24.5.1 (Christian et al., 2006). The thermal inertia of medium sized reptiles is real and measurable. However, body size is only one of several factors determining how a reptile interacts with its thermal environment (Stevenson, 1985b; Tracy et al., 1986), and this figure shows that the random movements of lizards at least as large as 10 kg negates the consequences of body size. Reptiles with a mass of 100 kg or more have substantial thermal lags that cannot be ignored. Thus, for the purpose of quantifying thermoregulation in an ecological context, we suggest that thermal inertia can be ignored for reptiles up to 10 kg. For reptiles with a mass between 10 and 100 kg, the decision to evaluate the consequences of thermal inertia or not would depend on the nature of the questions being asked, the complexity of the thermal environment, and the behaviour of the animals.

24.5.2 How big does an animal have to be before it is 'large'?

The fact that a large reptile would have substantial thermal inertia resulting in lags in its T_b as it moves between thermal environments is not surprising. Similarly, it is easy to imagine that a small lizard would have negligible thermal lags for the purposes of quantifying its thermal environment. But what about animals in between—how big does a reptile need to be before it is considered large enough to warrant using the modified thermal indices? Christian et al. (2006) addressed this question by modelling the $T_{b\ pred}$ of reptiles ranging in size, and the results are shown in Figure 24.3.

24.6 Conclusions

The techniques described here have yielded results that would not have been possible by simply measuring T_b. For example, the reasons for a species having different T_bs in different seasons cannot be explained by simply recording T_bs. Are differences due to seasonal limitations in what is possible for the animal to achieve regardless of its preferred T_b (Figure 24.2(d), Christian and Weavers, 1996), or do differences represent active shifts in preferred T_b due to ecological conditions other than the thermal environment, such as food availability (Christian et al., 1983; Christian and Bedford, 1995)? This can only be answered by quantifying the available thermal environment. Additionally, it may be possible to model the T_b of a particular animal when it is not possible to measure the T_b directly. For example, hatchling Galapagos Land Iguanas (*Conolophus pallidus*) are too small to carry temperature-sensitive radio transmitters, but it was possible to model their predicted T_bs to determine the relationship between their thermal ecology and their susceptibility to predation (Christian et al., 1983). The possibilities of modelling past or future scenarios was discussed in Section 24.3, and this characteristic of computational models represents a powerful tool for predicting and documenting the consequences of global climate change.

These techniques are not perfect or foolproof, and they come with assumptions that should be carefully examined before each application. Nevertheless, they represent the best tools available for quantifying the thermal biology of reptiles. Hertz et al. (1993) readily admitted that the use of the central 50% of T_bs to determine T_{set} was arbitrary and other ranges could be used. However, our experience suggests that, for whatever reason, it seems about right. Consider the case of *Varanus gouldii* (Figure 24.2(a)) and *V. panoptes* (Figure 24.2(b)), which can (and often do) live side by side in the same habitat in tropical Australia. Their body sizes overlap considerably (and were equal for the calculations in Figure 24.2), and their solar absorptivities differ only slightly. T_{set}, as originally defined, allows thermoregulation to be quantified in a way that was consistent with other observations and measurements of their thermal biology. Thus, although these two sympatric species had different thermoregulatory responses early and late in the day, the T_bs of both species reached a plateau in the middle of the day that was within their respective T_{set} ranges. The dry season T_{set} range of frillneck lizards changed in conjunction with a seasonal change in field T_bs, such that Ex was similar (75–78%) in the two seasons (Christian and Bedford, 1995). This suggests not only that this species is seasonally adjusting its preferred T_b consistently in the lab and in the field, but that the T_{set} range, as originally defined, is robust enough to document this consistency.

Field studies of the thermal biology of reptiles have taught us a great deal in the 70 years since Cowles and Bogert (1944) published their seminal paper. We know that some species thermoregulate carefully and others regulate less precisely. More importantly, we know that this is not an either–or distinction, but rather that thermoregulation and thermoconformity are two ends of a continuum, with many species in between. The techniques described here allow detailed quantifications and comparisons without pigeon-holing species into arbitrary categories.

Another pattern that has become clear over the years is that thermoregulation within a species can change in response to environmental circumstances, such as seasons and food availability (Christian et al., 1983; Christian and Bedford, 1995). Nevertheless, there appear to be strong species-specific characteristics in thermal biology, as illustrated by the sympatric *Varanus gouldii* and *V. panoptes* in Figure 24.2. This is not surprising given the overarching physiological and ecological importance of thermal biology

(Section 24.1), but it is worth noting, given the apparent flexibility and variety of thermal patterns both within and between species.

Understanding the consequences of climate change on reptiles is an example of when these tools will be critical. There have been suggestions that thermoconformers are at risk as a result of global warming (Huey et al., 2012). Questions about whether reptiles will respond behaviourally or will adapt evolutionarily (behaviourally, physiologically, or morphologically) in response to global warming are unanswered. The only certainty is that robust techniques that adequately account for the thermal environment will be required to answer these questions and to interpret the consequences of future environments.

If the past is a guide, the next 70 years will see great advances in the study of thermal biology. The inevitable technological advances in measuring T_b should not overshadow ways to quantify animal–environment interactions, which will hopefully also advance. It is possible to become enamoured by easy, inexpensive technology to the extent that it is used without considering the underlying principals; hence it is critically important to understand the fundamental processes described here, so that the technology is not used inappropriately.

References

Bakken, G.S. (1976). A heat transfer analysis of animals: unifying concepts and the application of metabolism chamber data to field ecology. *Journal of Theoretical Biology*, **60**, 337–84.

Bakken, G.S. (1981a). A two-dimensional operative-temperature model for thermal energy management by animals. *Journal of Thermal Biology*, **6**, 23–30.

Bakken, G.S. (1981b). How many equivalent black-body temperatures are there? *Journal of Thermal Biology*, **6**, 59–60.

Bakken, G.S. (1992). Measurement and application of operative and standard operative temperatures in ecology. *American Zoologist*, **32**, 194–216.

Bakken, G.S., and Angilletta, M.J., Jr. (2014). How to avoid errors when quantifying thermal environments. *Functional Ecology*, **28**, 96–107.

Bakken, G.S., and Gates, D. M. (1975). Heat-transfer analysis of animals: some implications for field ecology, physiology, and evolution. In D.M. Gates and R.B. Schmerl (eds) *Perspectives in Biophysical Ecology*. New York: Springer-Verlag, pp. 255–90.

Bartlett, P.N., and Gates, D.M. (1967). The energy budget of a lizard on a tree trunk. *Ecology*, **48**, 316–22.

Blouin-Demers, G., and Nadeau, P. (2005). The cost-benefit model of thermoregulation does not predict lizard thermoregulatory behavior. *Ecology*, **86**, 560–6.

Blouin-Demers, G., and Weatherhead, P.J. (2001). Thermal ecology of black rat snakes (*Elaphe obsoleta*) in a thermally challenging environment. *Ecology*, **82**, 3025–43.

Brattstrom, B.H. (1965). Body temperatures of reptiles. *American Midland Naturalist*, **73**, 376–422.

Brien, M.L. (2015). *Growth and survival of hatchling saltwater crocodiles (*Crocodylus porosus*) in captivity: the role of agonistic and thermal behaviour*. PhD Thesis, Charles Darwin University, Darwin, Australia.

Christian, K.A., and Bedford, G.S. (1995). Seasonal changes in thermoregulation by the frilled lizard, *Chlamydosaurus kingii*, in tropical Australia. *Ecology*, **76**, 124–32.

Christian, K.A., and Tracy, C.R. (1981). The effect of the thermal environment on the ability of hatchling Galapagos Land Iguanas to avoid predation during dispersal. *Oecologia*, **49**, 218–23.

Christian, K.A., and Tracy, C.R. (1985). Measuring air temperature in field studies: a comparison of methods. *Journal of Thermal Biology*, **10**, 55–6.

Christian, K.A., and Weavers, B.W. (1996). Thermoregulation of monitor lizards in Australia: an evaluation of methods in thermal biology. *Ecological Monographs*, **66**, 139–57.

Christian, K.A., Tracy, C.R., and Porter, W.P. (1983). Seasonal shifts in body temperature and use of microhabitats by the Galapagos Land Iguana. *Ecology*, **64**, 463–8.

Christian, K.A., Tracy, C.R., and Tracy, C.R. (2006). Evaluating thermoregulation in reptiles: an appropriate null model. *The American Naturalist*, **168**, 421–30.

Clusella-Trullas, S., van Wyk, J.H., and Spotila, J.R. (2009). Thermal benefits of melanism in cordylid lizards: a theoretical and field test. *Ecology*, **90**, 2297–312.

Cowles, R.B., and Bogert, C.M. (1944). A preliminary study of the thermal requirements of desert reptiles. *Bulletin of the American Museum of Natural History*, **83**, 261–96.

Dorcas, M.E., and Peterson, C.R. (1998). Daily body temperature variation in free-ranging rubber boas. *Herpetologica*, **54**, 88–103.

Dzialowski, E.M. (2005). Use of operative temperature and standard operative temperature models in thermal biology. *Journal of Thermal Biology*, **30**, 317–34.

Dzialowski, E.M., and O'Connor, M.P. (2001). Thermal time constant estimation in warming and cooling ectotherms. *Journal of Thermal Biology*, **26**, 231–45.

Heath, J.E. (1964). Reptilian thermoregulation: evaluation of field studies. *Science*, **146**, 784–5.

Hertz, P.E. (1992). Temperature regulation in Puerto Rican *Anolis* lizards: a field test using null hypotheses. *Ecology*, **73**, 1405–17.

Hertz, P.E., Huey, R.B., and Stevenson, R.D. (1993). Evaluating temperature regulation by field-active ectotherms: the fallacy of the inappropriate question. *The American Naturalist*, **142**, 796–818.

Huey, R.B. (1982). Temperature, physiology, and the ecology of reptiles. In C. Gans and F.H. Pough (eds) *Biology of the Reptilia, Volume 12. Physiology C: Physiological Ecology*. New York: Academic Press, pp. 25–91.

Huey, R.B., Kearney, M.R., Krockenberger, A., et al. (2012). Predicting organismal vulnerability to climate warming: roles of behaviour, physiology and adaptation. *Philosophical Transactions of the Royal Society of London B: Biological Sciences*, **367**, 1665–79.

Norris, K.S. (1967). Color adaptation in desert reptiles and its thermal relationships. In W.W. Milstead (ed) *Lizard Ecology: A Symposium*. Columbia, MO:University of Missouri Press, pp. 162–229.

Peterson, C.R. (1987). Daily variation in the body temperatures of free-ranging garter snakes. *Ecology*, **68**, 160–9.

Peterson, C.R., Gibson, A.R., and Dorcas, M.E. (1993). Snake thermal ecology: the causes and consequences of body-temperature variation. In R.A. Seigel and J.T. Collins (eds) *Snakes: Ecology and Behavior*. New York: McGraw-Hill, pp. 241–314.

Porter, W.P., and Gates, D.M. (1969). Thermodynamic equilibria of animals with environment. *Ecological Monographs*, **39**, 227–44.

Porter, W.P., and James, F.C. (1979). Behavioral implications of mechanistic ecology. II. The African rainbow lizard *Agama agama*. *Copeia*, **1979**, 594–619.

Porter, W.P., and Tracy, C.R. (1983). Biophysical analysis of energetics, time-space utilization, and distributional limits of. In R.B. Huey, E.R. Pianka, and T.W. Schoener (eds) *Lizard Ecology: Studies of a Model Organism*. Cambridge, MA:Harvard University Press, pp. 55–83.

Porter, W.P., Mitchell, J.W., Beckman, W.A., et al. (1973). Behavioral implications of mechanistic ecology. *Oecologia*, **13**, 1–54.

Price-Rees, S.J., Brown, G.P., and Shine, R. (2013). Habitat selection by bluetongue lizards (*Tiliqua*, Scincidae) in tropical Australia: a study using GPS telemetry. *Animal Biotelemetry*, **1**(7). Available at: http://www.animalbiotelemetry.com/content/1/1/7

Row, J.R., and Blouin-Demers, G. (2006). Thermal quality influences effectiveness of thermoregulation, habitat use, and behaviour in milk snakes. *Oecologia*, **148**, 1–11.

Sartorius, S.S., Vitt, L.J., and Colli, G.R. (1999). Use of naturally and anthropogenically disturbed habitats in Amazonian rainforest by the teiid lizard *Ameiva ameiva*. *Biological Conservation*, **90**, 91–101.

Shine, R., and Kearney, M. (2001). Field studies of reptile thermoregulation: how well do physical models predict operative temperatures? *Functional Ecology*, **15**, 282–8.

Spotila, J.R., Lommen, P.W., Bakken, G.S., et al. (1973). A mathematical model for body temperature of large reptiles: implications for dinosaur ecology. *The American Naturalist*, **107**, 391–404.

Stevenson, R.D. (1985a). Body size and limits to the daily range of body temperature in terrestrial ectotherms. *The American Naturalist*, **125**, 102–17.

Stevenson, R.D. (1985b). The relative importance of behavioral and physiological adjustments controlling body temperature in terrestrial ectotherms. *The American Naturalist*, **126**, 362–86.

Tracy, C.R. (1975). Water and energy relations of terrestrial amphibians: insights from mechanistic modeling. In D.M. Gates and R.B. Schmerl (eds) *Perspectives in Biophysical Ecology*. New York: Springer-Verlag, pp. 325–46.

Tracy, C.R. (1976). A model of the dynamic exchange of water and energy between a terrestrial amphibian and its environment. *Ecological Monographs*, **46**, 293–326.

Tracy, C.R. (1982). Biophysical modelling in reptilian physiology and ecology. In C. Gans and F.H. Pough (eds) *Biology of the Reptilia, Volume 12. Physiology C: Physiological Ecology*. New York: Academic Press, pp. 275–321.

Tracy, C.R., Turner, J.S., and Huey, R.B. (1986). A biophysical analysis of possible thermoregulatory adaptations in sailed pelycosaurs. In N.J. Hotton III, P.D. MacLean, J.J. Roth, et al. (eds) *The Ecology and Biology of Mammal-like Reptiles*. Washington, DC: Smithsonian Institution Press, pp. 195–206.

Vitt, L.J. and Sartorius, S.S. (1999). HOBOS, Tidbits, and lizard models: the utility of electronic devices in field studies of ectotherm thermoregulation. *Functional Ecology*, **13**, 670–4.

Waldschmidt, S., and Tracy, C.R. (1983). Interactions between a lizard and its thermal environment: implications for sprint performance and space utilization in the lizard *Uta stansburiana*. *Ecology*, **64**, 476–84.

Walsberg, G.E., and Wolf, B.O. (1996). A test of the accuracy of operative temperature thermometers for studies of small ectotherms. *Journal of Thermal Biology*, **21**, 275–81.

25

Genetics in field ecology and conservation

Nancy N. FitzSimmons and Joanna Sumner

25.1 Introduction

Valuable and intriguing insights into the migratory and mating behaviour of reptiles, their population dynamics and history, the relationships among populations, and the role of ecology in speciation have been provided through the use of genetic tools in reptilian studies. Genetic studies of reptiles have added to insights from field studies in the estimation of dispersal ability (Sumner et al., 2001); confirmed natal philopatry in breeding marine turtles (Meylan et al., 1990; FitzSimmons et al., 1997); and tested hypotheses about kinship and microgeographic population structure (Gibbs and Whitehead, 2001; Moore et al., 2008). Phylogeographic studies, which use DNA sequence variation to identify genetic lineages within a species and determine the geographic distribution of the lineages, have revolutionized our understanding of population history and colonization, regional barriers to gene flow, vicariate events, Pleistocene population expansion from refugia, speciation, and the presence of contact zones between lineages (Walker and Avise, 1998; Leavitt et al., 2007; Maldonado et al., 2012; Barlow et al., 2013; Spinks et al., 2014).

Studies of mating systems in reptiles using genetic markers reveal a range of behaviours from monogamy to promiscuity (Moore et al., 2009; Oliveira et al., 2014; Schofield et al., 2014), and have confirmed the operation of sperm storage across breeding seasons (Pearse et al., 2002; Booth and Schuett, 2011). Genetic studies have been used to address questions on the evolution of mating systems, parthenogenesis, sexual selection, and sex-biased dispersal (Gibbs and Whitehead, 2001; Johansson et al., 2008; Booth and Schuett, 2011). Multiple paternity as an important component of reptilian mating systems has been confirmed in genetic studies in an array of reptiles, with a wide range in the proportion of multiply sired clutches within a population and among populations (reviewed in Uller and Olsson, 2008).

Conservation genetic studies have assessed patterns of genetic diversity, determined the extent of genetic isolation in fragmented populations, estimated effective population sizes, and provided insights into behaviour in efforts to better inform conservation management in the designation of management units, establishment of conservation corridors, or reintroduction of individuals (Sumner et al., 2004; Koumoundouros et al., 2009; Meister et al., 2010; Maldonado et al., 2012; Michaelides et al., 2014). Genetic studies have been crucial for identifying cryptic species, understanding their origins,

Reptile Ecology and Conservation. Edited by C. Kenneth Dodd, Jr.
Published 2016 by Oxford University Press.

and assessing conservation needs (Leavitt et al., 2007; Oliver et al., 2009). The application of genetic tools has identified important natural hybrid zones in reptiles (Placyk et al., 2012; Vilaca et al., 2012; Haines et al., 2014), found human-mediated hybrids within natural populations (Fong and Chen, 2010), and identified rare 'species' to be hybrids of lesser conservation concern (Stuart and Parham, 2007). Captive breeding of reptiles for conservation has relied on genetic studies to identify pure-bred animals, determine the relatedness of potential breeders, and assess the extent of genetic diversity within possible founding individuals or populations. In wildlife forensic investigations of illegally trafficked reptiles, genetic markers have been a key component in identifying the origins of the trade and prosecuting offenders (Salinas et al., 2001; Siler et al., 2014). Genetic analyses of invasive reptile species have been used to understand colonization history (Silva-Rocha et al., 2012; Chapple et al., 2014), and to detect their presence from environmental DNA in water (Piaggio et al., 2014).

DNA research has advanced to the point where studies span the entire diversity of living reptiles (Pincheira-Donoso et al., 2013). The arrival of next-generation sequencing (NGS) for high throughput DNA sequencing and associated marker development means that there are now genomic resources available to develop hundreds or thousands of markers that can be sequenced for virtually any species. NGS refers to a range of techniques that allow sequencing of millions of DNA strands in a single run (massively parallel approaches) that can utilize minute amounts of DNA as template (Metzer, 2010). NGS tools continue to advance with developments such as nanopore technology (Jain et al., 2015).

Applications of genomics in ecology and conservation are being realized through approaches that allow for the screening of large numbers of genome-wide loci at a population level. Given the broad contributions that genetic studies can make to reptile research, consideration should be given to incorporating a genetic component when designing a research programme, even if just including a tissue sampling regime to facilitate future genetic analyses.

25.2 Genetic markers

Knowledge of the diversity of genetic markers available, their relevance for addressing particular questions, and their limitations are necessary for conducting meaningful research.

25.2.1 Allozymes and restriction fragment length polymorphisms

Early researchers did not have tools to directly investigate DNA variation, but instead used variation in protein allozymes sampled from blood or liquefied tissue as genetic markers. Protein variants are expected to be under selective pressure, so relationships between environmental variation or biological traits and genetic variability were studied (Salvidio et al., 1990; Salvi et al., 2009). However, allozymes are inappropriate for population genetic studies if neutral markers are needed for estimates of population divergence, gene flow, and population size (Richardson et al., 2012).

Early techniques of restriction fragment length polymorphism (RFLP) analysis targeted mutation sites in mtDNA sequences. These remain a useful method to screen samples for specific mtDNA variants after polymerase chain reaction (PCR) amplification of target sequences (FitzSimmons et al., 2002).

25.2.2 Mitochondrial DNA sequencing

Mitochondrial DNA (mtDNA) is found in most eukaryotic cells, including reptilian red blood cells. It is typically maternally inherited and generally does not undergo recombination (but see Ujvari et al., 2007), and so provides a unique perspective on the history and relatedness among populations. For species of reptiles in which an important focus of research and conservation is linked to female nesting sites, such as marine and freshwater turtles or crocodilians, mtDNA data are well suited to provide insights into the dynamics of breeding areas (Pearse et al., 2006; Jensen et al., 2013). A combination of mtDNA and nuclear markers allows testing for sex-biased dispersal, particularly when combined with field data (Ujvari et al., 2008), and important corroboration or contrast in phylogeographic and population genetic studies (Ferchaud et al., 2015, but see Zink and Barrowclough, 2008). For taxonomic questions, a combination of mtDNA and nuclear markers has become imperative (Galtier et al., 2009).

Direct (Sanger) sequencing of target regions, which has been the most common technique for studies using mtDNA as a genetic marker, involves amplification of the target region via PCR and visualization via capillary electrophoresis. Each gene and gene region in the mtDNA genome are under different selective pressures and have different mutation rates. For resolving a phylogeny within a genus or family, faster evolving genes are used (Godhino et al., 2005; Alfaro et al., 2008), whereas for higher taxonomy, slower evolving genes or the complete mtDNA genome (Dong and Kumazawa, 2005; Duchene et al., 2012) and nuclear genes (see Section 25.2.3) are used. Mutation rates vary across taxa and within the reptiles, crocodiles and turtles have lower rates of mutation than snakes (Jiang et al., 2007; Eo and DeWoody, 2010). Reptilian mtDNA genomes are complex and recognized for structural gene rearrangements and even a duplicated control region in one turtle, most snakes, and several lizards (Dong and Kumazawa, 2005; Parham et al., 2006; Jiang et al., 2007).

Another approach to species identification and biodiversity assessments has been 'DNA barcoding' of a standard region, typically the mitochondrial cytochrome c oxidase subunit 1 (CO1) gene for animals (Herbert et al., 2003). A DNA barcoding database can be used to supplement traditional taxonomic tools to recognize undescribed diversity, identify species during cryptic life stages, and to monitor the international pet trade (Nagy et al., 2012). Sequences submitted to a DNA barcoding database, such as BOLD (Barcode of Life Data Systems; http://www.boldsystems.org/), must meet required standards for a persistent linkage between the barcode sequence and the source voucher specimen. Cold Code is the international initiative to DNA barcode all amphibians and non-avian reptiles. Barcoding has had mixed success in reptiles and the approach may need updating as new techniques develop (Kress et al., 2015).

Limitations

The sole use of mtDNA data to address questions of population genetics or taxonomy is problematic due to reliance on a genome in which there is some evidence of selective sweeps where a variant of a particular mtDNA gene is selected for along with the rest of the linked genome (e.g. Rato et al., 2010). If this occurs, the relationships among lineages may not reflect the actual history of populations or species. In addition, polyphyly in mitochondrial gene trees has been demonstrated to be a common and taxonomically general occurrence among closely related species (Funk and Omland, 2003). The possibility of interspecific hybridization and subsequent introgression of alleles from a closely related species should be considered with paraphyletic and polyphyletic gene trees (McGuire et al., 2007). Another potential problem is sequencing of nuclear pseudogenes: fragments of mtDNA that have been duplicated within the nuclear DNA. This should be tested for when working on new taxa, particularly if highly divergent lineages are found within a species (see Calvignac et al., 2011; Hodges et al., 2014).

25.2.3 Nuclear gene sequencing

The nuclear genome contains protein-coding, RNA-coding, and non-coding regions, independent and unlinked markers that evolve at different rates. A range of nuclear DNA regions have been targeted for phylogenetically useful markers in reptile studies (Townsend et al., 2008, 2011; Spinks et al., 2010; Portik et al., 2012; Wiens et al., 2012). Researchers have sequenced faster evolving coding nuclear genes to study phylogeography, speciation, and hybridization in reptiles (Vilaca et al., 2012; Sanders et al., 2013; Tolley et al., 2013) and slower evolving markers to resolve deep phylogenetic questions (Pyron et al., 2013; Barley et al., 2014).

The arrival of NGS has resulted in a revolution in DNA sequence output and marker development with the amount of the data produced increasing exponentially and the cost per sequence decreasing dramatically. Large-scale, multi-locus data (i.e. hundreds to thousands of loci), combined with improved analytical tools for inferring gene trees, provide unprecedented opportunities for resolving species phylogenies. Obtaining homologous loci across the genome of an organism often requires reduction to a targeted genomic subset of interest. Some of the more common methods to reduce the genome include reduced representation libraries, restriction-site-associated DNA (RAD) sequencing, amplicon sequencing, and transcriptome sequencing (Good, 2011; McCormack et al., 2013). New methods continue to be developed for nuclear marker discovery (Bi et al., 2012; Lemmon and Lemmon, 2012; Peñalba et al., 2014).

As large-scale sequencing projects become more feasible, ecological and conservation-based studies will have greater capacity to understand the molecular basis of behaviour, physiology, selection, and adaptation (Primmer, 2009; Stapley et al., 2010). For example, nuclear genes that code for the major histocompatibility (MHC) proteins of the immune system show high levels of genetic variation that are under selection in reptiles (Jaratlerdsiri et al., 2012). Other emerging areas of research are the heat-shock proteins (Zatsepina et al., 2000) and the mechanisms of environmental sex determination (Quinn et al., 2007), both of which are relevant to species adaptability to climate change (Mitchell and Janzen, 2010). Hoffmann et al. (2015) argue that under climate change scenarios, genomics can help conservation managers decide when to increase gene flow and hybridization across climate zones to facilitate *in situ* evolutionary change. Thus, genomics can also help inform conservation priorities in maintaining genetically distinct populations and species or supporting processes of evolutionary change.

Building upon these new techniques are emerging fields of research such as landscape community genomics (LCG) focused on understanding the processes driving patterns of genomic variability in populations of interacting species in relation to landscape features and environmental gradients (Hand et al., 2015). For example, exon (the coding section of an RNA transcript, that is translated into a protein after non-coding introns are spliced out)-capture techniques allow capture of neutral and adaptive gene markers, and targeted sequencing of exons is also useful for LCG because multiple divergent species can be sequenced for the same exons.

Limitations

The huge amount of nuclear DNA data produced from NGS techniques requires considerable computing power and storage, which needs to be taken into consideration when planning projects using NGS data. Researchers and labs may struggle to keep up with emerging developments and will need to support training in genomic data acquisition and analyses, particularly as cost reductions make genomic approaches more feasible. Additionally, Shafer et al. (2015) argue that considerable advances need to be achieved before genomics provides 'real-world conservation potential', including better engagement of genome-focused researchers with applied conservation.

25.2.4 Nuclear microsatellites

A revolution in population genetics came in the mid-1990s with the advent of nuclear microsatellites as a tool for studies of mating systems and estimates of dispersal and relatedness. Microsatellites are regions of DNA characterized by short repetitive segments, in which the repeat unit is typically from 2–5 bp long. Microsatellite loci have very high mutation rates (up to 10^{-2} mutations/generation), alleles derived from both parents are visualized (thus individuals are either heterozygous or homozygous) and by combining data from several loci, individual genetic profiles are possible (Selkoe and Toonen, 2006). Techniques for developing microsatellite markers now utilize NGS methodology (Castoe et al., 2012). In population genetic studies, 10–20 (or more) loci are typically used to estimate genetic diversity, levels of inbreeding within populations, gene flow, and the extent of relatedness among populations, or to assign individuals to specific populations, including the identification of immigrants from other populations (Hale et al., 2012; Landguth et al., 2012; Colosimo et al., 2014). Use of five or more loci can provide extremely high (>99%) probabilities of detecting multiple paternity among reptilian clutches or in excluding paternity or maternity (Gonzalez et al., 2014). Microsatellite analyses have also dominated the world of forensics (Johnson et al., 2014). To minimize time and money spent, it is possible to amplify multiple microsatellite loci for a particular sample within the same PCR reaction (multiplexing), and to combine PCR products so that ~10 loci can be analysed for each sample in one run of fragment analysis.

Limitations

If useable microsatellite loci are not available for the species or genus of interest, then discovery, screening, and optimization of new loci requires expertise; however, several companies offer microsatellite development services. Allele length can vary by one or a few base pairs between electrophoretic runs, so controls of known length need to be run with each analysis. This can be problematic if a researcher wants to compare results with a previous study and the original samples are not available. The mutational mechanisms producing variation at microsatellite loci are still poorly understood (Oliveira et al., 2006) and this may affect the underlying assumptions of the analytical approaches used with microsatellite genotypes.

25.2.5 Single nucleotide polymorphisms

A single nucleotide polymorphism (SNP) is a DNA sequence variation occurring within a population and found throughout the genome in which there has been a single base pair substitution mutation, such as A to G, or T to A. They are increasingly the marker of choice in population genetics due to their greater genotyping efficiency, data quality, genome-wide coverage and analytical simplicity for modelling mutational dynamics. SNPs can be used in genotyping for a range of analyses, from parentage analysis and phylogeography to landscape genetics and forensics (Rosenblum and Novembre, 2007; Ogden, 2011; Spinks et al., 2014). Because each locus has only two alleles, more SNPs are needed than microsatellite loci to gain a similar level of statistical power (Helyar et al., 2011). SNPs are useful for species of conservation concern as sample sizes can be significantly reduced when using a large number of markers (Willing et al., 2012).

Discovery and development of SNPs involves screening sequence fragments from several individuals to find areas with single nucleotide substitutions. Several techniques have been developed for SNP discovery and with advances in NGS technology it is now possible to identify tens of thousands of SNPs (Seeb et al., 2011). Less expensive alternatives for non-model organisms have been developed (Ogden, 2011), such as RADseq and DArTseq, which can discover and analyse hundreds of thousands of SNPs across hundreds of individuals (James et al., 2008; Peterson et al., 2012).

Advances are also being made in the analysis of SNP loci and in statistical and software development (Helyar et al., 2011). One important advantage over microsatellite loci is that because a SNP genotype gives a definitive allele designation (e.g. A or T), researchers can easily compare results across studies. Importantly, the mutational dynamics of SNPs can be modelled and analysed using a phylogenetic framework in programs such as SNAPP (Bryant et al., 2012) and TREEMIX (Pickrell and Pritchard, 2012). Some software programs used for microsatellite analysis can be used for SNPs (such as Structure; Pritchard et al., 2000), and software designers have added extensions in other programs to manage the larger SNP data sets (Helyar et al., 2011; Catchen et al., 2013).

Limitations

For organisms with limited genomic data, the actual location of SNP loci may be unknown and it is expected that some loci will be located within coding or regulatory regions of the genome, and may thus be under selective pressure. Population genetic analyses that assume the use of neutral markers may thus be compromised unless this is tested for. Depending upon how SNPs are discovered, there may be ascertainment biases if a small number of individuals from a limited geographic range were originally screened and may influence measures of genetic diversity, although this can be corrected analytically (Rosemblum and Novembre, 2007). Start-up costs can be an impediment to the use of SNPs; however, several commercial labs now provide services in SNP discovery and analysis.

25.2.6 Whole genome research

NGS has propelled the expanding field of genomics towards a much greater understanding of evolutionary dynamics among species and individuals (Ellegren, 2014). Whole genome sequencing (WGS) projects for reptiles are becoming more common, and examples include three crocodilians (St John et al., 2012; Green et al., 2014), one lizard (Alföldi et al., 2011), two snakes (Castoe et al., 2013; Vonk et al., 2013), and three turtles (Wang et al., 2013; Shaffer et al., 2013). Genome 10K is a worldwide effort aiming to coordinate and help fund the sequencing of 10,000 vertebrate genomes, including reptiles, from across the diversity of life (Haussler et al., 2009). These genomic projects will allow tremendous insights into species and molecular evolution, as well as providing new genetic tools and insights relevant to field and conservation studies.

25.3 Initiating a genetic study

To begin, identify questions for genetic analysis that can provide the greatest insights beyond what is possible with field studies alone and that could not be fairly well predicted from theoretical population genetics. Discuss the questions you wish to address with someone who has appropriate population genetics expertise in the same or a related species.

Once you have formulated questions that can be best addressed with genetic analyses, you need to determine a range of genetic markers that could successfully address each question. Within these possibilities, the markers you select will be based on the finances, time and expertise available. Importantly, you need to determine if markers are available for the species of interest, and if not, what are the most closely related species for which they have been developed. Have they been tested on other species within the genus or other genera within the family for amplification success and variability?

Several journals publish on the development of genetic markers for use in wildlife studies, such as *Molecular Ecology Resources* and *Conservation Genetic Resources*. Look for taxon-specific reviews of genetic studies (FitzSimmons and Hart, 2007; King, 2009; Jensen et al., 2013) and available primers (Engstrom et al., 2007; Townsend et al., 2008; Spinks et al., 2010; Brandley et al., 2015).

Unless the development of genetic markers is a part of your research, it is extremely important to know if the genetic markers available for you will have the variability needed for your study. The ability to conduct population genetic studies is severely limited if there is not enough genetic variation to give the power to differentiate among populations. It is important, therefore, to conduct a pilot study and determine the extent of variation within a population by screening the potential genetic markers in 5–10 individuals, or if working on multiple populations, to test several individuals from across the range of the study.

25.4 Labwork

Before you begin any labwork, go back to basics, reread a good biology textbook to remind yourself about DNA structure and replication, and get a basic introduction to PCR and DNA sequencing.

Labwork usually begins with DNA extraction, of which there are many techniques, including phenol:chloroform extraction (produces high quality concentrated DNA but requires toxic chemicals that must only be used in a fume hood; see Sambrook and Russell, 2006), salting out (Miller et al., 1988), CHELEX® (Bio-Rad Laboratories, Inc.), and various kits that can be purchased and are increasingly popular due to their ease of use, limited labour, and ability to consistently produce high-quality DNA (e.g. Qiagen DNeasy® Blood and Tissue Kit). Doing good DNA extractions is crucial to all that follows, as you need DNA of sufficient concentration with a minimum of contaminants.

DNA extractions are checked for quantity and quality, typically by running a small amount through an agarose gel using electrophoresis and visualizing the size and quantity using a DNA stain and UV light. Fluorometric measurement of DNA and RNA are also possible with relatively inexpensive equipment like a Qubit® 3.0 (Life Technologies). When PCRing many samples, you will want consistent protocols and results, so DNA samples are diluted to have similar concentrations. Depending on the lab facilities available to you, you might only be doing DNA extractions and PCR reactions, with the PCR products being sent to another lab or company (e.g. Macrogen, Korea) for the final sequencing or fragment analysis for microsatellites.

Troubleshooting is an important skill in conducting labwork, so the better you understand the chemistry of the reactions you are performing, the better your chances of quickly solving problems. Most importantly, you need to work under the close supervision of someone familiar with all the techniques you will be doing until you have gained experience. One good way to do this is to volunteer in a lab to help someone else before you start doing your own labwork.

25.5 Sample design

In designing your sampling regime, it is important to consider the number of individuals you need to sample, the number of sampling localities, and the number of loci in the context of the type of analysis you will be undertaking. An excellent way to gauge sample size and sample design is to look at papers undertaking similar research that have been published recently in high quality journals.

As a general guide:

- *Population genetic* projects require a larger number of individuals (25–30; Hale et al., 2012) per site with the number of sites sampled dependent on the scale of the project. It is important to avoid sampling from within family groups as this may skew statistical analyses such as relatedness estimates.

- *Mating system* projects require sampling enough loci and offspring per clutch to ensure a high probability (>95%) of detecting multiple paternity and of as many breeding adults in the population as possible to maximize the chance of identifying parents.
- *Phylogeographic* projects require sampling of smaller numbers of individuals at multiple sites from across the geographic range of a species (Kidd and Ritchie, 2006). Consider nested geographic sampling that combines broad sampling with more intensive regional sampling. Include representatives of all subspecies, if applicable. You will need to sample outgroups for building the phylogeny. The choice of outgroups is extremely important and can affect the results, so consider them carefully (Albert et al., 2008).

25.6 Sample collection and storage

Early RFLP and allozyme studies required tissue that necessitated destructive sampling, such as from liver or muscle cells, but current techniques require less DNA, which allows for non-destructive sampling. This includes small skin samples, tail tips, toes, shed skin or sloughs, or buccal swabs (Poschadel and Möller, 2004; Gamble, 2014), plus reptiles have nucleated blood, which can be an excellent source of genome-wide DNA. When deciding what part of the animal to sample, consider the ecology of the species—for example, avoid taking toe tips from climbing species (McCarthy and Parris, 2004). As sequencing technology advances, new sampling methods and protocols for sampling vertebrate DNA are being developed (Wong et al., 2012). Newer NGS technologies allow the use of shorter fragments (~200 bp) of DNA relative to traditional approaches. This allows the use of degraded DNA samples such as from museum specimens (Bi et al., 2013).

25.6.1 Sampling considerations

It is extremely important to use sterile methods when tissue sampling to avoid contamination between samples and the possibility of passing diseases among the individuals being sampled. Clean scissors between uses with ethanol or hydrogen peroxide swabs or dip them into ethanol and flame. Scalpels should be replaced between uses if they can't be cleaned. Only place tissue samples from a single individual into each tube. Prior to embarking on field collections, consider the kind of information you will require for your analyses and create a data sheet to include that information and to keep track of tissue sampling. Pre-numbered vials can be made in advance and listed prior to sampling on data sheets. Ensure tubes are carefully labelled with unique labels so that each tissue sample can be traced back to the individual and location from which it came. Waterproof labels that can be placed within tubes are particularly helpful.

If you are tissue sampling and then releasing individuals, consider photographing each individual in case you need to verify species or individuals at a later date. Record photograph numbers in your data sheet, or include the vial number in the photograph to ensure you can match sample with image.

25.6.2 Sample preservation

Samples from which DNA is to be extracted can be preserved in 70–96% ethanol. If logistical constraints preclude the use of EtOH, several studies have used a solution of 20% dimethyl sulphoxide (DMSO) in a salt (NaCl)-saturated solution. There are also sample kits that can be purchased that use silica beads or specialized blotting paper (FTA® paper, Whatman). Samples can remain at room temperature for some time in a preservative, but should be refrigerated as soon as possible after collection to minimize degradation of tissue and DNA. Under hot field conditions, minimize the time that samples sit at higher temperatures. Samples from which RNA is to be extracted should be placed in RNAlater®; this is expensive to buy, but can be made in a laboratory quite cheaply. For best results, samples should be cut into <1 mm pieces before placing in RNAlater®, kept in a refrigerator for 24 hours, then placed in a −20°C to −80°C freezer. RNA degrades extremely quickly, so you must take great care to preserve tissue quickly if you wish to extract RNA from it.

25.6.3 Long-term storage

To minimize degradation of tissue samples or extracted DNA over time, store samples at the lowest temperature available. Liquid nitrogen dewars provide the best long-term storage but are not common and are expensive to maintain. Storage in a −80°C mechanical freezer can slow degradation down considerably. Storage in a −20°C freezer or 4°C fridge is less desirable for long-term storage, but can preserve tissue and DNA extracts for several years.

25.6.4 Sample curation

Tissue samples can be expensive to collect and are usually irreplaceable as they represent a species or population at a unique place and point in time. To maximize the possible uses that can be gained from tissue samples, it is important that tissues are carefully preserved and registered to an accessible tissue collection once your research is complete. Many publications in phylogenetic fields require a registration number for each tissue sampled that links to the voucher number of a museum specimen. Collection of voucher specimens as well as non-destructive tissue samples can be extremely important if your phylogenetic research is likely to uncover new species (Dubois and Nemésio, 2007; Clemann et al., 2014). By lodging voucher specimens in a museum, you will be able to describe any new species uncovered in your molecular research. If you have gaps in sample localities, you can contact museums or other institutions that have tissue collections to ask for samples to complement your research programme. Make sure that your samples are of known origin and provide that information in your publications.

25.7 Data analysis and management

Organized, efficient, and reliable data management is crucial to genetics projects. The size of data sets can grow quickly and there are numerous software programs to conduct genetic analyses, with each often requiring different data formats. Use a system in which the data can be easily understood by other people and one that could eventually be lodged on a shared website. Excel spreadsheets suffice for small data sets, but they are prone to human errors, particularly when sorting data across multiple columns. Database software in which rows of data cannot be disrupted accidently by sorting are more secure, as are programs that limit category entries or that allow easy linkage to additional data. The two most common database software programs are Microsoft Access (for Windows) and FileMaker Pro (for Mac and Windows) the latter of which supports linkage with multiple types of data (such as photos) and devices (iPad and iPhone). Some software for genetic analyses also includes linkages to multiple software programs and functions for organizing databases, such as in Geneious (Kearse et al., 2012).

How you will analyse your data should be determined at the start of your project. There are a plethora of software programs to conduct genetic analyses, and a review by Excoffier and Heckel (2006) was well named as a 'survival guide'. Understanding the theoretical basis of the genetic analyses will be some of the most challenging and rewarding aspects of your research and we encourage you to explore new analytical approaches as they are developed. Learn the limitations, and assumptions relevant to the various software programs. There are several websites listing relevant software such as: https://courses.washington.edu/popgen/Software.htm,

http://www1.montpellier.inra.fr/CBGP/?q=fr/content/populations-genetics-softwares, http://softlinks.amnh.org/popgen.html,

http://www.nature.com/nrg/journal/v7/n10/box/nrg1904_BX1.html, and

http://www.cmpg.iee.unibe.ch/content/softwares__services/computer_programs/index_eng.html.

The majority of journals require that sequence data be submitted to the GenBank database (http://www.ncbi.nlm.nih.gov/genbank) before publication. Alignments can be submitted to permanent online repositories like TreeBase (http://treebase.org/treebase-web) or Dryad (https://datadryad.org/).

Genetic analyses have provided tremendous insights into the behaviour of reptiles, their evolution, and the history and dynamics of populations. The rapidly expanding field of genomics is providing new genetic markers that will enhance our ability to address questions of ecology and conservation and, as always, the application of genetics will be an exciting endeavour.

References

Albert, E.M., Mauro, D.S., García-París, M., et al. (2008). Effect of taxon sampling on recovering the phylogeny of squamate reptiles based on complete mitochondrial genome and nuclear gene sequence data. *Gene*, **441**, 12–21.

Alfaro, M.E., Karns, D.R., Voris, H.K., et al. (2008). Phylogeny, evolutionary history, and biogeography of Oriental-Australian rear-fanged water snakes (Colubroidea: Homalopsidae) inferred from mitochondrial and nuclear DNA sequences. *Molecular Phylogenetics and Evolution*, **46**, 576–93.

Alföldi J., Di Palma, F., Grabherr, M., et al. (2011). The genome of the green anole lizard and a comparative analysis with birds and mammals. *Nature*, **477**, 587–91.

Barley, A.J., Spinks, P.Q., Thomson, R.C., et al. (2014). Fourteen nuclear genes provide phylogenetic resolution for difficult nodes in the turtle tree of life. *Molecular Phylogenetics and Evolution*, **55**, 1189–94.

Barlow, A., Baker, K., Hendry, C.R., et al. (2013). Phylogeography of the widespread African puff adder (*Bitis arietans*) reveals multiple Pleistocene refugia in southern Africa. *Molecular Ecology*, **22**, 1134–57.

Bi, K., Linderoth, T., Vanderpool, D., et al. (2013). Unlocking the vault: next-generation museum population genomics. *Molecular Ecology*, **22**, 6018–32.

Bi, K., Vanderpool, D., Singhal, S., et al. (2012). Transcriptome-based exon capture enables highly cost-effective comparative genomic data collection at moderate evolutionary scales. *BMC Genomics*, **13**, 403.

Booth, W., and Schuett, G.W. (2011). Molecular genetic evidence for alternative reproductive strategies in North American pitvipers (Serpentes,Viperidae): long-term sperm storage and facultative parthenogenesis. *Biological Journal of the Linnean Society*, **104**, 934–42.

Brandley, M.C., Bragg, J.G., Singhal, S., et al. (2015). Evaluating the performance of anchored hybrid enrichment at the tips of the tree of life: a phylogenetic analysis of Australian *Eugongylus* group scincid lizards. *BMC Evolutionary Biology*, **15**, 62.

Bryant, D., Bouckaert, R., Felsenstein, J., et al. (2012). Inferring species trees directly from biallelic genetic markers: bypassing gene trees in a full coalescent analysis. *Molecular Biology and Evolution*, **29**, 1917–32.

Calvignac. S., Konecny, L., Malard, F., et al. (2011). Preventing the pollution of mitochondrial datasets with nuclear mitochondrial paralogs (*numts*). *Mitochondrion*, **11**, 246–54.

Castoe, T.A., de Koning, A.P.J., Hall, K.T., et al. (2013). The Burmese python genome reveals the molecular basis for extreme adaptation in snakes. *Proceedings of the National Academy of Sciences of the USA*, **110**, 20645–50.

Castoe, T.A., Poole, A.W., de Koning, J.D., et al. (2012). Rapid microsatellite identification from Illumina paired-end genomic sequencing in two birds and a snake. *PLoS One*, **7**, e30953.

Catchen, J., Hohenlohe, P.A. Bassham, S., et al. (2013). Stacks: an analysis tool set for population genomics. *Molecular Ecology*, **22**, 3124–40.

Chapple, D.G., Whitaker, A.H., Chapple, S.N.J., et al. (2014). Biosecurity interceptions of an invasive lizard: origin of stowaways and human-assisted spread within New Zealand. *Evolutionary Applications*, **6**, 324–39.

Clemann, N., Rowe, K.M., Rowe, K.C., et al. (2014). Value and impacts of collecting vertebrate voucher specimens, with guidelines for ethical collection. *Memoirs of Museum Victoria*, **72**, 141–51.

Colosimo, G., Knapp, C.R., Wallace, L.E., et al. (2014). Inferred vs realized patterns of gene flow: an analysis of population structure in the Andros Island Rock Iguana, *PLoS One*, **9**, e106963.

Dong, S., and Kumazawa, Y. (2005). Complete mitochondrial DNA sequences of six snakes: phylogenetic relationships and molecular evolution of genomic features. *Journal of Molecular Evolution*, **61**, 12–22.

Dubois, A., and Nemésio, A. (2007). Does nomenclatural availability of nomina of new species or subspecies require the deposition of vouchers in collections. *Zootaxa*, **1409**, 1–22.

Duchene, S., Frey, A., Alfaro-Núñez, A., et al. (2012). Marine turtle mitogenome phylogenetics and evolution. *Molecular Phylogenetics and Evolution*, **65**, 241–50.

Ellegren, H. (2014). Genome sequencing and population genomics in non-model organisms. *Trends in Ecology & Evolution*, **29**, 51–63.

Engstrom, T.N., Edwards, T., Osentoski, M.F., et al. (2007). A compendium of PCR primers for mtDNA, microsatellite, and other nuclear loci for freshwater turtles and tortoises. *Chelonian Research Monographs*, **4**, 124–41.

Eo, S.H., and DeWoody, J.A. (2010). Evolutionary rates of mitochondrial genomes correspond to diversification rates and to contemporary species richness in birds and reptiles. *Proceedings of the Royal Society of London B: Biological Sciences*, **277**, 3587–92.

Excoffier, L., and Heckel, G. (2006). Computer programs for population genetics data analysis: a survival guide. *Nature Reviews Genetics*, 7, 745–58.

Ferchaud, A.L., Eudeline, R., Arnal, V., et al. (2015). Congruent signals of population history but radically different patterns of genetic diversity between mitochondrial and nuclear markers in a mountain lizard. *Molecular Ecology*, **24**, 192–207.

FitzSimmons, N.N., and Hart, K.M. (2007). Genetic studies of freshwater turtles and tortoises: a review of the past 70 years. *Chelonian Research Monographs*, **4**, 15–46.

FitzSimmons, N.N., Buchan, J.C., Lam, P.V., et al. (2002). Identification of purebred *Crocodylus siamensis* for reintroduction in Vietnam. *Journal of Experimental Zoology, Molecular and Developmental Evolution*, **294**, 373–81.

FitzSimmons, N.N., Goldizen, A.R., Norman, J.A., et al. (1997). Philopatry of male marine turtles inferred from mitochondrial markers. *Proceedings of the National Academy of Sciences of the USA*, **94**, 8912–17.

Fong, J.J., and Chen, T-H. (2010). DNA evidence for the hybridization of wild turtles in Taiwan: possible genetic pollution from trade animals. *Conservation Genetics*, **11**, 2061–6.

Funk, D.J., and Omland, K.E. (2003). Species-level paraphyly and polyphyly: frequency, causes, and consequences, with insights from animal mitochondrial DNA. *Annual Review of Ecology, Evolution, and Systematics*, **34**, 397–423.

Galtier, N., Nabholz, B., Glémin, S., et al. (2009). Mitochondrial DNA as a marker of molecular diversity: a reappraisal. *Molecular Ecology*, **18**, 4541–50.

Gamble, T. (2014). Collecting and preserving genetic material for herpetological research. *Society for the Study of Amphibians and Reptiles Herpetological Circular*, No. 41.

Gibbs, H.L., and Whitehead, P.J. (2001). Insights into population ecology and sexual selection in snakes through the application of DNA-based genetic markers. *Journal of Heredity*, **92**, 173–9.

Godhino, R., Crespo, E.G., Ferrand, N., et al. (2005). Phylogeny and evolution of the green lizards, *Lacerta* spp. (Squamata: Lacertidae) based on mitochondrial and nuclear DNA sequences. *Amphibia-Reptilia*, **26**, 271–85.

Gonzalez, E.G. Cerón-Souza, I., Mateo, J.A., et al. (2014). Island survivors: population genetic structure and demography of the critically endangered giant lizard of La Gomera, *Gallotia bravoana*. *BMC Genetics*, **15**, 121.

Good, J.M. (2011). Reduced representation methods for subgenomic enrichment and next-generation sequencing. In V. Orgogozo and M.V. Rockman (eds) *Molecular Methods for Evolutionary Genetics*. New York: Humana Press, pp. 85–103.

Green, R.E., Braun, E.L., Armstrong J., et al. (2014). Three crocodilian genomes reveal ancestral patterns of evolution among archosaurs. *Science*, **346**, DOI:10.1126/1254449.

Haines, M.L., Moussalli, A., Stuart-Fox, D., et al. (2014). Phylogenetic evidence of historic mitochondrial introgression and cryptic diversity in the genus *Pseudemoia* (Squamata: Scincidae). *Molecular Phylogenetics and Evolution*, **81**, 86–95.

Hale, M.L., Burg, T.M., and Steeves, T.E. (2012). Sampling for microsatellite-based population genetic studies: 25 to 30 individuals per population is enough to accurately estimate allele frequencies. *PLoS One*, 7, e45170.

Hand, B.K., Lowe, W.H., Kovach, R.P., et al. (2015). Landscape community genomics: understanding eco-evolutionary processes in complex environments. *Trends in Ecology & Evolution*, **30**, 161–8.

Haussler, D. Stephen, J., Ryder, O.A., et al. (2009). Genome 10K: a proposal to obtain whole-genome sequence for 10,000 vertebrate species. *Journal of Heredity*, **100**, 659–74.

Helyar, S.J., Hemmer-Hansen, J., Bekkevold, D., et al. (2011). Application of SNPs for population genetics of nonmodel organisms: new opportunities and challenges. *Molecular Ecology Resources*, **11**(Suppl. 1), 123–36.

Herbert, P.D.N., Ratnasingham, S., and deWaard, J.R. (2003). Barcoding animal life: cytochrome *c* oxidase subunit 1 divergences among closely related species. *Proceedings of the Royal Society of London B: Biological Sciences*, **270**(Suppl.), S96–9.

Hodges, K., Donnellan, S., and Georges, A. (2014). Phylogeography of the Australian freshwater turtle *Chelodina expansa* reveals complex relationships among inland and coastal bioregions. *Biological Journal of the Linnean Society*, **111**, 789–805.

Hoffmann, A., Griffin, P., Dillon, S., et al. (2015). A framework for incorporating evolutionary genomics into biodiversity conservation and management. *Climate Change Responses*, **2**, 1–24.

Jain, M., Fiddes, I.T., Miga, K.H., et al. (2015). Improved data analysis for the MinION nanopore sequencer. *Nature Methods* **12**, 351–6.

James, K.E., Schneider, H., Ansell, S.W., et al. (2008). Diversity arrays technology (DArT) for pan-genomic evolutionary studies of non-model organisms. *PLoS One*, **2**, e1682.

Jaratlerdsiri, W., Isberg, S.R., Higgins, D.P., et al. (2012). MHC class I of saltwater crocodiles (*Crocodylus porosus*): polymorphism and balancing selection. *Immunogenetics*, **64**, 825–38.

Jensen, M.P., FitzSimmons, N.N., and Dutton, P.H. (2013). Molecular genetics of sea turtles. In J. Wyneken, K.J. Lohmann, and J.A. Musick (eds). *The Biology of Sea Turtles, Volume 3*. Boca Raton, FL:CRC Press, pp 135–61.

Jiang, Z.J., Castoe, T.A., Austin, C.C., et al. (2007). Comparative mitochondrial genomics of snakes: extraordinary substitution rate dynamics and functionality of the duplicate control region. *BMC Evolutionary Biology*, 7, 123.

Johansson, H., Surget-Groba, Y., and Thorpe, R.S. (2008). Microsatellite data show evidence for male-biased dispersal in the Caribbean lizard *Anolis roquet*. *Molecular Ecology*, **17**, 4425–32.

Johnson, R.N., Wilson-Wilde, L., and Linacre A. (2014). Current and future directions of DNA in wildlife forensic science. *Forensic Science International: Genetics*, **10**, 1–11.

Kearse, M., Moir, R., Wilson, A., et al. (2012). Geneious Basic: an integrated and extendable desktop software platform for the organization and analysis of sequence data. *Bioinformatics*, **28**, 1647–9.

Kidd, D.M., and Ritchie, M.G. (2006). Phylogeographic information systems: putting the geography back into phylogeography. *Journal of Biogeography*, **33**, 1851–65.

King, R.B. (2009). Population and conservation genetics. In S.J. Mullin and R.A. Seigel (eds) *Snakes Ecology and Conservation*. Ithaca, NY:Cornell University Press, pp. 78–122.

Koumoundouros, T., Sumner, J., Clemann, N., et al. (2009). Current genetic isolation and fragmentation contrasts with historical connectivity in an alpine lizard (*Cyclodomorphus praealtus*) threatened by climate change. *Biological Conservation*, **142**, 992–1002.

Kress, W.J., García-Robledo, C., Uriarte, M., et al. (2015). DNA barcodes for ecology, evolution, and conservation. *Trends in Ecology & Evolution*, **30**, 25–35.

Landguth, E.L., Fedy, B.C., Oyler-McCance, S.J., et al. (2012). Effects of sample size, number of markers, and allelic richness on the detection of spatial genetic pattern. *Molecular Ecology Resources*, **12**, 276–84.

Leavitt, D.H., Bezy, R.L., Crandall, K.A., et al. (2007). Multi-locus DNA sequence data reveal a history of deep cryptic vicariance and habitat-driven convergence in the desert night lizard *Xantusia vigilis* species complex (Squamata: Xantusiidae). *Molecular Ecology*, **16**, 4455–81.

Lemmon, A.R., and Lemmon, E.M. (2012). High-throughput identification of informative nuclear loci for shallow-scale phylogenetics and phylogeography. *Systematic Biology*, **61**, 745–61.

Maldonado, S.P., Melville, J., Peterson, G.N., et al. (2012). Human-induced versus historical habitat shifts: identifying the processes that shaped the genetic structure of the threatened grassland legless lizard, *Delma impar. Conservation Genetics*, **13**, 1329–42.

McCarthy, M.A., and Parris, K.M. (2004). Clarifying the effect of toe clipping on frogs with Bayesian statistics. *Journal of Applied Ecology*, **41**, 780–6.

McCormack, J.E., Hird, S.M., Zellmer, A.J., et al. (2013). Applications of next-generation sequencing to phylogeography and phylogenetics. *Molecular Phylogenetics and Evolution*, **66**, 526–38.

McGuire, J.A., Linkem, C.W., Koo, M.S., et al. (2007). Mitochondrial introgression and incomplete lineage sorting through space and time: phylogenetics of crotaphytid lizards. *Evolution*, **61**, 2879–97.

Meister, B., Hofer, U., Ursenbacher, S., et al. (2010). Spatial genetic analysis of the grass snake, *Natrix natrix* (Squamata: Colubridae), in an intensively used agricultural landscape. *Biological Journal of the Linnean Society*, **101**, 51–8.

Metzer, M.L. (2010). Sequencing technologies—the next generation. *Nature Reviews Genetics*, **11**, 31–46.

Meylan, A.B., Bowen, B.W., and Avise, J.C. (1990). Genetic test of the natal homing versus social facilitation models for green turtle migration. *Science*, **248**, 724–7.

Michaelides, S., Cole, N., and Funk, S.M. (2014). Translocation retains genetic diversity of a threatened endemic reptile in Mauritius. *Conservation Genetics*, **16**, 661–72.

Miller, S.A., Dykes, D.D., and Polesky, H.F. (1988). A simple salting out procedure for extracting DNA from human nucleated cells. *Nucleic Acids Research*, **16**, 1215.

Mitchell, N.J., and Janzen, F.J. (2010). Temperature-dependent sex determination and contemporary climate change. *Sexual Development*, **4**, 129–40.

Moore, J.A., Daugherty, C.H., Godfrey, S.S., et al. (2009). Seasonal monogamy and multiple paternity in a wild population of a territorial reptile (tuatara). *Biological Journal of the Linnean Society*, **98**, 161–70.

Moore, J.A., Miller, H.C., Daugherty, C.H., et al. (2008). Fine-scale genetic structure of a long-lived reptile reflects recent habitat modification. *Molecular Ecology*, **17**, 4630–41.

Nagy, Z.T., Sonet, G., Glaw, F., et al. (2012). First large-scale DNA barcoding assessment of reptiles in the biodiversity hotspot of Madagascar, based on newly designed COI primers. *PLoS One*, 7, e34506.

Ogden, R. (2011). Unlocking the potential of genomic technologies for wildlife forensics. *Molecular Ecology Resources*, **11**(S1), 109–16.

Oliveira, D.P., Marioni, B., Farias, I.P., et al. (2014). Genetic evidence for polygamy as a mating strategy in *Caiman crocodilus. Journal of Heredity*, DOI: 10.1093/jhered/esu020.

Oliveira, E.J., Pádua, J.G., Zucchi, M.I., et al. (2006). Origin, evolution and genome distribution of microsatellites. *Genetics and Molecular Biology*, **29**, 294–307.

Oliver, P.M., Adams, M., Lee, M.S.Y., et al. (2009). Cryptic diversity in vertebrates: molecular data double estimates of species diversity in a radiation of Australian lizards (*Diplodactylus*, Gekkota). *Proceedings of the Royal Society of London B: Biological Sciences*, DOI: 10.1098/rspb.2008.1881.

Parham, J.F., Feldman, C.R., and Boore, J.L. (2006). The complete mitochondrial genome of the enigmatic bigheaded turtle (*Platysternon*): description of unusual genomic features and the reconciliation of phylogenetic hypotheses based on mitochondrial and nuclear DNA. *BMC Evolutionary Biology*, **6**, 11.

Pearse, D.E., Arndt, A.D., Valenzuela, N., et al. (2006). Estimating population structure under nonequilibrium conditions in a conservation context: continent-wide population genetic of the giant Amazon river turtle, *Podocnemis expansa* (Chelonia, Podocnemididae). *Molecular Ecology*, **15**, 985–1006.

Pearse, D.E., Janzen, F.J., and Avise, J.C. (2002). Multiple paternity, sperm storage, and reproductive success of female and male painted turtles (*Chrysemys picta*) in nature. *Behavioral Ecology and Sociobiology*, **51**, 164–71.

Peñalba, J.V., Smith, L.L., Tonione, M.A., et al. (2014). Sequence capture using PCR-generated probes: a cost-effective method of targeted high- throughput sequencing for nonmodel organisms. *Molecular Ecology Resources*, **14**, 1000–10.

Peterson, B.K., Weber, J.N., Kay, E.H., et al. (2012). Double digest RADseq: an inexpensive method for *De Novo* SNP discovery and genotyping in model and non-model species. *PLoS One*, 7, e37135.

Piaggio, A.J., Engeman, R.M., Hopken, M.W., et al. (2014). Detecting an elusive invasive species: a diagnostic PCR to detect Burmese python in Florida waters and an assessment of persistence of environmental DNA. *Molecular Ecology Resources*, **14**, 374–80.

Pickrell, J.K., and Pritchard, J.K. (2012). Inference of population splits and mixtures from genome-wide allele frequency data. *PLoS Genetics*, **8**, e1002967.

Pincheira-Donoso, D., Bauer, A.M., Meiri, S., et al. (2013). Global taxonomic diversity of living reptiles. *PLoS One*, **8**, e59741.

Placyk, J.S., Jr., Fitzpatrick, B.M., Casper, G.S., et al. (2012). Hybridization between two gartersnake species (*Thamnophis*) of conservation concern: a threat or an important natural interaction? *Conservation Genetics*, **13**, 649–63.

Portik, D.M., Wood, P.L., Jr., Grismer, J.L., et al. (2012). Identification of 104 rapidly-evolving nuclear protein-coding markers for amplification across scaled reptiles using genomic resources. *Conservation Genetic Resources*, **4**, 1–10.

Poschadel, J., and Möller, D. (2004). A versatile field method for tissue sampling on small reptiles and amphibians, applied to pond turtles, newts, frogs and toads. *Conservation Genetics*, **5**, 865–7.

Primmer, C.R. (2009). From conservation genetics to conservation genomics. *Annals of the New York Academy of Sciences*, **1162**, 357–68.

Pritchard, J.K., Stephens, M., and Donnelly, P. (2000). Inference of population structure using multilocus genotype data. *Genetics*, **155**, 945–59.

Pyron, R.A., Burbink, F.T., and Wiens, J.J. (2013). A phylogeny and revised classification of Squamata, including 4161 species of lizards and snakes. *BMC Evolutionary Biology*, **13**, 93.

Quinn, A.E., Georges, A., Sarre, S.D., et al. (2007). Temperature sex reversal implies sex gene dosage in a reptile. *Science*, **316**, 411.

Rato, C., Carranza, S., Perera, A., et al. (2010). Conflicting patterns of nucleotide diversity between mtDNA and nDNA in the Moorish gecko, *Tarentola mauritanica*. *Molecular Phylogenetics and Evolution*, **56**, 962–71.

Richardson, B.J., Baverstock, P.R., and Adams, M. (2012). *Allozyme Electrophoresis: A Handbook for Animal Systematics and Population Studies*. San Diego, CA:Academic Press.

Rosenblum, E.B., and Novembre, J. (2007). Ascertainment bias in spatially structured populations: a case study in the eastern fence lizard. *Journal of Heredity*, **98**, 331–6.

Salinas, M., Altet, L. Clavel, C., et al. (2001). Genetic assessment, illegal trafficking and management of the Mediterranean spur-thighed tortoise in Southern Spain and Northern Africa. *Conservation Genetics*, **12**, 1–13.

Salvi, D., Capula, M., Bombi, P., et al. (2009). Genetic variation and its evolutionary implications in a Mediterranean island endemic lizard. *Biological Journal of the Linnean Society*, **98**, 661–76.

Salvidio, S., Pasteur, G., Heulin, B., et al. (1990). Natural selection and geographical variation in a known sex-linked gene of the common lizard in Europe. Implications for chromosomal evolution. *Heredity*, **64**, 131–8.

Sambrook, J., and Russell, D.W. (2006). Purification of nucleic acids by extraction with phenol:chloroform. *Cold Spring Harbor Protocols*, **2006**, DOI: 10.1101/pdb.prot4455.

Sanders, K.L., Lee, M.S., Bertozzi, T., et al. (2013). Multilocus phylogeny and recent rapid radiation of the viviparous sea snakes (Elapidae: Hydrophiinae). *Molecular Phylogenetics and Evolution*, **66**, 575–91.

Schofield, J.A., Gardner, M.G., Fenner, A.L., et al. (2014). Promiscuous mating in the endangered Australian lizard *Tiliqua adelaidensis*: a potential windfall for its conservation. *Conservation Genetics*, **15**, 177–85.

Seeb, J.E., Carvalho, G., Naish, K., et al. (2011). Single-nucleotide polymorphism (SNP) discovery and application of SNP genotyping in nonmodel organisms. *Molecular Ecology Resources*, **11**(Suppl. 1), 1–8.

Selkoe, K.A., and Toonen, R.J. (2006). Microsatellites for ecologists: a practical guide to using and evaluating microsatellite markers. *Ecology Letters*, **9**, 615–29.

Shafer, A.B.A., Wolf, J.B.W, Alves, P.C., et al. (2015). Genomics and the challenging translation into conservation practice. *Trends in Ecology & Evolution*, **30**, 78–87.

Shaffer, H.B., Minx, P., Warren, D.E., et al. (2013). The western painted turtle genome, a model for the evolution of extreme physiological adaptations in a slowly evolving lineage. *Genome Biology*, **14**, R28.

Siler, C.D., Lira-Noriega, A., and Brown, R.M. (2014). Conservation genetics of Australasian sailfin lizards: flagship species threatened by coastal development and insufficient protected area coverage. *Biological Conservation*, **169**, 100–8.

Silva-Rocha, I., Salvi, D., and Carretero, M.A. (2012). Genetic data reveal a multiple origin for the population of the Italian wall lizard *Podarcis sicula* (Squamata: Lacertidae) introduced in the Iberian Peninsula and Balearic islands. *Italian Journal of Zoology*, **79**, 502–10.

Spinks, P.Q., Thomson, R.C., Barley, A.J., et al. (2010). Testing avian, squamate, and mammalian nuclear markers for cross amplification in turtles. *Conservation Genetics Resources*, **2**, 127–9.

Spinks, P.Q., Thomson, R.C., and Shaffer, H.B. (2014). The advantages of going large: genome-wide SNPs clarify the complex population history and systematics of the threatened western pond turtle. *Molecular Ecology*, **23**, 2228–41.

St John, J.A., Braun, E.L., Isberg, S.R., et al. (2012). Sequencing three crocodilian genomes to illuminate the evolution of archosaurs and amniotes. *Genome Biology*, **13**, 415.

Stapley, J., Reger, J., Feulner, P.G.D., et al. (2010). Adaptation genomics: the next generation. *Trends in Ecology & Evolution*, **25**, 705–12.

Stuart, B.L., and Parham, J.F. (2007). Recent hybrid origin of three rare Chinese turtles. *Conservation Genetics*, **8**, 169–75.

Sumner, J., Jessop, T., Paetkau, D., et al. (2004). The effect of anthropogenic habitat fragmentation on molecular diversity in a rainforest endemic skink, *Gnypetoscincus queenslandiae*. *Molecular Ecology*, **13**, 259–69.

Sumner, J., Rousset, F., Estoup, A., et al. (2001). 'Neighbourhood size', dispersal and density estimates in the prickly forest skink (*Gnypetoscincus queenslandiae*) using individual genetic and demographic methods. *Molecular Ecology*, **10**, 1917–27.

Tolley, K.A., Townsend, T.M., and Vences, M. (2013). Large-scale phylogeny of chameleons suggests African origins and Eocene diversification. *Proceedings of the Royal Society of London B: Biological Sciences*, **280**, DOI: 10.1098/rspb.2013.0184.

Townsend, T.M., Alegre, R.E., Kelley, S.T., et al. (2008). Rapid development of multiple nuclear loci from phylogenetic analysis using genomic resources: an example from squamate reptiles. *Molecular Phylogenetics and Evolution*, **47**, 129–42.

Townsend, T.M., Mulcahy, D.G., Noonan, B.P., et al. (2011). Phylogeny of iguanian lizards inferred from 29 nuclear loci, and a comparison of concatenated and species-tree approaches for an ancient, rapid radiation. *Molecular Phylogenetics and Evolution*, **61**, 363–80.

Ujvari, B., Dowton, M., and Madsen, T. (2007). Mitochondrial DNA recombination in a free-ranging Australian lizard. *Biology Letters*, **3**, 189–92.

Ujvari, B., Dowton, M., and Madsen, T. (2008). Population genetic structure, gene flow and sex-biased dispersal in frill neck lizards (*Chlamydosaurus kingii*). *Molecular Ecology*, **17**, 3557–64.

Uller, T., and Olsson, M. (2008). Multiple paternity in reptiles: patterns and processes. *Molecular Ecology*, **17**, 2566–80.

Vilaca, S.T., Vargas, S.M., Lara-Ruiz, P., et al. (2012). Nuclear markers reveal a complex introgression pattern among marine turtle species on the Brazilian coast. *Molecular Ecology*, **21**, 4300–12.

Vonk, F.J., Casewell, N.R., Henkel, C.V. et al., (2013). The king cobra genome reveals dynamic gene evolution and adaptation in the snake venom system. *Proceedings of the National Academy of Sciences of the USA*, **110**, 20651–6.

Walker, D., and Avise, J.C. (1998). Principles of phylogeography as illustrated by freshwater and terrestrial turtles in the southeastern United States. *Annual Review of Ecology and Systematics*, **29**, 23–58.

Wang Z., Pascual-Anaya, J., and Zadissa A. (2013). The draft genomes of the soft-shell turtle and green sea turtle yield insights into the development and evolution of the turtle-specific body plan. *Nature Genetics*, **45**, 701–6.

Wiens, J.J., Hutter, C.R. Mulcahy, D.G., et al. (2012). Resolving the phylogeny of lizards and snakes (Squamata) with extensive sampling of genes and species. *Biology Letters*, **8**, 1043–6.

Willing, E-M., Dreyer, C., and van Oosterhout, C. (2012). Estimates of genetic differentiation measured by F_{ST} do not necessarily require large sample sizes when using many SNP markers. *PLoS One*, 7, e42649.

Wong, B.Y., Wiley, E.O., Johnson, W.E., et al. (2012). Tissue sampling methods and standards for vertebrate genomics. *Giga Science*, **1**, 1–12.

Zatsepina, O.G., Ulmasov, K.H.A, Beresten, S.F., et al. (2000). Thermotolerant desert lizards characteristically differ in terms of heat-shock system regulation. *The Journal of Experimental Biology*, **203**, 1017–25.

Zink, R.M., and Barrowclough, G.E. (2008). Mitochondrial DNA under siege in avian phylogeography. *Molecular Ecology*, **17**, 2107–21.

Part 6
Trends Analysis and Conservation Options

26
Occupancy models

Darryl I. MacKenzie

26.1 Introduction

Occupancy models are a set of methods that have been developed to investigate factors affecting the presence and absence of species within an area, while explicitly accounting for the potential imperfect detection of the species. They have become increasingly popular since MacKenzie et al. (2002) described a flexible method for analysing appropriately collected data that can be used to examine the effect of possible predictor variables on both the presence (or occurrence) and detectability of a target species with logistic regression-like frameworks. It could also accommodate the practicalities of field sampling in terms of unequal sampling effort at different locations. In order to explicitly account for imperfect detection, a key data requirement of the MacKenzie et al. (2002) approach is that multiple surveys should be conducted at each location where the presence or absence of the species is being assessed. These surveys could be conducted in a single visit to a location, so the method does not necessarily require a much higher level of field effort. A number of similar, independently developed methods (Tyre et al., 2003; Stauffer et al., 2004; Wintle et al., 2004) and extensions (Royle and Nichols, 2003; MacKenzie et al., 2004) soon followed.

The above methods were developed to examine the patterns in species occurrence at a single point in time, which could be useful when as part of a monitoring programme (to quantify the current status of the species across a broad area), to investigate habitat requirements for a species, or to assess the current distribution of a species. These methods are often referred to as single-season or static occupancy models. MacKenzie et al. (2003) developed an extension that incorporated parameters representing the underlying dynamic processes of change in the presence of a species, which enabled changes in species occurrence through time to be considered. These multi-season or dynamic occupancy models are potentially much more relevant in terms of understanding the important factors for the persistence of a species in a region. Predictor variables for the dynamic parameters, probabilities of local colonization and extinction (or its complement, local persistence), can also be assessed in a logistic regression-like framework, and detectability can be allowed to change through time if required. Following estimation of the dynamic processes, the concepts behind the dynamic occupancy model can be used to make predictions about which locations are likely to be occupied in the future (i.e. make predictions about the future distribution of the species), making this a very relevant modelling approach to modern-day conservation.

In this chapter I briefly outline these methods, highlighting the most salient points of the modelling approaches, while noting some useful extensions. I also express my opinion on the most important advantages of these approaches, their disadvantages, and comment on some published criticisms of these methods.

26.2 Method overview

Occupancy models can be used to examine the probability of a target species being present in a sampling unit, to look at the factors that affect that probability resulting in occupancy being higher or lower in certain types of sampling units, and to assess the overall fraction of sampling units that are occupied in a region. They can also be used to investigate changes in occupancy across the set of sampling units through time. A very important, and necessary, first step is to clearly define the region of interest and what should be regarded as a *sampling unit*. The sampling unit for a particular study should be defined at the spatial scale at which the investigators wish to establish the presence or absence of the target species. There can often be a variety of choices for exactly how that should be defined, particularly in contiguous landscapes. A sampling unit may be a natural feature such as a pond, patch of vegetation, or rocky outcrop, or may be arbitrarily defined, such as a transect or grid cell of a specified shape and size. When an objective of the study is to produce species distribution maps, it may be useful to define the sampling unit according to the desired resolution of the map. The region of interest is then the collection of all sampling units that may potentially be selected for surveying, that is, all sampling units within the population of interest must have a non-zero probability of being included in the sample of units from which data are collected. If there are certain areas that have no chance of being selected for surveying (e.g. due to access issues or if they are considered 'non-habitat' for the species), those areas are excluded from the region of interest. Note that the statistical population of interest is not the population of individuals of the target species scattered on the landscape, but the collection of sampling or landscape units themselves. An important change of mind-set for those beginning with occupancy-based studies is that the objective of the study is not sampling individuals of the target species from a landscape (such as in mark–recapture studies), but sampling the landscape itself. The presence or absence of the target species is just a characteristic of a sampling unit, much like the type of vegetation cover, aspect, or elevation. Readers are directed to MacKenzie and Royle (2005) and MacKenzie et al. (2006) for more detailed discussions of sampling unit definition and study design considerations.

Once sampling units have been suitably defined and the region (or population) of interest identified, a probabilistic sampling scheme should be used to select a sample of units that are to be surveyed. A well-defined probabilistic sampling scheme is required to justify extrapolation of the results from the surveyed units to the unsurveyed units. If no probabilistic sampling scheme is used, any extrapolation has no statistical support and is typically based on untestable assertions. Clearly, such situations should be avoided.

For each time period within which the presence or absence of the species is to be assessed (often referred to as a 'season'), sampling units should be surveyed multiple times due to the (presumed) imperfect detection of the species. A 'survey' (in this context) is just a single opportunity to detect the target species, and, as noted earlier, the requirement for multiple surveys does not equate to multiple visits. There are a wide range of options for how the multiple survey information could be collected in practice, including, but not limited to, multiple visits, multiple observers, different survey methods, multiple surveys in a single visit, or surveying multiple sub-units within a larger sampling unit (e.g. smaller plots within a 4 ha cell). Rather than conducting multiple discrete surveys in a single visit, Garrard et al. (2008) and Guillera-Arriota et al. (2011) extended the modelling for situations where units are surveyed with a single continuous survey (which is a limiting case of having a large number of short repeat surveys). The repeat survey information enables separation of occupancy from detection probability which is otherwise confounded if only a single survey with a binary outcome is conducted, or alternatively, if multiple surveys are conducted but the resulting information is collapsed into only a single binary value for further analysis. Essentially, the repeat survey information quantifies the amount of effort expended to detect the species at least once during the repeat surveys. Furthermore, the true value of the repeat surveys is that it reduces the likelihood of a false absence due to imperfect detection, which can be considered as a form of measurement error with respect to species presence/absence (Welsh et al., 2013; MacKenzie 2014). As in any analysis, it is often desirable to reduce uncertainty due to measurement error to a suitable level.

Conceptually, the species may be present at sampling unit *i* with probability ψ_i and conversely, absent with probability $1 - \psi_i$. The probability of presence (or occupancy) at each unit cannot be estimated as completely independent values, and must vary either through some relationship with predictor variables or covariates (e.g. elevation, habitat, distance from nearest urban development), or by assuming the probabilities are random values from a well-defined statistical distribution. The basic occupancy models assume that if the species is absent from a sampling unit, the probability of detecting the species during a survey is 0, which effectively assumes there is no species misidentification. When the species is present at unit *i*, it may be detected in survey *j* with probability p_{ij}. As with occupancy probability, the detection probability cannot be estimated as a completely independent value for all units and surveys, and can only vary by way of covariates or random effects. However, there is a great deal of flexibility in the type of covariates that can be used, which may be characteristics of the units (e.g. habitat, elevation, distance to water) or information specific to the individual survey (e.g. time of day, date, air temperature, observer, survey method). Therefore, this assumption is not overly restrictive.

The resulting sequence of detections and non-detections from the repeat surveys of a sampling unit conducted within a sampling period, or season, can be represented as a sequence of 1s and 0s. This is commonly referred to as a detection history, and is analogous to a capture or encounter history in mark–recapture applications. For example, if three surveys were conducted at a sampling unit with the species being detected in surveys 1 and 2, but undetected in survey 3, that could be represented with the detection history 110. If the species was not detected in any of the three surveys, the detection history would be 000. Avoiding the technical side of how the data are actually used in the modelling procedure to estimate probabilities and effect sizes (for details see MacKenzie et al., 2002, 2006), the important point to note about the 'never detected' detection history is that there are two different biological options that would result in the same observed data: (1) the species may be genuinely absent from the sampling unit; or (2) it is present but undetected (i.e. a false absence). Occupancy models explicitly account for this, unlike other techniques such as standard logistic regression.

For longer-term studies, changes in occupancy can be defined in terms of the underlying dynamic processes. If the species is present at a unit in season 1, it may go locally extinct by season 2 and therefore be absent from the unit in season 2. Let ε_t represent the probability of the species going locally extinct between seasons t and $t + 1$, that is, the probability the occupancy status of a unit changes from the species being present to absent. If the species is absent from a unit in season 1, it may colonize the unit by season 2 and therefore be present there in season 2. Let γ_t represent the probability of the occupancy status of a unit changes from the species being absent in season t to present in season $t + 1$.

Using these dynamic processes, the presence and absence of the species over time can be modelled. For example, if the species was present at a unit in season 1, absent in season 2, and present again in season 3, the probability of that can be expressed in terms of the parameters defined above as $\psi_1\varepsilon_1\gamma_2$. However, in the face of imperfect detection, the non-detection of the species in season 2 does not necessarily equate to the species being absent. Unaccounted for, imperfect detection will cause estimates of dynamic parameters to be biased. MacKenzie et al. (2003) developed a method for accounting for imperfect detection in situations when multiple surveys are conducted in each season using similar logic to that used in MacKenzie et al. (2002). Namely, that in the face of imperfect detection, there will likely be multiple possible descriptions for any observed detection history. For example, suppose a unit is surveyed for three seasons, with two surveys being conducted each season, and the follow detection history observed (with groupings indicating the observations for each season): 10 00 11. As the species was detected at least once in the first and third seasons, it is known the species was present in those seasons. However, in the second season, the species may have been absent, but it may also have been present but undetected. In the former case, the species must have gone locally extinct between seasons 1 and 2, then (re)colonized between seasons 2 and 3, or in the latter case, the species continuously occupied the unit, that is, it did not go locally extinct between seasons 1 and 2 or seasons 2 or 3. The approach detailed by MacKenzie et al. (2003) explicitly accounts for these options and thereby reduces the bias in the dynamic parameter estimates.

It is also possible to use either the estimated dynamic parameters (and first season occupancy) to calculate other occupancy-related parameters that may be of interest (such as overall occupancy each season, or rate of change in occupancy), or to reparametrize the model of MacKenzie et al. (2003) to estimate these quantities directly (MacKenzie et al., 2006).

In these longer-term studies, researchers often express an interest in wanting to estimate the occupancy probability in each season, or whether there is a trend in occupancy. While these are interesting questions, focusing on the underlying dynamics is a much more interesting and useful construct as it will likely provide greater insight into why occupancy (or the species distribution) might be changing rather than just quantifying the size of any change. For example, the modelling approach of MacKenzie et al. (2003) provides a framework to better address questions of habitat suitability. By including habitat as a covariate for the dynamic processes, it is possible to identify whether certain habitats have lower rates of local extinction or higher rates of colonization, or both.

26.3 Grand Skink example

As an example of the methods of MacKenzie et al. (2003), data collected on the New Zealand Grand Skink (*Oligosoma grande*) are analysed to examine temporal and habitat-related effects on occupancy dynamics. A portion of these data have been previously analysed using the same methods by Seddon et al. (2011). The Grand Skink is endemic to New Zealand and currently ranked by the New Zealand Department of Conservation as Nationally Critically Endangered. It is the largest of New Zealand's skink species, growing to almost 300 mm. Like many New Zealand skinks, the Grand Skink is relatively long-lived, living up to 20 years in the wild, and gives birth to 2–3 live young per year. Their biology makes them highly susceptible to introduced mammalian predators, including cats (*Felis catus*; feral and domestic), ferrets (*Mustela furo*), stoats (*Mustela erminea*), and hedgehogs (*Erinaceus europaeus occidentalis*). In the early 2000s, it was thought that Grand Skinks were restricted to less than 10% of their historic range; they are currently found in two areas of Central Otago on New Zealand's South Island. The habitats in these areas are primarily sub-alpine native tussock grasslands interspersed with schist rock outcrops, upon which the skinks are typically found. During the late 1990s–early 2000s, the Department of Conservation conducted surveys each January of approximately 120 outcrops near MacRaes Flat in eastern Otago. In this area, the landscape consists of the native tussock grassland and modified sown pastures, and approximately half of the outcrops were located in each of these two predominant habitat types. Outcrops were surveyed up to three times per year to determine whether skinks were present on the outcrops. In total, 338 outcrops were surveyed at least once during the five years of data used here, although not all were surveyed in each year.

There are four parameter types in the model of MacKenzie et al. (2003); (1) first-year occupancy; (2) colonization; (3) extinction; and (4) detection. For first-year occupancy, two hypotheses were considered: skink occupancy of rocky outcrops was different depending on the surrounding habitat types (native tussock grassland or sown pasture, denoted as $\psi_1(H)$) or occupancy was the same for both habitat types ($\psi_1(\cdot)$). Four hypotheses for colonization were considered: colonization was different each year with a consistent difference between habitat types ($\gamma(Yr + H)$); colonization was different in each year only ($\gamma(Yr)$); colonization was different due to each habitat type ($\gamma(H)$); or colonization was the same in all years and both habitat types ($\gamma(\cdot)$). The same four hypotheses were considered for extinction probability (i.e. ($\varepsilon(Yr + H)$), ($\varepsilon(Yr)$), ($\varepsilon(H)$)

and ($\varepsilon(\cdot)$). Only one hypothesis was considered for detection: it was different for each outcrop type in each year, with a consistent difference depending on the surrounding habitats ($p(Yr + H)$). All combinations of these different hypotheses were considered resulting in 32 models being fit to the data (i.e. $2 \times 4 \times 4 \times 1$). The models were compared using Akaike's Information Criterion (AIC) and model averaging was used to combine the results from each model into a single set of estimates (Burnham and Anderson, 2002; MacKenzie et al., 2006).

Table 26.1 presents the ten highest-ranked models according to AIC. There is a lot of support for the hypothesis that first-year occupancy is different depending upon the habitat surrounding a rocky outcrop as all of the top-ranked models include $\psi_1(H)$. The five highest-ranked models all include habitat type for colonization probability, indicating greater support for the hypotheses that colonization probabilities are also different depending on the surrounding habitat. Models that include a year effect for colonization are ranked similarly to those that do not, indicating ambiguous support for a year effect on colonization. There is more support for a year effect *not* being important on extinction probability as year is not included in any of the four highest-ranked models. The two highest-ranked models include habitat for extinction probability, suggesting some support for different extinction probabilities depending upon the surrounding habitat. Hence, overall, the lack of important year effects on colonization and extinction would suggest the distribution of Grand Skinks was relatively stable over this time frame (MacKenzie et al. 2006), while the strength of the habitat effects would suggest that the surrounding habitat is an important factor for the dynamics of skink occupancy of outcrops.

The AIC model weights are spread across a number of different models suggesting there are a number of models that have a similar level of support, but with (potentially) different biological interpretations. Estimates from each of the supported models can be

Table 26.1 *Ten highest-ranked models according to AIC fit to the Grand Skink data. Models are denoted using the component used for each occupancy-related parameter. For all models detection probability varied by year with a consistent habitat effect (i.e. $p(H + Yr)$). Given for each model is the relative difference in AIC value (ΔAIC), the AIC model weight (w), twice the negative log-likelihood ($-2l$), and the number of estimated parameters (K).*

Model	ΔAIC	w	$-2l$	K
$\psi(H), \gamma(H), \varepsilon(H)$	0.00	0.23	1719.86	12
$\psi(H), \gamma(Yr + H), \varepsilon(H)$	0.18	0.21	1714.04	15
$\psi(H), \gamma(H), \varepsilon(\cdot)$	1.21	0.13	1723.07	11
$\psi(H), \gamma(Yr + H), \varepsilon(\cdot)$	1.44	0.11	1717.30	14
$\psi(H), \gamma(Yr + H), \varepsilon(Yr + H)$	2.54	0.06	1710.39	18
$\psi(H), \gamma(\cdot), \varepsilon(H)$	2.57	0.06	1724.43	11
$\psi(H), \gamma(H), \varepsilon(Yr + H)$	2.76	0.06	1716.62	15
$\psi(H), \gamma(Yr), \varepsilon(H)$	3.82	0.03	1719.68	14
$\psi(H), \gamma(Yr + H), \varepsilon(Yr)$	4.22	0.03	1714.08	17
$\psi(H), \gamma(H), \varepsilon(Yr)$	4.37	0.03	1720.23	14

examined and compared, although here model averaging has been used to combine the parameter estimates from all models to obtain overall estimates. Figures 26.1 and 26.2 present the model averaged colonization and extinction probabilities (respectively) for each habitat type and in each year. Naïve estimates, obtained from the raw data ignoring detection probability, are also given along with 95% confidence intervals.

Colonization probability is estimated to be generally low for both types of outcrops (<0.15), although it is approximately twice as high for those outcrops surrounded by the native tussock grassland, with some annual variation. The estimated colonization probabilities accounting for detection probability are notably lower than the naïve estimates. The probability of Grand Skinks going locally extinct from an outcrop are also estimated

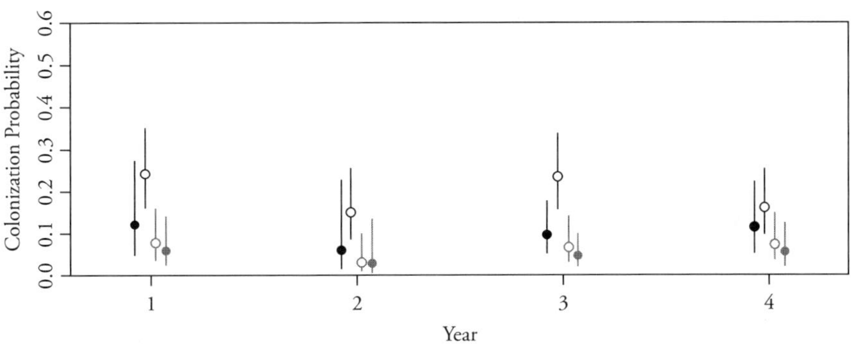

Figure 26.1 Model averaged colonization probability estimates for the Grand Skink example. Solid symbols indicate the estimates from the dynamic occupancy models with 95% confidence intervals indicated. Hollow symbols indicate naïve estimates. Black symbols relate to estimates for outcrops surrounded by native tussock grassland; grey symbols are for outcrops surrounded by sown pasture.

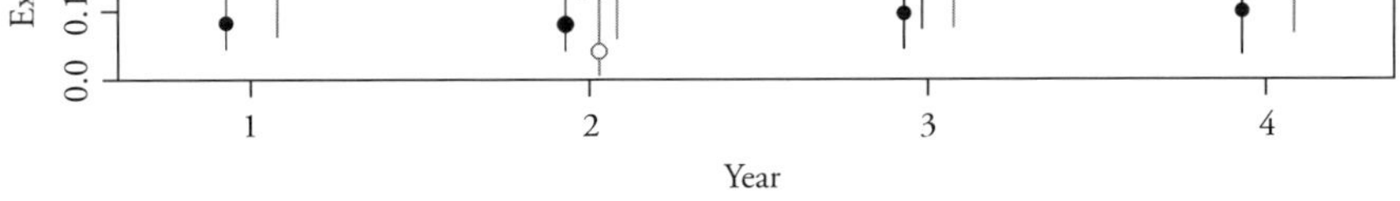

Figure 26.2 Model averaged extinction probability estimates for the Grand Skink example. Solid symbols indicate the estimates from the dynamic occupancy models with 95% confidence intervals indicated. Hollow symbols indicate naïve estimates. Black symbols relate to estimates for outcrops surrounded by native tussock grassland; grey symbols are for outcrops surrounded by sown pasture.

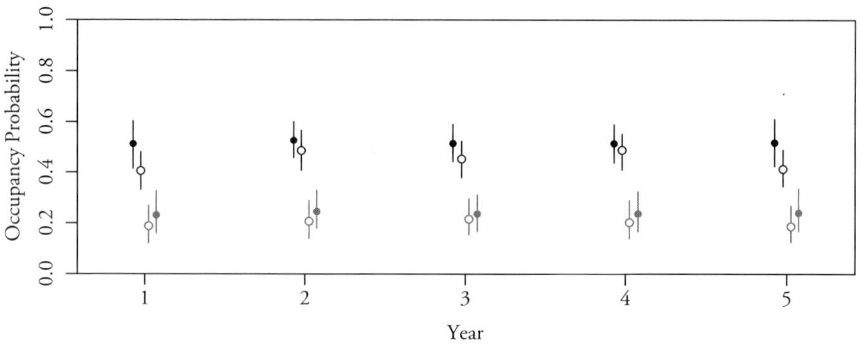

Figure 26.3 Model averaged occupancy probability estimates for the Grand Skink example. Solid symbols indicate the estimates from the dynamic occupancy models with 95% confidence intervals indicated. Hollow symbols indicate naïve estimates. Black symbols relate to estimates for outcrops surrounded by native tussock grassland; grey symbols are for outcrops surrounded by sown pasture.

to be generally low (<0.20), with estimates for outcrops surrounded by tussock being approximately half of the estimated extinction probability for outcrops surrounded by pasture. There is little annual variation. The naïve estimates for extinction are generally higher than those from the occupancy model where imperfect detection has been accounted for. Figure 26.3 presents the estimated occupancy probability in each habitat and year. Note that only the first-year occupancy estimates are estimated directly, with the estimates for years 2–5 being derived from the first-year occupancy estimates and estimates of colonization and extinction probabilities (MacKenzie et al., 2003, 2006). From Figure 26.3, the probability of an outcrop being occupied by Grand Skinks was consistently about 0.5 for those outcrops surrounded by tussock grassland during this timeframe, and about 0.25 for those outcrops surrounded by pasture. The naïve occupancy estimates are lower than those from the occupancy models that have accounted for imperfect detection, as would be expected. Overall, there is a clear indication that Grand Skinks are using the types of outcrops quite differently, with a preference (in some sense) for outcrops surrounded by the native grassland given the higher colonization and lower extinction probabilities, resulting in higher occupancy probabilities.

An important aspect of this example is that the naïve estimates of the dynamic parameters are greater than those obtained from accounting for detection. The naïve estimates would suggest that the system is much more dynamic with higher turnover rates than the estimates from the dynamic occupancy models that account for imperfect detection. This is because one effect of false absences (i.e. present but not detected) is that it increases the rate of apparent changes in occupancy.

26.4 Recent extensions

There have been a number of notable extensions of these methods since MacKenzie et al. (2006) that have been developed to relax some of the underlying assumptions, or to enable a broader range of biologically relevant questions to be addressed. In the following subsections, I highlight some of the more interesting ones, although a more comprehensive review of many of these extensions can be found in Bailey et al. (2014).

26.4.1 Multi-state occupancy

The initial models were developed for situations where interest is in the presence or absence of the species at each sampling unit, that is, there are two possible states for the status of the species. An obvious extension is for situations with more possible states. Royle and Link (2005) and Nichols et al. (2007) independently developed slightly different extensions for single-season applications, while MacKenzie et al. (2009) provided a unifying framework for the two approaches and an extension to multiple seasons. These multi-state occupancy models allow an additional level of resolution to the types of questions that can be asked at landscape-levels. For example, not only the whereabouts of a species, but within that distribution, where is breeding occurring, or where is the species infected by a particular disease or pathogen. Alternatively, the multiple states may represent different categories of abundance, for example, no individuals (species absence), some, or lots of individuals. An important identifying feature of multi-state occupancy models is that they allow for imperfect detection of the true state, recognizing that the evidence of breeding may not always be observed in a survey of a unit where breeding occurs, for example. As with regular occupancy models, having repeat surveys within each season allows for this imperfect detection to be accounted for.

26.4.2 Multi-scale occupancy

Nichols et al. (2008) developed an extension of the single-season model to look at occupancy at multiple spatial or temporal scales where there is a third level of sampling. For example, at each survey of each sampling unit, two observers independently survey for the species, or within each sampling unit, a number of sub-units are selected (e.g. three 100 m transects within a 4 ha unit) and each sub-unit is surveyed multiple times. This third level of information enables additional information to be extracted from the data and biologically relevant parameters estimated. An example of this is when the target species is only physically present at a unit at random within a season. This is technically a violation of the closure assumption of the regular models, but is reasonable if 'occupancy' is interpreted as a measure of which units are used by the species during a season (i.e. sometimes in the unit, but not always). In this case, using the regular models, the estimated 'detection' probability is a combination of the temporary presence, or availability, probability and detection given availability. These two components cannot be separated with only two levels of information. However, if there was an additional level of information such that multiple surveys were conducted within the time-scale when the species was actually present (e.g. at each visit, a surveyor conducts two 10 minute visual searches for the species), then the multi-scale occupancy models can separately estimate the probability of the species using a unit (spatial-scale occupancy), the probability of the species being present at the time of surveying given the unit is used (temporal-scale occupancy or availability), and the probability of detection given the species is present at the time of surveying (detection). Multiple-scale occupancy models can also be useful where occupancy is of interest at two spatial scales, for example, among and within watershed occupancy of ponds by the target species, while accounting for detection at the finer scale. As with all occupancy models, the framework used is flexible enough to incorporate covariates on the different model parameter and unequal survey effort.

26.4.3 Species misidentification

Species misidentifications can result if there are false detections, incorrectly recording the target species as detected when the detection was actually of another species. This will result in occupancy being overestimated. Importantly, this will happen even if the target species is detected perfectly otherwise. An obvious design consideration is to reduce the likelihood of species misidentifications through appropriate field crew training and protocols. It is also worth remembering that saying a species was not detected when it was really there is just a form of imperfect detection that can be dealt with, but having that and species misidentifications is more difficult to address. Hence, if in doubt, leave it out. Methods have been developed to account for misidentification, including those proposed by Royle and Link (2006), Miller et al. (2011), and Chambert et al. (2015). Chambert et al. (2015) outline a useful structure for how to conceptualize different forms of species misidentifications and the different available approaches for dealing with them. Readers are directed there for more details.

26.4.4 Correlated detections

The basic occupancy models assume that within-season surveys are independent, that is, the outcome of one survey has no effect on the outcome of another. This assumption can be violated in a number of ways, including if the temporal or spatial (when using spatially replicates) spacing between surveys is too close, particularly when the closure assumption is not fully met. Hines et al. (2010) extended the MacKenzie et al. (2002) model by including an additional underlying process to allow the species to be available for detection at each survey, similar to the multi-scale occupancy model, where the probability of being available can be different in each survey depending on whether the species was available to be detected in the previous survey. Unlike the multi-scale model, however, the correlated detection model does not require three levels of sampling. Note that if it was assumed the availability probability was the same regardless of whether the species was available in the previous survey, this would equate to the species being available at random. In a vagary of statistics, it is possible to estimate all the parameters with only two levels of sampling when the availability probabilities are different depending on the previous survey, but the parameters are not estimable when the probabilities are the same (i.e. for a simpler model).

The correlated detections model will be particularly useful when the repeat survey information is collected in a near continuous manner (either temporal or spatially), or when the closure assumption is not fully met and the species tends to be available for the duration of a number of consecutive survey occasions, then unavailable. When detections are collected in a continuous manner, the method detailed by Guillera-Arriota et al. (2011) can be used to account for correlated detections. In practice, it may be worth considering the correlated detection models when detections and non-detections at a unit tend to consistently occur in pulses rather than be more randomly spread throughout the season.

26.5 Response to criticisms

As with any statistical technique, there are a number of assumptions associated with occupancy models that are required to be met in order for final conclusions to be valid and well supported. There have been some papers (e.g. Lele et al., 2012; Welsh et al., 2013) and discussions in the blogosphere critical of these assumptions, arguing they are often biologically unreasonable and, therefore, these more complex methods should not be used in favour of simpler approaches that involve fewer assumptions, such as logistic regression. However, the 'simpler' methods are often actually making very similar assumptions, although they are not explicitly stated. Here I briefly comment on two key assumptions, closure and no heterogeneity in detection; readers should see Guillera-Arriota et al. (2014) for a more comprehensive response.

One of the key assumptions used by occupancy models is that sampling units are closed to changes in occupancy within a season, which is required to account for imperfect detection. That is, within a season, the species is either always present or always absent from a sampling unit over the period of the repeated surveys. This assumption is required so that when the species is detected at least once within a season at a unit, non-detection within a season is regarded as 'present but not detected' rather than as a possible absence. This closure assumption can be relaxed to some degree, such that a non-detection could be the result of a temporary absence (e.g. members of the species may be elsewhere within their home range), in which case occupancy should be regarded as a measure of what places are used by the species during the season (i.e. the species is present in the unit at some point within the season). Essentially the closure assumption assumes there is a period of time where the distribution of the species is relatively static, or stable, during which repeat surveys are conducted. Some have argued this closure assumption is unrealistic and therefore advocated the data be collapsed across the repeat surveys into a single detection–non-detection value for each sampling unit, or to only use a single survey per unit, and use logistic regression to analyse the data. However, the results of the logistic regression indicate the estimated probability of 'presence' associated with the observation. When the data are collapsed to a single value, the species is regarded as 'present' at the sampling unit if it was detected there at any point during the course of the repeat surveys, and 'absent' if it never was. This is identical to the relaxed version of the closure assumption where occupancy should be interpreted as use. In the case when only a single survey is conducted, the logistic regression results only apply to the duration of that single observation. For example, if data are from 10 minute visual encounter surveys, the estimated probabilities apply to where the species was detected during those 10 minutes. As soon as a researcher is willing to presume that where a species is found during the survey is indicative of where it will be outside the survey period (i.e. where the species will be for the next day, week, month), and similarly where it was not found is indicative that it will not be there over a longer period, they are assuming there is a period of time where the distribution of the species is unchanging. That is, they are implicitly assuming the system is closed to changes in occupancy over some timeframe in exactly the same manner as the explicit assumption of occupancy models.

Another criticism of occupancy models is that results can be sensitive to unmodelled detection heterogeneity, that is, among-unit variation in detection that is unaccounted for by the model (Welsh et al., 2013). Unmodelled heterogeneity typically leads to occupancy being underestimated, which has long been recognized; methods have been developed that can be used to account for it (e.g. Royle and Nichols, 2003; Royle, 2006; MacKenzie et al., 2006; Chapter 5). However, some have argued that the prospect of unmodelled heterogeneity is itself a cause to avoid using methods that account for detection, and again use simpler methods. Unfortunately, ignoring detection completely hardly lessens the potential impact of detection heterogeneity. Using logistic regression to model and estimate occupancy assumes that detection is perfect, or if the results are to be interpreted in a relative sense, the effect of detection is consistent across sampling units. If there is detection heterogeneity, then detection cannot be considered consistent and resulting estimates are likely to be at least as biased, or more likely have a greater bias, than estimates from an occupancy model (Guillera-Arroita et al., 2014). When detection heterogeneity is suspected of having an effect on results, then a responsible analyst will investigate ways to account for this, either with suitable covariates or extensions of the basic occupancy models.

It is worth noting that the effects of detection heterogeneity are greater when the average overall detection probability (i.e. the probability of at least one detection from the repeat surveys of a unit) is lower (Link, 2003; MacKenzie et al., 2006). The best way to deal with such heterogeneity is to design studies with higher overall detection probabilities through either increasing per-survey detection probabilities or using a greater number of repeat surveys.

26.6 Summary

Occupancy models are a useful addition to the researcher's toolbox for assessing questions about the patterns and dynamics of species occurrence. With the more recent extensions, it is possible to address a wider range of questions that are relevant both in terms of biological understanding and for conservation management. A key feature of them is the ability to explicitly account for imperfect detection that may create bias in estimates and affect sizes if left unaddressed. However, all the methods still work if detection is perfect, and in fact will work much better as detection no longer needs to be estimated. With perfect detection, the simple single-season occupancy model of MacKenzie et al. (2002) reduces to regular logistic regression, but that is not the case for the other modelling approaches discussed in this chapter (e.g. the dynamic occupancy model). In many cases, comparable modelling approaches have not been developed assuming perfect detection; therefore, this suite of methods is very relevant to researchers that do not believe they have false absences.

That there are a number of explicit assumptions associated with occupancy models is, in my view, a strength of these methods, as the researcher should take some time to consider whether the assumptions are likely to be met for their particular situation. Explicit assumptions clarify how the fieldwork, data, and modelling relate to each other, and therefore how estimates should be interpreted relative to the biological questions of interest. Where assumptions are not explicitly stated, errors of application and interpretation are likely to result.

The best time to consider potential assumption violations is well before fieldwork begins. Good design is paramount to the success of any study or monitoring programme. Objectives need to be clearly stated; likely methods of analysis should be identified; assumptions of the analysis methods need to be carefully considered; and the impact that those assumptions might have in terms of what data and information need to be collected while in the field should be addressed. There are a number of sources that provide useful advice with respect to designing occupancy-based studies while accounting for imperfect species detection (MacKenzie and Royle, 2005; MacKenzie et al., 2006; Guillera-Arroita et al., 2010).

As part of the design phase, careful attention should also be devoted to obtaining realistic expectations about the quality of the inferences that could be drawn from the study. I often say that these methods are statistical, not magical; they require a certain amount of information to work well. In my experience, situations when the methods do not work well are typically when the number of surveyed units, or number of repeat surveys conducted, is too small to provide adequate information relative to the researcher's questions of interest. A detailed assessment of their study design prior to the commencement of the fieldwork would have indicated that there were problems. That researchers decide the methods 'didn't work' is not necessarily a failure of the method, but simply an indication that they may have been trying to achieve too much with the resources they had available. Neither excuse should be used as justification for using simpler methods that ignore the practicalities of the fieldwork (Welsh et al., 2013), as this comes at the cost of compromising how the results should be interpreted (i.e. as a combination of biological and observation processes, rather than trying to separate the two as with occupancy models). Instead it should be a wake-up call to the researcher that it is unlikely that they will be able to reliably address their questions of interest with their current design, and that they will need to increase their field effort or modify their overall design. Alternatively, the study objectives may need to be refined such that they are likely achievable with the present design and available resources.

Finally, there are a range of freely available software packages that can be used for data analysis. Programs PRESENCE (http://www.mbr-pwrc.usgs.gov/software/presence.html) and MARK (http://warnercnr.colostate.edu/~gwhite/mark/mark.htm) are Windows-based packages that allow uses to fit a range of occupancy models to their data using (primarily) maximum-likelihood methods. There are also a range of R-packages https://www.r-project.org/) including unmarked, RPresence and RMark. The latter two are packages that enable PRESENCE and MARK to be used from within R, rather than via a Windows GUI (even on other operating systems). Bayesian methods of analysis can also be applied using OpenBUGS (http://www.openbugs.net/w/FrontPage) or JAGS (http://mcmc-jags.sourceforge.net/).

References

Bailey, L.L., MacKenzie, D.I., and Nichols, J.D. (2014). Advances and applications of occupancy models. *Methods in Ecology and Evolution*, **5**, 1269–79.

Burnham, K.P., and Anderson, D.R. (2002). *Model Selection and Multi-Model Inference*. 2nd ed. New York: Springer.

Chambert, T., Miller, D.A., and Nichols, J.D. (2015). Modeling false positive detections in species occurrence data under different study designs. *Ecology*, **96**, 332–9.

Garrard, G.E., Bekessy, S.A., McCarthy, M.A., et al. (2008). When have we looked hard enough? A novel method for setting minimum survey effort protocols for flora surveys. *Austral Ecology*, **33**, 986–98.

Guillera-Arroita, G., Lahoz-Monfort, J.J., MacKenzie, D.I., et al. (2014). Ignoring imperfect detection in biological surveys is dangerous: a response to 'Fitting and Interpreting Occupancy Models'. *PLoS One*, **9**, e99571.

Guillera-Arroita, G., Morgan, B.J.T., Ridout, M.S., et al. (2011). Species occupancy modeling for detection data collected along a transect. *Journal of Agricultural, Biological and Environmental Statistics*, **16**, 301–17.

Guillera-Arroita, G., Ridout, M.S., and Morgan, B.J.T. (2010). Design of occupancy studies with imperfect detection. *Methods in Ecology and Evolution*, **1**, 131–9.

Hines, J.E., Nichols, J.D., Royle, J.A., et al. (2010). Tigers on trails: occupancy modeling for cluster sampling. *Ecological Applications*, **20**, 1456–66.

Lele, S.R., Moreno, M., and Bayne, E. (2012). Dealing with detection error in site occupancy surveys: what can we do with a single survey? *Journal of Plant Ecology*, **5**, 22–31.

Link, W.A. (2003). Nonidentifiability of population size from capture-recapture data with heterogeneous detection probabilities. *Biometrics*, **59**, 1123–30.

MacKenzie, D.I. (2014). Occupancy models. In B.F.J. Manly and J.A. Navarro Alberto (eds) *Introduction to Ecological Sampling*. Boca Raton, FL: CRC Press.

MacKenzie, D.I., and Royle, J.A. (2005). Designing occupancy studies: general advice and allocating survey effort. *Journal of Applied Ecology*, **42**, 1105–14.

MacKenzie, D.I., Bailey, L.L., and Nichols, J.D. (2004). Investigating species co-occurrence patterns when species are detected imperfectly. *Journal of Animal Ecology*, **73**, 546–55.

MacKenzie, D.I., Nichols, J.D., Hines, J.E., et al. (2003). Estimating site occupancy, colonization, and local extinction when a species is detected imperfectly. *Ecology*, **84**, 2200–7.

MacKenzie, D.I., Nichols, J.D., Lachman, G.B., et al. (2002). Estimating site occupancy rates when detection probabilities are less than one. *Ecology*, **83**, 2248–55.

MacKenzie, D.I., Nichols, J.D., Royle, J.A., et al. (2006). *Occupancy Estimation and Modeling: Inferring Patterns and Dynamics of Species Occurrence*. Oxford: Elsevier.

MacKenzie, D.I., Nichols, J.D., Seamans, M.E., et al. (2009). Modeling species occurrence dynamics with multiple states and imperfect detection. *Ecology*, **90**, 823–35.

Miller, D.A., Nichols, J.D., McClintock, B.T., et al. (2011). Improving occupancy estimation when two types of observational error occur: non-detection and species misidentification. *Ecology*, **92**, 1422–8.

Nichols, J.D., Bailey, L.L., O'Connell, A.F., Jr., et al. (2008). Multi-scale occupancy estimation and modelling using multiple detection methods. *Journal of Applied Ecology*, **45**, 1321–9.

Nichols, J.D., Hines, J.E., MacKenzie, D.I., et al. (2007). Occupancy estimation and modeling with multiple states and state uncertainty. *Ecology*, **88**, 1395–400.

Royle, J. A. (2006). Site occupancy models with heterogeneous detection probabilities. *Biometrics*, **62**, 97–102.

Royle, J.A., and Link, W.A. (2005). A general class of multinomial mixture models for anuran calling survey data. *Ecology*, **86**, 2505–12.

Royle, J.A., and Link, W.A. (2006). Generalized site occupancy models allowing for false positive and false negative errors. *Ecology*, **87**, 835–41.

Royle, J.A., and Nichols, J.D. (2003). Estimating abundance from repeated presence–absence data or point counts. *Ecology*, **84**, 777–90.

Seddon, P.J., Roughton, C.M., Reardon, J., et al. (2011). Dynamics of an endangered New Zealand skink: accounting for incomplete detectability in estimating patch occupancy. *New Zealand Journal of Ecology*, **35**, 247–53.

Stauffer, H.B., Ralph, C.J., and Miller, S.L. (2004). Ranking habitat for marbled murrelets: new conservation approach for species with uncertain detection. *Ecological Applications*, **14**, 1374–83.

Tyre, A.J., Tenhumberg, B., Field, S.A., et al. (2003). Improving precision and reducing bias in biological surveys: estimating false-negative error rates. *Ecological Applications*, **13**, 1790–801.

Welsh, A.H., Lindenmayer, D.B., and Donnelly, C.F. (2013). Fitting and interpreting occupancy models. *PLoS One*, **8**, e52015.

Wintle, B.A., McCarthy, M.A., Parris, K.M., et al. (2004). Precision and bias of methods for estimating point survey detection probabilities. *Ecological Applications*, **14**, 703–12.

27
Estimating abundance

Chris Sutherland and J. Andrew Royle

27.1 Introduction

Fundamental to much of ecology, including species management and conservation, is the ability to reliably estimate population size, or *abundance*. This is particularly true for reptiles given the growing evidence of declines globally (Gibbons et al., 2000; Reading et al., 2010) and high levels of data deficiency (McDiarmid et al., 2011; Böhm et al., 2013). To understand the causes and consequences of population declines and to address data deficiencies requires monitoring, which is especially challenging because not all individuals in a population are encountered during a survey, and the resulting *counts*, that is, the number of individuals encountered, n, represents only some fraction of the true abundance, N. Although the distinction between counts and true abundance is an important one, the two are intuitively related and, by repeatedly sampling and marking individuals in a population, capture–recapture methods provide a formalization of this relationship and a framework for estimating true population size.

In this chapter, we provide a non-technical overview of 'closed population capture–recapture' models, a class of well-established models that are widely applied in ecology (Borchers et al., 2002; Williams et al., 2002) and regularly adopted for studies of reptiles (Mazerolle et al., 2007), to estimate abundance from counts of marked individuals while accounting for imperfect detection. We first describe some classic closed population models for estimating abundance (Otis et al., 1978), then consider some recent extensions that provide a spatial context for the estimation of abundance, and therefore density, D (spatial or spatially explicit capture–recapture; Efford, 2004; Royle et al., 2014), and finally provide an example of estimating abundance and density of reptiles using an artificial cover object survey of Slow Worms (*Anguis fragilis*).

27.2 Closed population capture–recapture

27.2.1 Sampling a population

The primary objective in a closed population capture–recapture study is to estimate the abundance of a population of interest. The population of N individuals is subjected to repeated sampling for a specified number of occasions, say K, where, in the first sampling occasion, all captured individuals are marked and released, and then at each subsequent sampling occasion the detection of marked individuals is recorded and unmarked individuals are marked. Identifying a focal population, the spatial extent of the population is implicitly defined and the method of capture depends on the species in question and available resources (see Chapter 10). For reptiles, survey methods that allow individuals to be captured and marked include, for example, visual searches within a defined area (Christy et al., 2010; Zylstra et al., 2010), cage traps (Tyrrell et al., 2009) or pitfall traps (Welsh, 1990), and the use of artificial cover objects (ACOs, Grant et al., 1992; Sutherland et al. 2016). Once captured, individuals can be uniquely identified using natural markings that can be used to determine individual identity (Sacchi et al., 2010), using tags or markings (Grant and Doherty, 2007), or by physical marking such as toe-clipping (Paulissen and Meyer, 2000; see Chapter 4).

Such repeated sampling results in individual encounter histories that, for each of the $i = 1, \ldots, n$ individuals encountered, describes whether or not they were detected in each of the K occasions. For example, in a $K = 4$ occasion capture–recapture study, an individual with an encounter history $y_i = (0101)$ was encountered 2 times, first in occasion 2, then again in occasion 4, and was not encountered in occasions 1 or 3. In Table 27.1 we provide an example of encounter history data for a $K = 4$ occasion capture–recapture study during which $n = 8$ individuals were captured.

Estimating abundance using encounter history data collected using the general sampling scheme we have described previously is basically the process of estimating how many individuals were *missed*, that is, how many individuals have encounter history $\sum y_i = 0$. The ability to do so requires that these basic assumptions are met:

1. The population is closed to demographic processes and to movement.
2. Individual marks can be identified unambiguously and are not lost.
3. Individuals are equally likely to be captured.

A 'closed' population is one that experiences no additions or subtractions for the duration of the study, and whose size is therefore assumed to be fixed during sampling. Defining a sampling period over which the assumption of closure can be

Table 27.1 *An example of an encounter history for a* K = 4 *occasion capture–recapture study during which* n = 8 *individuals were detected and* $\sum y$ *is the total number of captures.*

Individual	1	2	3	4	$\sum y$
1	0	1	0	1	2
2	1	0	0	1	2
3	0	1	1	1	3
4	1	1	0	0	2
5	1	0	0	1	2
6	0	0	1	1	2
7	0	0	1	0	1
$n = 8$	1	0	0	1	2

satisfied means that an individual detected at least once during the study was present for the entire study, and therefore, failure to detect that individual in any occasion was due to imperfect detection. This highlights the importance of the second assumption—that individuals are identified unambiguously—because misidentification would lead to erroneous encounter histories that don't reflect the true process of encountering individuals. The third assumption is less important as we will see later, but satisfying this assumption means that we can employ the simplest formulation of a capture–recapture model, model M_0. Failure to meet these assumptions can result in biased estimators of abundance as discussed in detail in Kendall (1999) and Williams et al. (2002).

27.2.2 Estimating abundance using model M_0

Under model M_0, the encounter probability for each individual, p_i, is assumed to be the same for all individuals in the population, that is, $p_i = p$. Then, whether or not we encounter an individual i = 1, 2,. . ., N during sampling occasion k, y_{ik}, is a Bernoulli trial (a 'coin flip') with constant probability p, which we can write formally as

$$y_{ik} \sim \text{Bernoulli}(p).$$

That is, there are no individual or temporal covariates that affect p. The basic idea of all closed population capture–recapture methods is that the pattern of detections in the encounter histories of individuals observed at least once provides information about individual detectability, or detection probability, p, which, in turn, can be used to estimate the number of individuals *not* encountered. The underlying concept can be understood by recognizing that the observed number of individuals n is related to the total population size N by the expression

$$E(n) = N\tilde{p},$$

where $E()$ denotes statistical expectation and $\tilde{p}$ is the probability that an individual is captured *at all* during the study. In a study of K survey occasions $\tilde{p}$ is directly related to the basic parameter p by the formula

$$\tilde{p} = 1 - (1 - p)^K.$$

This expression applies to model M_0 but the expression relating p to $\tilde{p}$ is different depending on the specific capture–recapture model being considered. In general, the parameter p can be estimated from the observed encounter histories and, in turn, this is used to estimate $\tilde{p}$ and then finally we estimate N by $\hat{N} = n/\tilde{p}$.

This estimator of N is typically called the 'conditional estimator' of N because the quantity $\tilde{p}$ is the probability of capture in at least one occasion, and the likelihood is formally constructed by conditioning on the event that an individual is encountered at all (i.e. at least once). The conditional estimator of population size N has only one estimated parameter, that being p, and N is a derived parameter. An alternative framework for inference about N is based on the 'full likelihood' in which N appears explicitly in the likelihood. The two approaches are both widely and there is little practical difference.

27.2.3 Variation in p: beyond M_0

The assumption of equal capture probability is a rather restricting one and there are many situations under which the capture probabilities would be expected to vary. For these situations, model M_0 is not appropriate and Otis et al. (1978) described a family of models that can be used to deal with most sources of variation in individual encounter probabilities:

M_0: Capture probability is the same for all individuals (Section 2.2).
M_t: Capture probability is the same for all individuals, but varies between sampling occasions (**t**ime) (note that subsequently we use *k* for the time index in our development).
M_b: Capture probabilities vary depending on whether or not individuals have been captured previously (**b**ehavioural response).
M_h: Capture probabilities vary among individuals (individual **h**eterogeneity).

Variations of these different models exist. For example, the usual application of model M_t involves occasion-specific parameters p_k, but we can also consider systematic variation in detection probability that results from explicit covariates. For example, we could model seasonal variation in p_k using a quadratic polynomial in Julian day, J_k, such as

$$logit(p_k) = \alpha_0 + \alpha_1 J_k + \alpha_2 J_k^2 ,$$

and estimate the parameters α_0, α_1, and α_2 instead of fully occasion specific parameters p_k. The behavioural response model is usually parameterized as a permanent change in p for individuals subsequent to their initial capture (i.e. p_{pre} and p_{post}, for capture probability *prior to* and *after* first capture, respectively). This could be the result of 'trap happiness' due to having baited traps (uncommon in most reptile studies) or it could be the result of 'trap shyness' due to aversion to handling. Sometimes a transient or ephemeral behavioural response may be sensible (Yang and Chao, 2005). Under this type of model the response to initial capture only lasts for a brief period after initial capture.

Model M_h has been an important model in capture–recapture because it has long been recognized that the existence of individual heterogeneity in capture probability will lead to under-estimation of N when it is not accounted for. Thus, much attention has been focused on developing more flexible classes of model M_h and it has many different variations. Norris and Pollock (1996) formulated the model in terms of a finite mixture or latent class model in which each individual in the population belongs to a finite (and small) number of classes represented by distinct values of p (see Pledger, 2000). Dorazio and Royle (2003) considered continuous formulations of model M_h, where individual heterogeneity is described by a continuous distribution such as the beta-binomial and logit-normal models (see Coull and Agresti, 1999).

27.2.4 Removal sampling

Removal sampling is unlike other capture–'recapture' methods in the sense that individuals are captured only once but then removed from the population and, hence, not available to be recaptured. The idea of removal sampling is that, by repeatedly removing individuals, we should realize a decrease in catch frequency, and this realized decrease is informative about detection probability that can then be used to obtain an estimate of the population size N as it was *prior* to the initiation of the sampling activity. Intuitively, we expect to capture $p \times N$ individuals during a single sample and, if we remove those individuals, we should expect to capture $p(N - pN) = p(1 - p)N$ individuals in the second bout of removal sampling. Note that the ratio of these two removal counts is $1 - p$ that produces a direct estimate of p from which N can be estimated by a suitable algebraic function of the counts.

In practice, removal sampling is done using 'temporary' removals in which an area is searched and individuals are temporarily housed in a bucket or cage during successive passes (Garden et al., 2007). At the end of the study, they would normally be released where they were initially captured. Of course earlier applications of removal sampling to fisheries involved permanent removal, typically harvest, but that is not practical in most studies of reptiles.

27.2.5 Hierarchical capture–recapture models

Closed population models are usually described in the context of sampling a single closed population. However in practice, it is almost always the case that multiple populations of individuals are sampled. For example, small mammal trapping studies might involve a number of live trapping grids, or reptile and amphibian studies might involve a number of distinct coverboard arrays (Sutherland et al. 2016). In such cases, we can think of the distinct populations being sampled by each array as spatial strata, with potentially stratum specific parameters such as p_g, N_g (g for 'grid') that could be estimated by applying closed population models to data collected from each stratum. However, analysing the data from these stratified populations 'one-at-a-time' can be statistically inefficient (Converse and Royle, 2012). Instead, there can be distinct advantages to combining all of the data into a single capture–recapture model but tying the different models together by imposing model structure on the detection probability or abundance parameters using a hierarchical modelling framework. For example, a simple model allowing for variation in population size among a collection of $g = 1,2,...,G$ trap arrays is $N_g \sim \text{Poisson}(\lambda)$. This model allows each population to have its own size, N_g, which is governed by an overall mean value, λ. Such structure builds in a weak stochastic dependence between the populations, so that all of the data are used to estimate the shared parameter λ. The end result is more precise to estimate the population-specific parameters by combining the information from all of the distinct populations. See Royle (2004), Dorazio et al. (2005), Royle et al. (2012), and Sutherland et al. 2016.

27.2.6 Individual covariate models and distance sampling

Model M_h is the standard closed population model when *unexplained* individual heterogeneity in capture probability exists. However, an important related class of models are models in which individual heterogeneity can be explained by explicit individual covariates. These are often called 'individual covariate models' but, in keeping with the classical nomenclature on closed population models, Kéry and Schaub (2012) referred to these models as 'model M_x', the x representing some explicit covariate (of course multiple covariates are allowed). Classical examples of covariates influencing detection probability are type of animal (juvenile/adult or male/female), a continuous covariate such as body mass, or a discrete covariate such as group or cluster size. For example, in a study of Timber Rattlesnakes (*Crotalus horridus*), Brown et al. (2007) modelled recapture probabilities as a function of 'colour morph', a discrete individual covariate ('yellow' or 'black').

The basic encounter model for model M_x is the same as our other closed models, the Bernoulli encounter model:

$$y_i \sim \text{Bernoulli}(p_i).$$

To model the covariate, we use a logit model for encounter probability of the form

$$\text{logit}(p_i) = \alpha_0 + \alpha_1 x_i,$$

where x_i is the covariate value for individual i and the parameters α_0 and α_1 are the parameters to be estimated.

Traditionally, estimation of N in model M_x is achieved using methods based on ideas of unequal probability sampling (i.e. Horvitz–Thompson estimation). This idea was developed independently by Huggins (1989) and Alho (1990). The estimator of N is given as a derived parameter:

$$\hat{N} = \sum_{i=1}^{n} \frac{1}{\tilde{p}_i},$$

where $\tilde{p}_i$ is the probability that individual i appeared in the sample. This is related to the more fundamental parameters α in the model for detection probability according to:

$$\tilde{p}_i = 1 - (1 - p_i)^K,$$

where p_i is a function of parameters α_0 and α_1 through the logit model defined above. In practice, parameters are estimated from the conditional-likelihood of the observed encounter histories.

An alternative formulation of model M_x is the 'full likelihood' which requires that we put a model on the individual covariate x allowing for the sample not only of the encounter histories but also of the covariate to be extrapolated to the population. For example, if we have a continuous trait measured on each individual, then we might assume that x has a normal distribution:

$$x_i \sim \text{Normal}(\mu, \sigma^2).$$

If the covariate was group size then, naturally, some discrete probability mass function would be needed. Inference for individual covariate models from the standpoint of the full likelihood is discussed widely, including by Royle (2009) and Kéry and Schaub (2012).

Individual covariate models are important in practice for the simple reason that heterogeneity exists in almost every capture–recapture study due to the spatial organization of traps and of individuals in the population (see next section). Thus, they were adopted historically to account for spatial structure in capture–recapture studies (Karanth and Nichols, 1998; Boulanger and McLellan, 2001). For this purpose, an individual covariate is created which describes *where* the individual is located in relation to the trapping array. This approach leads naturally to more recent spatial capture–recapture models described in the next section.

A very important and popular method for estimating abundance is distance sampling (Buckland et al., 2005; see also Rodda and Campbell, 2002, and Smolensky and Fitzgerald, 2010, for reptile applications). Unlike capture–recapture sampling, distance sampling requires only a single 'snap-shot' sample of the population. For each detected individual, distance from the observer is measured. Information about detection probability comes from an assumed model for the relationship between detection probability and distance to observer. Distance sampling is, formally, a special case of individual covariate models where there is only a single replicate sample ($K = 1$), and the individual covariate is distance. See Anderson et al. (2001) for an example of using distance sampling for reptiles.

27.2.7 Spatial capture–recapture

One of the main deficiencies with classical closed population models is that they do not permit direct estimation of animal *density* because, in almost all practical field applications, it is not possible to precisely define the area sampled by a set of trapping devices. This is because individuals being captured move and can be captured without the biologists knowing from whence those individuals originated or how much space they are using. Newly developed *spatial capture–recapture* (SCR) models (also called spatially explicit capture–recapture, SECR) provide a statistical framework for dealing with this problem (Efford, 2004; Borchers and Efford, 2008; Royle and Young, 2008; Royle et al., 2014). These models follow logically from model M_x, where x is the activity centre of the individual as we will see in the following paragraphs.

The sampling scheme for a SCR analysis is the same as that described previously, that is, there is a population of N individuals, but now we consider each individual having an *activity centre* that has X and Y coordinates ($\mathbf{s}_i = [s_{i,X}, s_{i,Y}]$). Now the goal is to estimate the number of individuals (or activity centres) within a region of interest which we refer to as the *state-space*, or S, which is to say we wish to estimate density: $D = N/\|S\|$, where $\|S\|$ is the area of S. We assume that these activity centres are distributed uniformly over S:

$$\mathbf{s}_i \sim \text{Uniform}(S).$$

As before, the population is subjected to sampling using some trapping devices (for convenience, we will refer to these as 'traps'). However, we explicitly acknowledge both how many traps there are: $j = 1,\ldots, J$ traps, and the locations of each of the traps, which we denote as x_j. The acknowledgment of the spatial structure of the traps means observations can be spatially indexed so encounter histories describe *who* (i), *when* (k), and, importantly, *where* (j) individuals were encountered, that is, $y_{i,j,k}$. Typically, these observations are assumed to be Bernoulli outcomes:

$$y_{i,j,k} \sim \text{Bernoulli}(p_{i,j,k}),$$

where $p_{i,j,k}$ is the probability of encountering individual i in trap j, and occasion k, which at a minimum depends on the distance between the trap location (x_j) and the individuals activity centre (s_i) as follows:

$$p_{i,j,k} = p_0 \times e^{-(1/2\sigma^2)d(x_j,s_i)^2}.$$

(It may also depend on sample occasion k in some fashion.) This is referred to as the half-normal encounter model where logit(p_0) = α_0 is the baseline encounter probability, which is the probability of encountering an individual at its activity centre, the parameter σ describes the rate at which detection probability declines as a function of distance, and d(x_j, s_i) is the Euclidean distance between trap j and the activity centre of individual i. In an SCR analysis, the parameters to be estimated are α_0 and σ in addition to population size N. As in model M_h, the additional parameter σ accommodates individual heterogeneity in p but, unlike model M_h, the parameter represents an explicit source of heterogeneity, that due to distance between individual activity or home range centres and trap locations.

SCR models address the density estimation problem directly by parameterizing the model directly in terms of individual activity centres s_i and prescribing the state-space S. The inference problem then reduces to estimating the number of such activity centres in the well-defined area S, that is, density. Density, D, is simply a transformation of N: $D = N/\text{Area}(S)$.

27.3 Software

Non-spatial closed population capture–recapture models can be fit using both classical (frequentist) and Bayesian methods, and here we outline some of the common software options and material sources for doing so.

By far the most widely used software for fitting such models is the Windows-based program MARK (White and Burnham, 1999), a free, user friendly and extremely well-documented application for fitting most of the standard closed population models using maximum likelihood (http://www.phidot.org/software/mark). In an effort to facilitate easier model development, increased reproducibility and to allow for a more organized and automated work flow, Laake and Rexstad (2008) developed the R package RMark, an interface between R and MARK that can be used to construct input files and extract output that can manipulated in the R environment (http://www.phidot.org/software/mark/rmark/). The R package unmarked (Fiske and Chandler, 2011) can also be used to fit hierarchical versions of some standard closed population capture–recapture models using maximum likelihood (see Kéry and Royle, 2015, Chapter 7). In addition to detailed documentation, both MARK (http://www.phidot.org/forum/index.php) and unmarked (https://groups.google.com/forum/#!forum/unmarked) have supporting web based forums.

SCR models cannot (yet) be implemented in MARK, and, although a Windows-based spatial equivalent exists (DENSITY: Efford et al., 2004), it is no longer in development and likelihood analysis of SCR models is typically conducted using the R package secr (Efford 2011). The R package secr implements a wide variety of SCR methods, including spatial versions of the models described earlier and in Otis et al. (1978), all within the R environment (the SECR forum can be found here: https://groups.google.com/forum/#!forum/secrgroup).

Often situations arise when an analysis requires a 'non-standard' approach (e.g. an integrated population model: Schaub and Abadi, 2011; demographic metapopulation models: Sutherland et al., 2014; models for transience and dispersal: Royle et al., 2016). In these cases, available likelihood-based methods cannot be used and instead a Bayesian analysis is needed. Bayesian methods, specifically the use of Markov chain Monte Carlo (MCMC) using the BUGS language, offer a great deal of flexibility and modelling freedom (Kéry and Schaub, 2012).

Bayesian analysis of capture–recapture models is done using a method related to classical 'data augmentation' from the statistics literature (e.g. Tanner and Wong, 1987). This is a general concept in statistics but, in the context of capture–recapture models, where N is unknown, it has a consistent implementation across classes of capture–recapture models and one that is convenient from the standpoint of doing MCMC (Royle et al., 2007; Royle and Dorazio, 2012). Chapter 6 of Kéry and Schaub (2012) provides an accessible and complementary development of Bayesian analysis of non-spatial closed population models. The tremendous benefit of formulating SCR models in the BUGS language is that it is relatively trivial to extend form the non-spatial model to the formulation on the closed population model. In their book, Royle et al. (2014) provide a thorough treatment of SCR and Bayesian methods for analysing SCR models that has an associated forum (https://groups.google.com/forum/#!forum/spatialcapturerecapture).

27.4 Example: population size and density estimation of Slow Worms

We provide an example using a study of a legless lizard, the Slow Worm (*Anguis fragilis*), conducted in Mueterschwanderberg, Canton Nidwalden, Switzerland (Meier, 2012). A more detailed SCR analysis of the data from that study is given by Meier et al. (unpublished data). Here we analyse a portion of the data from one of the artificial cover object arrays which used 23 cover objects (Figure 27.1). This ACO array was operated over 59 days and it produced encounter histories on 44 unique individuals, 23 captured once, 4 capture twice, 4 thrice, 5 four times, 3 five times, 1 six times, and 4 seven times.

An immediately obvious 'feature' of this ACO array is the irregular and uneven distribution of cover objects. While we may apply capture–recapture models to the data from this study to estimate population size N, it is clear that the irregular ACO array should be a hindrance to a clear interpretation of this estimate because there is no obvious way to compute a sample area for this grid. Moreover, the spatial heterogeneity in trap density is almost certain to induce heterogeneity in detection probability of individuals depending on the location of individuals relative to traps. While we can use model M_h to account for some heterogeneity, that model is not an explicit model of heterogeneity due to spatial proximity of individuals to traps. Nevertheless, we fit both models M_0 and M_h to these data in order to see how they compare. We will also fit the SCR model to see how we can convert an estimate of N to density.

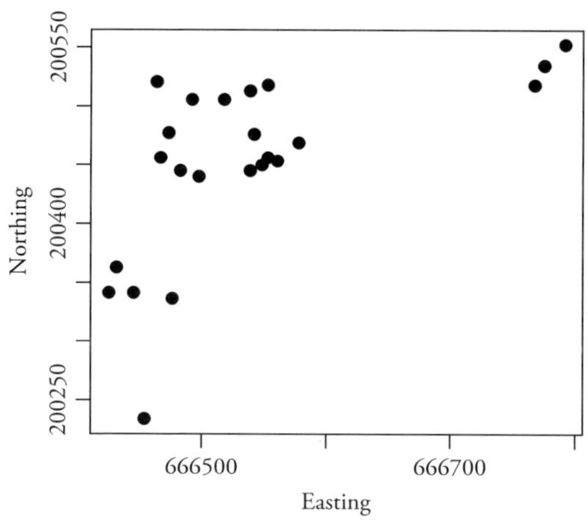

Figure 27.1 Artificial cover object (ACO) array from the Slow Worm study of Meier (2012). Sampling from this array over 59 days produced encounter histories of 44 individuals.

We fit the logit-normal variety of model M_h (Coull and Agresti, 1999; Dorazio and Royle, 2003) which assumes that the logit-transformation of individual detection probability p_i has a normal distribution with variance θ^2:

$$logit(p_i) \sim \text{Normal}(\mu, \theta^2).$$

Therefore this model has three parameters (N, μ, and θ). The additional parameter θ accommodates over-dispersion or individual heterogeneity in p. For the analysis of the SCR model we defined the state-space by creating the minimum area rectangle around the trap array and then buffering that by 40 m, creating a state-space having a buffered area of 17.58 ha (Figure 27.1). The summary results from fitting these three models to the Slow Worm data are given in Table 27.2. We note that the Akaike Information Criterion (AIC) is *not* comparable between SCR and ordinary closed models because the models are fitted to different data sets: SCR models use the individual, trap and occasion-specific encounter data whereas ordinary capture–recapture models aggregate over all traps to produce a simpler reduced-information individual by occasion-specific encounter history. A key point of this comparison is that the estimates of n_0 (the number of uncaptured individuals), and hence of N, are radically different. Model M_h and the SCR model both accommodate heterogeneity in encounter probability which is indicated in these data (θ = 1.114 under the logit-normal model M_h, which is favoured strongly over model M_0 by AIC). Thus, the estimated N is increased substantially in comparison to model M_0. In addition, the SCR model produces an estimate of N that applies to a specific, well-defined region (the state-space) having an area of 17.58 ha. Thus we can directly compute an estimate of the *density* of Slow Worms, which is 14.68 Slow Worms/ha. With model M_0 and model M_h there's really no telling what area should be associated with those estimates and, therefore, no basis for comparison with the estimate obtained by the SCR model.

Table 27.2 *The results of fitting models M_0 and M_h and an ordinary SCR model to the Slow Worm data. Model M_h has one extra parameter compared to model M_0, the individual heterogeneity parameter θ. The SCR model has one additional parameter, σ, the scale parameter of the encounter probability model relating individual p to distance between individual activity centre and trap location.*

Model	p or p_0	n_0	Extra parameter	N	D (/ha)	AIC
M_0	0.0396	3.97		47.97	–	−73.42
M_h	0.0149	30.88	1.114	74.88	–	−89.21
SCR	0.0501	–	20.971	155.77	8.86	1086.587

27.5 Summary

Capture–recapture methods have been the standard for estimating population size and density for many decades and provide a framework for sampling populations and estimating key parameters, namely population size or abundance. However, we have also argued that capture–recapture is inherently spatial because of the way we typically conduct sampling (e.g. ACOs, pitfall, or cage traps), and yet traditionally this has almost never been addressed in the application of capture–recapture methods. These non-spatial methods have a number of deficiencies that would appear to limit their usefulness in practice (but, strangely, have not). For example, they do not allow the direct estimation of density; they do not account for the spatial organization of trapping arrays or of individuals within the population being studied. Moreover, despite capture–recapture studies often involving explicit questions about spatial ecology, space is largely ignored, opting instead for a 'fishbowl' view of systems.

We demonstrated that the source of heterogeneity that is likely to be an issue in almost all CR studies, that is, spatial distribution of individuals relative to traps, can be addressed using spatial extensions of capture–recapture models, namely 'spatial capture–recapture (SCR)'. The obvious benefit of SCR is that the spatial region for which abundance is estimated is explicitly defined and that *absolute* density can be computed directly. Perhaps of greater importance, however, is the explicit focus on the spatial processes giving rise to encounter data which enable researchers to study many aspects of spatial ecology from individual encounter history data, including resource selection or space usage (Royle et al. 2013b), landscape connectivity (Royle et al., 2013a; Sutherland et al., 2015, Fuller et al., 2016), spatial variation in density (Borchers and Efford, 2008; Royle et al., 2013a), and movement or dispersal (Schaub and Royle, 2014; Ergon and Gardner, 2014; Royle et al., 2016).

As discussed throughout this book, reptiles represent an important component of many communities, are a group of animals in global decline, and can be extremely difficult to monitor (McDiarmid et al., 2011). Many of the methods described in earlier chapters for monitoring, capturing and identifying individuals for reptilian populations lend themselves naturally to analysis using capture–recapture methods and should be preferred over indices of abundance like raw counts. Moreover, many of these sampling procedures are naturally spatial (e.g. trapping grids, line transect) in which case we recommend analysis using SCR methods in order to arrive at meaningful, spatially referenced abundance estimates as we demonstrated with the Slow Worm example.

References

Alho, J.M. (1990). Logistic regression in capture–recapture models. *Biometrics*, **46**, 623–35.

Böhm, M., Collen, B., Baillie, J.E., et al. (2013). The conservation status of the world's reptiles. *Biological Conservation*, **157**, 372–85.

Anderson, D.R., Burnham, K.P., Lubow, B.C., et al. (2001). Field trials of line transect methods applied to estimation of desert tortoise abundance. *Journal of Wildlife Management*, **65**, 583–97.

Borchers, D.L., Buckland, S.T., and Zucchini, W. (2002). *Estimating Animal Abundance. Closed Populations*. Statistics for Biology and Health, Vol. 13. New York: Springer Science & Business Media.

Borchers, D.L., and Efford, M.G. (2008). Spatially explicit maximum likelihood methods for capture–recapture studies. *Biometrics*, **64**, 377–85.

Boulanger, J., and McLellan, B. (2001). Closure violation in DNA-based mark–recapture estimation of grizzly bear populations. *Canadian Journal of Zoology*, **79**, 642–51.

Brown, W.S., Kéry, M., and Hines, J.E. (2007). Survival of timber rattlesnakes (*Crotalus horridus*) estimated by capture–recapture models in relation to age, sex, color morph, time, and birthplace. *Copeia*, **2007**, 656–71.

Buckland, S.T., Anderson, D.R., Burnham, K.P., et al. (2005). *Distance Sampling*. Hoboken, NJ:John Wiley & Sons.

Chris Sutherland, David J. Muñoz, David A. W. Miller, and Evan H. Campbell Grant (2016). Spatial Capture–Recapture: A Promising Method for Analyzing Data Collected Using Artificial Cover Objects. Herpetologica (In-Press). doi: http://dx.doi.org/10.1655/HERPETOLOGICA-D-15-00027.

Christy, M.T., Yackel Adams, A.A., Rodda, G.H., et al. (2010). Modelling detection probabilities to evaluate management and control tools for an invasive species. *Journal of Applied Ecology*, **47**, 106–13.

Converse, S.J., and Royle, J.A. (2012). Dealing with incomplete and variable detectability in multi-year, multi-site monitoring of ecological populations. In R.A. Gitzen, J. J. Millspaugh, A.B. Cooper, et al. (eds) *Design and Analysis of Long-term Ecological Monitoring Studies*. Cambridge: Cambridge University Press, pp. 426–42.

Coull, B.A., and Agresti, A. (1999). The use of mixed logit models to reflect heterogeneity in capture–recapture studies. *Biometrics*, **55**, 294–301.

Dorazio, R.M., and Royle, J.A. (2003). Mixture models for estimating the size of a closed population when capture rates vary among individuals. *Biometrics*, **59**, 351–64.

Dorazio, R.M., Jelks, H.L., and Jordan, F. (2005). Improving removal-based estimates of abundance by sampling a population of spatially distinct subpopulations. *Biometrics*, **61**, 1093–101.

Efford, M. (2004). Density estimation in live-trapping studies. *Oikos*, **106**, 598–610.

Efford, M.G. (2011). secr: spatially explicit capture–recapture models. R package ver. 2.0. Available at: http://www.r-project.org/.

Efford, M.G., Dawson, D.K., and Robbins, C.S. (2004). DENSITY: software for analysing capture–recapture data from passive detector arrays. *Animal Biodiversity and Conservation*, **27**, 217–28.

Ergon, T., and Gardner, B. (2014). Separating mortality and emigration: modelling space use, dispersal and survival with robust-design spatial capture–recapture data. *Methods in Ecology and Evolution*, **5**, 1327–36.

Fiske, I., and Chandler, R.B. (2011). Unmarked: an R package for fitting hierarchical models of wildlife occurrence and abundance. *Journal of Statistical Software*, **43**, 1–23.

Fuller, A. K., Sutherland, C. S., Royle, J. A., & Hare, M. P. (2016). Estimating population density and connectivity of American mink using spatial capture-recapture. *Ecological Applications*. (in press). http://dx.doi.org/10.1890/15-0315.1.

Garden, J.G., McAlpine, C.A., Possingham, H.P., et al. (2007) Using multiple survey methods to detect terrestrial reptiles and mammals: what are the most successful and cost-efficient combinations? *Wildlife Research*, **34**, 218–27.

Gibbons, J.W., Scott, D.E., Ryan, T.J., et al. (2000). The global decline of reptiles, déjà vu amphibians. *Bioscience*, **50**, 653–66.

Grant, B.W., Tucker, A.D., Lovich, J.E., et al. (1992). The use of coverboards in estimating patters of reptile and amphibian biodiversity. In D.R. McCullough and R.H. Barrett (eds) *Wildlife 2001: Populations*. London:Elsevier Science, pp. 379–403.

Grant, T.J., and Doherty, P.F. (2007). Monitoring of the flat-tailed horned lizard with methods incorporating detection probability. *Journal of Wildlife Management*, **71**, 1050–6.

Huggins, R.M. (1989). On the statistical analysis of capture experiments. *Biometrika*, **76**, 133–40.

Karanth, K.U., and Nichols, J.D. (1998). Estimation of tiger densities in India using photographic captures and recaptures. *Ecology*, **79**, 2852–62.

Kendall, W.L. (1999). Robustness of closed capture–recapture methods to violations of the closure assumption. *Ecology*, **80**, 2517–25.

Kéry, M., and Royle, J.A. (2015). *Applied Hierarchical Models*. San Diego, CA:Academic Press.

Kéry, M., and Schaub, M. (2012). *Bayesian Population Analysis using WinBUGS: A Hierarchical Perspective*. San Diego, CA:Academic Press.

Laake, J., and Rexstad, E. (2008). RMark: an alternative approach to building linear models in MARK. Program MARK. Available at: http://www.phidot.org/software/mark/docs/book/pdf/app_3.pdf.

Mazerolle, M.J., Bailey, L.L., Kendall, W.L., et al. (2007). Making great leaps forward: accounting for detectability in herpetological field studies. *Journal of Herpetology*, **41**, 672–89.

McDiarmid, R.W., Foster, M.S., Guyer, C., Jr., et al. (eds) (2011). *Reptile Biodiversity. Standard Methods for Inventory and Monitoring*. Berkeley, CA:University of California Press.

Meier, A. (2012). *Reptilienaufwertungskonzept am Mueterschwandenberg*. Bachelorarbeit, Zurich University of Applied Sciences, Wädenswil, Switzerland.

Norris, J.L., III, and Pollock, K.H. (1996). Nonparametric MLE under two closed capture–recapture models with heterogeneity. *Biometrics*, **52**, 639–49.

Otis, D.L., Burnham, K.P., White, G.C., et al. (1978). Statistical inference from capture data on closed animal populations. *Wildlife Monographs*, **62**, 3–135.

Paulissen, M.A., and Meyer, H.A. (2000). The effect of toe-clipping on the gecko *Hemidactylus turcicus*. *Journal of Herpetology*, **34**, 282–5.

Pledger, S. (2000). Unified maximum likelihood estimates for closed capture–recapture models using mixtures. *Biometrics*, **56**, 434–42.

Reading, C.J., Luiselli, L.M., Akani, G.C., et al. (2010). Are snake populations in widespread decline? *Biology Letters*, **6**, 777–80.

Rodda, G.H., and Campbell, E.W. (2002). Distance sampling of forest snakes and lizards. *Herpetological Review*, **33**, 271–4.

Royle, J.A. (2004). Generalized estimators of avian abundance from count survey data. *Animal Biodiversity and Conservation*, **27**, 375–86.

Royle, J.A. (2009). Analysis of capture–recapture models with individual covariates using data augmentation. *Biometrics*, **65**, 267–74.

Royle, J.A., and Dorazio, R.M. (2012). Parameter-expanded data augmentation for Bayesian analysis of capture–recapture models. *Journal of Ornithology*, **152**, 521–37.

Royle, J.A., and Young, K.V. (2008). A hierarchical model for spatial capture–recapture data. *Ecology*, **89**, 2281–9.

Royle, J.A., Chandler, R.B., Gazenski, K.D., et al. (2013a). Spatial capture–recapture models for jointly estimating population density and landscape connectivity. *Ecology*, **94**, 287–94.

Royle, J.A., Chandler, R.B., Sollmann, R., et al. (2014). *Spatial Capture–Recapture*. San Diego, CA:Academic Press.

Royle, J.A., Chandler, R.B., Sun, C.C., et al. (2013b). Integrating resource selection information with spatial capture–recapture. *Methods in Ecology and Evolution*, **4**, 520–30.

Royle, J.A., Converse, S.J., and Link, W.A. (2012). Data augmentation for hierarchical capture–recapture models. arXiv preprint arXiv:1211.5706.

Royle, J.A., Dorazio, R.M., and Link, W.A. (2007). Analysis of multinomial models with unknown index using data augmentation. *Journal of Computational and Graphical Statistics*, **16**, 67–85.

Royle, J.A., Fuller, A.K., and Sutherland, C. (2016). Spatial capture–recapture models allowing for Markovian transience or dispersal. *Population Ecology*, **58**, 53–62.

Sacchi, R., Scali, S., Pellitteri-Rosa, D., et al. (2010). Photographic identification in reptiles: a matter of scales. *Amphibia-Reptilia*, **31**, 489–502.

Schaub, M., and Abadi, F. (2011). Integrated population models: a novel analysis framework for deeper insights into population dynamics. *Journal of Ornithology*, **152**, 227–37.

Schaub, M., and Royle, J.A. (2014). Estimating true instead of apparent survival using spatial Cormack–Jolly–Seber models. *Methods in Ecology and Evolution*, **5**, 1316–26.

Smolensky, N.L., and Fitzgerald, L.A. (2010). Distance sampling underestimates population densities of dune-dwelling lizards. *Journal of Herpetology*, **44**, 372–81.

Sutherland, C.S., Elston, D.A., and Lambin, X. (2014). A demographic, spatially explicit patch occupancy model of metapopulation dynamics and persistence. *Ecology*, **95**, 3149–60.

Sutherland, C., Fuller, A.K., and J. A. Royle, J.A. (2015). Modelling non-Euclidean movement and landscape connectivity in highly structured ecological networks. *Methods in Ecology and Evolution*, **6**, 169–77.

Sutherland, C., Brambilla, M., Pedrini, P., Tenan, S., (2016). A multi-region community model for inference about geographic variation in species richness. Methods in Ecology and Evolution (In Press). DOI: 10.1111/2041-210X.12536.

Tanner, M.A., and Wong, W.H. (1987). The calculation of posterior distributions by data augmentation. *Journal of the American Statistical Association*, **82**, 528–40.

Tyrrell, C.L., Christy, M.T., Rodda, G.H., et al. (2009). Evaluation of trap capture in a geographically closed population of brown treesnakes on Guam. *Journal of Applied Ecology*, **46**, 128–35.

Welsh, H.H., Jr. (1990). Relictual amphibians and old-growth forests. *Conservation Biology*, **4**, 309–19.

White, G.C., and Burnham, K.P. (1999). Program MARK: survival estimation from populations of marked animals. *Bird Study*, **46**, S120–39.

Williams, B.K., Nichols, J.D., and Conroy, M.J. (2002). *Analysis and Management of Animal Populations*. San Diego, CA:Academic Press.

Yang, H.C., and Chao, A. (2005). Modeling animals' behavioral response by Markov chain models for capture–recapture experiments. *Biometrics*, **61**, 1010–17.

Zylstra, E.R., Steidl, R.J., and Swann, D.E. (2010). Evaluating survey methods for monitoring a rare vertebrate, the Sonoran desert tortoise. *Journal of Wildlife Management*, **74**, 1311–18.

28

Collecting biological samples for disease monitoring

Elliott R. Jacobson

28.1 Introduction

Health and disease form a continuum and are inextricably linked to the ecology of the population(s) being studied. As in other animals, pathogens and resulting infectious diseases have surfaced as inescapable threats to populations of wild non-avian reptiles (hereafter simply called reptiles). Viral, bacterial, fungal, and parasitic diseases have been identified in all major groups of reptiles, with more being studied in captive reptiles than their wild counterparts. Examples of infectious diseases that have been well studied in wild reptiles are fibropapillomatosis of marine turtles (Jacobson, 2007a), a worldwide disease, an upper respiratory tract disease of North American and European tortoises (Jacobson, 2007b), and *Ranavirus* infection of Gopher Tortoises (*Gopherus polyphemus*), exotic tortoises, and Eastern Box Turtles (*Terrapene carolina*) (Johnson et al., 2008). A fungal skin disease has been seen at multiple sites in different species of snakes in the eastern half of the USA (Cheatwood et al., 2003; Allender et al., 2013). Non-infectious diseases have also been seen in wild reptiles and include metabolic/nutritional, endocrine disruptors, traumatic, and neoplastic diseases. For marine turtles, plastic and tar ingestion and oil from spills have caused significant morbidity and mortality.

Along with the collection of samples for disease diagnosis, this chapter includes discussion of: (1) ethics and animal welfare considerations; (2) the role of Institutional Animal Care and Use Committees; (3) pain; (4) analgesia and anaesthesia; (5) several of the most important diseases/pathogens for the major taxonomic groups of reptiles; and (6) biosecurity.

28.2 Ethics and animal welfare

Ethical issues need to be considered when preparing to do research and collect samples from animals. The decision to use animals in research requires critical thought, judgment, and consideration of animal welfare and ethical issues. Ethical issues are distinguished from those of animal welfare in that the former are not fact based, whereas the latter is fact-oriented and scientifically based. Animal welfare arose from ethical concerns raised by scientists regarding the way we treat animals (Palmer and Sandøe, 2011). As pointed out by Warwick et al. (1995), 'to act cautiously and preventatively is responsible and humane and "costs nothing"'. To proceed with activities that may affect animal welfare simply because harmful effects are unknown is, arguably, abuse. Therefore assigning the 'benefit of the doubt' to animals may be integral to humane, ethical, and scientific consideration.

In the United States, the Laboratory Animal Welfare Act, the first version of what is now known as the Animal Welfare Act (AWA) (U.S. Department of Agriculture, 2013), was passed by Congress in 1966 establishing the legal standards for laboratory animal care and use in this country. A number of amendments to the AWA have led to regulations that include animal transportation, keeping marine mammals in captivity, and animals in the research laboratory. However, the regulations currently exclude reptiles along with common laboratory rats and mice, birds, and farm animals used in production agriculture research. Still, ethical and animal welfare issues apply equally to all groups of vertebrates (and certain invertebrates). Although there is sufficient evidence indicating that reptiles feel pain (see Section 28.4), the level of pain being experienced by an animal due to collection of different samples is most often subjectively assessed by the investigator collecting the samples. Bad habits are easy to perpetuate, especially if the animal survives the procedure. But this cannot be condoned. When projects are being defined, the investigator needs to ask the question: Is this project going to provide more harm than good?

28.3 Institutional Animal Care and Use Committees

In the United States, the term IACUC (Institutional Animal Care and Use Committee) was formally introduced in 1986 with an amendment to the AWA and corresponding changes in Public Health Service (PHS) policy. As such, they are a federally mandated committee. The function of the IACUC is to ensure humane use of animals being studied in the laboratory and in field research. For in-depth information on IACUCs in the United States, see the *Institutional Animal Care and Use Committee Guidebook* (2002).

Although reptiles are not covered under the AWA or the PHS policy, investigators in the United States doing research or teaching in academic, zoological, private, and conservation organizations (especially U.S. federally funded institutions) on live reptiles (and other vertebrates), or their tissues, typically need to submit and have an IACUC protocol approved before the research begins. Professional field biologists in organizations devoted to the study of fish, amphibians, reptiles, birds, and mammals have prepared separate or additional guidelines for fieldwork with these groups. For reptiles, see *Guidelines for the Use of Live Amphibians and Reptiles in Field and Laboratory Research* (Herpetological Animal Care and Use Committee, 2004). Other countries may have more or less stringent requirements for conducting research on reptiles and the reader of this chapter should adhere to those requirements.

28.4 Pain

Pain and suffering are serious considerations for projects involving live animals. The definition of pain by the International Association for the Study of Pain (IASP) is 'an unpleasant sensory and emotional experience associated with actual or potential tissue damage, or described in terms of such damage' (http://www.iasp-pain.org/Taxonomy#Pain). A term often used interchangeably with pain is nociception, defined by IASP as 'the neural process of encoding noxious stimuli'. A workshop on pain in 2002 developed a consensus statement indicating that operationally, all vertebrates and some invertebrates experience pain (Paul-Murphy et al., 2004). Because reptiles have appropriate neuroanatomic and neuropharmacologic components, display the appropriate behaviour in response to a painful stimulus, and possess antinociceptive mechanisms to modulate pain, pain perception in these species is therefore likely to be analogous to that of mammals (Machin, 2001) and should be accordingly managed if a procedure likely results in pain (Mosley, 2011).

28.5 Analgesia and anaesthesia

Whereas many diagnostic samples can be obtained using manual restraint alone, biopsies generally require either a local injectable analgesic or central acting injectable or gas anaesthetic. Numerous anaesthetics have been used in reptiles indicating that no single 'best' anaesthetic exists for all age groups of all reptiles under all conditions. Biologists should consult wildlife veterinarians in order to determine the best drug(s) to be used when collecting samples that may cause more than momentary or slight pain or distress. Since collection and physical restraint of an animal can result in distress, samples should be collected using the least amount of restraint in the shortest period of time. Animals anaesthetized in the field should be fully recovered before release.

In the following subsections, I briefly mention the most commonly used injectable and gas anaesthetics used in reptiles. For more in depth reviews of anaesthesia in reptiles, see DeNardo (2012) and West et al. (2014). For anesthesia and analgesia of wildlife in the field see Chinnadurai et al. (2016).

28.5.1 Local analgesics and central acting injectable anaesthetics

For collecting biopsies from soft components of the integument, the author routinely uses 2% lidocaine that is injected as a ring-block around the lesion to be biopsied using a 1 ml syringe fitted with a 25-gauge needle. The volume used will vary with the size of the animal and size of the lesion to be removed. If the biopsy is for the identification of an infectious agent, sterile water should be used for cleansing the biopsy site. Depending upon the size of the lesion removed, the site may need to be sutured or closed using a surgical glue.

The most commonly used injectables that have central nervous effects are the alpha antagonists xylazine and medetomidine (which are sedatives) that are often administered intramuscularly along with the dissociative anaesthetic ketamine. The alpha antagonist effects of xylazine and medetomidine can be reversed with the specific antagonist atipamezole. Another drug that has been used to reach a plane of anaesthesia for biopsies and minor surgical procedures is telazol, a combination drug consisting of tiletamine (related to ketamine) and zolazepam (sedative; member of the benzodiazepines). However, in some animals it may take a full day to recover from an anaesthetic dose of this drug combination. An injectable agent that has been used in all major groups of reptiles is propofol, a rapid-acting and relatively short-lasting hypnotic having sedative and anaesthetic qualities. But propofol needs to be administered intravenously and has no analgesic properties. Thus it needs to be administered with an analgesic if a painful procedure is performed. Another injectable that has become popular in reptile medicine is alfaxalone, a fast and short-acting drug, that similar to propofol, has minimal to no analgesic properties. For specific doses, see Carpenter and Marion (2013).

28.5.2 Inhalant anaesthetics

Inhalant anaesthetics are generally delivered to reptiles in pure oxygen, with or without nitrous oxide, using an anaesthesia machine having a precision vaporizer. Such machines can be modified for use in the field. The two inhalant anaesthetics most commonly used in reptiles are isoflurane and sevoflurane. Due to its low solubility in blood, sevoflurane has a shorter induction time and faster recovery time then isoflurane (West et al., 2014). Since reptiles will generally stop breathing under a surgical plane of gas anaesthesia, positive pressure ventilation will be needed, at approximately six breaths per minute. One of the best methods of monitoring an animal's recovery is a Doppler unit (Parks Medical Electronics Sales, Inc., Las Vegas, NV, USA), with an ultrasonic flat or pencil probe placed over the heart or a major vessel.

28.5.3 Post-surgical analgesia

Investigators should consult a veterinarian for recommendations on appropriate post-surgical analgesics to be used. The most commonly used analgesics in reptiles include the opioids (morphine, hydromorphone, butorphanol, tramadol), non-steroidal anti-inflammatory drugs (meloxicam, caprofen, ketoprofen), and local analgesics (lidocaine, bupivicaine). A review of the literature and specific doses can be found elsewhere (Whiteside, 2014).

28.6 Major infectious and non-infectious diseases

28.6.1 Infectious diseases

Numerous infectious agents (viruses, bacteria, fungi, parasites), and the diseases they cause, have been reported in captive reptiles, and to a lesser degree, in wild populations (Table 28.1). Herpesviruses, adenoviruses, poxviruses, iridoviruses, paramyxoviruses, reoviruses, flaviruses, and nidoviruses are a few of the viral groups known to infect reptiles (Jacobson, 2007a; Stenglein et al., 2014.). Certain wild reptiles may serve as reservoir hosts for the eastern equine encephalomyelitis virus (EEE) and western encephalomyelitis virus (Thomas and Eklund, 1962; Bingham et al., 2012). Numerous bacteria have been isolated from lesions in reptiles (Jacobson, 2007b). Of the various bacteria known to cause disease in reptiles, the mycoplasmas of chelonians and crocodilians, *Chlamydophila*, which has been identified in all major groups of reptiles

Table 28.1 Major infectious agents of disease in reptiles.

Disease agent	Pathogen group	Reptile host	Major system affected	Diagnostic testing
Herpesviruses	DNA virus	Turtles, tortoises Lizards	Skin, respiratory tract, oral cavity	Cell culture TEM, PCR
Adenoviruses	DNA virus	All reptile groups	Brain, liver, blood vessels, GI tract	Cell culture TEM, PCR
Iridoviruses	DNA virus	Turtles, tortoises, snakes, lizards	Red blood cells, oral cavity, liver, spleen	TEM, PCR
Poxviruses	DNA virus	Crocodilians, lizards	Skin, macrophages, spleen	Cell culture TEM, PCR
Paramyxoviruses	RNA virus	Snakes, lizards, tortoises	Lung, pancreas, skin	TEM, PCR
Reoviruses	RNA virus	Lizards, snakes	Brian, lung, pancreas	Cell culture TEM, PCR
Arenaviruses	RNA virus	Boas and pythons	Multiple tissues, brain	Cell culture TEM, PCR
Flaviviruses	RNA virus	American Alligator	Multiple tissues, blood	PCR
Nidoviruses	RNA virus	Ball Pythons, Indian Python	Lung, spleen	Cell culture, PCR, serology
Mycoplasma agassizii and *M. testudineum*	Bacteria	Tortoises	Nasal cavity	PCR, TEM
Mycoplasma crocodyli and *M. alligatoris*	Bacteria	Crocodilians	Multiple tissues	Culture and PCR
Chlamydophila	Bacteria	Sea turtles, crocodilians, lizards, snakes	Multiple tissues	Culture and PCR
Devriesea agamarum	Bacteria	Bearded Dragons, Spiny-tailed Lizards	Skin	Immunohistochemistry, PCR
Nannizziopsis	Fungus	Lizards	Skin	Culture and PCR
Ophidiomyces	Fungus	Snakes	Skin	Culture, PCR
Paranannizziopsis	Protozoa	Lizards and Tuataras	Skin	Culture and PCR
Cryptosporidium	Protozoa	Tortoises, lizards, snakes,	GI tract, inner ear	Histology, faecal exam, PCR
Caryospora	Protozoa	Marine turtles	Viscera and brain	Histology, faecal exam, PCR

Unidentified intranuclear coccidia	Protozoa	Tortoises	Multiple visceral tissues	Histology, PCR
Entamoeba invadens	Protozoa	Turtles, tortoises, lizards, and snakes	GI tract, liver	Histology, PCR, faecal exam
Spirorchis spp	Spirorchiid trematode	Emydine freshwater turtles	Multiple visceral tissues and brain	Morphology of parasite, PCR
Neospirorchis, Laeredius, Hapalotrema, Carettacola	Spirorchiid trematodes	Marine turtles	Multiple visceral tissues and brain	Morphology of parasite, PCR
Sulcascaris	Ascarid nematode	Marine turtles	GI tract	Morphology of parasite, PCR
Dujardinascaris	Ascarid nematode	Crocodilians	GI tract	Morphology of parasite, PCR
Ophidascaris, Hexametra	Ascarid nematodes	Snakes and lizards	GI tract and other viscera	Morphology of parasite, PCR
Macdonaldius	Filarid nematode	Snakes	Vascular system	Morphology of parasite, PCR
Foleyella	Filarid nematode	Lizards	Subcutaneous tissues, coelomic cavity	Morphology of parasite, PCR

(Jacobson, 2007b), and *Devriesea agamarum* of agamid lizards (Hellebuyck et al., 2009) stand out as a few of the more important bacterial pathogens.

The *Chrysosporium* anamorph of *Nannizziopsis vriesii* (CANV), and other *Chrysosporium* spp., have been reported as a major cause of fungal dermatitis in a wide variety of squamate reptiles (Paré and Jacobson, 2007). Molecular studies revealed that CANV isolates from reptiles belong to three lineages (Sigler et al., 2011). One lineage represents the genus *Nannizziopsis* from chameleons and geckos, crocodiles, agamid and iguanid lizards, and humans, and two lineages comprise the genus *Ophidiomyces* with the species *Ophidiomyces ophiodiicola* occurring only on snakes, and *Paranannizziopsis* gen. nov., with three new species from squamates and tuataras.

Free-ranging reptiles are infected and infested with a great diversity of endo- and ectoparasites (Jacobson, 2007c). The coccidian *Caryospora* was responsible for mortality in captive (Leibovitz et al., 1978) and wild (Gordon et al., 1993) marine turtles. The author has also identified amoebiasis, cryptosporidiosis, and various helminthes in wild reptiles found ill in the field. An unclassified intranuclear coccidian has been reported as a pathogen in multiple species of tortoises in private collections in the United States (Garner et al., 2006). Members of the trematode family Spirorchiidae utilize freshwater turtles and sea turtles as their definitive hosts where the adults or their eggs may result in considerable pathology (Gordon et al., 1998; Jacobson et al., 2006; Stacy et al., 2010). Of nematodes, ascarids may cause lesions either as larvae migrating through visceral structures or as adults embedded within the gastrointestinal mucosa. The Pentastomida are a group of worm-like arthropods that infect all major groups of reptiles. Ectoparasites of reptiles include mites and ticks in terrestrial reptiles and leeches in fresh and salt-water sea turtles and crocodilians.

28.6.2 Non-infectious diseases

Various non-infectious diseases have been reported to cause illness and death in wild populations of reptiles. Cutaneous dyskeratosis, a shell disease of the keratinaceous layer of shell, was observed in multiple populations of Desert Tortoises *(Gopherus agassizii)* in the Mojave and Colorado Deserts of California, USA (Jacobson et al., 1994). The lesion was suggestive of either a deficiency disease or toxicosis. The author has seen severe osteopenia (thinning of the dermal bone of the shell), especially of the plastron, in Gopher Tortoises on Sanibel Island, Florida, USA. Lack of quality nutrition was considered the ultimate cause. Urolithiasis refers to an accumulation of uric acid calculi anywhere in the urinary system. It has been seen sporadically in the bladder of captive and wild juvenile and adult Desert Tortoises (Homer et al., 1998). Gout is an accumulation of uric acid crystals in visceral organs, such as the kidney, liver, and pericardial sac, and/or skeletal joint (articular) spaces. This has been reported sporadically in Desert Tortoises in the Mojave Desert of the southwest USA (Homer et al., 1998). Renal oxalosis was identified in wild Desert Tortoises in the Mojave Desert, where crystals of calcium oxalate obstructed renal tubules (Jacobson et al., 2009). This may have resulted from Desert Tortoises feeding upon invasive plants having higher oxalate content than native plants.

Drought and subsequent dehydration and starvation are considered responsible for die-offs of Desert Tortoises in the southwest United States, and have been implicated in low growth rates of young Desert Tortoises, reduced egg production and reproductive effort in female tortoises, reduced activity levels and movement, and low metabolic rates (Berry et al., 2002).

Gopher Tortoises submitted to the Zoological Medicine Service, University of Florida, are most commonly hit by a car or traumatized by dogs. Sea turtles worldwide are threatened by a variety of anthropogenic influences including incidental catch and death in fisheries, entanglement, trauma from hooks, exposure to spilled oil and tar, ingestion of plastics, artificial lighting, and habitat degradation (Shigenaka, 2003). Natural occurring mortality events have been reported due to toxicosis from harmful algal blooms and trauma from shark predation.

Environmental contaminants known as endocrine-disrupting contaminants (EDCs) have been studied in a wide variety of wildlife including American alligators, turtles, and lizards. They have been shown to have oestrogenic, androgenic, antiandrogenic, and antithyroid actions in different species (Guillette, 2006). As an example, the antiandrogenic actions of *p,p′*-DDE, the major bioaccumulated metabolite of DDT, is thought to operate on androgen receptors, ultimately resulting in an 'oestrogenic' feminizing or demasculinizing response.

28.7 Collecting samples for disease diagnostics

When assessing disease status of wildlife populations, always keep in mind that novel pathogens may be involved. More than likely, the vast majority of pathogens that exist in populations of reptiles are yet to be identified. Investigators working on outbreaks of disease need to be ready and capable of collecting multiple samples for diagnostic purposes. Rarely do investigators stumble upon an outbreak of disease as it is happening. Animals showing clinical signs of illness that are euthanatized and those recently dead should be submitted for a full necropsy. This will necessitate proper overnight shipping of a whole animal on ice (not frozen), frozen tissues and plasma samples on dry ice, and microbial swabs on ice packs to a laboratory capable of doing the specific testing. Tissues in neutral buffered 10% formalin (NBF) should not be frozen, with shipment in heat sealed plastic bags, either by normal or overnight mail. If the samples are initially fixed in NBF, after 24 hours of fixation they can be wrapped in paper towels soaked in 70% ethanol and shipped in a heat sealed plastic bag. Data sheets and health profile forms should be used for each animal (Berry and Christopher, 2001) and copies should accompany the material. A complete set of digital images of the animal should be obtained.

28.7.1 Equipment

For the field, plastic tool and tackle boxes are ideal for holding all the following items needed for collecting blood and other samples: field and necropsy forms, camera, sharpies, syringes of different size, blood tubes and vacutainer tubes of different volumes including those with and without anticoagulants (lithium heparin is the anticoagulant of choice), needles and vacutainer needles of different size, butterfly catheters, vials of heparin, heparinized microhaematocrit tubes, haematocrit tube sealer, plastic transfer pipettes, callipers of different sizes, measuring tape, containers and bags for holding small reptiles, sterile and disposable latex or nitrile gloves (see Chapter 6 when working with NBF), hand sanitizer, disposable booties for covering shoes/boots, digital electric scale, AC/DC converter for plugging into cigarette lighter in vehicle, sterile culturettes (plus special media depending on organism being surveyed), sterile twirl bags, cryotubes, cryotube marking pens, sterile urine cups, microscopic slides and coverslips for making blood films, container of 100% methanol to fix blood films on either coverslips or glass slides, plastic containers with formalin, a rotary power tool with diamond wheel (Dremel Mototool, Dremel Mfg. Co., Racine, WI, USA), sterile instrument packs, free forceps and haemostats, absorbable suture material, biopsy punches tissue cassettes, packs of gauze sponges, alcohol swabs, and disinfectant such as chlorhexidine and sodium hypochlorite. A dry shipper liquid nitrogen tank should be used for transporting plasma/ serum samples. If not possible, use a Styrofoam or plastic cooler with ice. A small electric microhaematocrit/microtainer centrifuge and generator may be needed.

28.7.2 Blood collection and handling

Blood is a tissue that is commonly collected in the field for biological and health assessment studies. Blood is a fragile tissue and is easily abused from the time of collection to time of testing. Either the tube needs to be placed on ice and worked up within a few hours of collection or the blood needs to be centrifuged and the plasma removed and placed in a cryotube for transport on dry ice or in liquid nitrogen. If a complete blood count is to be performed, it is best for blood films to be made in the field.

Blood sampling represents an invasive procedure, and as such, there is associated pain and the risk of microbial infection, either locally or systemically. An aseptic technique is necessary, with cleansing of the sampling site using several alternating applications (typically three) of 2% chlorhexidine and 10% isopropyl alcohol or 70% ethanol. When the needle is removed, moderate steady pressure should be applied to the site to prevent haematoma formation.

The total blood volume of reptiles varies between species but as a generalization is approximately 5–8% of total body weight. Thus, a 100 g reptile has an estimated blood volume of 5–8 ml. Since clinically healthy reptiles can acutely lose 10% of their blood volume without any detrimental consequences, from a reptile weighing 100 g, 0.5–0.8 ml of blood can be withdrawn safely.

Depending on reptile group, size, and species, blood can be collected from several different sites (Heard et al., 2015). For chelonians, blood can be collected from the heart, jugular vein, brachial vein, ventral coccygeal vein, dorsal coccygeal vein, cervical sinus, and subcarapacial vein. Blood sampling from trimmed toenails has been utilized, but is not recommended due to the associated pain, risk of subsequent infection, and the small amounts that can be obtained from this site. In crocodilians, blood samples can be obtained from the supravertebral vessel located caudal to the occiput, the heart, and ventral and lateral tail veins. In large lizards, blood is easily obtained from the ventral tail vein. In snakes, blood samples can be obtained from a variety of sites, including the palatine veins, ventral tail vein, and via cardiocentesis. In snakes >100 g, the author prefers cardiocentesis to other methods. Some muscular snakes such as large boas and pythons are difficult to manually restrain and bleed from the heart. Venomous snakes are also problematic, and the heart may not be the ideal sampling site unless the animal is anaesthetized. The most important points to remember when collecting blood for haematological and biochemical evaluations are: (1) try and utilize the same blood collection technique at all times; (2) when sampling from peripheral vessels, haemodilution with lymph may occur due to accidental puncture or sampling from adjacent lymphatics, and when sampling from the heart, haemodilution can occur with pericardial fluid; (3) handle the blood in a consistent fashion; (4) use the same anticoagulant and try and add the same volume of blood to the collection tube; (5) centrifuge the blood immediately following collection and remove the plasma immediately following centrifugation; (6) freeze plasma/serum following collection, preferably on dry ice, in liquid nitrogen, or in an ultra-cold freezer at –70°C; (7) the sample should be transported frozen to the laboratory, preferably on dry ice; (8) try and use the same clinical pathology laboratory utilizing the same machine.

28.7.3 Serology

Serological tests are available to determine exposure of reptiles to a range of potential pathogens including herpes virus (tortoises), iridovirus (chelonians), paramyxovirus (lizards and snakes), *Mycoplasma agassizii* (tortoises), *Cryptosporidium* (different species of reptiles), and spirorchiid trematodes (marine turtles) (Jacobson, 2007d). Ideally, a portion of the plasma/serum from blood of reptiles sampled in the field should be stored in a properly marked cryotube for possible use at a later date. Positive serology at the time of sampling only provides evidence of previous exposure. Further diagnostic testing is needed to establish infection.

28.7.4 Biopsies

The collection of biopsies often is necessary to diagnose disease problems in reptiles. In the field, biopsies are most commonly obtained from the integumentary system. Reptiles are often observed with skin disease, and a great variety of infectious and non-infectious disease problems can result in pathological changes in the integument. It is best to include samples from the interface of normal and abnormal tissue. In collecting biopsy specimens, a minimum of two samples should be obtained for: (1) histopathology; and (2) microbiology (both culture and molecular diagnostics such as polymerase chain reaction (PCR)). Additional samples can be collected for cytology and electron microscopy (Jacobson and Samuelson, 2007).

Of all the reptiles, chelonians present the greatest challenge for biopsy, especially when lesions involve the shell. The reptile shell is a very hard biological structure that makes biopsy somewhat difficult. If a pathogen is suspected, the site should be cleansed with sterile saline or water so not to suppress growth of microbes on artificial media or denature them for identification by PCR. If a non-infectious process is suspected, the site can be cleansed with chlorhexidine or 70% ethanol. The shell is highly innervated below the epidermal component, so some degree of general anaesthesia may be necessary. While under anaesthesia, a rotary power tool and bone elevator can be used to obtain a small wedge out of the shell. The epidermal component of the shell (scute) is relatively thin and the depth of the thickness of the biopsy should only be a few millimetres. Ideally a veterinarian should be consulted when collecting such samples. Depending on the size of the sample removed, the sampling site may need to be cauterized and covered with sterile plastic that is sealed to the edges of the site with a methacrylate resin (Cyanoveneer, Ellman International Mfg., Inc., Hewlett, NY, USA).

For biopsy of soft tissue, the area around the biopsy should be infiltrated with a 2% lidocaine solution. A full-thickness biopsy may be difficult in those areas of crocodilian integument having osteoderms. Small crocodilians and most lizards can be manually restrained, whereas large crocodilians and large monitors may need to be chemically immobilized. Snakes are ideally suited for skin biopsy. Harmless species can be manually restrained, and venomous species can be guided into a plexiglass tube for restraint or anaesthesia. Affected skin can be removed with a scalpel blade or a biopsy punch. In such cases, the area around the lesion should be infiltrated with 2% lidocaine hydrochloride.

28.7.5 Pathological evaluations

Necropsy techniques for reptiles can be found elsewhere (Terrell and Stacy, 2007). For histopathology, NBF is the most commonly used fixative. The NBF volume to tissue ratio should be at least 10:1. For best penetration, tissue should not exceed 6 mm in thickness. A fixative that the author routinely uses for doing light and electron microscopy is Trump's solution (McDowell and Trump, 1976), which is a combination of 4% formaldehyde and 1% glutaraldehyde. This solution needs to be refrigerated prior to use and is stable at refrigeration temperatures for several months. When working with NBF, it is essential to use appropriate gloves such as neoprene or nitrile gloves (see Chapter 6).

28.7.6 Cytodiagnostics

Examination of touch impressions and wet mounts of various lesions on glass slides is extremely helpful in diagnosing disease problems in reptiles. Because of the ease and rapidity of processing these samples, much information can be gathered in a short time. For integumentary lesions, samples can be collected with relative ease. Depending upon the size and nature of the reptile, manual restraint alone may be all that is needed. In larger more fractious reptiles, or in those patients with subcutaneous masses, sedation or anaesthesia will be required.

The most common stain used for initial evaluation of smears is the Wright–Giemsa stain. Prior to staining, the smear is fixed in absolute methanol for approximately 10 seconds. Similar quick staining techniques are commercially available (Dif-Quik, American Scientific Products, McGraw Park, IL, USA) and allow staining of smears in a few seconds. Other stains that are commonly used in evaluating smears of lesions from reptiles include Gram's stain for bacteria, acid-fast stain for mycobacteria, and new methylene blue for fungi.

28.7.7 Microbiology

Swab specimens, aspirates, and biopsy specimens can be collected and submitted for microbial isolation attempts including those for: (1) viruses; (2) aerobic bacteria; (3) anaerobic bacteria; (4) special bacterial organisms such as *Chlamydophila* and *Mycoplasma*; and (5) fungi. Proper collection technique is a prerequisite for successful recovery of microorganisms responsible for an infectious disease. While isolation of aerobic organisms is relatively inexpensive and fairly rapid to do, isolation of the other groups of pathogens listed requires special techniques and is far more costly. Viruses, *Chlamydophila*, and *Mycoplasma* require special media and conditions for isolation. Fluids, aspirates, and biopsy specimens can be frozen on dry ice or in an ultrafreezer at −70°C until a decision is made as to specific isolations that will be attempted.

If a laboratory is available to attempt viral isolation, samples should be submitted either: (1) as fresh tissue samples in a sterile container on wet ice; (2) as minced pieces in a transport media such as Eagle's minimal essential medium; or (3) as fresh frozen samples transported on dry ice or in liquid nitrogen. For isolation of aerobic bacteria, a variety of sterile swabs, of different sizes, with associated transport media are commercially available. Once the sample is collected, the swab should be returned to its protective holder, placed in a plastic zip-lock bag, and sent to the laboratory on ice as rapidly as possible. Samples taken in the field from reptiles can be placed in a cryotube containing tryptic soy broth and frozen in a tank containing liquid nitrogen.

Two species of *Mycoplasma, M. agassizii and M. testudineum*, have been identified in nasal exudate of Desert Tortoises, Gopher Tortoises, and other tortoises with chronic upper respiratory tract disease (Jacobson et al., 2014). A special enrichment media, SP4 media, is used both as a broth and agar in culturing for this organism in chelonians (Brown et al., 2002). A syringe containing sterile phosphate buffered saline or SPF4 media is placed over the external nares of a tortoise and is flushed in and out of the nares several times. The sample can be either placed in an empty cryotube or a cryotube containing SPF4 media, frozen on dry ice, and shipped to an appropriate diagnostic laboratory. SP4 media is commercially available (Remel, Lenexa, KS, USA).

As mentioned previously, fungal diseases are commonly encountered in reptiles, particularly in cutaneous or subcutaneous lesions in lizards and snakes. Biopsies of suspected mycotic disease are preferable to swab specimens or scrapings (Paré and Jacobson, 2007). Biopsy specimens can be minced or if the sample is heavily keratinized, gently ground in a sterile tissue grinder containing sterile saline and antibiotics such as gentamicin or amikacin. Minced or ground samples can be cultured on an appropriate mycotic media such as Sabouraud's agar containing antibiotics to suppress bacterial growth.

28.7.8 Molecular diagnostics

Molecular diagnostics, most commonly PCR, conventional and quantitative, are being used routinely in many diagnostic laboratories to determine the presence of gene sequences of numerous pathogens in tissues of living or dead animals. Certain pathogens can also be identified in paraffin embedded tissue specimens. For viruses, DNA viruses are better preserved for testing than RNA viruses. For interpretation of findings, histopathology is needed to correlate identification of a pathogen on or in tissues with histopathological changes indicating a disease process. The presence of a microbial agent in a lesion or on a tissue does not necessarily incriminate the microbe as the causative agent of a disease or lesion. A Zoo Medicine Infectious Disease Testing Laboratory has been established at the University of Florida to test samples obtained from reptiles and other animals by PCR. For information of submission of samples see: http://labs.vetmed.ufl.edu/sample-requirements/zoo-med-infections/. Samples can be sent as swabs of the lesions or biopsies/necropsy specimens can be sent fresh on ice or frozen on dry ice. For *Mycoplasma* diagnostics in the United States, contact Dr. Mary Brown at: mbbrown@ufl.edu. Commercial diagnostic laboratories have limited ability to perform many of these tests, whereas research laboratories tend to focus on one or a few pathogens. From the literature, the reader of this chapter can identify those laboratories capable of performing the needed diagnostic testing.

28.7.9 Preserving ecto- and endoparasites for identification

Both ecto- and endoparasites are encountered on and within field-collected reptiles. Expect to find mites and ticks on all terrestrial reptiles. Detailed methods of collection and preservation can be found elsewhere (Gardner et al., 2012). For mites, the reptile can be placed over a white sheet and mites can be brushed off the host using a fine paint brush. Ticks generally will have to be removed using a forceps. Care will be needed to make sure the mouth parts of the tick remain intact. Leeches also can be mechanically removed from aquatic turtles and crocodilians. For tortoises and freshwater turtles, the animal may need to be sedated to remove ectoparasites. For reptiles, a jeweller's loupe should be used, especially for mites and ensuring that mouth parts of ticks are intact. Specimens from different parts of the body should be placed in individual vials containing 70% ethanol that include labels marked with a permanent ink pen. A portion of the collected parasites can be placed in cryotubes and frozen on dry ice or liquid nitrogen for future PCR testing. For blood parasites, blood films should be made and fixed for 2–5 minutes in absolute methanol. Endoparasites are generally found at necropsy or identified in faeces. For identifying coccidia, faeces can be placed in vials half filled with 2% potassium dichromate. Helminths that may be encountered include cestodes, trematodes, nematodes, and spiny-headed worms. Cestode cysts can be placed intact in NBF and several can be opened and placed in vials of NBF and 95% ethanol. Cestodes, trematodes, and acanthocephalans should first be placed in distilled or tap water for relaxation and killing the parasite. Once relaxed, the parasites should be placed in NBF. For helminths that cannot be seen, intestinal contents can be placed in a container of NBF and the contents examined under a dissecting microscope in the laboratory. Avoid placing nematodes in distilled water since they will subsequently burst. If nematodes are placed in glacial acetic acid for a few minutes, they will uncoil and straighten and then can be transferred to and stored in either NBF or 70% ethanol. Pentastomids should be placed in cold NBF. Several pentastomids should be placed in containers and frozen in an ultrafreezer for future PCR testing.

28.8 Biosecurity: preventing pathogen transmission

When collecting samples from a population of animals, fresh clean clothing need to be used. While some pathogens are fragile and not easily transferred from clothing, boots, and equipment to other sites and animals, there are many pathogens that can remain in a viable state on clothing, equipment and vehicles for extended periods of time. Each animal should be handled with a fresh pair of disposable gloves and all reusable equipment that comes in contact with the animal should be disinfected immediately after use. First, equipment and boots should be washed with soap and water to physically remove contaminants. Next, the equipment should be disinfected with chemical disinfectants, such as chlorhexidine (0.75%) and sodium hypochlorite (0.175–0.5%; regular Clorox bleach is 6.15% sodium hypochlorite), two of the most common disinfectants used in the field (Miller and Gray, 2009). Still, the disinfectant used will vary with the specific pathogen that needs to be controlled. For instance, for certain fungi, chlorine disinfectants have been shown to be the most effective disinfectants that were evaluated (Gupta et al., 2002). The disinfectant should remain in or on the equipment for several minutes to have an effect. The diluted disinfectants should be changed every few days and the stock solutions should be protected from heat and sunlight. Vehicles should be washed (preferably in a car wash) between sites to reduce the possibility of transferring pathogens from site to site. For further information on disinfection and disinfectants, see Friend and Franson (1999) and Rutala and Weber (2013). To prevent transmission of pathogens between study plots, field workers should not travel directly from one site to another without bathing and changing clothes and shoes.

Pathogen transmission is of extreme importance when animals are being transported from one site to another. This includes those in captive breeding programmes being returned to the wild and those being translocated from one site to another. Protocols need to be established that ensure that such animals are screened for specific pathogens and overall health.

28.9 Conclusions

Of all the factors relevant to the structure of wild populations of reptiles, the impacts of diseases have been poorly studied. Of reptiles, more diseases have been reported in wild chelonians compared to other groups, and overall, more diseases and pathogens have been reported in captive reptiles than those in the wild. Wildlife biologists, who conduct most of the research on populations of wild reptiles, have become more aware of the importance of reporting health problems and collecting samples from the populations being studied. Sample collection for diagnostic purposes is often opportunistic, and biologists need to be prepared to collect samples from ill animals as they are encountered. When performing surveys of populations for specific pathogens, the numbers of animals that need to be sampled will depend on the prevalence of the pathogen in the population being studied. An epidemiologist should be consulted for advice on determining statistical appropriate sample sizes and best methods for sampling.

References

Allender, M.C., Dreslik, M.J., Wylie, D.B., et al. (2013). Ongoing health assessment and prevalence of *Chrysosporium* in the Eastern Massasauga (*Sistrurus catenatus catenatus*). *Copeia*, **2013**, 97–102.

Berry, K.H., and Christopher, M.M. (2001). Guidelines for the field evaluation of desert tortoise health and disease. *Journal of Wildlife Diseases*, 37, 427–50.

Berry, K.H., Spangenberg, E.K., Homer, B.L., et al. (2002). Deaths of desert tortoises following periods of drought and research manipulation. *Chelonian Conservation and Biology*, **4**, 436–48.

Bingham, A.M., Graham, S.P., Burkett-Cadena, N.D., et al. (2012). Detection of eastern equine encephalomyelitis virus RNA in North American snakes. *American Society of Tropical Medicine and Hygiene*, **87**, 1140–4.

Brown, D.R., Schumacher, I.M., McLaughlin, G.S., et al. (2002). Application of diagnostic tests for mycoplasmal infections of desert and gopher tortoises, with management recommendations. *Chelonian Conservation and Biology*, **4**, 497–507.

Carpenter, J.W., and Marion, C.J. (2013). *Exotic Animal Formulary*. 4th ed. St. Louis, MO: W.B. Saunders.

Cheatwood, J.L., Jacobson, E.R., May, P.G., et al. (2003). An outbreak of fungal dermatitis and stomatitis in a free-ranging population of pigmy rattlesnakes (*Sistrurus miliarius barbouri*) in Florida. *Journal of Wildlife Disease*, **39**, 329–37.

Chinnadurai, S.K., Strahl-Heldreth, D., Fiorello, C.V., et al. (2016). Best-practice guidelines for field-based surgery and anesthesia of free-ranging wildlife. I. Anesthesia and analgesia. *Journal of Wildlife Diseases*, **52**, S14–S27.

DeNardo, D. (2012). Anesthesia in reptiles. In R.W. McDiarmid, M.S. Foster, C. Guyer, et al. (eds) *Reptile Biodiversity: Standard Methods for Inventory and Monitoring*. Berkeley, CA: University of California Press, pp. 134–8.

Friend, M., and Franson, J.C. (1999). *Field Manual of Wildlife Diseases. General Field Procedures and Diseases of Birds*. Information and Technology Report 1999–001, U.S. Geological Survey.

Gardner, S.L., Fisher, R.N., and Barry, S.J. (2012). Collecting and preserving parasites during reptile biodiversity surveys. In R.W. McDiarmid, M.S. Foster, C. Guyer, et al. (eds) *Reptile Biodiversity, Standard Methods for Inventory and Monitoring*. Berkeley, CA: University of California Press, pp. 114–21.

Garner, M.M., Gardiner, C., Wellehan, J.F., et al. (2006). Intranuclear coccidiosis in tortoises: nine cases. *Veterinary Pathology*, **43**, 311–20.

Gordon, A.N., Kelly, W.R., and Cribb, T.H. (1998). Lesions caused by cardiovascular flukes (Digenea: S pirorchidae) in stranded green turtles (*Chelonia mydas*). *Veterinary Pathology*, **35**, 21–30.

Gordon, A.N., Kelly, W.R., and Lester J.G. (1993). Epizootic mortality of free-living green turtles, *Chelonia mydas*, due to coccidiosis. *Journal of Wildlife Disease*, **29**, 490–4.

Guillette, L.J. (2006). Endocrine disrupting contaminants: beyond the dogma. *Environmental Health Perspectives*, **114**(Suppl. 1), 9–12.

Gupta, A.K., Ahmad, I., and Summerbell, R.C. (2002). Fungicidal activities of commonly used disinfectants and antifungal pharmaceutical spray preparations against clinical strains of *Aspergillus* and *Candida* species. *Medical Mycology*, **40**, 201–8.

Heard, D., Stacy, N.I., and Wellehan, J.F.X. (2016). Diagnostic sampling and laboratory tests. In S. Girling and P. Raiti (eds) *BSAVA Manual of Reptiles*, 3rd ed. British Small Animal Veterinary Association. In Press.

Hellebuyck, T., Martel, A., Chiers, K., et al. (2009). *Devriesea agamarum* causes dermatitis in bearded dragons (*Pogona vitticeps*). *Veterinary Microbiology*, **134**, 267–71.

Herpetological Animal Care and Use Committee. (2004). *Guidelines for the Use of Live Amphibians and Reptiles in Field and Laboratory Research*. 2nd ed rev. American Society of Ichthyologists and Herpetologists. Available at: http://www.asih.org/files/hacc-final.pdf (accessed 31 March 2015).

Homer, B.L., Berry, K.H., Brown, M.B., et al. (1998). Pathology of diseases in desert tortoises from California. *Journal of Wildlife Disease*, **34**, 508–23.

Institutional Animal Care and Use Committee Guidebook Guidebook. (2002). Applied Research Ethics National Association (ARENA), Office of Laboratory Animal Welfare (OLAW), National Institutes of Health. Available at: http://grants.nih.gov/grants/olaw/GuideBook.pdf. (accessed 31 March 2015).

Jacobson, E.R. (2007a). Viruses and viral diseases of reptiles. In E.R. Jacobson (ed) *Infectious Diseases and Pathology of Reptiles: A Color Atlas and Text.* Boca Raton, FL: CRC Press, pp. 395–460.

Jacobson, E.R. (2007b). Bacterial diseases of reptiles. In E.R. Jacobson (ed) *Infectious Diseases and Pathology of Reptiles: A Color Atlas and Text.* Boca Raton, FL: CRC Press, pp. 461–526.

Jacobson, E.R. (2007c). Parasites and parasitic diseases of reptiles. In E.R. Jacobson (ed) *Infectious Diseases and Pathology of Reptiles: A Color Atlas and Text.* Boca Raton, FL: CRC Press, pp. 571–665.

Jacobson, E.R. (2007d). Serodiagnostics. In E.R. Jacobson (ed) *Infectious Diseases and Pathology of Reptiles: A Color Atlas and Text.* Boca Raton, FL: CRC Press, pp. 381–94.

Jacobson, E.R., and Samuelson, D.A. (2007). Identifying reptile pathogens using electron microscopy. In E.R. Jacobson (ed) *Infectious Diseases and Pathology of Reptiles: A Color Atlas and Text*. Boca Raton, FL: CRC Press, pp. 299–349.

Jacobson, E.R., Berry, K.H., Stacy, B., et al. (2009). Oxalosis in wild desert tortoises, *Gopherus agassizii. Journal of Wildlife Disease*, **45**, 982–8.

Jacobson, E.R., Brown, M.B., Wendland, L.D., et al. (2014). Mycoplasmosis and upper respiratory tract disease of tortoises: a review and update. *The Veterinary Journal*, **201**, 257–64.

Jacobson, E.R., Homer, B., Greiner, E., et al. (2006). A neurological disease of loggerhead sea turtles (*Caretta caretta*) in coastal waters of Florida, USA. *Diseases of Aquatic Organisms*, **70**, 139–54.

Jacobson, E.R., Wronski, T., Schumacher, J., et al. (1994). Cutaneous dyskeratosis in free ranging desert tortoises, *Gopherus agassizii*, in the Colorado Desert of southern California. *Journal of Zoo and Wildlife Medicine*, **25**, 68–81.

Johnson A.J., Pessier, A.P., Wellehan, J.F.X., et al. (2008). *Ranavirus* infection of free-ranging and captive box turtles and tortoises in the United States. *Journal of Wildlife Disease*, **44**, 851–63.

Leibovitz, L., Rebell, G., and Boucher, G.C. (1978). *Caryospora cheloniae* sp. n.: a coccidial pathogen of mariculture-reared green sea turtles (*Chelonia mydas mydas*). *Journal of Wildlife Disease*, **14**, 269–75.

Machin, K.L. (2001). Fish, amphibian, and reptile analgesia. In D. Heard (ed) *Veterinary Clinics of North America: Exotic Animal Practice 4*. St. Louis, MO: W.B. Saunders, pp. 19–33.

McDowell, E.M., and Trump, B.F. (1976). Historical fixatives suitable for diagnostic light and electron microscopy. *Archives of Pathology and Laboratory Medicine*, **100**, 405–14.

Miller, D.L., and Gray, M.J. (2009). *Disinfection of Field Equipment and Personal Gear.* Information Sheet #10, Disease, Pathogens, and Parasites Task Team, Southeastern Partners in Amphibian and Reptile Conservation. Available at: http://fwf.ag.utk.edu/mgray/SEPARC/SEPARCDisinfectingGuidelinesFinal.pdf.

Mosley, C. (2011). Pain and nociception in reptiles. *Veterinary Clinics of North America: Exotic Animal Practice*, **14**, 45–60.

Palmer, C., and Sandøe, P. (2011). Animal ethics. In M.C. Appleby, J.A. Mench, I.A. Olsson, et al. (eds) *Animal Welfare*. 2nd ed. Wallingford, UK: CABI, pp. 1–12.

Paré, J.A., and Jacobson,E.R. (2007). Mycotic diseases of reptiles. In E.R. Jacobson, (ed) *Infectious Diseases and Pathology of Reptiles: A Color Atlas and Text.* Boca Raton, FL: CRC Press, pp. 527–70.

Paul-Murphy, J., Ludders, J.W., Roberston, S.A., et al. (2004). The need for a cross-species approach to the study of pain in animals. *Journal of the American Veterinary Medical Association*, **224**, 692–7.

Rutala, W.A., and Weber, D.J. (2013). Disinfection and sterilization: an overview. *American Journal of Infection Control*, **41**, S2–25.

Shigenaka, G. (2003). *Oil and Sea Turtles: Biology, Planning, and Response*. Washington, DC: National Oceanic and Atmospheric Administration.

Sigler, L., Hambleton, S., and Parée, J.A. (2011). Molecular characterization of reptile pathogens currently known as members of the *Chrysosporium* Anamorph of *Nannizziopsis vriesii* complex and relationship with some human-associated isolates. *Journal of Clinical Microbiology*, **51**, 3338–57.

Stacy, B.A., Foley, A.M., Greiner, E., et al. (2010). Spirorchiid trematode infections in stranded loggerhead (*Caretta caretta*) and green turtles (*Chelonia mydas*) in Florida (USA): host pathology and significance. *Diseases of Aquatic Organisms*, **89**, 237–59.

Stenglein, M.D., Jacobson, E.R., Wozniak, E.J., et al. (2014). Ball python Nidovirus: A candidate etiologic agent for severe respiratory disease in *Python regius*. *mBio*, 5(5):e01484–14. doi:10.1128/mBio.01484–14.

Terrell, S.P., and Stacy, B.A. (2007). Reptile necropsy techniques. In E.R. Jacobson (ed) *Infectious Diseases and Pathology of Reptiles: A Color Atlas and Text*. Boca Raton, FL: CRC Press, pp. 219–56.

Thomas, L.A., and Eklund C.M. (1962). Overwintering of western equine encephalomyelitis virus in garter snakes experimentally infected by *Culex tarsalis*. *Proceedings of the Society of Experimental Biology and Medicine*, **109**, 421–4.

U.S. Department of Agriculture. (2013). *Animal Welfare Act and Animal Welfare Regulations*. U.S. Code, 7 U.S.C., Sections 2131–59. Available at: http://www.nal.usda.gov/awic/legislat/awa.htm (accessed 31 March 2015).

Warwick, C., Frye, F.L., and Murphy, J.B. (1995). *Health and Welfare of Captive Reptiles*. 1st ed. London: Chapman and Hall.

West, G., Heard, D., and Caulkett, N. (2014). *Zoo Animal and Wildlife Immobilization and Anesthesia*. 2nd ed., Hoboken, NJ: John Wiley & Sons.

Whiteside, D.P. (2014). Analgesia. In G. West, D. Heard, and N. Caulkett (eds) *Zoo Animal and Wildlife Immobilization and Anesthesia*. 2nd ed. Ames, IA: John Wiley & Sons, pp. 83–108.

29
Conservation management

David A. Pike

29.1 Introduction

Reptiles have extremely diverse life histories and ecologies, which is reflected in the wide range of management strategies being used to conserve species and their habitats. Conservation management requires setting objectives that incorporate the life-history constraints of focal species, protection of existing habitats and ecosystem functioning, and monitoring population responses to management. Here, I briefly summarize the overarching principles governing conservation management and review approaches that have been used to successfully manage reptile populations worldwide. Detailed overviews of management strategies are available for crocodilians (Grigg and Kirshner, 2015), turtles (Klemens, 2000; Moll and Moll, 2004), Tuatara (Cree, 2014), and snakes (Mullin and Seigel, 2009). Detailed overviews of management strategies for reptile habitats are available for Australia (Webb et al., 2015), Europe (Edgar et al., 2010), and some regions of the United States (Bailey et al., 2006).

Management of reptiles requires an integrated landscape approach that incorporates habitats and connections among these at multiple spatial scales, rather than a species-oriented focus. Reptile assemblages not only depend upon multiple habitat features within landscapes, but are influenced by multiple stressors, necessitating management approaches that influence all aspects of habitat. In some cases it may be relatively easy to reintroduce natural disturbances, such as fire or grazing (Figure 29.1(a, b)), to help maintain landscapes. Many reptiles, especially rare species, are unable to persist in highly modified landscapes. Thus, conserving and managing existing habitats is preferable to restoration, when the option is available. An integrated landscape approach requires long-term funding for management activities and follow-up monitoring so that the project objectives can be continually assessed and evaluated. This feedback is used to ensure that project objectives are being met and, if not, that management activities are altered to reflect new information. Management plans should incorporate environmental change, including stochastic disturbances and more directional processes, such as temperature increases due to anthropogenic climate change. The environments within and surrounding protected landscapes is continually changing, and management should reflect that. The primary objectives of management should always focus on the species or community of concern, and not on peripheral or extended objectives, such as political motivations. There is no single management approach that is suitable in all

Reptile Ecology and Conservation. Edited by C. Kenneth Dodd, Jr. © Oxford University Press 2016.
Published 2016 by Oxford University Press.

Figure 29.1 Habitat management strategies often increase the availability of important reptile habitats in the landscape, which can enhance monitoring efforts by increasing detectability, and provide structures that are important for aspects of life history (e.g. shelter, thermoregulation). Traditional habitat management approaches include application of (a) prescribed fire (longleaf pine savannah habitat in Ocala National Forest, Florida USA) and (b) livestock grazing (eucalypt woodland at Wambiana Cattle Station, Queensland Australia), both of which can reduce woody vegetation encroachment and maintain heterogeneity in natural habitat structure. The timing, intensity, and return interval of these disturbances can profoundly influence the quality of habitat for reptiles, as well as community composition, population dynamics, and individual fitness. Moderate disturbance regimes applied in a heterogeneous manner throughout the landscape are typically most effective at maintaining ecosystems. In some cases, the application of natural disturbances is ineffective or impossible and more targeted approaches are needed. Examples include (c) the manual removal of woody vegetation that is encroaching reptile habitats, which can enhance basking sites (woody vegetation encroaching bog habitat, which is overgrown due to suppression of large herbivores, North Carolina, USA). Another strategy involves increasing the availability of thermally suitable retreat sites, such as by (d) using artificial rocks that mimic the thermal properties of real rocks (used by endangered snakes and their lizard prey, Morton National Park, New South Wales, Australia). These are only some of the management options that have been used to manage and conserve reptiles and their habitats; more are listed in Table 29.1. Photographs by (a) E. Roznik, (b-d) D. Pike.

instances and for all taxonomic groups. Thus, experimentation, monitoring, and adaptation are necessary to ensure management effectiveness.

29.1.1 Statutory protection

The simplest approach to reptile conservation is to enact legislative protection for species or habitats. In some cases, this can be extremely effective at reducing or controlling take and adverse trade, preventing habitat alteration or destruction, and ensuring that management and monitoring are being undertaken regularly. Legislative approaches work best when threats to reptiles or crucial habitats are direct and identifiable, where the protective regulations will be enforced, and when penalties are sufficient to deter potential offenders. Statutory protection can also lead to measures designed to decrease or eliminate threats to reptiles, such as mandated mitigation requirements for the disturbance of individuals, populations, or habitats, and directives to develop species or habitat recovery plans. Statutory protection should be accompanied by continual enforcement, monitoring, and management practices that keep pace with scientific advances.

29.1.2 Habitat protection

Habitat protection requires control over habitats where imperilled reptile species or communities reside. This can be achieved directly through outright purchase by governments or private organizations, by changing zoning regulations to prevent degradation, or indirectly by protecting private lands via conservation easements or agreements, tax incentives, or payment subsidies (Michael et al., 2014). Purchased lands can be incorporated into existing parks, preserves, refuges, or wildlife management units, and in some cases can be used by a wide range of stakeholders for recreational activities, as long as this does not conflict with the conservation objectives outlined in the purchase plan. Agreements can ensure that landowners can access and use their land as long as broader conservation objectives are being met. Management plans must ensure that land use occurs in accordance with initial agreements, and that long-term research and monitoring are in place to gauge the effectiveness of protection measures. Land acquisition and alternative conservation programmes require early, careful fiscal planning in order to ensure long-term continuity of protection, management, and assessments of how well objectives are being achieved (Michael et al., 2014).

The people residing adjacent to protected lands should be incorporated into long-term habitat protection efforts to help garner local stakeholder support for protection. Engaging local communities in habitat protection is particularly effective in biodiverse regions because of the economic incentives provided by tourism, which extend to employment and educational opportunities, as well as outlets for local arts. Examples of well-run habitat protection efforts that incorporate ecotourism include UNESCO World Heritage Sites (Great Barrier Reef in Australia, Galápagos Islands in Ecuador) and species-oriented programmes, such as sea turtle nesting beach tours. Incorporating local stakeholders into ecotourism provides local economic incentives for protecting and maintaining habitats, and helps protect against further habitat degradation.

29.1.3 Managing reptile populations

Five underlying principles should guide the development of reptile management programmes (Klemens, 2000; Moll and Moll, 2004; Mullin and Seigel, 2009; Cree, 2014; Grigg and Kirshner, 2015; Webb et al., 2015):

- Understand life histories. Life history and demography form the biological constraints upon which conservation and management efforts are based. Conservation and management only proceed with knowledge of reptile biology, including habitat requirements, spatial use of habitats, reproduction, diet, causes of mortality, and population biology.
- Address causes of decline. Imperilled species or populations can only be successfully managed or recovered once the causes of decline or need for conservation are first understood, and then corrected.

- Consider human-based constraints. Finances, personnel, time, logistics, local sensitivities and collaboration, and administrative policy must be incorporated into management programmes. Programmes that will succeed take into account human-based constraints and biological requirements of the species (Table 29.1).
- Set achievable biological objectives, management plans, and measures of success. Objectives that reflect species biology should be clearly stated and management efforts directed at achieving specific goals. Management plans are guidelines that should be updated regularly to accommodate new biological insights; they should not be altered to meet political objectives.
- Monitor management outcomes. Management effectiveness is evaluated based on the results of monitoring efforts that determine population status and habitat quality. Management should be viewed as long-term fundamental research programmes designed to address specific biological questions that are revised as new information becomes available in an adaptive management framework.

29.1.4 Monitoring populations

Monitoring programmes provide the foundation for statutory protection of species and their habitats, and guide population management. Monitoring provides the knowledge used to estimate changes in the distribution or abundance of populations or species, determine the impacts of habitat management, assess the efficacy of protected area networks, and document the spread of invasive species and impacts of climate change. Monitoring programmes are often quantitative, recording information such as search locations and effort, and details of the numbers and types of reptiles observed. Monitoring is usually done by trained biologists, but in some instances the general public can be recruited to help. Citizen scientists can be trained (usually via websites) to identify species and record important biological information and submit these observations directly to researchers using the internet or cell phone applications. This approach builds conservation support by engaging the public in monitoring and research, and also provides data at much broader spatial and temporal scales than would otherwise be possible. Citizen scientists may be especially effective for detecting elusive species over large spatial scales and monitoring spatiotemporal changes in distribution.

29.2 Habitats

29.2.1 Contiguous habitats, buffer zones, and edge effects

Reptile populations and communities are influenced by habitat connectivity, the prevalence of anthropogenic edges, the degree of landscape connectivity, the structure and heterogeneity of modified areas, and ecological connectivity (Fischer and Lindenmayer, 2007). Generally, local extinction is less likely when there are large areas of contiguous habitat, when landscapes have high native vegetation cover and connectivity, and when intensive land uses and habitat modifications are few (Fischer and Lindenmayer, 2007). Landscape planning is essential to ensure that habitat, species, and ecological connectivity are maintained. Often, land use changes extend into natural areas due to

Table 29.1 *Conservation management actions that have been successfully implemented for reptiles, and the taxonomic groups in which these actions have been tested successfully. When available, I provide citations of review articles that synthesize a large body of primary literature on the topic, rather than individual studies focusing on only one or a few species.*

Goal	Management action and outcome	Crocodilians	Turtles	Tuatara	Snakes	Lizards	Sources
Habitat addition	Creation of artificial wetland habitats (storm-water retention ponds, canals, etc.) and regulating hydroperiod	X	X		X	X	Klemens (2000), Moll and Moll (2004), Mitchell et al. (2008), Mullin and Seigel (2009), McDougall et al. (2015)
Habitat addition	Creation of nesting areas, including adding substrate to existing habitats (e.g. to ensure that sandy substrates are available for nesting), or other artificial habitats that provide oviposition sites, such as large rock walls containing deep crevices		X		X	X	Klemens (2000), Moll and Moll (2004), Shine and Bonnet (2009), Wnek et al. (2013), Roosenberg et al. (2014)
Habitat addition	Adding features to the landscape that reptiles and/or their prey use, such as shelter sites including rocks, coarse woody debris, leaf litter, hibernacula, artificial burrows, and other artificial shelters including cover boards				X	X	Moll and Moll (2004), Souter et al. (2004), Shine and Bonnet (2009), Burger and Zappalorti (2011), Christie et al. (2013), Roosenberg et al. (2014), Webb et al. (2015)
Habitat connectivity	Maintain or enhance connections among habitats, this can be achieved using underpasses that funnel animals beneath transportation corridors, or include corridor habitats that provide shelter and safe passage, such as hedgerows, ditches and ditch banks, stone walls, meadows, orchards, field margins, ponds, and manure heaps	X	X		X	X	Klemens (2000), Dodd et al. (2004), Moll and Moll (2004), Mullin and Seigel (2009), Shine and Bonnet (2009), Edgar et al. (2010), Grigg and Kirshner (2015)
Habitat maintenance	Using mechanical canopy removal to maintain or increase basking sites by maintaining open microhabitats				X	X	Mullin and Seigel (2009), Pike et al. (2011)

continued

Table 29.1 *Continued*

Goal	Management action and outcome	Crocodilians	Turtles	Tuatara	Snakes	Lizards	Sources
Habitat maintenance	Using prescribed fire regimes to maintain or increase basking sites by maintaining open microhabitats		X		X	X	Webb et al. (2015)
Habitat maintenance	Manage grazing herbivores to limit shrub development and maintain open herbaceous patches that allow reptiles to thermoregulate		X		X	X	Edgar et al. (2010), Webb et al. (2015)
Habitat retention	Retain bare sand for fossorial species or egg-layers (e.g. walking paths), which recreational users help maintain via their disturbance		X		X	X	Edgar et al. (2010)
Habitat retention	Protect and effectively manage remnant linear strips, and edge habitats, especially vegetation structural attributes	X	X		X	X	Klemens, 2000; Moll and Moll (2004), Edgar et al. (2010), Jellinek et al. (2014), Grigg and Kirshner (2015)
Pest reduction	Screening for invasive species being transported in cargo, including canine detection	X	X	X	X	X	Rodda et al. (1999)
Pest reduction	Reducing the abundance of reptile predators that are pests, such as inflated populations of mammalian meso-predators (e.g. raccoons, foxes, rats)		X	X	X		Klemens (2000), Moll and Moll (2004), Mullin and Seigel (2009), Cree (2014), Webb et al. (2015)
Pest reduction	Reducing the abundance of invasive reptiles, which can be predators and/or competitors of native fauna		X		X	X	Rodda et al. (1999), Mullin and Seigel (2009)
Reduce human disturbance	Statutory protection and enforcement of habitat closure from human disturbance. This may include sensitive habitats or areas with large aggregations of animals, such as important nesting areas	X	X	X	X	X	Klemens (2000), Moll and Moll (2004), Mullin and Seigel (2009), Grigg and Kirshner (2015)
Regulate harvest of wild populations	Statutory regulation regulating harvest of eggs or adults from the wild by setting take limits combined with seasonal closures, can include selective harvest that targets life stages which already have high mortality (eggs, juveniles)	X	X		X		Klemens (2000), Moll and Moll (2004), Mullin and Seigel (2009), Shine and Bonnet (2009), Grigg and Kirshner (2015)

Relocation, repatriation, translocation	Assessing microhabitat suitability of release sites prior to translocation (including nesting/basking sites, shelters, habitat structure)	X	X	X	X	X	Klemens (2000), Moll and Moll (2004), Mullin and Seigel (2009), Cree (2014), Jarvie et al. (2014), Grigg and Kirshner (2015)
Relocation, repatriation, translocation	Head-starting to increase survival of released animals	X	X	X	X	X	Klemens (2000), Moll and Moll (2004), Mullin and Seigel (2009), Cree (2014), Grigg and Kirshner, (2015)
Relocation, repatriation, translocation	Penning prior to release at RRT sites		X	X		X	Klemens (2000), Mullin and Seigel (2009), Knox and Monks (2014), McCoy et al. (2014)

edge effects, such as increased solar radiation. This can degrade habitats, and must be taken into account when planning for conservation by area. Providing additional 'buffer zones' around core habitats will help ensure that individual animals and populations can move normally while remaining protected. The sizes of these areas should reflect the mobility and habitat requirements of focal species or communities (Fischer and Lindenmayer, 2007), and incorporate the potential for future environmental changes, including anthropogenic climate change, stochastic disturbances, and their interactions with anthropogenic effects (see Box 29.1).

29.2.2 Habitat connectivity

Individual animals often move among habitats throughout their lives. Maintaining connectivity among habitats helps buffer against local catastrophe by facilitating gene flow

Box 29.1 Managing for environmental change

The environment is continually changing over daily, seasonal, and annual timescales, and this can influence habitat quality and population status. Some environmental changes are unpredictable and stochastic (e.g. events such as major storms that alter habitat structure or flood/drought), whereas others are regular and more predictable (e.g. plant community succession, increased air temperatures from anthropogenic climate change). Management should be placed in an adaptive framework that is reinforced by hypotheses that are tested and refined using the results of monitoring. This approach will ensure that management activities take into account changes in the environment as they happen. Increases in temperatures as a result of anthropogenic climate change has the potential to affect many reptile populations by altering hatchling sex ratios in species with temperature-dependent sex determination (Jarvie et al., 2014), changes in geographic range size (Penman et al., 2010), alterations in gene flow (Dubey et al., 2012), and activity patterns, which can influence thermoregulation and therefore growth rates, reproduction, and survival (Sinervo et al., 2010).

Changes in land use often degrades habitat, and could magnify the impacts of environmental changes. Thus, maintaining connectivity in complex landscapes (corridors between fragments, crossing corridors) and ensuring that these connect long-term refugia from climatic and other disturbances is a current challenge. In addition, some species may be able to be used to protect others—ecosystem engineers can provide shelters for a wide range of species within communities, and these could be thermal refugia under climate change (Pike and Mitchell, 2013). Therefore, maintaining the distribution and abundance of these species will be important, as is the location and distribution of suitable microhabitats (Christie et al., 2013). New kinds of studies will be required to investigate the complex issues involving the responses of reptiles to environmental changes. These include developing new approaches in macrophysiology and biophysical modelling, both of which are becoming powerful tools to predict the responses of animals to climate change (Webb et al., 2015). The information derived from such studies will facilitate the successful conservation management of reptiles in changing environments.

and metapopulation structure. Protecting these connections, when possible, should be guided by the ecology and behaviour of the focal species. Existing connections should be identified and maintained when possible, but these can also be restored by planting vegetation that reptiles use, or even created using artificial means (e.g. underpasses beneath roads). Examples of species that regularly migrate include many pond turtles, which undertake terrestrial migrations to locate nest sites or to move among water bodies, and sea turtles, which migrate between foraging areas and nesting sites, which can be hundreds or even thousands of kilometres apart. Some freshwater turtles, notably in Australia, aestivate in terrestrial habitats when seasonal or unpredictable droughts reduce water levels and food supplies in ponds. Dams can help to regulate water flow for human purposes—water retention, flood reduction, power generation—but can substantially alter reptile populations by impeding movements of riverine turtles or altering hydrological cycles in ways that disrupt portions of the life cycle (Klemens, 2000; Moll and Moll, 2004; McDougall et al., 2015). Protecting the linkages among habitats is essential for these animals to complete their life cycles. In some cases migratory pathways can be protected and maintained by incorporating them into existing conservation planning. In other instances, management can be used to facilitate movement through the landscape, for example, by providing safe passages through otherwise unsuitable habitats. This could include creation of wetland stepping-stones, tunnels or passageways that facilitate movement beneath or over obstacles such as transportation corridors.

Habitat connectivity should draw on the theory of biogeography, and incorporate knowledge of reptile ecology and habitat requirements. Along with providing connectivity comes the responsibility of maintaining the quality of these habitats by managing for native communities of plants and animals and limiting or reducing invasive species. Invasive plant species can alter the microhabitat structure preferred by reptiles, and invasive reptile species can negatively impact native reptile communities (Mitchell et al., 2008; Webb et al. 2015). Maintaining or enhancing habitat connectivity can be encouraged using statutory measures that require enforcement, and through incentives to landowners for maintaining or restoring natural habitats. Understanding connectivity among spatially disparate resources, and how individual animals and populations migrate between them, will ensure the conservation management of all life stages.

29.2.3 Crossing transportation corridors

Transportation corridors (e.g. roadways, railway lines, pipelines) represent formidable obstacles to reptiles. Whereas some species regularly cross transportation corridors, others are less willing or able to cross them successfully, resulting in networks of landscape barriers. When transportation corridors divide important migratory pathways or habitats, reptile populations can experience substantial mortality that affects population viability (Klemens, 2000; Mitchell et al., 2008; Mullin and Seigel, 2009; Van der Ree et al., 2015). The ability for some reptiles to cross these corridors successfully may depend on traffic volume and the distance needed to travel. Proper planning can ensure that transportation corridors have minimal impacts on reptile populations by placing them in areas that are infrequently used by reptiles or by facilitating movement beneath or around these obstacles. Barrier walls or fences along transportation corridors can prevent reptiles from crossing and help funnel animals along them to underpasses, culverts, or tunnels which can help ferry them safely to the other side (Van der Ree et al., 2015). The structures should be designed to facilitate movement of target fauna, in terms of size, shape, and attractiveness, and designed to reduce depredation of animals using them. Experimentation is necessary to determine the efficacy of these structures in terms of reductions in crossing mortality. Continual maintenance of underpasses is necessary to ensure that barriers remain effective and animals are successfully moving through these structures (Van der Ree et al., 2015).

29.2.4 Microhabitats

Within the broader spatial area used by individual reptiles are microhabitats, the specific sites actually used by animals that directly influence life history. These include nesting sites, basking sites, and shelters such as woody debris piles, rocks, burrows, and hibernacula. In many reptiles these sites are within home range areas used by individuals of the species, but some species migrate seasonally to nesting sites, foraging sites, or hibernacula. Sea turtles may migrate hundreds or even thousands of kilometres between breeding and foraging areas, and aquatic turtles often travel overland during the reproductive season in search of terrestrial nesting sites (Klemens, 2000; Moll and Moll, 2004). Determining the microhabitat features that may limit reptile populations requires detailed ecological study of habitat selection and use by individual animals, often achieved via radiotelemetry (see Chapter 9). Once identified, these microhabitats can often be manipulated in biologically meaningful ways to aid in population recovery and enhance detectability for monitoring (Table 29.1). Management can include limiting human disturbance of microhabitats, especially important nesting sites or key habitat features that are easily destroyed (rocks, crevices, burrows). Directly manipulating the distribution and abundance of microhabitat features in the landscape is rapidly becoming an effective conservation strategy applicable to a wide range of reptile populations, species and communities (Box 29.2; Table 29.1).

29.3 Human-altered habitats

Habitat disturbance, alteration, and destruction are primary threats to reptiles worldwide; of 1500 species assessed (Böhm et al., 2013), 74% were threatened by agriculture, 64% by biological resource use (silviculture, agriculture), 34% by urban development, and 25% by modifications to natural systems (changes in fire regimes, hydrological cycles).

29.3.1 Agricultural landscapes

Agricultural landscapes can disrupt natural habitat connectivity, change microhabitat structure, and alter prey availability for reptiles, all of which can alter community composition in ways that disadvantage rare species (Jellinek et al., 2014; Zeng et al., 2014; Ebrahimi and Bull, 2015). Some reptiles may thrive in agricultural landscapes (some snakes, which eat rodents attracted to crops), whereas species that require habitat features only present in natural habitats may be unable to persist. Heterogeneous networks

Box 29.2 Restoring microhabitat features

Microhabitats selected by reptiles often have very distinct thermal or structural properties that differ from similar, but unused sites in the landscape. Microhabitat features are often relatively easy to manipulate for conservation management over large spatial scales, can be monitored to assess the outcome of their occupancy and use by focal animals, and when used as a standardized sampling tool may aid in monitoring population trends. Targeted management intervention can manipulate specific habitat features, such as canopy cover (Pike et al., 2011), coarse woody debris for individual species (Christie et al., 2013) or communities (Márquez-Ferrando et al., 2009), or increases in the availability of additional habitat features in the landscape, such as hibernacula, leaves, or compost piles that provide insulation (Shine and Bonnet, 2009; Burger and Zappalorti, 2011), artificial rocks that serve as shelters for endangered snakes and their lizard prey (Webb et al., 2015), artificial burrows that provide shelters for lizards (Souter et al., 2004), and nesting beaches for turtles (Roosenburg et al., 2014). In addition to supplementing additional habitats or habitat features for one or more life stages, the entire ecosystem can be manipulated. Examples include regulating water flow to reduce nest flooding of riverine turtles (McDougall et al., 2015) or using a levee to reduce flooding of snake hibernacula (Shine and Bonnet, 2009). In some cases, ecosystems and reptile microhabitats are regulated by large herbivores, such as elephants that create shelters for lizards by partially stripping bark off trees (Pringle, 2008) or kangaroos and cattle that help maintain open woodland habitats (Webb et al., 2015). Sex ratios can be manipulated in artificial hatcheries and natural nesting areas by shading nests using artificial means or manipulating natural vegetation cover (Klemens, 2000; Moll and Moll, 2004; Cree, 2014). Natural microhabitats should not be destroyed and replaced by artificial ones, but instead artificial microhabitats can be used to enhance degraded landscapes and increase monitoring success of elusive species.

of agricultural areas interspersed with natural habitats retain more biodiversity than do monocultures. Connectivity among natural habitats can be retained in agricultural landscapes using hedgerows, ditches and ditch banks, stone walls, meadows, orchards, field margins, ponds, and manure heaps (Edgar et al., 2010).

29.3.2 Silviculture

Silviculture poses threats from selective logging, rotational clearcutting or stand thinning, and plantation forestry practices. Reptiles are directly affected by machinery, which can crush animals and alter their microhabitats by compacting the soil, collapsing underground burrows, and crushing other microhabitat features. Species with restricted distributions or specific habitat preferences that are targeted for extraction, such as large trees, may be most at risk. Silviculture often protects large tracts of land, including habitats that are not used for extraction, and disturbances are usually targeted and infrequent (e.g. selective removal of individual trees, or rotational logging of small tracts of land). Retaining a mosaic of standing dead and live trees of all sizes, coarse woody debris on the ground, and leaf litter provides important microhabitat features that may aid in local persistence (Todd and Andrews, 2008).

29.3.3 Urban environments

Urban environments are rapidly becoming dominant features in landscapes worldwide. Urban design can incorporate reptile populations by maintaining and increasing habitat connectivity, avoiding key habitats or habitat features, and ensuring that ecosystem management continues to occur (e.g. prescribed fire in close proximity to houses and other infrastructure; Mitchell et al., 2008). Connectivity can be achieved at the landscape level by providing corridors or stepping-stone habitats, such as greenways and parks with native vegetation communities, or aquatic connections, such as drainage ditches and storm-water ponds. Urbanization brings additional management concerns in terms of increased human disturbance to reptiles and their habitats, unwanted human–reptile interactions (venomous snakebite, encounters with crocodilians), environmental contaminants, and the accidental or deliberate introduction of non-native species. Non-native reptile species can increase competition, predation, and pathogen transmission to native species and communities, and non-native birds and mammals may act as reptile predators.

29.3.4 Environmental contaminants

Reptiles face threats from chemical contamination and pollution, which are rapidly emerging as threats to air, soil, and water quality (Sparling et al., 2010). Environmental contaminants are sometimes linked to a specific point source, such as leaking pipelines, but in other cases the source may be much more difficult or impossible to detect. Environmental contaminants include pesticides and herbicides, metals, plastics, petroleum, pharmaceuticals, and radioactive waste. Contaminants can have severe genetic effects and reduce fitness by increasing baseline metabolic rates in individual animals, leaving less energy available for growth and reproduction. Pollution may directly reduce survival, which can be accompanied by suffering (e.g. sea turtles asphyxiating after ingesting plastic bags mistaken as food or drowning in abandoned nets; snakes overheating after becoming entangled in landscape netting; aquatic turtles drowning in abandoned traps). Long-lived species may be at particular risk from the effects of contaminants, which often accumulate in the body. Environmental contaminants have been associated with reduced gonadal development and altered sex hormones in crocodilians, and mothers can pass some contaminants to offspring, with higher contaminant levels in larger progeny. Negative effects from environmental contaminants are likely much more widespread than is currently realized, due to the difficulty associated with detecting contaminants in reptiles (Sparling et al., 2010).

29.4 Intensive manipulation of individuals

29.4.1 Captive breeding

Captive breeding of imperilled reptiles is a high-technology option of last resort; few programmes have been successful and all are extremely costly. The goal of these programmes may be to establish 'assurance colonies', which maintain species in captivity that are threatened with extinction in the wild; release progeny into the wild to supplement or establish wild populations; and/or provide animals for the pet or food trades. Although many reptiles have attributes suitable for captive breeding, such as short generation times and high fecundity, many turtle and crocodilian species do not have those characteristics and may be difficult to breed successfully in large numbers. In other cases, some species can be farmed in numbers large enough to satisfy global demand; overexploitation is the main reason of the collapse of wild Asian turtle populations, which are widely used as food and pets in Asian countries. Combined with statutory protection of endangered turtle species and increased security at border crossings, turtle farming at large scales is reducing the pressure on wild populations (Klemens, 2000; Moll and Moll, 2004; Mitchell et al., 2008).

Prior to undertaking captive breeding for conservation or reintroduction, a careful plan must be developed that takes into consideration housing, cleanliness, disease screening, specialized cover, diet and biophysical requirements, and the potential effects of long-term captivity and inbreeding on fitness. Captive breeding programmes led by zoos, aquariums, or museums often use educational displays to inform visitors about these efforts and the plight of reptiles in the wild. Maintaining captive breeding populations should be a preventative measure that accompanies other attempts to reduce threats in the wild. In some cases, providing a legal supply of captive-bred individuals for the pet or food trades can substantially reduce exploitation of wild populations (Klemens, 2000; Moll and Moll, 2004; Grigg and Kirshner, 2015). Conservation is best carried out *in situ* unless the threats become so dire that extinction is imminent without intervention and removal to a secure location.

29.4.2 Relocation, repatriation, translocation (RRT)

Moving animals out of harm's way (relocation), into areas currently or previously occupied (repatriation), or even to areas not previously occupied (translocation) are popular management strategies. Despite their popularity, many RRT programmes have been unsuccessful or monitoring has been insufficient to determine the ultimate outcome (Dodd and Seigel, 1991; Klemens, 2000; Mullin and Seigel, 2009; Cree, 2014; Ewen et al., 2014; Jarvie et al., 2014; Knox and Monks, 2014; McCoy et al., 2014; Ebrahimi et al., 2015). The main concerns are that animals may not maintain fidelity to their release site, numbers of animals released may be too low to establish wild populations, and survival or reproduction will be insufficient to maintain population viability. To increase the likelihood of RRT success, managers should clearly outline all relevant objectives and constraints, rank their importance, and understand their relevance (Ewen et al., 2014).

RRT has substantial control over the demography of animals released. Various life stages have been used in RRT programmes, which can be sourced from wild populations or captive breeding programmes. Adult reptiles are often used for RRT because these animals are more easily captured than juveniles and because adults will be able to initiate breeding in the wild more quickly than can juveniles. The juvenile stage is particularly vulnerable to mortality, especially in long-lived species with delayed maturation.

Therefore, many RRT programmes with turtles, crocodilians, and tuatara raise animals in captivity to increase body size and condition prior to being released (Klemens, 2000; Moll and Moll, 2004; Cree, 2014). Sometimes the nests of high-profile species, such as sea turtles, are relocated to protect them from direct threats, such as predators or flooding due to tidal overwash.

Prior to moving animals, monitoring of the recipient site should take place, including population censuses, assessment and preparation of suitable habitat at recipient sites, disease and genetic screening, predator removal, and public education. In addition, it is essential to ensure that the recipient site is protected from threats and that habitats contain all of the features necessary for all life stages (e.g. shelters, foraging areas, nesting/gestation sites). For many species, assessment of habitat quality includes an evaluation of the thermal suitability of possible release sites, for both adults and in terms of nesting sites (Wnek et al., 2013; Cree, 2014; Jarvie et al., 2014), and possibly how thermal suitability could be affected by anthropogenic climate change.

Approaches designed to increase site retention, such as penning animals inside enclosures at the release site for a period of time prior to release (termed a 'soft release') are more effective than releasing animals and allowing them to move and disperse freely (Knox and Monks, 2014; McCoy et al., 2014). Penning provides animals the opportunity to establish local territories, which increases site fidelity and reduces mortality associated with animals dispersing away from recipient sites (Klemens, 2000; Knox and Monks, 2014; McCoy et al., 2014; Ebrahimi et al., 2015). Most RRT programmes use a staggered release through time, releasing individuals over several months or years, rather than all at once. This helps to ensure that individuals can become established in stochastic environments.

29.4.3 Pest reduction

Some reptiles are considered pests, usually because they have unwanted effects on humans or native ecosystems, or because they are invasive. Invasive reptiles can reach high densities, and may negatively impact native species through competition for resources (space, food) or by acting as predators. For example, the Red-eared Slider (*Trachemys scripta elegans*) invades many parts of the world, and may outcompete local turtle species (Klemens, 2000; Moll and Moll, 2004). Management of native reptile pests usually includes removal of unwanted individual animals, such as crocodilians habituated to humans or venomous snakes, whereas management of invasive species usually attempts to reduce population size or even completely eradicate these species from islands or other discrete areas (Rodda et al., 1999; Grigg and Kirshner, 2015). Many introduced reptile populations are difficult or impossible to eradicate because of their secretive nature, high fecundity, and difficulty capturing all life stages.

29.4.4 Biosecurity and disease

Diseases and parasites, including bacteria (*Mycoplasma agassizii*), viruses (*Herpesviridae, Ranavirus*), fungi (*Ophidiomyces ophiodiicola*), enterobacteria (*Salmonella*), and trematodes, may seriously affect individual reptiles and populations (see Chapter 28). Biosecurity protocols are fundamental to preventing the spread of these pathogens and parasites at local scales, throughout a landscape, and among continents. Pathogens and parasites may move as organisms or propagules, and can be transported in water, air, or on or within human and wildlife vectors. Humans play an important role in transporting pathogens and parasites by moving animals, such as when wild reptiles are sold as pets, deliberately released in new locations, or inadvertently moved with vehicles, boats, and other forms of human transportation. In some instances, statutory protection may prohibit pathogens or parasites and mandate screening for infected reptiles. Shipments of reptiles (especially those imported from other regions or continents) should be screened for pathogens and parasites (e.g. ticks), and the entire contents of the shipment (including reptiles, containers, substrate, and water) should be treated to reduce likelihood of spread. These types of testing and quarantine procedures are often mandated for amphibians, mammals, and birds because of concerns that animal pathogens could have negative impacts on local diversity and/or pose risks to human health. Biosecurity risks are especially high when pathogens or parasites threaten remaining populations of critically endangered species, during relocation, repatriation, and translocation efforts, and when non-indigenous species are released or become established outside their native ranges (see Chapter 28).

29.5 Conclusion

Reptile conservation can be extremely effective when sound science is used to implement statutory protection, habitat protection and management, population monitoring, and in some cases, manipulation of habitats or individuals. Habitat alteration is the single most widespread threat to reptiles worldwide; protecting and connecting remaining natural habitats are essential to maintaining healthy reptile communities, as is restoration of degraded environments. Habitat management and restoration actions can range from traditional approaches that mimic natural ecosystem disturbances (prescribed fire, grazing to mimic extinct megaherbivores) to artificial habitat replacement that directly targets microhabitat features used by reptiles (shelters, basking sites, and nesting sites). Management should seek to use approaches that are likely to be most effective in particular cases and experimentally test their effectiveness locally, rather than relying on the results of earlier studies. Although reptiles are in decline worldwide, there is increasing public interest in these animals, which can provide the momentum necessary for effective protection, management, and conservation in changing environments.

References

Bailey, M.A., Holmes, J.N., Buhlmann, K.A., et al. (2006). *Habitat Management Guidelines for Amphibians and Reptiles of the Southeastern United States*. Technical Publication HMG-2. Montgomery, AL: Partners in Amphibian and Reptile Conservation.

Böhm, M., Collen, B., Baillie, J.E.M., et al. (2013). The conservation status of the world's reptiles. *Biological Conservation*, **157**, 372–85.

Burger, J., and Zappalorti, R.T. (2011). The northern pine snake (*Pituophis melanoleucus*) in New Jersey: its life history, behavior and conservation. In K.J. Baker (ed) *Reptiles: Biology, Behavior and Conservation*. Hauppauge, NY:Nova Science Publishers, pp. 1–56.

Christie, K., Stokes, V.L., Craig, M.D., et al. (2013). Microhabitat preference of *Egernia napoleonis* in undisturbed Jarrah Forest, and availability and introduction of microhabitats to encourage colonization of restored forest. *Restoration Ecology*, **21**, 722–8.

Cree, A. (2014). *Tuatara: Biology and Conservation of a Venerable Survivor*. Christchurch, New Zealand:Canterbury University Press.

Dodd, C.K., Jr., and Seigel, R.A. (1991). Relocation, repatriation, and translocation of amphibians and reptiles: are they conservation strategies that work? *Herpetologica*, **47**, 336–50.

Dodd, C.K., Jr., Barichivich, W.J., and Smith, L.L. (2004). The effectiveness of a barrier wall and underpasses in reducing wildlife mortality on a heavily traveled highway in Florida. *Biological Conservation*, **118**, 619–31.

Dubey, S., Pike, D.A., and Shine, R. (2012). Predicting the impacts of climate change on genetic diversity in an endangered lizard species. *Climatic Change*, **117**, 319–27.

Ebrahimi, M., and Bull, C.M. (2015). Behavioural changes in an endangered grassland lizard resulting from simulated agricultural activities. *Journal of Arid Environments*, **113**, 102–7.

Ebrahimi, M., Ebrahimie, E., and Bull, C.M. (2015). Minimizing the cost of translocation failure with decision-tree models that predict species' behavioral response in translocation sites. *Conservation Biology*, in press.

Edgar, P., Foster, J., and Baker, J. (2010). *Reptile Habitat Management Handbook*. Bournemouth, UK:Amphibian and Reptile Conservation.

Ewen, J.G., Soorae, P.S., and Canessa, S. (2014). Reintroduction objectives, decisions and outcomes: global perspectives from the herpetofauna. *Animal Conservation*, **17**, 74–81.

Fischer, J., and Lindenmayer, D.B. (2007). Landscape modification and habitat fragmentation: a synthesis. *Global Ecology and Biogeography*, **16**, 265–80.

Grigg, G., and Kirshner, D. (2015). *Biology and Evolution of Crocodylians*. Ithaca, NY:Cornell University Press.

Jarvie, S., Besson, A.A., Seddon, P.J., et al. (2014). Assessing thermal suitability of translocation release sites for egg-laying reptiles with temperature-dependent sex determination: a case study with tuatara. *Animal Conservation*, **17**, 48–55.

Jellinek, S., Parris, K.M., McCarthy, M.A., et al. (2014). Reptiles in restored agricultural landscapes: the value of linear strips, patches and habitat condition. *Animal Conservation*, **17**, 544–54.

Klemens, M.W. (ed). (2000). *Turtle Conservation*. Washington, DC:Smithsonian Institution Press.

Knox, C.D., and Monks, J.M. (2014). Penning prior to release decreases post-translocation dispersal of jewelled geckos. *Animal Conservation*, **17**, 18–26.

Márquez-Ferrando, R., Pleguezuelos, J.M., Santos, X., et al. (2009). Recovering the reptile community after the mine-tailing accident of Aznalcóllar (southwestern Spain). *Restoration Ecology*, **17**, 660–7.

McCoy, E.D., Osman, N., Hauch, B., et al. (2014). Increasing the chance of successful translocation of a threatened lizard. *Animal Conservation*, **17**, 56–64.

McDougall, A.J., Espinoza, T., Hollier, C., et al. (2015). A risk assessment approach to manage inundation of *Elseya albagula* nests in impounded waters: a win–win situation? *Environmental Management*, **55**, 715–24.

Michael, D.R., Wood, J.T., Crane, M., et al. (2014). How effective are agri-environment schemes for protecting and improving herpetofaunal diversity in Australian endangered woodland ecosystems? *Journal of Applied Ecology*, **51**, 494–504.

Mitchell, J.C., Jung Brown, R.E., and Bartholomew, B. (eds). (2008). *Urban Herpetology*. Salt Lake City, Utah: Society for the Study of Amphibians and Reptiles.

Moll, D., and Moll, E.O. (2004). *The Ecology, Exploitation, and Conservation of River Turtles*. New York: Oxford University Press.

Mullin, S.J., and Seigel, R.A. (eds). (2009). *Snakes: Ecology and Conservation*. Ithaca, New York: Comstock Publishing Associates.

Penman, T.D., Pike, D.A., Webb, J.K., et al. (2010). Predicting the impact of climate change on Australia's most endangered snake, *Hoplocephalus bungaroides*. *Diversity and Distributions*, **16**, 109–18.

Pike, D.A., and Mitchell, J.C. (2013). Burrow-dwelling ecosystem engineers provide thermal refugia throughout the landscape. *Animal Conservation*, **16**, 694–703.

Pike, D.A., Webb, J.K., and Shine, R. (2011). Removing forest canopy restores a reptile assemblage. *Ecological Applications*, **21**, 274–80.

Pringle, R.M. (2008). Elephants as agents of habitat creation for small vertebrates at the patch scale. *Ecology*, **89**, 26–33.

Rodda, G.H., Sawai, Y., Chiszar, D., et al. (eds) (1999). *Problem Snake Management: The Habu and the Brown Treesnake*. Ithaca, NY:Comstock Publishing Associates.

Roosenburg, W.M., Spontak, D.M., Sullivan, S.P., et al. (2014). Nesting habitat creation enhances recruitment in a predator-free environment: *Malaclemys* nesting at the Paul S. Sarbanes Ecosystem Restoration Project. *Restoration Ecology*, **22**, 815–23.

Shine, R., and Bonnet, X. (2009). Reproductive biology, population viability, and options for field management. In S.J. Mullin and R.A. Seigel (eds) *Snakes: Ecology and Conservation*. Ithaca, New York: Cornell University Press, pp. 172–200.

Sinervo, B., Méndez-de-la-Cruz, F., Miles, D.B., et al. (2010). Erosion of lizard diversity by climate change and altered thermal niches. *Science*, **328**, 894–9.

Souter, N.J., Michael Bull, C., and Hutchinson, M.N. (2004). Adding burrows to enhance a population of the endangered pygmy blue tongue lizard, *Tiliqua adelaidensis*. *Biological Conservation*, **116**, 403–8.

Sparling, D.W., Linder, G., Bishop, C.A., et al. (eds). (2010). *Ecotoxicology of Amphibians and Reptiles*. Pensacola, Florida: Society of Environmental Toxicology and Chemistry Press.

Todd, B.D., and Andrews, K.M. (2008). Response of a reptile guild to forest harvesting. *Conservation Biology*, **22**, 753–61.

Van der Ree, R., Smith, D.J., and Grilo, C. (eds). (2015). *Handbook of Road Ecology*. Oxford: Wiley-Blackwell.

Webb, J.K., Harlow, P.S., and Pike, D.A. (2015). Australian reptiles and their conservation. In A. Stow, N. Maclean, and G.I. Holwell (eds) *Austral Ark: The State of Wildlife in Australia and New Zealand*. Cambridge: Cambridge University Press, pp. 354–81.

Wnek, J.P., Bien, W.F., and Avery, H.W. (2013). Artificial nesting habitats as a conservation strategy for turtle populations experiencing global change. *Integrative Zoology*, **8**, 209–21.

Zeng, Z.-G., Bi, J.-H., Li, S.-R., et al. (2014). Effects of habitat alteration on lizard community and food web structure in a desert steppe ecosystem. *Biological Conservation*, **179**, 86–92.

30
Education and outreach

Brian Gratwicke, Matthew Neff, Lindsay Renick Mayer, Sharon Ryan, and Jennifer Sevin

30.1 Introduction

Those of us who work in the natural sciences come from every corner of the globe, but we tend to have at least one thing in common: nearly all of us had formative interactions with nature that instilled in us a basic love for wildlife and wilderness. That love eventually led each of us to a career in the natural sciences. While no single interaction with nature is necessarily life changing, the cumulative effect of spending time outdoors is a deep love of and respect for nature (Dietz et al., 2004). For many environmental education programmes, facilitating connections to nature is the end goal, which may also have benefits both for students and for the natural areas they are exploring. Exposure to nature can be one of the best ways to hone skills in observation and critical thinking about the world around us. Some argue that in our increasingly urban modern world, a deficit in interactions with nature is actually detrimental to a child's development (Louv, 2008).

An education in natural history certainly instills conservation ethics, but the more we learn about the challenges wildlife face, the more we realize that it is necessary to change people's behaviour in ways that solve environmental problems. As Aldo Leopold put it:

> One of the penalties of an ecological education is that one lives alone in a world of wounds. Much of the damage inflicted on land is quite invisible to laymen. An ecologist must either harden his shell, and make believe that the consequences of science are none of his business, or he must be the doctor who sees the marks of death in a community that believes itself well and does not want to be told otherwise. (Leopold, 1989)

Conservation education projects come in many different forms. There are small-scale conservation outreach efforts such as the production and distribution of posters to inform the public. These low-cost actions require little effort, but the effects are usually so indirect that it is very difficult to evaluate their conservation benefit (Gratwicke et al., 2007). The key to making environmental education into a conservation action is to establish a solid foundation of natural history knowledge and then take it a level further, changing people's behaviour in a way that will mitigate a conservation threat (Butler et al., 2013). If, for example, you are working on reptiles that are experiencing high levels of mortality as the result of roadkill, it is not sufficient to simply make people

Reptile Ecology and Conservation. Edited by C. Kenneth Dodd, Jr. © Oxford University Press 2016.
Published 2016 by Oxford University Press.

aware of reptiles and their ecological values. That is a crucial first step, but it is also necessary to convince drivers to slow down for turtles and snakes crossing the road, or, at the very least, to refrain from deliberately hitting them on the tarmac. A truly effective education intervention must evaluate whether the campaign successfully changed driving behaviour sufficiently to reduce reptile mortality. In reality, it is extraordinarily difficult to change people's attitudes and behaviours, so many environmental managers instead choose other kinds of conservation interventions that manage the species directly, such as designing road culverts (Chapter 29).

The challenge of developing effective conservation education projects for reptiles is further compounded by the fact that many people fear them, especially snakes and crocodilians. A Gallup poll found that about half of Americans fear snakes (Brewer, 2001). Why do people fear snakes? Most people can't explain their fear. Some have postulated that humans pointed their finger before developing spoken language because of a deeply ingrained fear of snakes, resulting in a positive evolutionary selection force that was actually a behavioural precursor to the evolution of language (Isbell, 2011; Van Le et al., 2013). People retain an enhanced ability to detect dangerous taxa even if they have grown up in urban areas and were never exposed to them (Penkunas and Coss, 2013). Similarly, captive orangutans that have never been exposed to venomous snakes are very fearful of non-venomous Eastern Ratsnakes (*Pantherophis alleghaniensis*) that occasionally find their way into the outdoor primate enclosures at a zoo (Murphy et al., 2014). In some cases, an innate fear of snakes can be overcome through exposure and positive reinforcement, but in some people who are truly ophiophobic, no amount of rational instruction will alleviate their irrational fear (Burghardt et al., 2009).

This chapter will explore the various environmental educational tools at our disposal and reptile-specific conservation education challenges reptiles pose.

30.2 Setting goals

The most important part of any environmental project is setting goals. Every situation is likely to be extremely context specific, so it is critical to think through all the elements of your problem. What is the conservation threat you are trying to address? What action could you take that will mitigate the threat? Once you have defined your goals, you must define your target audience and evaluate your existing capacity to reach them. The next phase involves developing multiple creative activities, messages, images, or lesson plans that might invoke your desired response and that you can test on small focus groups. It is important to establish sensible indicators that help you evaluate the results, and to discard activities that don't work. This customer-centric approach is routine in the marketing world, but many non-profits focus exclusively on promotional marketing of their organization without taking the extra steps to maximize their impact (Chao et al., 2009).

30.3 Campaigns and constituency-building

A strategic and comprehensive campaign over a sustained period of time can do more than just tell a good story about your project—it can be instrumental in shaping the way people think and talk about wildlife, biodiversity, and conservation. Communications campaigns rely on a number of channels to reach a few target audiences. These channels include earned media (stories in newspapers, on the radio, on TV, and in professional blogs), digital media (social media, e-newsletters, websites, and organizational blogs) and community outreach (e.g. sponsorships, tabling at events, and film screenings). While a single story in the *New York Times* may not change the hearts and minds of its readers, the repetition over time of targeted messages through compelling stories and beautiful images can encourage an audience to see the intrinsic value of the natural world and its connection to nature.

30.3.1 Audience

The first step in building an effective campaign for your project is identifying your target audiences. Are you trying to reach like-minded individuals who already identify as environmentalists? Are you trying to reach individuals who may never have considered the value of the wildlife we share this planet with? Do you want to target a demographic by age or geography? By profession or income? Once you've identified your key audiences, next you have to consider the communications channels that will most likely reach them and the conservation action you'd ultimately like them the take, even if that action is to give money to build capacity for an organization to conduct field conservation programmes. Building a constituency for your campaign usually requires reaching individuals a number of times through a number of different channels until they recognize both the name of your project, and an anecdote related to the project's work. There are huge differences in communications channels, from 140-character tweets to 1000-word *New York Times* cover stories, but one common element unifies effective campaigns: compelling storytelling.

30.3.2 Story

It may be easy to identify the main messages you'd like to push out to build a constituency, but those messages need to be delivered in the form of a story. With some exception, telling a story that people will remember requires incorporating three or four elements of the nine elements of a good story:

1. Timeliness: If it is happening now, it's news. If it happened yesterday, it's probably not.
2. Superlatives: The biggest, the first, the only, or a new finding.
3. Relevancy: It's related to a common thread in the news.
4. Conflict: There's a conflict that is being resolved in an innovative or inspiring way. Stories about conservation tend to highlight doom and gloom, but the stories that stick are those that offer solutions and give hope.
5. Prominence: The story involves someone famous or someone in power.
6. Consequence: The outcome of the work has a direct and immediate impact on people's lives.
7. Proximity: The work has a tangible and immediate impact on people geographically closest to where the work is happening.

8. Human interest: The interesting person doing the work becomes the focus of the story; or there's an anthropological aspect.
9. Quirkiness: Unexpected, offbeat details that make the story engaging for that reason.

If you are working in earned media, it is important to remember that while the media outlet has an audience that you need access to, they do not necessarily share your mission. Snakes and crocodilians in particular are prime candidates for cheap sensationalism that may not be consistent with your message. For example, Discovery Channel's episode 'Eaten Alive', which featured naturalist Paul Rosolie dressed in a snake-proof suit being squeezed by a 5.8-m anaconda was controversial. It was, by the star's own account, an unmitigated disaster for snake conservation messaging (Rosolie 2014). Steve Irwin is probably one of the most colourful celebrities who developed a devoted following by focusing on reptiles and their conservation in Australia. Despite being a charismatic storyteller and an experienced crocodile handler, Steve's cavalier interactions with wildlife on camera endangered him and set a poor example for his audience. This behaviour arguably led to his untimely death while interacting with a stingray on the Great Barrier Reef, tragically undermining his own message and legacy (Bradshaw et al., 2007). The lesson learned here is that you should have a compelling story to tell that doesn't need to be spun into something different, pick the target of your earned media pitches carefully, and never do anything silly or harmful to reptiles on camera.

30.3.3 Constituency-building

Assuming your recent front-page story on the *New York Times* has successfully piqued the interest of some like-minded supporters, it is important to figure out how to continue to engage them. In the last decade, this has been a rapidly changing field and many conservation non-profits have moved from traditional hard-copy mailings to online and social media campaigns. The Orianne Society is a contemporary example of a membership-based reptile conservation group that has an excellent social media footprint on Facebook and Twitter, with an engaging online presence. Best practices will surely evolve along with the rapid changes in digital communications, but being engaged in the broader community on digital platforms is an important way to stay abreast of the best practices. David Steen (@AlongsideWild) has become a Twitter sensation by identifying snakes for people on the popular social media platform. Despite the fact that these snakes are usually dead and misidentified, he exhibits restraint in his communications, politely answering the question and encouraging the person asking the question not to kill snakes in the future. However, don't limit your search for potential models just to the herpetology world. Sharks, for example, may be an analogous taxon to snakes and crocodiles, given people's innate fear of them. On Twitter, marine biologist David Schiffman (@whysharksmatter) is an outstanding example of someone who has managed to grow a large audience and insert himself—and his passion—into the popular discourse. Reptile conservation needs to scale up and we can learn a great deal by observing and emulating the efforts of other organizations that have successfully built their own constituencies.

30.3.4 Community conservation

Turtles are an example of a taxon that is gravely threatened by insatiable demand for turtles as food and medicine (Van Dijk et al., 2000). Promoting public awareness of the plight of turtles in order to reduce demand is a priority conservation action vital to tackling the illegal wildlife trade (Dutton et al., 2013). Even though there are many very good zoo exhibits about threatened turtles that are being maintained in captivity as ex situ assurance colonies, these are unlikely to reduce the demand for turtle meat and medicine unless they reach the intended audience of consumers, who are unlikely to be patrons of Western zoos and aquariums. In order to mount an effective community-based campaign, you have to become a part of it. One example is the 'I don't eat turtle eggs' campaign in Nicaragua. About 50% of Nicaraguans are now familiar with the campaign. The campaign coordinator explains that she experienced a very hostile reception in the fish markets initially, but she is now a familiar face around the country and receives a warm welcome at fish markets (Abarca, 2012). Similarly, the Brazilian sea turtle conservation project TAMAR-IBAMA directly involves 400 fishermen by hiring them to patrol beaches to prevent poaching of nests. They charge tourists who pay a small fee to witness turtles nesting on the beach. These fishermen have alternative livelihoods and personally benefit from healthy turtle populations. They become ambassadors for the project within their local communities (Marcovaldi et al., 2005). Community conservation can be extremely difficult, and poacher-turned-protector programmes do not always work well in the short term (Kiester et al., 2013), but there is a growing body of evidence from multidisciplinary research that over the longer term, knowledge, attitude, interpersonal communications and removal of barriers lead to behaviour change, threat reduction and a positive conservation result (Butler et al., 2013).

30.4 Nature centres, museums, and exhibits

Nature centres offer a unique opportunity to use first-hand experiences to change people's attitude about and behaviour towards reptiles. Museums such as the Florida Museum of Natural History (http://www.flmnh.ufl.edu/herpetology/fl-snakes) are places where people can learn about local species and view positive displays of these animals (see also Box 30.1). The exhibits inform visitors about the local reptiles that are native to their areas, educate them about the threats reptiles face, and shed light on misperceptions about the threats they pose.

One of the most important issues a nature centre can tackle is that of mistaken identity. The majority of snakes are non-venomous, and the majority of human interactions with snakes are likely to be with non-venomous species. For example, there is only one venomous snake in the Washington, DC metropolitan area, the Northern Copperhead (*Agkistrodon contortrix*). The remaining 18 snake species are non-venomous and harmless. Despite this, snakes are often misidentified as venomous and killed. The Long Branch Nature Center in Virginia focuses much of its outreach on the myths surrounding snakes and educating the public about the snakes of the area. One common

Box 30.1 The Corn Snake ambassador

The Smithsonian's National Zoo Reptile Discovery Center uses a Corn Snake (*Pantherophis guttatus*) as an ambassador animal. This species was chosen because it is non-venomous, it is known to have a calm disposition, and it is usually tolerant of being handled. This demonstration animal was bred at the zoo in 2005 and has been handled since it was a hatchling and so is accustomed to human handling. Trained staff members regularly handle the snake, which is kept in an exhibit in a separate room where small groups of visitors can be brought into a calm environment to ask questions and get up close to the animal. Seeing another person respectfully interact with a snake can help reduce negative or hysterical reactions (Figure 30.1) and significantly affects people's attitudes toward snakes (Ballouard et al., 2012). When people see this, many zoo visitors want to participate and touch the snake, which they can do one at a time while the trained staff keeper holds the animal. This close interaction with a snake is much more memorable for visitors than simply viewing a snake behind glass, and has been shown to increase the time visitors spend with the animals (Burghardt et al., 2009). This will often be the first time any of the zoo visitors have touched a live snake, and it is a great jumping-off point for an engaging dialogue about the native snakes in the area, venomous vs. non-venomous snakes, the value of snakes in ecosystems, and how to conserve wild snakes. Anyone who touches the animal has access afterward to a hand-washing station. This ambassador animal programme was developed using the American Association of Zoos and Aquariums' ambassador animal programme guidelines (AZA, 2014).

Figure 30.1 Matt Neff, a keeper at the National Zoo's Reptile Discovery Center, handles a captive-bred Corn Snake as part of the zoo's ambassador animal programme to engage visitors in reptile conservation discussions. Photo courtesy of Alan Peters, Smithsonian's National Zoo.

misconception around Washington, DC is that Cottonmouths live in the creeks where kids like to swim, but Cottonmouths are a venomous snake native to the James River drainage near Richmond and south-eastern Virginia, with a range extending southwards to Texas. What people are actually encountering are non-venomous Northern Water Snakes (*Nerodia sipedon*), which are usually quick to flee when they encounter people. Unfortunately, visitors occasionally bring in a dead snake to be identified to make sure that it really is a copperhead or rattlesnake, yet the unfortunate victims are almost always non-venomous—mostly Eastern Ratsnakes.

Non-venomous snakes can be a desirable form of pest control in your garden, but there are many places where living alongside venomous snakes is an everyday reality. Cape Snake Conservation is a group working in South Africa whose primary objective is to empower people with knowledge. When venomous snakes such as puff adders and cobras take up residence under people's patios, they put homeowners in touch with a network of volunteer herpetologists that will help to identify, and if appropriate, remove venomous snakes in a way that is safe for the snake, the herpetologist, and the homeowner.

30.5 Citizen science

Citizen science is an appealing tool for any conservation project. Citizen scientists can be viewed as a volunteer, almost infinitely scalable labour force that can effectively collect data over a wide area. The experience of being a citizen scientist can also inspire people to become conservation advocates (Ryan et al., 2001). The most successful projects such as eBird can achieve remarkable resolution and generate real-time bird distribution maps throughout the year in the United States that would simply not be achievable any other way (Sullivan et al., 2014). Successful citizen science programmes should have an explicit goal or hypothesis, well-designed data collection methods, clearly articulated assumptions, data that can be validated, and they should provide volunteers with feedback (Silvertown, 2009). Recent advances in technology such as environmental DNA and digital photography can serve as convenient non-destructive records that can be verified by third parties even without traditional museum specimens (Minteer et al., 2014).

In practice, however, volunteers are not free. Recruiting, training, and retaining volunteers is challenging and requires a real investment of financial and human resources to be successful (Silvertown et al., 2013). It is important to remember that citizen-science projects involve a trade-off. More complex tasks have higher costs of training, increased probability of bias or inaccuracies, and reduced participation rates. Simple, straightforward tasks generating data that can be validated regardless of the differential field skills of volunteers tend to make the most powerful citizen science projects.

One of the biggest challenges of creating a scalable, usable citizen science programme is developing a user-friendly platform that allows users to share their observations. iNaturalist.org is one platform that allows users to upload photos or sounds as observations of a particular species, provide a date, a locality, and identification (see Box 30.2). In essence, the photographic observation is the curated record, and for cryptic species, multiple photos that include key identification features with each record. Other members of the community can assist with the identification process, which becomes

Box 30.2 The Global Reptile Bioblitz

iNaturalist.org created a special project called the Global Reptile Bioblitz in partnership with the Reptile Database, HerpNET, Society for the Study of Amphibians and Reptiles, the Center for North American Herpetology, California Academy of Sciences, and several IUCN/SSC Reptile Specialist Groups. The goal of the project is for volunteers to eventually observe and photograph all of the world's 9500 reptiles *in situ* and to share their observations. The primary constituency for this project is experienced herpetologists, which is not a huge target audience, but participation is open to everyone. In the four years since the project's launch in 2011, about 700 users submitted 16,000 observations covering about 30% of the world's described reptile species. Users can share their observations on an app on their phones or upload pictures from their computers. The records are immediately displayed on a range map of the species, along with other user observations (Figure 30.2). Any user can create a similar project with the option of restricting submissions to a user-defined polygon that geographically restricts the records. For example, *Herps of Texas* is a very active project that generates more than 1000 observations per month. The project is actively curated by six professional herpetologists and has generated citizen science records for nearly all of the herp species in the state. iNaturalist provides a flexible, free, user-friendly platform that removes technological limitations for some of the most popular types of citizen science projects requiring locations, date, and verifiable record. This eliminates the need to reinvent the technology wheel, allowing users to focus on the challenge of recruiting and engaging volunteers.

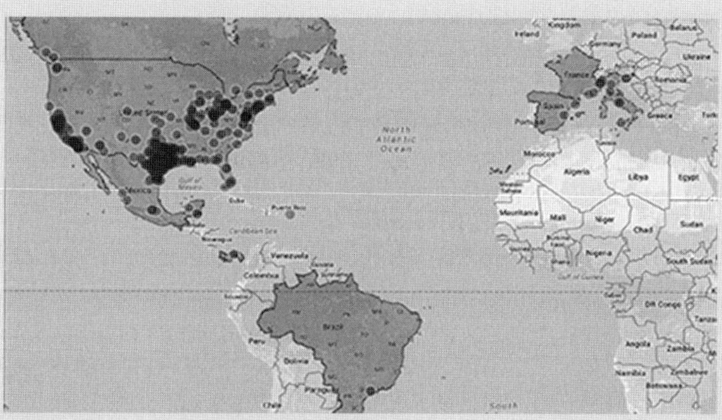

Figure 30.2 iNaturalist users created this map, which displays every submitted record of Red-eared Sliders (*Trachemys scripta*) as grey dots. Native to the south-eastern United States, this species is a popular pet and has been released in many places around the world. iNaturalist provides verifiable citizen records that can be evaluated by curators who may assign the observations a research grade status and reveal a global snapshot of this species' expanding range that is not feasible to generate in any other way. The darker representation of records in Texas is undoubtedly an indication of how effectively the curators of the *Herps of Texas* project have been at mobilizing observation submissions in that state. Image courtesy of S. Loarie, iNaturalist.org.

a truly rewarding learning experience for the user and acts as a verification step (Loarie et al., 2011). Users can modify their own records based on input from others and retain copyright to their images. In addition, endangered species localities are automatically obscured at a fine scale to protect that information, but still provide accurate landscape-level distributions (Bowser et al., 2014).

30.6 Engaging teachers and schools

Many teachers are interested in bringing science and other curricula to life through hands-on, local opportunities for their students to learn critical thinking and problem-solving skills. Teachers also seek opportunities for professional growth, particularly those that align with the lessons and concepts they are required to deliver in the classroom.

By talking to teachers in your community, you can often identify areas of intersection between both of your educational goals. Where possible, try to connect your programme content with the needs of teachers and schools. Support for teachers, for example, through professional development workshops, educational materials, or involving them in creating lessons, has been shown to positively influence teaching practice and effectiveness (Paul and Volk, 2002; Penuel et al., 2007; Niesche and Jorgensen, 2010; Polly and Hannafin, 2011; Zeegers et al., 2012). Thus, you might offer a workshop for teachers in which you work together to develop a lesson on the natural history of reptiles found in the community. The lesson could link information on reptile biology and their role in ecosystems to the science curricula and standards of local schools, an approach that is more likely to engage teachers and, therefore, reach students (Shepardson et al., 2002; Meichtry and Smith, 2007).

A classroom activity or lesson could be combined with a field trip to a local nature centre where students learn to identify reptiles and learn the differences between venomous and non-venomous snakes first hand, and what to do if they encounter one in the wild. This kind of 'place-based learning' engages youth and teachers in local, real-world environmental issues that directly impact them (Meichtry and Smith, 2007; Ernst, 2009). Place-based learning is widely viewed as an effective method of environmental education (Gruenewald, 2005), and can positively affect teachers' confidence, attitudes, and classroom practice when teaching about the environment (Meichtry and Smith, 2007).

Working hand in hand with teachers as you develop activities and materials can really strengthen your education programme while benefiting teachers and reaching students. Remember, while you are a subject matter expert, teachers have experience with curriculum and they will know students' needs and their age limitations better than you.

30.7 Tips for designing educational programmes for schoolchildren

1. Work closely with local teachers to pilot activities and materials linked to local school curriculum and standards.
2. Emphasize place-based activities rooted in the natural history and concerns of the community.

3. Create opportunities for teachers and students to work with scientists solving real-world problems, for example, projects or field trips to nature centres.
4. If offering a field trip or visit, follow up with activities and materials for use back in the classroom and support their implementation.
5. Suggest ideas for supplementary activities that can be modified for a variety of age and ability levels.
6. Make a plan for distributing your materials to teachers, and for follow up.
7. Think through learning outcomes you could measure directly or through follow-up surveys to evaluate your success (here's where teachers can help!).
8. Suggest ideas for how interested students might help you achieve your mission, like volunteering or fund-raising for conservation.

30.8 Leadership development

It is important for every effective conservationist to reflect on whether they have the right skills and training to accomplish their goals. No matter one's role in reptile conservation, traditional discipline-specific knowledge and skills learned in herpetology, zoology, and ecology alone do not necessarily prepare you to be a leader in conservation. Conservationists must have leadership attributes to inspire and influence others, they must be effective listeners and communicators, have strong interpersonal skills, and the ability to negotiate and resolve conflict (Dietz et al., 2004). In today's world, effective conservation leaders must recruit and manage board members, define organizational structures, fund raise, manage programmes, communicate with stakeholders, influence policy, and perform a myriad of other needed tasks that require strong leadership skills (Blickley et al., 2013; Dietz et al., 2004; Newing, 2010). These are extensively taught in financial and business contexts, but are seldom considered in conservation (Dietz et al., 2004).

Some people have innate leadership abilities, while others need to develop their leadership potential. The types of skills that you need to work on are dependent upon the context of your situation and career goals, but it can be challenging to figure out what areas you need to work on. A mentor can often help you figure out what areas of personal growth you need to work on, but reading posted job advertisements, reviewing performance plans, and talking to your colleagues in other similar positions about leadership skills are good ways to reflect on developing your own leadership potential.

Fortunately there is a growing recognition that leadership development is needed on an ongoing basis and there are several formal and informal development opportunities now tailored to help ensure application to the field of conservation (Manolis et al., 2009). Leadership skills are now incorporated into many conservation-related undergraduate and graduate academic programmes (Manolis et al., 2009; Blickley et al. 2013). Early career conservationists who have shown potential can apply to participate in highly competitive, well-known leadership development programmes that are open to applicants from around the world. These vary from two-week intensive courses such as the Conservation Leadership Program, to one month programmes like the Kinship fellowships or two-year part-time programmes such as the Emerging Wildlife Conservation Leaders programme. The David H. Smith Conservation Fellowships are full-time fellowships targeted at post-doctoral emerging conservation leaders with specific leadership training and mentoring opportunities integrated into the programme. Other short courses, workshops, conferences, continuing education programmes, books, and online tutorials are offered in leadership development, but it is worth choosing carefully as you bear in mind that experiential learning is often the most effective method in this field (McCall, 2010). On-the-job training, temporary duty assignments, project-based assignments, volunteering, internships, and fellowships are great means of experiential learning. Having a mentor or participating in a mentor-type programme is also recognized as an important component of leadership development (Dietz et al., 2004).

30.9 Summary

Taking a broader view of education, building interdisciplinary partnerships, and exploring all the different educational tools at your disposal can be a fun and powerful endeavour. While education is often seen as an inherently valued activity, not all educational activities are equally effective at changing people's behaviour in a way that makes a difference for conservation. While most education efforts do have indirect value for conservation, the most powerful reptile conservation efforts are those that have a clear understanding of how their education actions directly affect conservation outcomes.

References

AZA. (2014). Recommendations for developing an institutional ambassador animal policy. American Association of Zoos and Aquariums. Available at: https://www.aza.org/Education/detail.aspx?id=2451.

Abarca, G. (2012). I don't eat turtle eggs—do you? Fauna and Flora International. Available at: http://www.fauna-flora.org (blog post 30.11.11).

Ballouard, J., Provost, G., Barré, D., et al. (2012). Influence of a field trip on the attitude of schoolchildren toward unpopular organisms: an experience with snakes. *Journal of Herpetology*, **46**, 423–8.

Blickley, J.L., Deiner, K., Garbach, K., et al. (2013). Graduate student's guide to necessary skills for nonacademic conservation careers. *Conservation Biology*, **27**, 24–34.

Bowser, A., Wiggins, A., Shanley, L., et al. (2014). Sharing data while protecting privacy in citizen science. *Interactions*, **21**, 70–3.

Bradshaw, J.A., Brook, B.W., and McMahon, C.R. (2007). Dangers of sensationalizing conservation biology. *Conservation Biology*, **21**, 570–1.

Brewer, C. (2001). Cultivating conservation literacy: 'trickle-down' education is not enough. *Conservation Biology*, **15**, 1203–5.

Burghardt, G.M., Murphy, J.B., Chiszar, D., et al. (2009). Combating ophiophobia: origins, treatment, education, and conservation tools. In S.J. Mullin and R.A. Seigel (eds) *Snakes: Ecology and Conservation*. Ithaca, NY: Cornell University Press, pp. 262–80.

Butler, P., Green, K., and Galvin, D. (2013). *The Principles of Pride: The Science behind the Mascots*. Arlington, VA: Rare. Available at: http://www.rare.org/sites/default/files/Principles%2520of%2520Pride%25202013%2520lo%2520res.pdf.

Chao, P., Dolnicar, S., and Lazarevski, K. (2009). Marketing in non-profit organizations: an international perspective. *International Marketing Review*, **26**, 275–91.

Dietz, J.M., Aviram, R., Bickford, S., et al. (2004). Defining leadership in conservation: a view from the top. *Conservation Biology*, **18**, 274–8.

Dutton, A.J., Gratwicke, B., Hepburn, C., et al. (2013). Tackling the unsustainable wildlife trade. *Key Topics in Conservation Biology*, **2**, 74–91.

Ernst, J. (2009). Influences on US middle school teachers' use of environment-based education. *Environmental Education Research*, **15**, 71–92.

Gratwicke, B., Seidensticker, J., Shrestha, M., et al. (2007). Evaluating the performance of a decade of Save the Tiger Fund's investments to save the world's last wild tigers. *Environmental Conservation*, **34**, 255–65.

Gruenewald, D.A. (2005). Accountability and collaboration: institutional barriers and strategic pathways for place-based education. *Ethics, Place and Environment*, **8**, 261–83.

Isbell, L.A. (2011). *The Fruit, the Tree, and the Serpent*. Cambridge, MA: Harvard University Press.

Kiester, R.A., Mandimbihansina, A.R., Lewis, R.E., et al. (2013). Conservation of the angonoka (ploughshare tortoise), *Astrochelys ynphora*. *Chelonian Research Monographs*, **6**, 162–70.

Leopold, A. (1989). *A Sand County Almanac, and Sketches Here and There*. New York: Oxford University Press.

Loarie, S.R., Kahn, T.R., Gratwicke, B., et al. (2011). Global amphibian bioblitz. *FrogLog*, **97**, 48–50.

Louv, R. (2008). *Last Child in the Woods: Saving Our Children from Nature-Deficit Disorder*. Chapel Hill, NC: Algonquin Books.

Manolis, J.C., Chan, K.M., Finkelstein, M.E., et al. (2009). Leadership: a new frontier in conservation science. *Conservation Biology*, **23**, 879–86.

Marcovaldi, M.A., Patiri, V., and Thomé, J. (2005). Projeto TAMAR-IBAMA: twenty-five years protecting Brazilian sea turtles through a community-based conservation programme. *Maritime Studies*, **3**, 39–62.

McCall, M.W. (2010). Recasting leadership development. *Industrial and Organizational Psychology*, **3**, 3–19.

Meichtry, Y., and Smith, J. (2007). The impact of a place-based professional development program on teachers' confidence, attitudes, and classroom practices. *The Journal of Environmental Education*, **38**, 15–32.

Minteer, B.A., Collins, J.P., Love, K.E., et al. (2014). Avoiding (re)extinction. *Science*, **344**, 260–61.

Murphy, J., Stomberg, E., Rosenthal, E., et al. (2014). Orangutans and gorillas just do not like snakes. *Herpetological Review*, **45**, 723–8.

Newing, H. (2010). Interdisciplinary training in environmental conservation: definitions, progress and future directions. *Environmental Conservation*, **37**, 410–18.

Niesche, R., and Jorgensen, R. (2010). Curriculum reform in remote areas: the need for productive leadership. *Journal of Educational Administration*, **48**, 102–17.

Paul, G., and Volk, T.L. (2002). Ten years of teacher workshops in an environmental problem-solving model: teacher implementation and perceptions. *The Journal of Environmental Education*, **33**, 10–20.

Penkunas, M.J., and Coss, R.G. (2013). A comparison of rural and urban Indian children's visual detection of threatening and nonthreatening animals. *Developmental Science*, **16**, 463–75.

Penuel, W.R., Fishman, B.J., Yamaguchi, R., et al. (2007). What makes professional development effective? Strategies that foster curriculum implementation. *American Educational Research Journal*, **44**, 921–58.

Polly, D., and Hannafin, M.J. (2011). Examining how learner-centered professional development influences teachers' espoused and enacted practices. *The Journal of Educational Research*, **104**, 120–30.

Rosolie, P. (2014). Edited reality: what I learned from filming Eaten Alive. *Mongabay*. Available at: http://news.mongabay.com/2014/1222-rosolie-rps-eaten-alive.html.

Ryan, R.L., Kaplan, R., and Grese, R.E. (2001). Predicting volunteer commitment in environmental stewardship programmes. *Journal of Environmental Planning and Management*, **44**, 629–48.

Shepardson, D.P., Harbor, J., Cooper, B., et al. (2002). The impact of a professional development program on teachers' understandings about watersheds, water quality, and stream monitoring. *The Journal of Environmental Education*, **33**, 34–40.

Silvertown, J. (2009). A new dawn for citizen science. *Trends in Ecology & Evolution*, **24**, 467–71.

Silvertown, J., Buesching, C.D., Jacobson, S.K., et al. (2013). Citizen science and nature conservation. *Key Topics in Conservation Biology*, **2**, 124–42.

Sullivan, B.L., Aycrigg, J.L., Barry, J.H., et al. (2014). The eBird enterprise: an integrated approach to development and application of citizen science. *Biological Conservation*, **169**, 31–40.

Van Le, Q., Isbell, L.A., Matsumoto, J., et al. (2013). Pulnivar neurons reveal neurobiological evidence of past selection for rapid detection of snakes. *Proceedings of the National Academy of Sciences of the USA*, **110**, 19000–5.

Van Dijk, P.P., Stuart, B.L., and Rhodin, A.G.J. (eds). (2000). *Asian Turtle Trade: Proceedings of a Workshop on Conservation and Trade of Freshwater Turtles and Tortoises in Asia*. Chelonian Research Monographs No. 2. Lunenburg, MA: Chelonian Research Foundation.

Zeegers, Y., Paige, K., Lloyd, D., et al. (2012). 'Operation Magpie': inspiring teachers' professional learning through environmental science. *Australian Journal of Environmental Education*, **28**, 27–41.

Index

Figures and tables are indexed in bold.

D

M

N

O

U

V